Advanced Manufacturing Technologies: Development and Prospect

Advanced Manufacturing Technologies: Development and Prospect

Guest Editors

Tibor Krenicky
Juraj Ruzbarsky
Maros Korenko

MDPI

Basel • Beijing • Wuhan • Barcelona • Belgrade • Novi Sad • Cluj • Manchester

Guest Editors

Tibor Krenicky
Department of Technical Systems Design and Monitoring
Technical University of Košice
Prešov
Slovakia

Juraj Ruzbarsky
Department of Technical Systems Design and Monitoring
Technical University of Košice
Prešov
Slovakia

Maros Korenko
Institute of Design and Engineering Technologies
Slovak University of Agriculture in Nitra
Nitra
Slovakia

Editorial Office
MDPI AG
Grosspeteranlage 5
4052 Basel, Switzerland

This is a reprint of the Special Issue, published open access by the journal *Applied Sciences* (ISSN 2076-3417), freely accessible at: www.mdpi.com/journal/applsci/special_issues/AM_technologies.

For citation purposes, cite each article independently as indicated on the article page online and as indicated below:

Lastname, A.A.; Lastname, B.B. Article Title. *Journal Name* **Year**, *Volume Number*, Page Range.

ISBN 978-3-7258-4044-1 (Hbk)
ISBN 978-3-7258-4043-4 (PDF)
https://doi.org/10.3390/books978-3-7258-4043-4

Contents

Preface

The main aim of this Special Issue was to present the current state of research on the subjects of the theory, modelling, monitoring, and control of the operation of technology systems and processes, along with the research and diagnostics of manufacturing systems and process operation. The contributions focused on the manufacturing research, operation reliability, and diagnostics of machines; the inspection, measurements, evaluation, and diagnostics of production quality in technologies of standard and progressive machining; reversible engineering; 3D printing; pressure die casting; injection moulding; EDM; AWJ cutting; etc., which are used for the advanced processing of materials and various kinds of technological applications.

The Guest Editors are thankful to all authors whose contributions were accepted, as well as those whose contributions were not included in this Special Issue—the latter were mainly declined not as a result of a lack of scientific rigour but rather because they did not closely align with this Special Issue's focus. The authors who demonstrated a commendable response rate, effectively implementing the constructive feedback given by reviewers and editors, deserve the deepest gratitude. Their cooperative and conscientious efforts were instrumental in the advancement of the articles, thereby making a substantial contribution to the field. In addition, gratitude is extended to all reviewers who assumed this significant responsibility and contributed to this Special Issue. Finally, the Applied Sciences Editorial Team merits extra recognition for their indispensable assistance in this Special Issue.

Tibor Krenicky, Juraj Ruzbarsky, and Maros Korenko
Guest Editors

applied sciences

MDPI

Editorial

Advanced Manufacturing Technologies: Development and Prospect

Tibor Krenicky

Department of Technical Systems Design and Monitoring, Faculty of Manufacturing Technologies, Technical University of Kosice, Bayerova 1, 080 01 Presov, Slovakia; tibor.krenicky@tuke.sk

Abstract: The main aim of this Special Issue was to present the current state of the research on the subjects of theory, modeling, monitoring, and control of the operation of technology systems and processes, along with research and diagnostics of manufacturing systems and processes operation. The contributions have focused on manufacturing research, operation reliability, and diagnostics of machines; inspection, measurements, evaluation, and diagnostics of production quality in technologies of standard and progressive machining; reversible engineering; 3D printing; pressure die casting; injection molding; EDM; AWJ cutting; etc., which are used for advanced processing of materials and various kinds of technological applications.

Keywords: manufacturing systems; modeling; manufacturing technology; computer-aided manufacturing; automation; production quality; reliability; advanced machining; surface properties; environmental aspects of manufacturing; additive manufacturing; manufacturing management; risk analysis

1. Introduction

The manufacturing industry is currently experiencing a paradigm shift, driven by the integration of advanced technologies. These innovations have been shown to enhance efficiency, precision, and sustainability, thereby transforming production processes and supply chains. Emerging trends in the field, including automation, artificial intelligence (AI), additive manufacturing [1,2], and smart manufacturing systems, are collectively driving a paradigm shift in traditional manufacturing practices. Concomitantly, the mounting emphasis on sustainability has precipitated the emergence of eco-friendly materials and energy-efficient production techniques [3,4].

The evolution of manufacturing technologies has paved the way for increased efficiency, customization, and sustainability [5,6]. The integration of automation [7], additive manufacturing [8], and digital technologies [9,10] has transformed traditional manufacturing into a more intelligent [11] and adaptive industry. In order to maintain competitiveness in the global market, it is imperative that industries continue to embrace innovation as these technologies advance further.

The theme of this Special Issue in the field of advanced manufacturing technologies has been met with a positive response from the scientific community. The significant engagement of the scientific community is evidenced by the numerous views on the articles, but more importantly by the rich citation response. To date, the articles in this Special Issue have been cited a total of 107 times, with an average of 5.1 citations per article, and this figure continues to increase. The success of the response can be attributed to the universality of the challenges and perspectives in this field, which are being addressed

Citation: Krenicky, T. Advanced Manufacturing Technologies: Development and Prospect. *Appl. Sci.* **2025**, *15*, 4597. https://doi.org/10.3390/app15094597

Received: 31 March 2025
Accepted: 17 April 2025
Published: 22 April 2025

Citation: Krenicky, T. Advanced Manufacturing Technologies: Development and Prospect. *Appl. Sci.* **2025**, *15*, 4597. https://doi.org/10.3390/app15094597

based on familiar approaches in various settings. The positive response to the initiative to collate and disseminate the latest findings in the field resonated and was confirmed, somewhat surprisingly, across a broad base of research areas. Conversely, the objectives of all the contributions are closely related.

2. Overview of Published Articles

In their study, Ramirez et al. (1) investigated the implementation of a 5G digital twin (DT) in warehouse logistics, emphasizing dynamic network monitoring and cybersecurity. Their study highlighted the role of network function virtualization (NFV) in securing IoT devices through advanced encryption standards (AES). The utilization of simulation tools such as Altair Feko and GNS3 demonstrated the potential of digital twins to enhance communication capabilities within warehouse environments. Matos et al. (2) propose a model that integrates electricity cost consumption and power demand into aggregate production planning, addressing market uncertainties and the rising costs of electricity. Applied in the food industry, the model utilized the Holt–Winters forecasting method and successfully optimized power demand, reducing workers labor and achieving cost savings while minimizing overall production expenses.

A methodology for determining the optimal number of grinding operations in cold rolling, with the objective of reducing labor intensity and improving processing efficiency, was developed by Bratan et al. (3). The findings of the present study contribute to the refinement of industrial grinding techniques. Mascenik and Coranic's (4) investigation into the coefficient of friction in screw joints revealed that lubrication significantly reduces friction. The study found that copper-based grease yielded the most significant reduction in friction, with a recorded 47.64% decrease. This outcome demonstrates the efficacy of copper-based grease in enhancing the efficiency and stability of the joints.

Ružbarský (5) conducted an investigation into the difficulties of using laser profilometry for roughness measurements on glossy surfaces. The study concluded that while non-contact laser methods provide useful data, they are less reliable for high-gloss materials like stainless steel and aluminum alloys. Iakovets et al. (6) explored the role of mobile applications in supporting SMEs' transition to Industry 4.0. Their study highlighted how digital tools facilitate workforce adaptation, process optimization, and sustainable business development in manufacturing enterprises. Kuznetsov et al. (7) explore the optimization of combined wind–solar power plants for sustainable smart city development. To enhance energy efficiency and stability, the study proposes using autonomous renewable energy sources instead of traditional power grids. The research highlights the benefits of integrating a vertical Darier rotor with photovoltaic converters, improving heat dissipation and maximizing energy output. Through digital aerodynamic modeling, the study identifies optimal design parameters for wind turbines and a photovoltaic module tilt for a particular location, ensuring efficient energy utilization in urban environments.

Surface quality is a pivotal factor in determining the functionality and lifespan of machined components. Palová et al. (8) proposed an efficient method for evaluating surface roughness using a secondary surface roughness standard. Their study identified a uniform distribution law for measurement uncertainty assessment and optimized measurement parameters such as evaluation standards, measurement speed, and cut-off filters. The findings of this study are instrumental in reducing measurement uncertainty while minimizing financial costs. In their study, Chang et al. (9) proposed an incremental learning model using the YOLO framework to improve tire bubble defect detection via digital shearography. The efficacy of this approach was evidenced by the attainment of 98% detection accuracy, thereby addressing the limitations of conventional quality control methodologies and providing a foundation for automation in the context of tire manufacturing.

Mitaľ et al. (10) analyze the impact of technological parameters on the tribological properties of 3D-printed polyetherimide (PEI) models using the fused filament fabrication (FFF) technique. By varying build orientation and fiber deposition strategy, the study examines abrasive wear resistance through tribometer testing per ASTM standards. Results indicate that wear and friction force continuity depend on model orientation. Notably, samples with fiber orientation in the Z direction exhibited greater weight loss than those in the X direction, highlighting the influence of deposition strategy on wear performance.

Rimar et al. (11) proposed a methodology of successive approximation for the modeling of heat transfer and gas dynamics in combustion equipment. Their approach has been demonstrated to simplify complex mathematical relationships, offering practical solutions with controlled error rates of 5–7%, which remain acceptable for engineering applications. A hybrid PFMEA–DEMATEL–ERPN model with the objective of enhancing defect management in post-communist economies was introduced by Bujna et al. (12), and the efficacy of this approach was demonstrated by its superior performance in reducing defect occurrence and associated costs when compared with conventional PFMEA methods. Gulyaev et al. (13) present a method for monitoring nanoparticle and agglomerate sizes using a scanning probe microscope. The study outlines an image analysis process involving segmentation, particle allocation, and overlap correction to ensure accurate size measurements. Existing commercial microscopes and software were found to have limitations in complex image processing. To address this, the authors propose a curvature-based method for particle segmentation and size determination, along with the use of sample displacement sensors for improved image stitching, enhancing measurement precision. Ságová et al. (14) conducted an examination of friction anisotropy in gear transmissions, presenting a tensor-based model with the objective of enhancing the accuracy of friction loss calculations. The findings of the present study are of crucial importance for improving efficiency in mechatronic systems by refining mechanical models of frictional anisotropy.

Efficient data exchange in computer-aided design (CAD) and computer-aided manufacturing (CAM) systems is essential for maintaining accuracy in digital modeling. Kuryło et al. (15) examined system compatibility issues and neutral format effectiveness, focusing on 2D and 3D object transmission. Their research highlighted the need for ongoing development of data exchange standards to minimize errors in geometry transversions. The integration of Industry 4.0 technologies has had a profound impact on the realm of quality control in manufacturing. Saif et al. (16) introduced an Internet of Things (IoT)-based computer vision system for roundness measurement, employing image processing techniques to enhance accuracy. The system they developed offers a non-contact method capable of inspecting multiple rounded components, thereby significantly improving the efficiency of automated quality control processes. Deburring is of paramount importance in the domain of aerospace manufacturing, particularly in the context of jet engine component production. In their study, Falandys et al. (17) conducted an analysis of the automation of edge deburring and its impact on production parameters. Their findings indicate that automated systems provide enhanced repeatability, improved working conditions, and reduced labor fatigue in comparison to manual deburring, thereby significantly enhancing process efficiency.

Liu et al. (18) developed a bi-level optimization model for industrial park layouts, with the objective of minimizing association risks while maximizing rental profit. The model incorporated risk assessment calculations and an enhanced genetic algorithm for layout design, thereby demonstrating its efficacy through a real-world case study. Holubek and Vagaš (19) proposed a vision system for estimating center of gravity coordinates using an overall brightness average from a 3D vision system. Designed for advanced manufacturing and robotic applications, the system utilizes two CCD cameras and an RS 232C communication interface for real-time data exchange. By integrating 1D photogrammetric calibration and advanced image processing techniques, the system achieves a correlation factor of 84–100% for object recognition. Successful data transfer to a robotic system demonstrates its feasibility as a cost-effective alternative for industries seeking high-performance vision solutions without significant investment in cutting-edge technology.

The semiconductor industry is contingent on the transportation of wafers that are both highly precise and free from contamination. In this context, Ha et al. (20) developed a pioneering magnetic levitation-based wafer handling system that eliminates friction and particle generation. The study demonstrated the effectiveness of this system in achieving dust-free, high-speed transport and precise positioning, which is critical for ultra-fine semiconductor production. Gearing accuracy exerts a direct influence on the performance of mechanical systems. Mascenik et al. (21) investigated the influence of angular settings in hobbing mills on gear tooth profile modification. Their research demonstrated that varying the angular settings affects final gear geometry and identified optimization strategies for improving precision in gear manufacturing.

3. Conclusions

The evolution of manufacturing technologies has paved the way for increased efficiency, customization, and sustainability. The integration of automation, additive manufacturing, and digital technologies has transformed traditional manufacturing into a more intelligent and adaptive industry. In order to maintain competitiveness in the global market, it is imperative that industries continue to embrace innovation as these technologies advance further. Future research and investments in smart manufacturing will further enhance the capabilities of production systems, rendering them more resilient and environmentally responsible.

Consequently, the primary objective of the Special Issue has been accomplished. The papers presented have the potential to contribute to the reduction of the cost of manufacturing processes, to the promotion of simulation tools, and to the design of key elements and components to achieve exceptional quality and reliability in the manufacturing field.

The latest advancements in manufacturing technologies, drawing insights from recent research and its impact on industrial applications, provide visions for the future research and investments in smart manufacturing that will further enhance the capabilities of production systems, rendering them more resilient and environmentally responsible.

Funding: This work was funded by the Ministry of Education, Science, Research, and Sport of the Slovak Republic within projects VEGA 1/0509/23, KEGA 052TUKE-4/2024 and APVV-18-0316.

Acknowledgments: The Guest Editors wish to extend their gratitude to all authors whose contributions were accepted, as well as to those whose contributions were not included in this Special Issue. The latter were declined not as a result of a lack of scientific rigor, but rather because they did not strongly align with the Special Issue focus. The authors who demonstrated a commendable response rate, effectively implementing the constructive feedback provided by reviewers and editors, are deserving of the deepest gratitude. Their cooperative and conscientious efforts have been instrumental in the advancement of the articles, thereby making a substantial contribution to the field. In addition, gratitude is extended to all reviewers who have assumed this significant responsibility and have contributed to the success of this Special Issue. Finally, the *Applied Sciences* Editorial Team merits particular recognition for their indispensable assistance in the successful realization of this Special Issue.

Conflicts of Interest: The author declares no conflicts of interest.

List of Contributions

1. Ramirez, R.; Huang, C.-Y.; Liang, S.-H. 5G Digital Twin: A Study of Enabling Technologies. *Appl. Sci.* **2022**, *12*, 7794. https://doi.org/10.3390/app12157794

2. Matos, C.; Sola, A.V.H.; Matias, G.d.S.; Lermen, F.H.; Ribeiro, J.L.D.; Siqueira, H.V. Model for Integrating the Electricity Cost Consumption and Power Demand into Aggregate Production Planning. *Appl. Sci.* **2022**, *12*, 7577. https://doi.org/10.3390/app12157577
3. Bratan, S.; Ságová, Z.; Sága, M.; Yakimovich, B.; Kuric, I. New Calculation Methodology of the Operations Number of Cold Rolling Rolls Fine Grinding. *Appl. Sci.* **2023**, *13*, 3484. https://doi.org/10.3390/app13063484
4. Mascenik, J.; Coranic, T. Experimental Determination of the Coefficient of Friction on a Screw Joint. *Appl. Sci.* **2022**, *12*, 11987. https://doi.org/10.3390/app122311987
5. Ružbarský, J. The Difficulty of Measuring the Roughness of Glossy Surfaces Using the Triangulation Principle. *Appl. Sci.* **2023**, *13*, 5155. https://doi.org/10.3390/app13085155
6. Iakovets, A.; Balog, M.; Židek, K. The Use of Mobile Applications for Sustainable Development of SMEs in the Context of Industry 4.0. *Appl. Sci.* **2023**, *13*, 429. https://doi.org/10.3390/app13010429
7. Kuznetsov, P.; Rimar, M.; Yakimovich, B.; Kulikova, O.; Lopusniak, M.; Voronin, D.; Evstigneev, V. Parametric Optimization of Combined Wind-Solar Energy Power Plants for Sustainable Smart City Development. *Appl. Sci.* **2021**, *11*, 10351. https://doi.org/10.3390/app112110351
8. Palová, K.; Kelemenová, T.; Kelemen, M. Measuring Procedures for Evaluating the Surface Roughness of Machined Parts. *Appl. Sci.* **2023**, *13*, 9385. https://doi.org/10.3390/app13169385
9. Chang, C.-Y.; Su, Y.-D.; Li, W.-Y. Tire Bubble Defect Detection Using Incremental Learning. *Appl. Sci.* **2022**, *12*, 12186. https://doi.org/10.3390/app122312186
10. Mitaľ, G.; Gajdoš, I.; Spišák, E.; Majerníková, J.; Jezný, T. An Analysis of Selected Technological Parameters' Influences on the Tribological Properties of Products Manufactured Using the FFF Technique. *Appl. Sci.* **2022**, *12*, 3853. https://doi.org/10.3390/app12083853
11. Rimar, M.; Yeromin, O.; Larionov, G.; Kulikov, A.; Fedak, M.; Krenicky, T.; Gupalo, O.; Myanovskaya, Y. Method of Sequential Approximation in Modelling the Processes of Heat Transfer and Gas Dynamics in Combustion Equipment. *Appl. Sci.* **2022**, *12*, 11948. https://doi.org/10.3390/app122311948
12. Bujna, M.; Lee, C.K.; Kadnár, M.; Korenko, M.; Baláži, J. New Possibilities of Using DEMATEL and ERPN in the New PFMEA Hybrid Model. *Appl. Sci.* **2023**, *13*, 3627. https://doi.org/10.3390/app13063627
13. Gulyaev, P.; Krenicky, T.; Shelkovnikov, E.; Korshunov, A. Particle and Particle Agglomerate Size Monitoring by Scanning Probe Microscope. *Appl. Sci.* **2022**, *12*, 2183. https://doi.org/10.3390/app12042183
14. Ságová, Z.; Tarasov, V.V.; Klačková, I.; Korshunov, A.I.; Sága, M. Study of Anisotropic Friction in Gears of Mechatronic Systems. *Appl. Sci.* **2022**, *12*, 11021. https://doi.org/10.3390/app122111021
15. Kuryło, P.; Frankovský, P.; Malinowski, M.; Maciejewski, T.; Varga, J.; Kostka, J.; Adrian, Ł.; Szufa, S.; Rusnáková, S. Data Exchange with Support for the Neutral Processing of Formats in Computer-Aided Design/Computer-Aided Manufacturing Systems. *Appl. Sci.* **2023**, *13*, 9811. https://doi.org/10.3390/app13179811
16. Saif, Y.; Rus, A.Z.M.; Yusof, Y.; Ahmed, M.L.; Al-Alimi, S.; Didane, D.H.; Adam, A.; Gu, Y.H.; Al-masni, M.A.; Abdulrab, H.Q.A. Advancements in Roundness Measurement Parts for Industrial Automation Using Internet of Things Architecture-Based Computer Vision and Image Processing Techniques. *Appl. Sci.* **2023**, *13*, 11419. https://doi.org/10.3390/app132011419
17. Falandys, K.; Kurc, K.; Burghardt, A.; Szybicki, D. Automation of the Edge Deburring Process and Analysis of the Impact of Selected Parameters on Forces and Moments Induced during the Process. *Appl. Sci.* **2023**, *13*, 9646. https://doi.org/10.3390/app13179646
18. Liu, X.; Huang, G.; Gao, X.; Li, H.; Ou, S.; Hezam, I.M. Large-Scale 3D Multi-Story Enterprise Layout Design in a New Type of Industrial Park in China. *Appl. Sci.* **2022**, *12*, 8110. https://doi.org/10.3390/app12168110
19. Holubek, R.; Vagaš, M. Center of Gravity Coordinates Estimation Based on an Overall Brightness Average Determined from the 3D Vision System. *Appl. Sci.* **2022**, *12*, 286. https://doi.org/10.3390/app12010286

20. Ha, C.-W.; Jung, S.; Park, J.; Lim, J. Development of a Magnetic Levitation Wafer Handling Robot Transfer System with High-Accuracy and High-Cleanliness: Experimental Evaluation. *Appl. Sci.* **2023**, *13*, 9482. https://doi.org/10.3390/app13169482
21. Mascenik, J.; Coranic, T.; Krenicky, T. Options on Tooth Profile Modification by Hob Adjustment. *Appl. Sci.* **2023**, *13*, 10646. https://doi.org/10.3390/app131910646

References

1. Liu, Y.; Li, Q.; Ren, Z.; Jiang, Z.; Luo, H.; Zhang, X. Effect of Process Parameters on Stress Field of Laser Additive Manufacturing. *Machines* **2022**, *10*, 1197. [CrossRef]
2. Lu, Y.; Xu, W.; Leng, J.; Liu, X.; Xu, H.; Ding, H.; Zhou, J.; Cui, L. Review and Research Prospects on Additive Manufacturing Technology for Agricultural Manufacturing. *Agriculture* **2024**, *14*, 1207. [CrossRef]
3. Li, X.; Li, G.; Lei, Y.; Gao, L.; Zhang, L.; Yang, K. Influence of Laser Surface Texture on the Anti-Friction Properties of 304 Stainless Steel. *Machines* **2023**, *11*, 473. [CrossRef]
4. Zhang, B.; Shi, W.; Lin, Y.; Jiang, L.; Wang, L.; He, K. Effect of CeO_2 Content on Microstructure and Wear Resistance of Laser-Cladded Ni-Based Composite Coating. *Lubricants* **2024**, *12*, 227. [CrossRef]
5. Sommer, M.; Stjepandić, J.; Stobrawa, S.; von Soden, M. Automated generation of Digital Twin for a built environment using scan and object detection as input for production planning. *J. Ind. Inf. Integr.* **2023**, *33*, 100462. [CrossRef]
6. Zhong, D.; Xia, Z.; Zhu, Y.; Duan, J. Overview of predictive maintenance based on digital twin technology. *Heliyon* **2023**, *9*, e14534. [CrossRef] [PubMed]
7. Skardžius, J.; Gargasas, J. Alternative Real-Time Part Quality Monitoring Method by Using Stamping Force in Progressive Stamping Process. *J. Manuf. Mater. Process.* **2025**, *9*, 104. [CrossRef]
8. Zou, D.; Li, C.; Yang, Y.; Jin, X.; Liu, S.; Zhang, H.; Zhang, N. Using the Ethaline Electropolishing Method on the Internal Surface of Additive Manufactured Tubes. *Materials* **2024**, *17*, 4915. [CrossRef] [PubMed]
9. Natarajan, S.; Thangamuthu, M.; Gnanasekaran, S.; Rakkiyannan, J. Digital Twin-Driven Tool Condition Monitoring for the Milling Process. *Sensors* **2023**, *23*, 5431. [CrossRef] [PubMed]
10. Li, Y.; Ding, F.; Wang, D.; Tian, W.; Zhou, J. Predicting the Dynamic Parameters for Milling Thin-Walled Blades with a Neural Network. *J. Manuf. Mater. Process.* **2024**, *8*, 43. [CrossRef]
11. Oščádal, P.; Huczala, D.; Bém, J.; Krys, V.; Bobovský, Z. Smart Building Surveillance System as Shared Sensory System for Localization of AGVs. *Appl. Sci.* **2020**, *10*, 8452. [CrossRef]

applied sciences

MDPI

Article

5G Digital Twin: A Study of Enabling Technologies

Ramiro Ramirez [1], Chien-Yi Huang [1] and Shu-Hao Liang [2,*]

[1] Department of Industrial Engineering and Management, National Taipei University of Technology, Taipei 106344, Taiwan; t109379402@ntut.edu.tw (R.R.); jayhuang@mail.ntut.edu.tw (C.-Y.H.)
[2] Industry 4.0 Implementation Center, National Taiwan University of Science and Technology, Taipei 106335, Taiwan
* Correspondence: shuhaoliang@mail.ntust.edu.tw; Tel.: +886-2-27333141 (ext. 5207)

Abstract: 5G networks require dynamic network monitoring and advanced security solutions. This work performs the essential steps to implement a basic 5G digital twin (DT) in a warehouse scenario. This study provides a paradigm of end-to-end connection and encryption to internet of things (IoT) devices. Network function virtualization (NFV) technologies are crucial to connecting and encrypting IoT devices. Innovative logistical scenarios are undergoing constant changes in logistics, and higher deployment of IoT devices in logistic scenarios, such as warehouses, demands better communication capabilities. The simulation tools enable digital twin network implementation in planning. Altair Feko (WinProp) simulates the radio behavior of a typical warehouse framework. The radio behavior can be exported as a radio simulation dataset file. This dataset file represents the virtual network's payload. GNS3, an open-source network simulator, performs data payload transmission among clients to servers using custom NFV components. By transmitting data from client to server, we achieved end-to-end communication. Additionally, custom NFV components enable advanced encryption standard (AES) adoption. In summary, this work analyzes the round-trip time (RTT) and throughput of the payload data packages, in which two data packages, encrypted and non-encrypted, are observed.

Keywords: 5G; radio access network (RAN); network function virtualization (NFV); cybersecurity; digital twin

Citation: Ramirez, R.; Huang, C.-Y.; Liang, S.-H. 5G Digital Twin: A Study of Enabling Technologies. *Appl. Sci.* **2022**, *12*, 7794. https://doi.org/10.3390/app12157794

Academic Editors: Juraj Ruzbarsky, Maros Korenko and Tibor Krenicky

Received: 27 June 2022
Accepted: 31 July 2022
Published: 3 August 2022

1. Introduction

The quantity of user equipment (UE) devices connected to the internet has increased rapidly since 2007 due to the introduction of the iPhone, and the IoT "birth" during 2008 fueled this increase [1]. More and more connected devices require higher network capabilities. Ericsson responded to these needs and launched a white paper on the 5G concept to meet those demands in 2014 [2].

A definition of the main system architecture requirements is needed to provide a straightforward approach. The 3rd Generation Partnership Project (3GPP) performs a task in cooperation with the European Telecommunication Standard Institute (ETSI), which released details in 3GPP TS 23.501 Release 16 [3]. Based on this, radio access network (RAN), user equipment (UE), and access management function (AMF) are the minimum components. The introduction of network function virtualization (NFV) and software-defined network (SDN) allows additional network functions, creating a more feasible and even safer network.

Network function virtualization (NFV) is the virtualization of routers, firewalls, and advanced packet cores. These components can execute software applications and connect to the central infrastructure. Software-defined network (SDN) is an approach to networking that provides tools for programming, traffic control, or network slicing. These functions help to achieve higher network performance. The technologies above enable most use cases of 5G networks, and the primary use cases are enhanced mobile broadband capabilities (eMBB), ultra-reliable low-latency communications (URLLCs), and massive machine-type communications (mMTCs).

Industrial applications with a high density of devices need the implementation of mMTC solutions. There is an increasing demand for Smart logistic solutions such as tracking and fleet management. Supply chain and logistics companies are willing to adopt this kind of solution [4]. Smart logistics needs real-time data to track current operations, create predictions and increase the overall equipment efficiency (OEE). Migration from previous wireless networks to 5G meets this need. However, the adoption rate is still moderate due to the high complexity of the technologies involved.

5G test-beds are crucial to define, test, analyze and deploy solutions in different mobile network scenarios. Digital twin solutions allow the use of virtual equipment for its testing stage. Additionally, the digital environment can be connected to its physical counterpart. Adopters can test different networking or security settings in virtual environments by developing digital models.

5G hardware, software, and training methods are unfamiliar to most non-academic users contributing to the slow 5G adoption. The use of 5G digital twins can enable radio and network simulation to help adopters to justify investment decisions. Regarding network simulation, Network function virtualization (NFV) technologies represent a cost-effective approach. By simulating network equipment, adopters can estimate network performance and understand the amount of required equipment for their particular case [5]. Implementation of NFV solutions reduces the energy consumption and hardware cost of the network.

The academic community provides an extensive collection of frameworks, reviews, and surveys about digital twins in 5G scenarios. Our study complements the research method introduced by the previous research authors to provide a practical application of a basic digital twin in a 5G scenario with data information. The selected testing environment is a warehouse at the Industry 4.0 Implementation Center at the National Taiwan University of Science and Technology. The electromagnetic software, Altair Feko (WinProp), simulates radio coverage testing in the environment, generating a radio dataset in comma-separated values (.CSV). Later, the network transmission simulation used the radio dataset file as the network payload.

With an emphasis on end-to-end communication, this study conducted data transmission and security. GNS3, the selected network simulator in this study, contains Ubuntu NFV solutions for data transmission. Ubuntu virtual machines work as clients and servers. Wireshark, an open-source packet analyzer, has been selected to monitor the data transmission between client and server. In addition, cybersecurity encryption can be implemented in Ubuntu devices to avoid security risks.

Digital twin simulation allows basic testing of 5G technologies, and radio coverage graphics visualize the interaction between 5G radio and its surrounding environment. The combination of NFV technologies and additional modules allows end-to-end connection between client and server. Implementing this set of tools can help speed up the adoption process of 5G technologies for non-familiar adopters.

2. Related Work

Norbert Wiener, the author of the book "cybernetics", suggested the combination of physical processes, computation, and communication, and he coined the term in 1948 [6]. The term evolved into cyber-physical systems (CPS), which expands the integration from process to complete systems. Its main goal is seamless integration between the practical world and its digital counterpart. In contrast, digital twins (DTs) are focused applications that allow providing high-fidelity virtual models. For this reason, DT can be understood as a focused application of CPS [7].

The origin of the digital twin (DT) concept began in the 1960s, and during this period, NASA developed "twinning" methods for its Apollo Space Program. Constant data connection provided telemetry data to an array of 15 simulators. This system could compute many simulated equipment failure scenarios or tested communication failures [8]. Nowadays, it is possible to define a DT as an advanced system that can provide high-fidelity models [9].

The connection between DT and its physical counterpart is crucial to generating high-fidelity models. The two-way data connection enables constant updates, and the main goal is to achieve a higher correlation between simulated and actual data tested in the field. Due to its modularity, digital twin solutions combine many technologies. As an example, Nvidia Omniverse combines telemetry data from sensors with augmented reality (AR) solutions.

Digital twin industrial applications are presented in manufacturing, energy, industrial assets, or architecture structures. Intelligent manufacturing solutions (CPS) are currently listed as the top priority in manufacturing of government-level strategies, such as Industrial Internet (US), Industry 4.0 (Germany), or Made in China 2025 (China), reflecting this priority [10]. Early adopters such as BMW, General Electric, Microsoft, or Siemens are developing their digital twin technologies.

The logistics sector needs signal monitoring systems to avoid unexpected disruptions [11]. The digital supply chain twin (DSCT) is a digital dynamic simulation model [12]. It has three scopes: network level, site level, and asset level.

Smart logistics belongs to the site level, covering warehouses and manufacturing scenarios. IoT solutions in ports [13], smart contracts for tracing supply chain parts [14], or new frameworks for next-generation ports and warehouses are examples of smart logistics applications [15].

The implementation of IoT solutions in ports requires the deployment of a massive number of devices. Therefore, the adoption of massive machine-type communication (mMTC) is required. Data encryption and end-to-end performance are required to test and validate mMTC applications. In this regard, 5G digital twins could perform network simulations for these two specific applications [16].

Digital twin integration with 5G technologies is gaining popularity among the research community. The research efforts from Dimitris et al. (2021) help to provide a generalized architecture of an IoT smart manufacturing scenario. In this case, the study highlights using digital twins to apply advanced technologies based on 5G wireless communications [17].

According to Qi et al. (2021), "Many researchers and participators in engineering are not clear which technologies and tools should be used". Their work provided an extensive summary of enabling technologies for digital twins and established general research directions [18]. The definition of clear research directions is part of the work of Zao et al. (2022), where the authors proposed a framework for digitalized museums. Digitalized museums require combining digital twins, artificial intelligence (AI), and 5G technologies. The implementation of the proposed framework was simulated using MATLAB [19].

According to Hu et al. (2021), "There is a lack of consideration of the environmental coupling, which results in the inaccurate representation of the virtual components in existing DT models". The authors suggested a strong focus on real-world data, including physical parameters [20].

Research from Zeb et al. (2022) provided the expected features beyond 5G/6G networks. The main list of features includes network services (eMBB, mMTC, and URLLC), network slicing (SDN, NFV), cloud-native deployments (Kubernetes, KubeFlow), federated learning (FL), age of information (AoI), and green communication (power saving, energy harvesting) [21].

Based on the previous considerations, our study provides the needed components to simulate and test a small-scale 5G digital twin. The study of enabling technologies provides a straightforward approach to using software tools. Our study considers the influence of real-world materials in the data generation section to obtain accurate radio representations. In addition, the use of NFV technologies in the data transmission section allows end-to-end communication between client and server. There is a need to provide an empirical analysis of the software tools to test and validate the previous list of features. This study's result section analyzes package data size, TCP streams, round-trip time, average network throughput, and capture of data packages. The discussion and conclusion sections summarize the study and future research directions.

2.1. Available Test-Beds

To test and evaluate the performance of a 5G network, academic and corporate institutions use the 5G test-bed to research, develop, and predict its future implementation under a wide variety of scenarios [22].

Currently, many network test-bed solutions can fulfill network slice selection function (NSSF) needs. Norwegian University of Science and Technology researchers conducted an extensive review of a small-scale 5G test-bed in this regard. In the review, the team provided a valuable summary of a wide variety of test-bed for network slicing in 5G [23]. A common trend was combining SDN and NFV technologies (the BlueArch, shown in 19 of the 21 test-bed evaluated). On the other hand, radio access technologies (RATs) had a lower adoption rate (the SliceNet, shown in 13 of the 21 test-bed). RAT combines radio networks, Wi-Fi, long-term evolution (LTE), and 5G new radio). Radio propagation is sensitive to its physical environment. The presence of materials such as metal or brick can cause unexpected blind spots.

Radio datasets are the basis for distance estimation or location fingerprint techniques [24]. Smart logistic solutions such as fleet management and tracking solutions require distance estimation and location fingerprint. The use of simulators with data generation capabilities can provide customized radio datasets.

The lower adoption rate of RATs in the 5G small-scale test-beds review represents a research opportunity. Combining radio simulators with other technologies, such as NFV or SDN, could provide tools to test and validate essential 5G technologies for non-familiar adopters.

2.2. Additional Requirements

5G relies on the latest technology to perform equipment virtualization, network slicing, and traffic control that exposes the network to cyber-attacks [25]. The next-generation networks must include cybersecurity tools that can provide safe connections. Data encryption, blacklist listing, and machine learning network analysis are some examples of cybersecurity tools.

3. Materials and Methods

This section describes the materials and methods used in this study. When adopting new technologies, decision-makers want to evaluate the system before its acquisition. Using simulators helps estimate coverage and the main parameters of a 5G deployment scenario.

In general, physical objects affect the coverage of the 5G network in an industrial environment where heavy machinery, production lines, or compound materials are ubiquitously in working fields. Antenna position, frequency band, or transmitting power also affect radio coverage. Antenna position or transmitting power changes can improve or deteriorate the radio signal coverage. Radio simulators can provide visual representation features to understand the effect of these changes.

Additionally, network simulators allow data transmission and analysis with virtual equipment representing either a valid user or a non-authorized user. Virtual equipment can provide security measures to guarantee safety across the network, where the simulation of firewalls or encryption techniques can help to increase network security.

Using 5G digital twins allows radio and network behavior simulation to represent common issues about 5G technologies. This study aims to provide a practical application scenario of enabling 5G technologies, providing an initial testing approach to non-academic adopters.

3.1. Main Subjects of the 5G Digital Twin

This section describes the materials and methods used in the central parts of this study, data collection, transmission, and methods. Feko-WinProp (Altair) simulates a deterministic radio signal for data collection, and GNS3 represents the whole network and its data transmission. Finally, the method part depicts the step-by-step approach of this research. The main subjects conducted are listed in Table 1.

Table 1. The portions in the practice of the 5G digital twin.

Data Generation	Data Transmission	Analysis Methods
List of radio parameters	Data encryption	Round-trip time (RTT)
Building CAD file	TCP	Throughput
Data generation (.CSV)	Wireshark	Packet analyzer

3.2. Radio Access Network (RAN)

Radio access network (RAN) is the implementation of both legacy and new radio technologies. New features such as multiple input, multiple output (MIMO), beamforming, and new dedicated radio bands (1–24 GHz) are available. These features enhance the radio performance of 5G. Due to its higher operating frequencies, 5G tends to be vulnerable to its environment. This issue is significant in warehouses or manufacturing sites, which tend to have higher penetration losses. To test the feasibility of a project, it is essential to study radio coverage analysis.

The set of bands available for 5G new radio (NR) has been defined in the technical specification 3GPP TS 38.101-1 Release 15 [26]. According to this technical specification, there are three major frequency bands: low (700 MHz), high (3.1–4.9 GHz), and very high (26 GHz). The use of each specific band depends on the geographical area. Table 2 represents Taiwan's current 5G frequency bands for individuals and private users.

Table 2. Commercially Available 5G Frequency Bands in Taiwan.

Band	f (MHz)
N1	2100
N3	1800
N28	700
N41	2500
N78	3500
N79 [1]	4700

[1] N79 is a dedicated band for private users and requires special authorization.

Frequency bands under 2.7 GHz have lower performance due to a higher concentration of cellular networks (3G and 4G) across the ultra-high-frequency band (UHF). In contrast, the C-band frequency range (3.4–3.6 GHz) is highly available. During the early phase of the 5G deployment in Taiwan, the central band provided at that time was N1. Nowadays, N78 is the most popular frequency band. N1 is the second most adopted frequency band in Taiwan.

Currently, the leading mobile network operators (MNOs) in Taiwan offer five frequency bands, as shown in Table 3. The current bands offered cover low- to high-frequency bands.

MNOs provide wireless network coverage maps to evaluate the service availability in a given location. However, this can differ from the network experience of each user. OpenSignal, a private analytics company, collects network parameters such as signal strength, available networks, or data speed of each application user. Thus, users can visualize the radio coverage map in their current location [27].

Radio coverage simulators are needed to predict the coverage of a custom wireless network. In the case of custom wireless networks, equipment parameters such as transmitting power or radiation patterns affect the result of the coverage. In this regard, VE2DBE is a free online radio simulator, merging the antenna parameters with location data from geographic databases such as OpenStreetMap, Bing Road, Bing Satellite, USGS Topo, USGS Sat, Open TopoMap, and OSM gray. The coverage file is a geo-referenced map of the antenna coverage in a selected area. This file contains multiple images, such as a radio coverage map (.PNG), terrain relief map (.JPG), land cover map (.PNG), and population density map (.PNG). VE2DBE can be used for radio coverage simulation in large-scale areas [28].

Table 3. List of frequency bands available in Taiwan by MNO.

Band/MNO	N1	N3	N28	N41	N78
Chunghwa Telecom	•	•			•
Taiwan Mobile	•	•	•		•
Far East Tone	•	•		•	•
T-Star Telecom	•	•	•		•
Asia Pacific Telecom			•	•	

Physical obstacles and reflective materials affect the coverage of custom wireless networks notorious in environments such as warehouses or manufacturing scenarios. In these environments, accurate radio coverage requires both antenna parameters and obstacles' material parameters. In this regard, advanced electromagnetic simulators such as Altair Feko (winProp) enable radio network planning for small and large-scale areas [29]. Small-scale areas include buildings, warehouses, or manufacturing sites. Large-scale areas include urban and rural environments.

The electromagnetic radio simulator provides a graphical representation of the radio coverage in a selected area with real-world physical data. In addition, the simulated radio coverage values can be exported as a comma-separated value (.CSV) file. The file contains the space coordinates and power values (expressed in dBm).

This work focuses on the 5G NR specification provided by the local MNO, Chunghwa Telecom, the largest telecommunications company in Taiwan. Using Altair, WinProp conducted the radio coverage analysis in a warehouse environment. By simulating the physical characteristics of the space, it is possible to estimate the maximum range of coverage of the radio signal. This helps to speed up the validation time of 5G use cases.

3.3. Testing Environment

The primary experiment environment includes a warehouse, outdoor space, and buildings as the testing ground for the DSCT Site Level. Figure 1A shows the layout of the testing ground in scale, which is the facilities of the Industry 4.0 Implementation Center at the National Taiwan University of Science and Technology. The layout space uses computer-aided design (CAD) to represent the geometric location of the building and antenna. The warehouse space is about 10×10 square meters, and the radio waves transmission range is about 30×40 m around the warehouse, peripheral space, and building. Figure 1B shows the location of the antenna at the testing site. The antenna sits at the top of the Industry 4.0 Implementation Center building, with a relative height of 5.3 m from the ground.

The testing environment layout allows radio coverage simulation using an electromagnetic radio simulator, an initial approach for radio coverage analysis and network deployment. The generated radio coverage simulation data can compare with practical signal measurements in a further plan. Acquiring dedicated hardware devices such as industrial boards with 5G capabilities or software-defined radio (SDR) modules is required to obtain practical measurements. The acquisition of dedicated hardware is currently under consideration and included in our future research development plan.

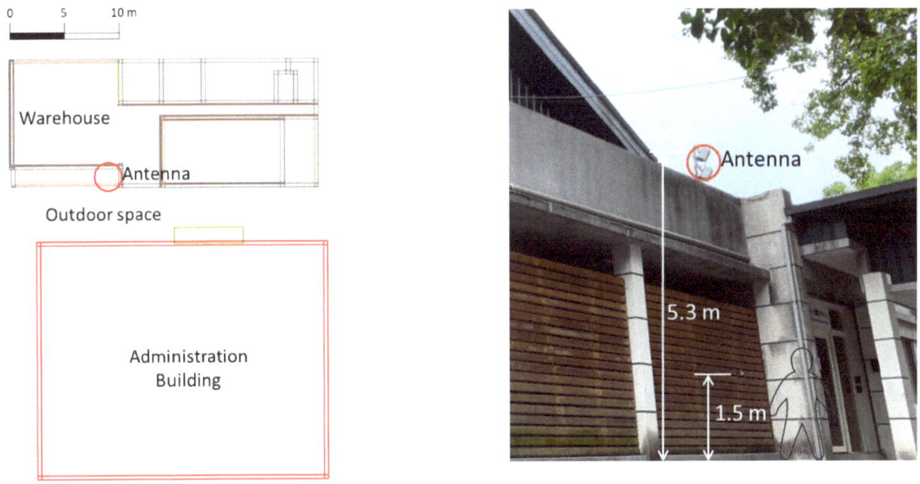

(**A**) Layout (including peripherals) (**B**) Antenna (warehouse outside)

Figure 1. Testing environment. (**A**) Layout (CAD file). (**B**) Antenna at the testing site.

3.4. Data Generation

Chunghwa Telecom provides service for three frequency bands, N1, N3, and N78. N78 is located in the C-Band frequency range and has an operating frequency of 3500 MHz. The following radio coverage analysis focuses on implementing the N78 frequency band. Supported channel Bandwidths for the N78 band are divided into different channels (10–100 MHz). The most popular channel bandwidth for N78 is 100 MHz, the maximum supported channel bandwidth. Smaller channel widths are prevalent in early N78 deployment before spectrum auctions.

Regarding the location of the antenna, it sat at the top of the Industry 4.0 Implementation Center building, with a relative height of 5.3 m from the ground. The azimuth and down-tilt values of the antenna are 150° and 35°. Additionally, its transmitting power is 40 dBm (10 W). The height of the user equipment (UE or client) is assumed to be 1.5 m above the ground. Table 4 represents the 5G network coverage simulation parameters.

Table 4. 5G parameters for radio simulation.

Parameters	Value
Frequency band	N78
Operating frequency	3500 MHz
Bandwidth	100 MHz
BS antenna height	5.3 m
Azimuth	150°
Down-tilt	35°
BS transmit power	40 dBm
Receiver height	1.5 m
Receiver power	23 dBm

Material penetration losses are proportional to the frequency value. A slight frequency increment can generate different radio propagation results. Advanced electromagnetic radio simulators contain material libraries for accurate environment representation. In this study, these values are in the WinProp Material library, which includes an extensive catalog of common materials for indoor and outdoor spaces.

The radio coverage simulation utilizes deterministic software (WinProp, Altair Software Inc., Troy, MI, USA), which is in charge of data modeling. Figure 2 shows 5G radio signal coverage contours in the CAD file layout. This figure combines simulation parameters and test environment materials listed in Tables 4 and 5. The WinProp software can export the data generated by the simulation process with the parameters we mentioned in the last subsections.

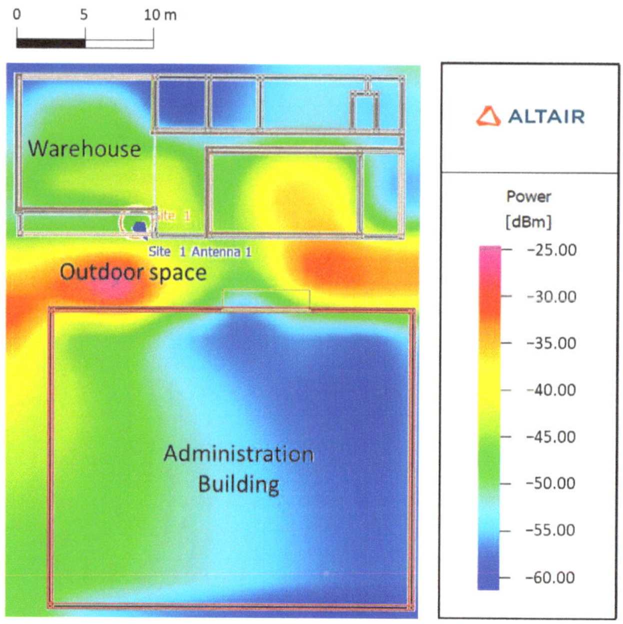

Figure 2. Predicted radio coverage on the test environment map.

Table 5. Test environment material parameters.

Material	Size (mm)	Transmission Loss (dB)	Reflection Loss (dB)
Concrete	300	28.95	7.51
Brick	300	20.11	9.52
Wood	150	6.18	11.43
Metal	5	750.2	0.05
Glass	5	1.72	7.53

The generated data contains power and distance values (expressed in the cartesian coordinate system) and can be saved as a dataset that consists of 1283 samples, as Table 6 shown. The first column represents the horizontal plane (x plane), the second column represents the vertical plane (y plane), and the last column stores the power value expressed in dBm.

Table 6. Simulation values in WinProp.

X	Y	dBm
30,945.269	4874.697	−48.047
30,946.269	4874.697	−48.079
30,947.269	4874.697	−48.811
30,948.269	4874.697	−50.194
[. . .]	[. . .]	[. . .]

This data can be saved in a comma-separated values (.CSV) file. The generated radio simulation dataset represents the payload of the network. Payload transmission between client and server allows network performance analysis. The following section will study the different components required for data transmission.

Altair Feko (WinProp), as an electromagnetic simulator for wireless networks, generated the values in Table 6 and the radio coverage map in Figure 2. Please note that both antenna and receiver have been simulated with the parameters indicated in Table 4. Therefore, no hardware is implemented in this study except the antenna provided by the local MNO. For future research, the acquisition of hardware components, such as industrial boards with 5G capabilities or software-defined radio (SDR) modules, is currently under consideration.

3.5. Data Transmission

Network software programs allow the generation of virtual networks. This network uses different components such as routers, switches, or similar. By using said components, data transmission is possible. The data package selected for this task is the dataset generated before. The following points explain the different components of the data transmission part.

3.5.1. Network Management User Interface

Graphical Network Simulator-3 (GNS3) is an open-source network software written in Python that utilizes to present the network topology. GNS3 can host Docker containers, network switch, firewalls, or Ubuntu servers. A two-way data connection is possible via telnet. This combination of components allows the simulation of complex network environments [30].

3.5.2. Network Functionality Virtualization (NFV)

NFV is the total virtualization of multiple network functions, such as routers or firewalls. Networking hardware companies such as Cisco [31], Juniper [32], Huawei [33], or FortiNet [34] are providing kernel-based virtual machines (KVMs), which are digital versions of their equipment. For the implementation of user equipment (guest users), companies such as Canonical [35] also provide a collection of KVMs. KVM solutions provide similar functions to their physical counterparts, a feature highly valuable for equipment testing, which helps reduce the error rate during deployment. In addition, KVM enables the connection between the simulated environment and the natural world (two-way connection), providing a hybrid combination of hardware and software components.

3.5.3. Software-Defined Network (SDN)

Dynamic network control is necessary to achieve higher network performance. Improving routing processes and data flow requirements is possible with SDN [36]. SDN controllers can be present in both front-haul and back-haul links. In this regard, Ryu is an SDN controller with a well-defined API written in Python [37]. Located at the node level, Ryu can communicate with the Mininet controller.

Based on the study from NUST, there are over 21 test-beds currently available. Due to the vast offer and proven effectiveness, we did not plan to apply SDN solutions in this study.

3.5.4. Packet Analyzer (Wireshark)

Wireshark is an open-source packet analyzer used for data interception over computer networks. The software allows the graphical representation of data packets and supports many internet protocols, according to the internet protocol suite (TCP/IP model) [38]. Table 7 depicts a brief list of supported protocols used for IoT applications. We study the packet round-trip time (RTT) and throughput in this work.

Table 7. Wireshark supported protocols.

TCP/IP	Protocol
Application	HTTP, MQTT, Telnet
Transport	TCP, UDP
Network	IPv4, IPv6
Network interface	Ethernet, 802.11 b/g/n (Wi-Fi), LTE

3.5.5. AES Encryption

The use of the advanced encryption standard (AES) in this study aims to fulfill the additional requirements stated. AES is a block cipher encryption method, where all the data are encrypted in blocks, and the length of each block is 128 bits. In addition, it is a symmetric algorithm requiring the same key for encryption and decryption of data. The key size would determine how many rounds of encryption are required [39].

From a performance point of view, symmetric algorithms are less resource-intensive and faster than asymmetric algorithms. The radio simulation dataset is the selected payload to encrypt in this study.

3.6. Dynamic NFV Components

This study requires the adoption of dynamic components to perform data transmission and encryption. This type of component runs kernel modules. The OS modules allow script file implementation using Python3 as the programming language.

3.6.1. Ubuntu-Docker (NFV)

Ubuntu is the selected operating system (OS) to simulate the behavior of user equipment (client) and server. Using a container image-hosted operating system can decrease the setup time. Container images are standalone executable packages of software with essential components to run applications. Docker is an open-source container service. Linux command line interface (CLI) commands are needed to interact with the system. Default modules of the system can be updated at will. The extra modules in Table 8 are necessary for data transmission and encryption.

Table 8. Docker-Ubuntu extra modules for data transmission and encryption.

Module	Version	Description
socket	3.8.13	TCP socket
pycryptodome	3.14.1	AES
tqdm	4.30.0	Progress bar

3.6.2. FortiGate (NFV)

FortiGate is a KVM solution provided by FortiNet. The software module includes a graphic user interface (GUI) and a CLI. The module setting allows the data transmission between client and server via ethernet port. For this reason, its behavior in this research can be passive, similar to an ethernet switch. Dynamic testing of FortiOS is under consideration for future work.

3.7. List of Network Components

Testing of 5G technologies included NFV and RAN technologies in the research. To summarize the previous points, Table 9 contains the list of components necessary to conduct data transmission. Physical equipment and its virtual counterpart are both listed. GNS3 allows the connection of the previous list of components to the virtual network via multiple protocols such as TCP or UDP [40]. The connection between the physical and virtual worlds can be considered for future work.

Table 9. List of network components.

Technology	Virtual Equipment	Physical Equipment	Reference Name
NFV	Docker-Ubuntu	5G UE	Ubuntu-DockerGuest-1
NFV, RAN	Ethernet Switch	Ericsson 5G Small Cell	5G AP
NFV	Ethernet Switch	Cisco 2959X	L2_Switch
NFV	FortiGate 6.45	FortiGate 81E	FortiGate6.4.5-1
NFV	Ethernet Switch	Cisco 3850	L3_Switch
NFV	Docker-Ubuntu	Server	Ubuntu-DockerGuest-2

3.8. 5G Network Architecture

This study implements kernel-based virtual machines (KVMs), digital equipment versions. KVM solutions provide similar functions to their physical counterparts, a feature highly valuable for equipment testing. In this study, FortiGate 6.45 and Docker-Ubuntu are dynamic NFV components capable of emulating actual networking hardware activities. FortiGate KVM requires the acquisition of its physical hardware. Therefore, its use is limited to authorized users. In contrast, Ubuntu-Docker is an open-source container service.

A 5G network is composed of two sections: front-haul and back-haul. The front-haul section refers to the network's wireless access (5G RAN) and establishes a radio connection between the user equipment (UbuntuDockerGuest-1) and the 5G Access Point. In contrast, back-haul access relies on a fiber to the X (FTTX) connection and connects the different network equipment components.

The network representation of the Industry 4.0 Implementation Center has been conducted with GNS3, as shown in Figure 3. Its architecture consists of public and private networks. The blue highlighted block (Figure 3, left) is the "known" part of the network (private). This part of the network uses NFV components to send data. The connection between Ericsson's 5G access point (Small Cell) and L2 Switch (OSI Layer 2) relies on the FTTX connection. A firewall is after the L2 Switch to guarantee network security.

The red block on the right side of Figure 3 is the highlighted public network, which interconnects to private networks via the L3 switch (OSI Layer 3). The primary function of this part is to guarantee access to the internet.

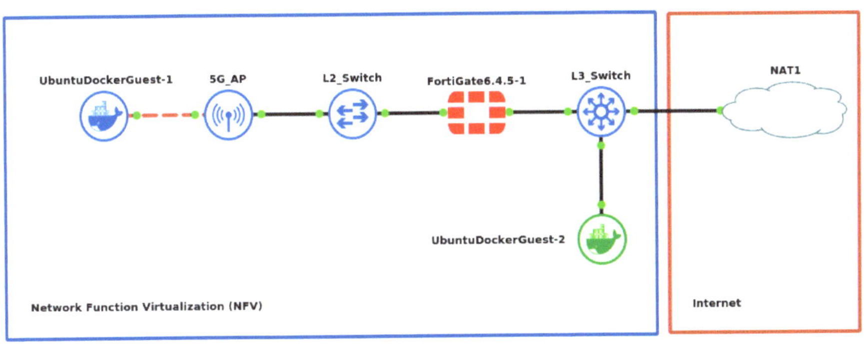

Figure 3. 5G network architecture.

Figure 3 represents the network layout of the private network. Therefore, the image represents a dynamic network simulator system rather than graphic icons. The components included in Figure 3 are KVM solutions (UbuntuDocker-1, UbuntuDocker-2, and FortiGate 6.4.5-1) and open-source NFV components (ethernet switch). Linux command line interface (CLI) commands are required to interact with the system to control each component.

3.9. Method

The radio parameters needed for simulation vary in each country. These 5G radio parameters are required for the deterministic radio simulator. In this case, the operating frequency band in this study is N78.

The simulator requires a CAD file representation to simulate radio coverage. A material library provided by WinProp allows material attributes to the CAD file. GNS3 is the network simulation host of the virtual network, and Ubuntu-Docker allows data transmission and encryption. The generated dataset contains cartesian coordinates and their predicted power value.

Data encryption relies on the AES algorithm, and it is an optional feature. To send data between the client (UbuntuDockerGuest-1) and server (UbuntuDockerGuest-2), Ubuntu-Docker uses TCP protocol. The selected port for data transmission is 21,880. After establishing a TCP connection, data transmission is possible. The packet analyzer (Wireshark) is required to observe data transmission. Figure 4 shows the workflow of the execution process of the 5G digital twin.

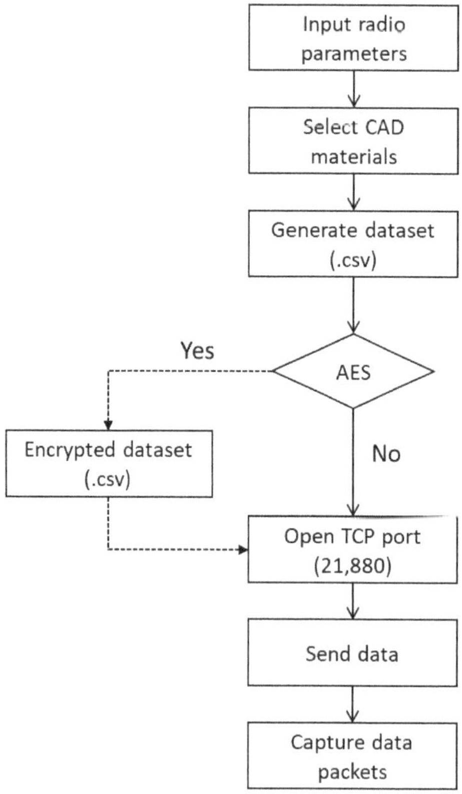

Figure 4. 5G digital twin workflow.

4. Results

Wireshark is the selected packet analyzer for this study, which provides a graphical representation of TCP streams, round-trip time (RTT), and throughput. The TCP protocol is executed in both client and server to establish communication. After the server-side data transmission port opens, the client sends data to the selected IP address and its respective port. The following points can explore the network performance differences between the data packages.

4.1. Data Size Comparison

In this study, two data packages are available: non-encrypted and encrypted packets, and both save in comma-separated value (.CSV) files. Non-encrypted package refers to the generated radio dataset containing raw data. Encrypted package refers to the generated dataset after encryption with the AES algorithm. After encryption, the new file is 34 percent larger than the non-encrypted dataset. Table 10 represents the difference in file size among these two data packages.

Table 10. Size comparison between encrypted and non-encrypted data packages.

Data Package	Size (kB)
Non-encrypted	32.7
Encrypted	43.7

4.2. TCP Streams

The first step of establishing a TCP session is the client's and server's handshake. The data transmission begins after the connection is established. Figure 5 represents both handshake (green) and data transmission (red).

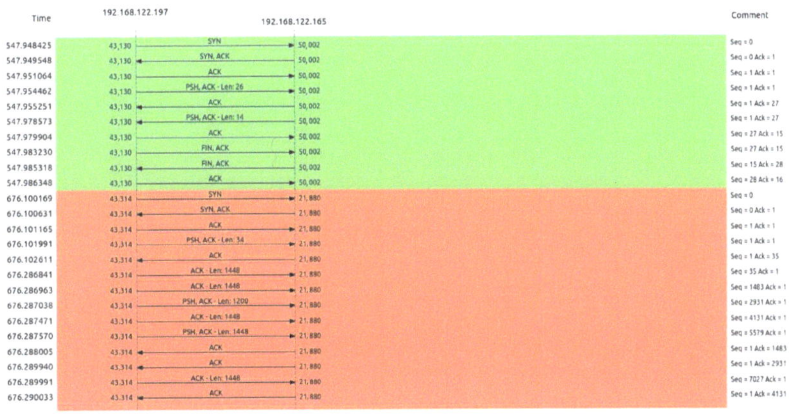

Figure 5. TCP stream graph.

4.3. Round-Trip Time

The time between request and response from the client to server, measured in milliseconds, represents the round-trip time of the network. RTT can be affected by distance, transmission medium, number of network hops, traffic levels, or server response time.

In this study, the main difference between encrypted and non-encrypted file transmission is the size of the document. Figures 6 and 7 represent the different RTT times of non-encrypted and encrypted files, respectively.

4.4. Average Network Throughput

Network throughput is the amount of data transmitted from client to server in a given period, and the measurement unit is the bit per second (bits/s). The horizontal axis represents the time in seconds for each data section. Enforced limitations, network congestion, latency or packet loss, and errors can be affected throughput.

Non-encrypted file requires 0.02 s to achieve its highest throughput, 260,000 (bits/s). In comparison, the encrypted file requires 0.025 s to achieve a higher throughput of 350,000 bits/s. Figures 8 and 9 represent both non-encrypted and encrypted files.

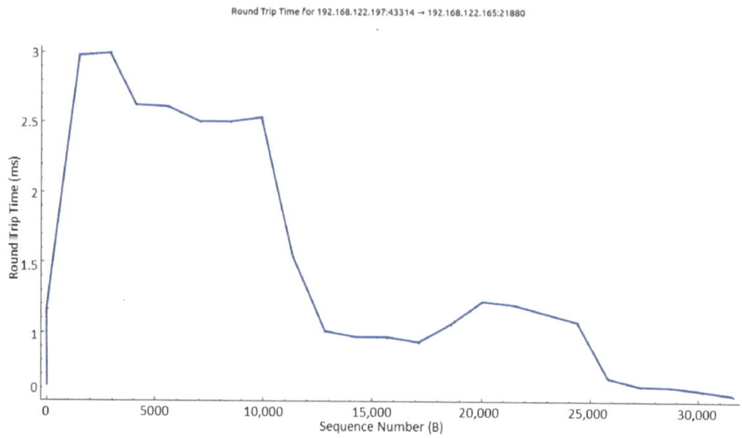

Figure 6. Non-encrypted file RTT.

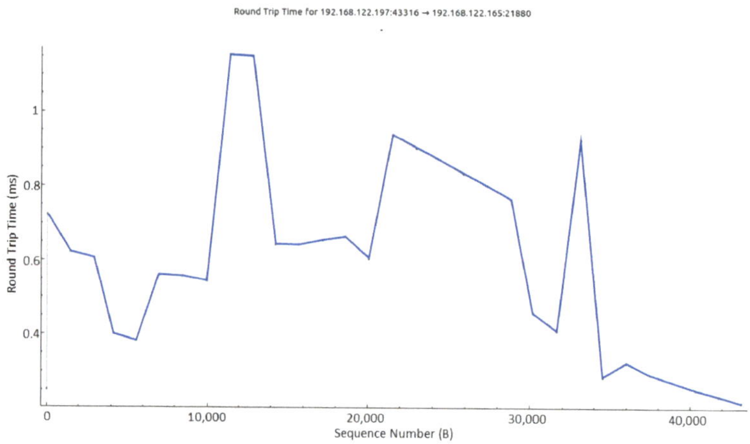

Figure 7. Encrypted file RTT.

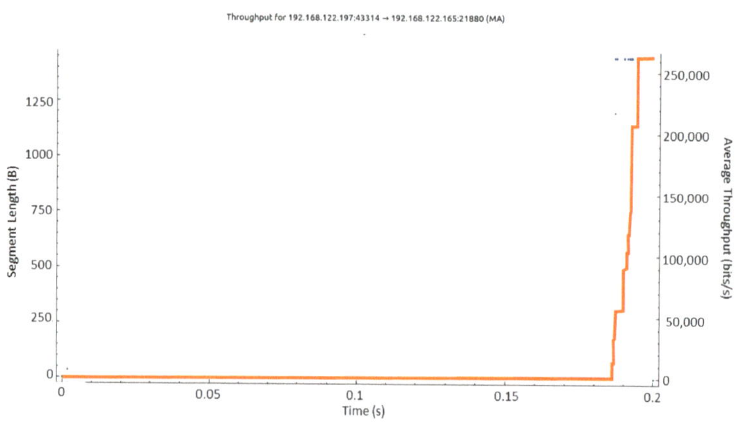

Figure 8. Non-encrypted file throughput.

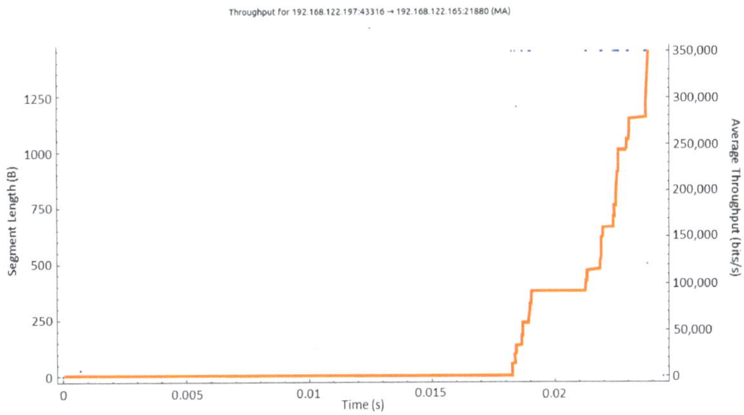

Figure 9. Encrypted file throughput.

4.5. Data Package Analyzer

Wireshark allows packet sniffing, capturing the data stream from client to server. After selecting the target package, Wireshark provides a representation in both HEX and UTF-8 values. Without data encryption, raw values are exposed to both white-hat and black-hat hackers. Figure 10 displays the compromised values on its right side, which are identical to the dataset obtained in Table 6.

```
0000  22 40 97 d9 2c ec 5e 17  ab 92 20 ec 08 00 45 00   "@  , ^       E
0010  05 dc 56 11 40 00 40 06  68 4f c0 a8 7a c5 c0 a8     V @·@  hO  z
0020  7a a5 a9 32 55 78 bb db  3d 13 d4 56 b5 46 80 10   z  2Ux   =  V F
0030  01 f6 cb 97 00 00 01 01  08 0a 35 b1 14 31 9e 01              5  1
0040  11 4b 33 30 39 34 35 2e  32 36 39 09 34 38 37 34    K30945. 269 4874
0050  2e 36 39 37 09 2d 34 38  2e 30 34 37 0a 33 30 39   .697 -48 .047 309
0060  34 36 2e 32 36 39 09 34  38 37 34 2e 36 39 37 09   46.269 4 874.697
0070  2d 34 38 2e 30 37 39 0a  33 30 39 34 37 2e 32 36   -48.079  30947.26
0080  39 09 34 38 37 34 2e 36  39 37 09 2d 34 38 2e 38   9 4874.6 97 -48.8
0090  31 31 0a 33 30 39 34 38  2e 32 36 39 09 34 38 37   11 30948 .269 487
00a0  34 2e 36 39 37 09 2d 35  30 2e 31 39 34 0a 33 30   4.697 -5 0.194 30
00b0  39 34 39 2e 32 36 39 09  34 38 37 34 2e 36 39 37   949.269  4874.697
00c0  09 2d 35 31 2e 37 34 38  0a 33 30 39 35 30 2e 32    -51.748  30950.2
00d0  36 39 09 34 38 37 34 2e  36 39 37 09 2d 35 33 2e   69 4874. 697 -53.
00e0  36 34 39 0a 33 30 39 35  31 2e 32 36 39 09 34 38   649 3095 1.269 48
00f0  37 34 2e 36 39 37 09 2d  35 34 2e 32 33 35 0a 33   74.697 - 54.235 3
0100  30 39 35 32 2e 32 36 39  09 34 38 37 34 2e 36 39   0952.269  4874.69
```

Figure 10. Non-encrypted data package capture.

The AES algorithm requires the use of the same key for encryption and decryption, and it can also provide excellent security if the key is not compromised. In contrast, Figure 11 represents the ciphertext values on its right side, which are extremely hard to decode.

```
0000  22 40 97 d9 2c ec 5e 17  ab 92 20 ec 08 00 45 00   "@·  , ^      E
0010  05 dc fa 0e 40 00 40 06  c4 51 c0 a8 7a c5 c0 a8       @ @   Q  z
0020  7a a5 a9 34 55 78 32 7f  a6 27 8f fe 2d 93 80 10   z  4Ux2    '  -
0030  01 f6 d3 e1 00 00 01 01  08 0a 35 cd 56 16 9e 1d              5 V
0040  53 d7 32 45 61 35 72 77  4e 4a 38 79 54 62 55 68   S 2Ea5rw NJ8yTbUh
0050  74 79 7a 54 43 50 35 42  4c 59 6b 54 41 6d 69 5a   tyzTCP5B LYkTAmiZ
0060  6f 58 50 68 4a 36 53 79  6f 79 35 63 49 55 4d 58   oXPhJ6Sy oy5cIUMX
0070  7a 70 44 65 6c 56 4c 36  74 37 72 33 7a 72 6c 6b   zpDelVL6 t7r3zrlk
0080  67 73 45 32 55 76 33 68  6f 2b 30 65 78 32 41 70   gsE2Uv3h o+0ex2Ap
0090  6f 52 42 65 79 37 34 30  65 58 4e 65 50 30 54 4e   oRBey740 eXNeP0TN
00a0  47 4e 74 73 51 73 71 2f  47 42 43 6e 76 69 53 6d   GNtsQsq/ GBCnviSm
00b0  74 6a 53 35 37 55 72 31  55 65 37 33 6e 71 58 67   tjS57Ur1 Ue73nqXg
00c0  31 41 7a 56 6b 46 76 38  67 72 52 36 55 36 79 74   1AzVkFv8 grR6U6yt
00d0  48 64 76 73 77 59 43 67  57 72 78 48 39 4c 38 7a   HdvswYCg WrxH9L8z
00e0  6d 52 58 45 38 69 53 6c  31 38 36 54 56 4c 77 65   mRXE8iSl 186TVLwe
00f0  4b 35 41 79 67 65 6e 6c  37 4b 6f 57 53 46 45 4f   K5Aygenl 7KoWSFEO
0100  45 59 6e 42 68 48 38 52  78 70 6f 69 68 71 2f 36   EYnBhH8R xpoihq/6
```

Figure 11. Encrypted data package capture.

5. Discussion

The implementation of the different technologies presented in this study facilitates the simulation of 5G digital twins. The network simulator (GNS3) allows the combination of NFV components and data transmission across a virtual network. The deterministic radio simulator provides a coverage map of the warehouse scenario. This data can be exported as a comma-separated value file (.CSV file), generating a 5G radio coverage dataset with cartesian coordinates and power values. The use of dynamic network components is required for sending data. In this regard, Ubuntu-Docker KVM is quite helpful. This component allows extra modules in a given scripting language (Python3). The TCP protocol enables client and server connection by denoting a specific port (21880). The selected transmission mode for this study is simplex. This communication mode is unidirectional between client to server. The transmission of files is individual. Wireshark is the software used to monitor data transmission.

Data security is a constant challenge in data communication networks. The packet analyzer allows the capture of packages across the network. Capturing data is known as packet sniffing, and it could represent a security risk when dealing with more sensitive data. The AES algorithm has been adopted in this study to hedge this risk.

This study compares the difference between encrypted and non-encrypted files across the network. RTT time between files is slightly different, ranging from 1.2 to 3 milliseconds. Throughput measures the amount of data transmitted from client to server in a given period. Therefore, when the data package increases, the throughput also tends to increase. In this case, the file size increases an extra 34 percent after encryption. The encrypted file takes an extra 0.005 s to achieve its maximum throughput, which is 90,000 bits/s higher than its counterpart. The extra 0.005 s represent a 25 percent increase in time. In other words, the extra size of the file significantly influences time and throughput.

Packet capture exposes the vulnerability of non-encrypted files. The raw data can be decoded and easily read by an attacker. In contrast, data encrypted with the AES algorithm can be captured but not easily decoded.

6. Conclusions

Three main issues are common when implementing a digital twin. To begin, digital twin needs a clear definition of tools and technologies. In addition, the applications need to consider real-world physical data. Finally, digital twins must adopt the expected features beyond 5G/6G networks. This study provides the following solutions to solve the main issues. Firstly, this study is a practical analysis of Altair Feko and GNS3 software tools. Secondly, our study considers the influence of real-world materials in the radio section. Finally, this study adopts network function virtualization (NFV) technologies for 5G networks.

In 5G scenarios, using simulation software for radio enables the adoption of essential digital twin technologies. The generated dataset represents the payload for transmission between client and server. Network monitoring and advanced security features depend on implementing NFV technologies. In this regard, communication between virtual network components has been achieved. This study adopted a specific encryption method (AES). This method solved packet sniffing issues by encrypting the dataset in this study.

Filtering mechanisms in firewalls can alter the network's performance. FortiGate KVM functions can simulate virtual firewalls, which is a topic of further study. Finally, our future goal is to enable a two-way connection that allows telnet communication between hardware and software in the GNS3 platform. Two-way connections allow interaction between physical and digital environments. The acquisition of hardware components, such as industrial boards with 5G capabilities or software-defined radio (SDR) modules, is currently under consideration and included in our future research development plan.

Author Contributions: Methodology and software, R.R. and S.-H.L.; measurements and data curation, R.R.; and writing—original draft preparation, C.-Y.H. All authors have read and agreed to the published version of the manuscript.

Funding: This research was funded by MOST 110-2218-E-011-011, the funded project of the Ministry of Science and Technology (MOST) of Taiwan. The experiment field and equipment support were mainly provided by the Center for Cyber-Physical System Innovation (CPSi), National Taiwan University of Science and Technology (NTUST), Taiwan.

Institutional Review Board Statement: Not applicable.

Informed Consent Statement: Not applicable.

Data Availability Statement: The data presented in this study are available on request from the corresponding author. The data are not publicly available due to their use in future research activities.

Conflicts of Interest: The authors declare no conflict of interest.

References

1. Dave, E. *The Internet of Things: How the Next Evolution of the Internet Is Changing Everything*; White paper; Cisco Internet Business Solution Group: San Jose, CA, USA, 2011.
2. ERICSSON. *"5G: What Is it?"*; White paper; ERICSSON: Kista, Stockholm, Sweden, 2014.
3. 3GPP TS 23.501, 5G.; System Architecture for the 5G System (5GS) (3GPP TS 23.501 Version 16.6.0 Release 16). Available online: https://portal.3gpp.org/desktopmodules/Specifications/SpecificationDetails.aspx?specificationId=3144 (accessed on 13 June 2022).
4. Godfrey, A.A.; Bruno, J.S.; Gerhard, P.H.; Adnan, M.A.-M. A Survey on 5G Networks for the Internet of Things: Communication Technologies and Challenges. *IEEE Access* **2018**, *6*, 3619–3647. [CrossRef]
5. Hilary, F.; Carlos, C.-M.; Assis, K.D.R.; Shuangyi, Y.; Dimitra, S. Techno-Economic Analysis of 5G Non-Public Network Architectures. *IEEE Access* **2022**, *10*, 70204–70218. [CrossRef]
6. Norbert, W. *Cybernetics: Or Control and Communication in the Animal and the Machine*, 2nd ed.; Hermann & Cie: Cambridge, MA, USA, 1961; MIT Press: Paris, France, 1948; ISBN 978-0-262-73009-9.
7. Fei, T.; Meng, Z.; Nee, A.Y.C. Chapter 12—Digital twin, cyber-physical system, and internet of things. In *Digital Twin Driven Smart Manufacturing (Science Direct)*; Academic Press: Cambridge, MA, USA, 2019; pp. 243–256. [CrossRef]
8. Guo, J.; Lv, Z. Application of Digital Twins in multiple fields. *Multimed. Tools Appl.* **2022**, *81*, 26941–26967. [CrossRef] [PubMed]
9. Michael, G.; John, V.; (Florida Institute of Technology, Melbourne, FL, USA). Origins of the Digital Twin Concept. Personal Communication, 2016. [CrossRef]
10. Fei, T.; Qinglin, Q.; Ang, L.; Andrew, K. Data-driven smart manufacturing. *Manuf. Syst.* **2018**, *48 Pt C*, 157–169. [CrossRef]
11. Jessica, O.A.; Alejandro, V.S. Supply Chain Resilience Roadmaps for Major Disruptions. *Logistics* **2021**, *5*, 78. [CrossRef]
12. Gerlach, B.; Zarnitz, S.; Nitsche, B.; Straube, F. Digital Supply Chain Twins—Conceptual Clarification, Use Cases and Benefits. *Logistics* **2021**, *5*, 86. [CrossRef]
13. Lars, H.; Jan, H.; Christian, R. Challenges of wifi-enabled and solar-powered sensors for smart ports. In Proceedings of the 4th International Workshop on Energy Harvesting and Energy-Neutral Sensing Systems (ENSsys'16), New York, NY, USA, 14 November 2016; pp. 13–18. [CrossRef]
14. Manoshi, D.T.; Mohammad, M.K.; Manjit, K.; Atef, Z. Smart Supply Chain Management Using the Blockchain and Smart Contract. *Sci. Program.* **2021**, *2021*, 6092792. [CrossRef]
15. Haobin, L.; Xinhu, C.; Pankaj, S.; Loo, H.L.; Ek, P.C. Framework of O2DES.NET digital twins for next generation ports and warehouse solutions. In Proceedings of the Winter Simulation Conference (WSC), Orlando, FL, USA, 14 December 2020; pp. 3188–3199. [CrossRef]
16. Huang, X.N.; Ramona, T.; Duc, T.; Mallik, T. Digital Twin for 5G and Beyond. *IEEE Commun. Mag.* **2021**, *59*, 10–15. [CrossRef]
17. Mourtzis, D.; Angelopoulos, J.; Panopoulos, N. Smart Manufacturing and Tactile Internet Based on 5G in Industry 4.0: Challenges, Applications and New Trends. *Electronics* **2021**, *10*, 3175. [CrossRef]
18. Qi, Q.; Tao, F.; Hu, T.; Answer, N.; Liu, A.; Wei, Y.; Wang, L.; Nee, A.Y.C. Enabling technologies and tools for digital twin. *J. Manuf. Syst.* **2021**, *58*, 3–21. [CrossRef]
19. Jin, Z.; Lei, G.; Yueqiao, L. Application of Digital Twin Combined with Artificial Intelligence and 5G Technology in the Art Design of Digital Museums. *Wirel. Commun. Mob. Comput.* **2022**, *2022*, 8214514. [CrossRef]
20. Hu, W.; Zhang, T.; Deng, X.; Liu, Z.; Tan, J. Digital twin: A state-of-the-art review of its enabling technologies, applications and challenges. *J. Intell. Manuf. Spec. Equip.* **2021**, *2*, 1–34. [CrossRef]
21. Zeb, S.; Mahmood, A.; Hassan, S.A.; Piran, M.J.; Gidlund, M.; Guizani, M. Industrial digital twins at the nexus of nextG wireless networks and computational intelligence: A survey. *J. Netw. Comput. Appl.* **2022**, *200*, 103309. [CrossRef]
22. Bjoern, H.; Arne, S.; Anders, E.; Ranvir, C.; Paulo, M.; Henrik, A. 5G NR test-bed 3.5 GHz coverage results. In Proceedings of the 2018 IEEE 87th Vehicular Technology Conference (VTC Spring), Porto, Portugal, 3 June 2018; pp. 1–5. [CrossRef]

23. Ali, E.; Katina, K. Small-Scale 5G Testbeds for Network Slicing Deployment: A Systematic Review. *Wirel. Commun. Mob.Comput.* **2021**, *2021*, 665216. [CrossRef]
24. Martin, K.W.; Dirk, P. A Bayesian approach for RF-based indoor localisation. In Proceedings of the 4th International Symposium on Wireless Communication Systems, Trondheim, Norway, 17 October 2007; pp. 133–137. [CrossRef]
25. Stanislav, V.; Alberto, M.; Antonio, P.; Diego, R.L. A digital twin network for security training in 5G industrial environments. In Proceedings of the IEEE 1st International Conference on Digital Twins and Parallel Intelligence (DTPI), Beijing, China, 15 July 2021; pp. 395–398. [CrossRef]
26. 3GPP TS 38.101-1, 5G NR.; User Equipment (UE) Radio Transmission and Reception Part 1: Range 1 Standalone (3GPP TS 38.101-1 version 15.5.0 Release 15). Available online: https://portal.3gpp.org/desktopmodules/Specifications/SpecificationDetails.aspx?specificationId=3283 (accessed on 13 June 2022).
27. OpenSignal. Available online: https://www.opensignal.com/apps#section-os-app (accessed on 21 July 2022).
28. VE2DBE. Available online: https://www.ve2dbe.com/rmonline_s.asp (accessed on 21 July 2022).
29. Altair Feko. Available online: https://www.altair.com/feko-applications (accessed on 21 July 2022).
30. GNS3. Getting Started with GNS3. Available online: https://docs.gns3.com/docs/ (accessed on 22 June 2022).
31. Cisco. Cisco UCS Manager Administration Management Guide 3.1 (KVM Console). Available online: https://www.cisco.com/c/en/us/td/docs/unified_computing/ucs/ucs-manager/GUI-User-Guides/Admin-Management/3-1/b_Cisco_UCS_Admin_Mgmt_Guide_3_1/b_Cisco_UCSM_GUI_Admin_Mgmt_Guide_3_1_chapter_01111.html (accessed on 17 June 2022).
32. vMX Getting Started Guide for KVM. Available online: https://www.juniper.net/documentation/us/en/software/vmx/vmx-getting-started/index.html (accessed on 17 June 2022).
33. Architecture Kunpeng BoostKit for Virtualization (Technical White Paper, Virtualization Architecture). Available online: https://support.huaweicloud.com/intl/en-us/twp-kunpengcpfs/kunpengcpfs_19_0005.html (accessed on 17 June 2022).
34. KVM Administration Guide. Available online: https://docs.fortinet.com/document/fortigate-private-cloud/6.4.0/kvm-administration-guide/706376/about-fortigate-vm-on-kvm (accessed on 17 June 2022).
35. Ubuntu. KVM Hypervisor: A Beginners' Guide. Available online: https://ubuntu.com/blog/kvm-hyphervisor (accessed on 17 June 2022).
36. Afolabi, I.; Taleb, T.; Samdanis, K.; Ksentini, A.; Flinck, H. Network Slicing and Softwarization: A Survey on Principles, Enabling Technologies, and Solutions. *IEEE Commun. Surv. Tutor.* **2018**, *20*, 2429–2453. [CrossRef]
37. Lucas, N.T.; Cleverson, N.; Silvia, L.; Aldebaro, K. Middleware implementation for RYU SDN controller to manage switches in a C-RAN scenario. In Proceedings of the 48th Integrated Software and Hardware Seminar, Porto Alegre, Brazil, 23 July 2021. [CrossRef]
38. Internet Protocol Family. Available online: https://wiki.wireshark.org/InternetProtocolFamily (accessed on 22 June 2022).
39. Dworkin, M.; Barker, E.; Nechvatal, J.; Foti, J.; Bassham, L.; Roback, E.; Dray, J. *Advanced Encryption Standard (AES), Federal Inf. Process. Stds. (NIST FIPS)*; National Institute of Standards and Technology: Gaithersburg, MD, USA, 2001; pp. 9–51. [CrossRef]
40. Petrovic, R.; Simic, D.; Drajic, D.; Cica, Z.; Nikolic, D.; Peric, M. Designing Laboratory for IoT Communication Infrastructure Environment for Remote Maritime Surveillance in Equatorial Areas Based on the Gulf of Guinea Field Experiences. *Sensors* **2020**, *20*, 1349. [CrossRef] [PubMed]

applied sciences

MDPI

Article

Measuring Procedures for Evaluating the Surface Roughness of Machined Parts

Katarína Palová, Tatiana Kelemenová * and Michal Kelemen

Faculty of Mechanical Engineering, Technical University of Kosice, Letna 9, 04200 Kosice, Slovakia; katarina.palova@student.tuke.sk (K.P.); michal.kelemen@tuke.sk (M.K.)
* Correspondence: tatiana.kelemenova@tuke.sk; Tel.: +421-556022361

Abstract: This article deals with the problems that arise during the evaluation of the surface of the components, which can have a fundamental impact on the functionality and service life of the component as well as the entire product. A fast method for estimating the condition and maximum error of the surface tester used for assessing the surface roughness of components was proposed. Using the secondary surface roughness standard, a uniform distribution law of the measured data was experimentally identified as a suitable model, according to which it is then possible to determine the coverage factor for determining the measurement uncertainty. The design of surface roughness measurement methodology for Ra determination is critical when selecting optimal settings of conditions and measurement parameters for various types of component surface treatment technologies. This is mainly due to the effect of the selection of the evaluation standard, the effect of the measurement speed, and the effect of setting the cut-off filter for the evaluation of the quantity Ra. Etalon samples of surface roughness realized by different technologies were used for experimental verification of these phenomena. Large sets of measurements were carried out on selected samples from each technology, and, accordingly, a methodology was proposed for determining the optimal number of measurements for determining the surface roughness of the investigated component, which represents an important factor for achieving measurement results with minimal measurement uncertainties at low financial costs.

Keywords: surface roughness; measurement; uncertainty of measurement; stylus; technology

Citation: Palová, K.; Kelemenová, T.; Kelemen, M. Measuring Procedures for Evaluating the Surface Roughness of Machined Parts. *Appl. Sci.* **2023**, *13*, 9385. https://doi.org/10.3390/app13169385

Academic Editors: Tibor Krenicky, Maros Korenko and Juraj Ruzbarsky

Received: 31 July 2023
Revised: 12 August 2023
Accepted: 12 August 2023
Published: 18 August 2023

1. Introduction

Surface roughness is a key parameter in many industrial devices and significantly affects the quality and equipment composed of several machined components; therefore, reliable measurement of surface roughness plays a fundamental role in ensuring the functionality and reliability of the final product. However, the measurement of surface roughness is not without problems, especially when it comes to quantifying the uncertainty associated with these measurements. This estimate of uncertainty is necessary to evaluate the reliability and comparability of the surface roughness data and, to the same extent, the measurement surface roughness data obtained.

There are several measuring methods and devices for investigating surface roughness, and generally, we can categorize them into two main groups, contact and non-contact methods and devices.

There are a number of devices available for evaluating surface roughness, and they are probably best divided according to individual methods of measuring and evaluating surface roughness.

The device for contact stylus-based contact measurements is one of the most widely used methods for measuring surface roughness, where the contact stylus profilometer is probably the only representative. The disadvantages of the contact-based measurement include mainly the stylus tip wear caused due to the contact between the sharp tip of the stylus and the surface roughness. The sample's surface will be scratched because of the movement of the stylus tip on the surface during the measurement, and the measurement is limited to the radius of the stylus tip.

Devices for non-contact methods of assessing surface roughness are more expensive. There are many devices and methods for non-contact methods, such as laser triangulation, optical interferometry, confocal microscopy, and atomic force microscopy, which is the most accurate of these methods. The scan quality of these measuring instruments is mainly dependent on the optical properties of the materials tested.

One of the common methods for determining surface roughness is the contact pen method or stylus profilometers, which provides relatively quick measurement results at a favorable price than other methods that require relatively expensive equipment. For production practice, the contact method is thus an ideal tool for evaluating the surface of machined parts, but this contact method is associated with a number of problems because the very implementation of this method depends on the correct setting of the measurement conditions and the correct implementation of the measurement and evaluation of surface roughness values.

Experience from different experiments and work with this contact method shows that inappropriate setting of the wavelength limit, also known as cut-off, inappropriate stylus speed, and inappropriate standard for evaluation, causes diametrically different results of surface roughness assessment and thus, the whole process of surface roughness assessment is then completely devalued. Another task is to understand and quantify the uncertainty of the measurement and evaluation of surface roughness values for each instrument quality, calibration standards, measurement conditions, and the way the surface of the part is produced, which can significantly validate the results and evaluations. For common practice, it is necessary to create a methodology for the optimal process of evaluating the surface of machined parts and determining the measurement uncertainty estimate.

In this article, we will delve into the topic of surface roughness measurement with a focus on the methodology of implementing the contact method and determining the estimate and evaluation of measurement uncertainty. Investigating and understanding specific issues is important to the operator in obtaining reliable results from the surface roughness evaluation process. This issue is very extensive, so for this article, we will deal with quantities—the arithmetic mean deviation of the assessed roughness profile and the quality of the surface of the components is relevant for the assessment.

Author Whitehouse [1] provides a definition for the term "surface metrology", which aims to optimize the functions of the production process and its tools to support effective management. This optimization directly affects the quality of products, making surface metrology crucial in predicting outcomes within the production process.

In the paper by Vorburger [2], a comprehensive review of various methods for measuring surface texture is presented, accompanied by a classification scheme for these methods.

Kumar's work [3] focuses on the measurement and uncertainty analysis of surface roughness, specifically addressing the influence of cutting conditions during turning. The study investigates the arithmetic mean deviation of surface roughness (*Ra*) and Material Removal Rate (MRR).

In the study conducted by Beno [4], a practical methodology for measuring roughness when turning with tool inserts featuring variable tool faces is introduced. The research emphasizes the importance of adopting a unified roughness measurement methodology to compare the capability of different tool inserts. The study evaluates measured data such as *Ra*, *Rq*, *Rz*, *Rt*, and *Rp*.

Davinci [5] conducts a detailed experimental study to assess the individual and combined effects of scanning speed and spacing on surface roughness parameters. The study focuses on a reference specimen, selecting profile parameters (*Ra*, *Rq*, and *Rz*) and surface parameters (*Sa*, *Sq*, and *Sz*) as response variables. The findings highlight the significant impact of scanning speed on *Ra*, *Rz*, *Sa*, and *Sz* measurements, while spacing affects all profile and surface measurements.

Surface roughness measurements are primarily performed using stylus roughness meters for machined surfaces. However, atomic force microscopes (AFM) are also utilized for surface roughness evaluation, including the assessment of uncertainty for surface roughness parameter *Ra* [6].

Surface topography measurements are becoming more and more popular and complement the 2D analysis of surface texture [7–10]. The selection of the measurement area is not yet included in the standards, and the size of this area affects the values of the determined parameters. The article [11] presents the results of research on determining the measurement area based on the smooth-rough crossover scale (SCR) and mean profile element spacing (*RSm*) parameters. Other works also address the measurement of the surface roughness of components [12–24].

The aim of the article is to determine the uncertainty estimate of the measurement of arithmetic mean deviation of surface roughness *Ra* for the used measuring device of surface roughness quantities and to verify the measurement methodology from the point of view of setting the measurement parameters. The problem is that the manufacturer usually does not specify the maximum permissible error of the measuring device, and even for surface roughness standards (calibration etalon), the maximum error of determining its nominal value is not declared either, and therefore, values must be determined by calibration.

A fundamental problem in this measurement is the appropriate measurement conditions, and the measurement and evaluation of surface roughness values are characterized by the fact that it is sensitive to the correct selection of measurement conditions. Although there are recommendations that are in some available standards, they may not always take into account the actual measurement conditions, and it is advisable to verify them for specific conditions. There is no suitable methodology for the correct setting and selection of measurement conditions, which would ensure the achievement of the correct measurement result with minimal measurement uncertainty for various technologies for processing parts' surfaces.

It is obvious from common metrological practice that one measurement is insufficient to determine the value of the measured quantity, so the unknown value of the measured quantity is approximated by the mean value (arithmetical mean deviation), determined as the arithmetic mean of repeated measurements. The question is, however, what is the sufficient number of measurements needed for determining the value of the measured quantity, and what effect does this number of measurements have on the reported measurement uncertainty? The seemingly best solution is to perform as many measurements as possible in order to minimize the measurement uncertainty. However, here it is also necessary to take into account the economic point of view, as the economic costs and the time required to carry out the measurements are also important criteria when determining the number of measurements. Therefore, it is necessary to find the optimal number of measurements so that the measurement uncertainty is sufficient, and, at the same time, it is advantageous from an economic and time point of view.

Based on the above, the research criteria are defined:

Criterion 1: Identification of suitable measurement conditions to achieve minimal measurement uncertainty.

Criterion 2: Determination of the optimal number of repeated measurements to achieve minimum measurement uncertainty.

Criterion 3: Identifying the possibility of rapid measurement uncertainty estimation.

Criterion 4: Identification or verification of the distribution law of a random measured variable.

Three main contributions of this work can be summarized as follows:

(1) Determination of the estimate of the condition of the used surface tester for assessing the surface roughness values of the parts.

(2) Proposal of a methodology for measuring surface roughness for determination of Ra when determining the optimal settings of conditions and measurement parameters for various types of technologies for processing parts' surfaces. This is mainly the effect of the

selection of the evaluation standard, the effect of the measurement speed, and the effect of setting the cut-off filter for the evaluation of the quantity *Ra*.

(3) Determination of the optimal number of measurements to determine the roughness of the surface of the examined part, which represents an important factor for achieving measurement results with minimal measurement uncertainties at low financial costs.

The rest of this paper is organized as follows. Section 2 describes the methodology used in this article. Section 3 is focused on the presentation of experiments for solving defined aspects of the surface roughness evaluation process. In the last part of the work, there is a discussion of the results in Section 4. Section 5 concludes this paper. Appendix A defines the basic terminology and describes the methodology employed in this study. Appendix B presents experiments carried out on samples with a non-periodic surface roughness profile. Appendix C contains the results of experiments on samples with a periodic surface roughness profile.

2. Methodology

Surface roughness is a geometric property of the surface, and there are no direct methods to measure it. Surface roughness has a decisive influence on the coefficient of friction of contacting surfaces, on adhesion between surfaces, and various other properties and phenomena; it also has direct effects on the material properties of components [25–32] but also has a key role in other areas, where it significantly affects the function of various devices. Terminology according to the standards and used methods of measurement and data processing is processed in Appendix A.

Profile curves are usually measured, and appropriate characteristics and quantities are evaluated from them, which are considered criteria for assessing surface roughness. Surface roughness parameters and values are defined in standards EN ISO 4287 and EN ISO 5436. These standards were later replaced by standards ISO 21920-1, ISO 21920-2, ISO 21920-3 [33–40].

Current engineering practice is characterized by an increase in requirements for precision and reproducibility of production in the period of production of sophisticated production. The optimal quality of the surface of the parts is one of the prerequisites for the correct function of products (machines, devices, tools) and significantly affects the service life of the parts and even the entire machines. The roughness of the surface represents a collection of irregularities with relatively small distances from a perfect and ideally smooth surface, while these irregularities arise as a result of the used production technology or wear due to use (STN ISO 4287, ISO 21920-1, ISO 21920-2, and ISO 21920-3) [33–36].

In the following sections, we will distinguish the so-called periodic and non-periodic surface roughness profiles (for more details, see Appendix A).

The ISO 21920-3 [39] standard introduces no distinction between periodic and non-periodic profiles compared to the edition of ISO 4288:1996 [34]. Nevertheless, the following part is focused on the analysis in the course of the choice of cut-off filter for individual technologies, which were used to produce individual samples. Just like the other methods, the stylus-based method encounters a whole range of limiting factors that result mostly from the requirements of its specific implementation, and sometimes there are conflicting requirements that require a compromise solution. This method makes it possible to determine the numerical values of normalized and non-normalized characteristics of surface roughness and then use them for the most modern statistical and spectral assessment of surface unevenness. The implementation of this method is ensured by the stylus-based surface roughness tester, which consists of a mechanical and electronic part (Figure 1), while the mechanical deflection, generated by the sensing tip, which monitors the surface irregularities of the measured surface, is transformed into an electrical signal in the induction position transducer, which it is further processed and interpreted as a numerical value of the selected surface roughness characteristic, or as a graphic record of the profile of surface roughness.

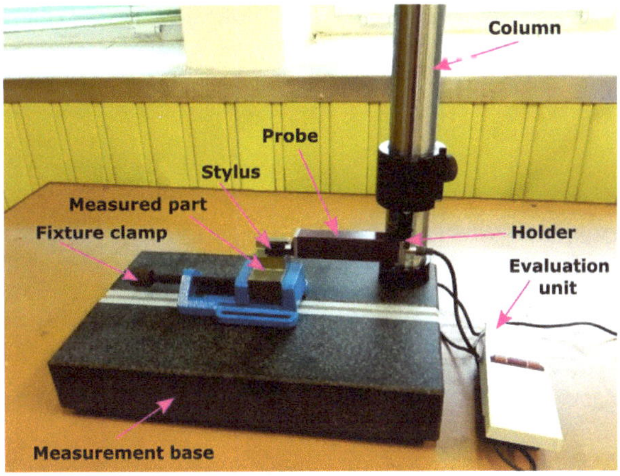

Figure 1. Stylus-based surface roughness measurement using a surface tester.

A touch profilometer is a contact profilometer that brings the stylus tip into direct contact with the measuring surface and tracks or moves along a desired path to determine the topography of the surface. The working principle of a touch profilometer is to analyze the topography of a surface using a probe that physically moves over the surface to obtain surface properties such as height.

An unspecified (due to commercial reasons) stylus-based surface roughness tester was used for experimental measurements. The device allows you to set four different values of the cut-off filter parameter (0.08 mm; 0.25 mm; 0.8 mm; 2.5 mm). However, the manufacturer of the surface tester does not define the maximum permissible error or measurement uncertainty, so it is necessary to determine this uncertainty in some way for measurement purposes.

The surface tester used has a manufacturer-defined resolution for the *Z*-axis, i.e., for measuring height characteristics with a value of 0.02 μm. This data applies to the largest measurement range that this device has available. Since the device has an automatic selection of the measurement range, for reporting measurement uncertainty, we will consider this resolution value as a potential source of measurement uncertainty.

The resolution is not defined for the *X*-axis, but this data can be determined from the sampling that the device performs at individual settings (Table 1). For individual setting values of the cut-off filter parameter, the sampling for the *X*-axis also changes.

Table 1. Surface tester sampling and resolution at different settings.

Cut-Off Filter (mm)	Evaluation Length (mm)	Number of Samples	*X*-Axis Resolution (μm)
0.08	0.4	800	0.5
0.25	1.25	2500	0.5
0.8	4	8000	0.5
2.5	12.5	8330	1.5

This device for evaluating individual quantities uses the course of the surface roughness profile curve with resolution values for individual measurement axes (*X*-axis in the horizontal plane and *Z*-axis in the vertical plane). These are longitudinal measurements, from which the values of individual surface roughness values are then indirectly determined by calculation in accordance with the relevant standards. Thus, determining the measurement uncertainty for *Ra*, for example, is a rather complicated task beyond the scope of this article. In this article, we will attempt at least an approximate estimation of the measurement uncertainty because, in our opinion, each measurement must have an associated measurement uncertainty in order to be able to evaluate the reliability of the mentioned experimental measurements [41–49].

Therefore, if we consider the differences in measurement in both directions as a significant source of uncertainty, then we can state them as the standard uncertainty determined by method B:

$$u_{BX} = 1.5\ \mu\text{m}$$
$$u_{BZ} = 0.02\ \mu\text{m}. \quad (1)$$

If we consider that the digital device and the measured quantities follow the uniform distribution law of random values, then we will consider the extended measurement uncertainty with a coverage coefficient with a value of $\sqrt{3}$ at the significance level of 0.95. Then we will consider the expanded uncertainties of the displacement measurement:

$$u_{BX} = u_{BX} \cdot k_{\text{cov}} = 2.6\ \mu\text{m}$$
$$u_{BZ} = u_{BZ} \cdot k_{\text{cov}} = 0.035\ \mu\text{m}. \quad (2)$$

It should be emphasized that this is only a very rough estimate, and for a more precise determination of the measurement uncertainty, it is also necessary to consider the uncertainty resulting from the geometry of the measuring tip, the uncertainty resulting from its elastic deformation during measurement, the uncertainty resulting from lateral movements during measurement, the uncertainty resulting from the noise of the device. It is very complicated to identify all components of measurement uncertainty when using such a complex device.

The surface tester used has the function of automatic calibration using secondary surface roughness standards. Secondary surface roughness standards for surface roughness testers, which is a calibration etalon artifact with unidirectional irregular profile of type D1 (according to ISO 5436-1:2000) [36], is recommended for the calibration of the gauge. For this artifact, the value of *Ra* and *Rz* is defined. In this work, this function was used, and the calibration process was started before each batch of measurements; after the calibrations, a control measurement was always carried out to verify the calibration process. Likewise, at the end of the measuring dose, a control measurement was carried out to verify the stability of the surface tester calibration. Secondary surface roughness standards for surface roughness testers (Figure 2) were verified by a superior standard in a certified laboratory, and the uncertainty of this used calibration specimen was determined.

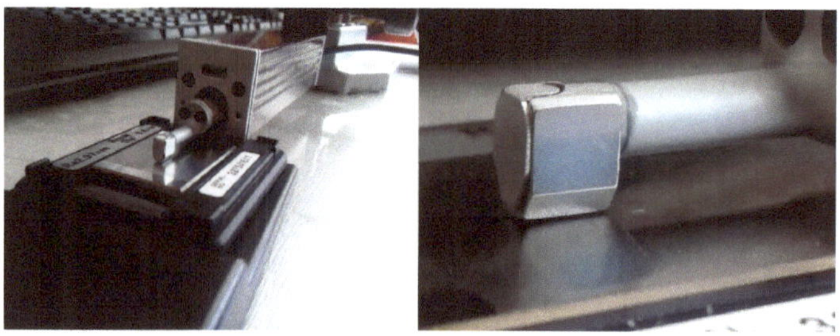

Figure 2. Surface tester calibration using secondary surface roughness standards.

One way to determine the measurement uncertainty is to look at the gauge as a black box and determine the measurement uncertainty from the statistical evaluation of the data measured on the secondary surface roughness standards. For this purpose, 100 measurements were carried out at the same place under unchanged measurement conditions on the secondary surface roughness standard (measured to the standard—ISO 1997 [33]; 0.25 mm/s; Lc = 2.5 mm; X = 5; Gauss). From the measured values, histograms of the frequency of individual intervals of arithmetic mean height of the roughness and standard deviations of the measurement were created.

The measurement graph (Figure 3) shows the variability and range of measured values of the quantity—*Ra* for measurement under the same conditions. From these data, it is possible to create a histogram of the frequency of individually measured data (Figure 4), from which the nature of the distribution of the measured data is visible, so in further considerations, we will assume a uniform law of distribution of the measured data.

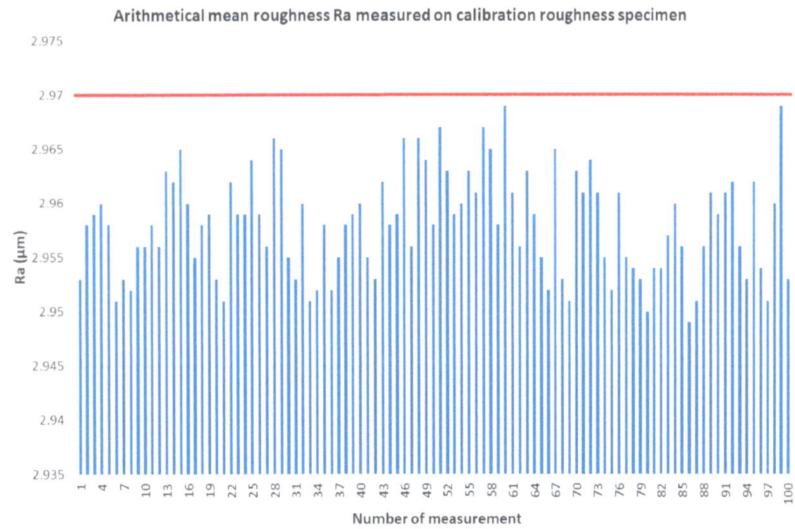

Figure 3. Arithmetical mean height of the roughness *Ra* values for 100 measurements at the same place under unchanged conditions (measured on Standard—ISO 1997 [33]; 0.25 mm/s; *Lc* = 0.8; *X* = 5; Gauss).

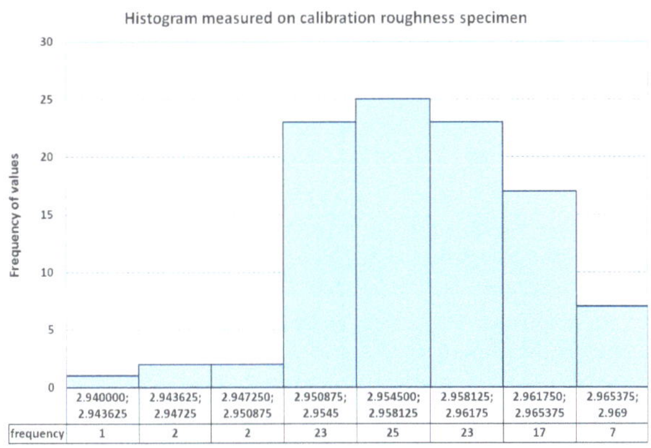

Figure 4. Histogram for values of arithmetic mean height of the roughness *Ra* for 100 measurements at the same place under unchanged conditions.

To assess the number of measurements as an indicator, it will be appropriate to use the standard deviation of a series of measurements. The value of these measurements depends on the dispersion of the evaluated values around the mean value. So, the smaller the standard deviation, the smaller the dispersion of the data. The evaluation of the standard deviation is therefore carried out in such a way that the standard deviation of a series of measurements is gradually determined, which is recalculated after the addition of each additional measurement. Therefore, we designated it as the cumulative standard deviation (Figure 5). At the moment when the cumulative standard deviation stops changing, it means that the number of evaluated values has been reached, which will no longer improve the uncertainty of the measurement result. At the same time, these standard deviations give us the answer to the course of the standard deviation with the increasing number of measurements, and thus it is possible to obtain the maximum standard deviation of a series of measurements (Figure 5).

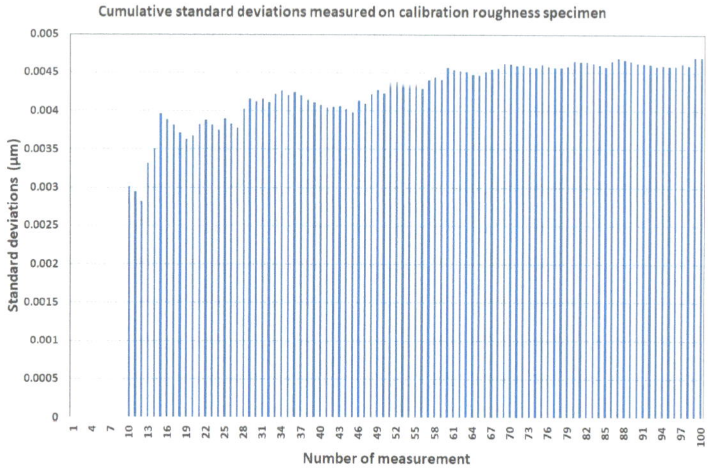

Figure 5. Cumulative standard deviations for 100 measurements at the same place under unchanged conditions (measured to standard—ISO 1997 [33]; 0.25 mm/s; Lc = 2.5 mm; X = 5; Gauss).

From the stated values of standard deviations, it is possible to consider the maximum value Z_{max} = 0.005 μm for the standard uncertainty of measurement and evaluation of the quantity Ra for this device. However, the Secondary Surface Roughness Standard itself also contributes to this uncertainty. However, this uncertainty is still very small. This means that the device itself will not show a large measurement uncertainty, and the dispersion of the measured data will instead be related to the dispersion of values as a result of the implemented technology for the surface treatment of the component.

3. Experiments

Quantities of surface roughness are precisely defined in the standard, but obtaining the correct value of these quantities by measurement depends on the correct choice of measurement parameters. These include cut-off filter settings, evaluation standards, number of section lengths, and measuring speed.

Important factors to consider when assessing surface roughness using a stylus-based surface roughness tester:

- The size of the measured irregularities must be appropriate to the measuring range of the sensor (vertical stroke); irregularities exceeding its size of the maximum value of vertical lift cannot be measured.
- The set value of the sensed length should include 5—a multiple of the section length plus the start-up and run-down path of the sensor; if it is not possible to meet this requirement (e.g., due to the size of the measured area), it is necessary to take into account that the result of the measurements will be affected by the failure to meet the requirement.
- The object to be measured must be clamped, orientated spatially, and the sensor must be placed appropriately on the measured surface; for this, it is necessary to choose the appropriate shape of the sensor with regard to its diameter, length, the possibility of fitting the bottom of the groove, the possibility of measuring on roundness, etc.; in case of inadmissibility of the place of measuring the surface roughness of the profilometer

sensor, replicas (imprints) of irregularities can be used in plastic film (or other suitable material) and the profile of the irregularities can be taken from this film (negative of the irregularities).

- Corresponding working conditions must be set, including measuring range, sensed length, filter type, basic length (cut-off) of the filter, and speed of the probe.

The measurement conditions during the measurements were as follows:

The surface tester was equipped with a stylus with a sliding tip. The tip of the stylus rides in a line across the surface, moving vertically over the peaks and valleys of the evaluated surface. Changes in the stylus' height are registered electrically and tracked against the position as the stylus moves, creating a measured profile. The shape of the stylus is conical with a spherical tip with a radius of 2 μm and a tip angle of 60°, and it is pressed with a measuring force of 0.75 mN. The skid force is less than 400 mN. The cone angle and tip radius determine the smallest features that a stylus can trace.

The relative humidity was in the interval (65 ± 5)%, and the temperature was (20 ± 2) °C.

The measurement methodology encounters several problems that need to be defined—here, it must be remembered that even the setting of the measurement conditions can have an impact on the result of the evaluation of the arithmetical mean deviation of the assessed roughness profile. It is, therefore, necessary to give an answer to the question of what this effect is and how to choose individual measurement conditions when assessing surface roughness.

The following measurement conditions are investigated:

- Choice of cut-off filter λc (or textual equivalent *Lc*): 0.08 mm; 0.25 mm; 0.8 mm; 2.5 mm; 8 mm (the surface tester, which is available for experiments, does not have the setting of the value *Lc* = 8 mm).
- Choice of standard for evaluation: ISO 1997 [33–40], JIS 1982 [50], JIS 1994 [51], JIS 2001 [52–58], ANSI Gauss [59–62], ANSI PC 75 [59–62], VDA, [63–65].
- Selection of speed of the sensing tip (stylus): 0.25 mm/s; 0.5 mm/s; 0.75 mm/s.

3.1. The Influence of the Choice of Standard and the Measuring Speed of the Surface Tester's Sensing Tip on the Evaluation of the Arithmetic Mean Height of the Roughness Ra

In this part, the influence of the selection of the measurement speed of the sensor tip displacement and the influence of the selection of the standard for the evaluation of the surface roughness values are assessed. Experimental measurements were carried out on the surface roughness standard (Figures 6 and 7) for grinding with nominal values of *Ra* = 1.6 μm and 3.2 μm. For each combination of the selected speed of the sensor and the evaluation standard, ten measurements were carried out at the same place in order to exclude the variability of the standard sample.

The influence of the selection of the evaluation standard and the selection of the sensor movement speed has an effect on the resulting *Ra* value. It can be concluded that for some standards, this influence is negligible, but for some standards, it is visible that increasing the speed of the sensor movement tends to reduce the resulting *Ra* value, and this influence is visible mainly for samples with a larger *Ra* value. It seems like a good choice to use the speed of movement of the sensor with a value of 0.5 mm/s, and the standard of evaluation can be selected according to the country where the evaluation is carried out. For our conditions, we will prefer evaluation according to the ISO 1997 [33–40] standard in further experiments.

The variability of the measured and evaluated *Ra* data obtained under the same conditions is also an interesting indicator for assessing the influence of the selection of the evaluation standard and the selection of the speed of the sensor movement.

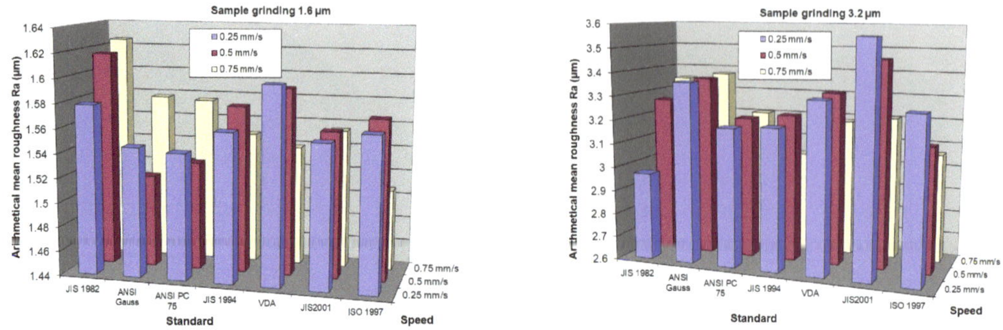

Figure 6. The influence of the selection of the evaluation standard [33–40,51–65] and the selection of the sensor movement speed for 1.6 µm and 3.2 µm samples in spatial imaging (conditions: *Lc* = 0.25 mm; *N* = 5).

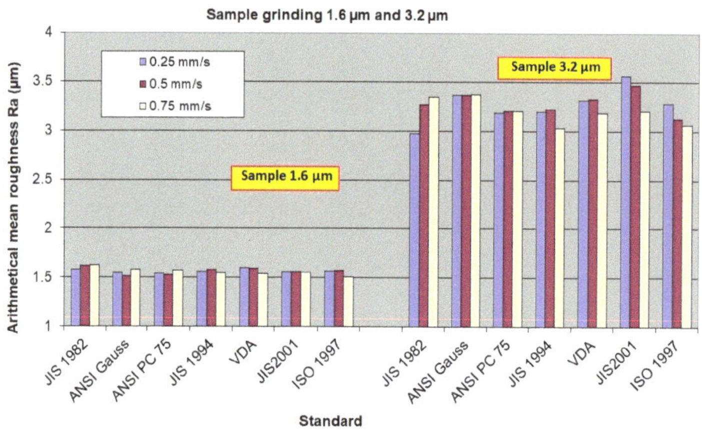

Figure 7. The influence of the selection of the evaluation standard [33–40,51–65] and the selection of the sensor movement speed for the 1.6 µm and 3.2 µm samples in the column plotting graph for comparison (conditions: *Lc* = 0.25 mm; N = 5).

The variability of the data, depending on the selection of the evaluation standard and the speed of the sensor movement, is expressed in percentage terms with respect to the average evaluated values of *Ra* and is shown in the following figure. The evaluation shows that the variability of the *Ra* data is in percentage terms up to a maximum of 18% (Figure 8).

Evaluations according to JIS1982 [50,51] and JIS2001 [52–58] standards have the greatest variability of Ra, and evaluations according to ANSI and VDA standards have the smallest variability. The ISO 1997 [33] and JIS1994 [51] standards are somewhere in the middle of these percentage expressions of variability. From these experiments, it can be concluded that the evaluations according to the JIS1982 [50] and JIS2001 [52–58] standards had the greatest sensitivity to the change in the movement speed of the sensor. However, this result may also depend on the specific sample on which the experiments were carried out (Figure 9).

In measurements where the speed of the sensor's movement did not change, and the standard of evaluation was changed, it can be seen that at a speed of 0.25 mm/s, the evaluations of *Ra* are most sensitive to the choice of the standard according to which the evaluation took place. For speeds of 0.5 mm/s and 0.75 mm/s, this effect was smaller.

Based on the above results, we decided to use a speed setting of 0.5 mm/s in further experiments, and we will use ISO 1997 [33–40] as the evaluation standard.

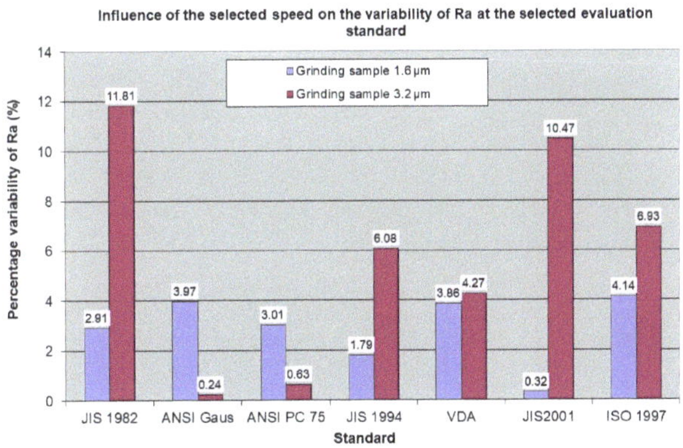

Figure 8. Percentage expression of the variability of the evaluated data *Ra* if the measurements were carried out for the given standard [33–40,51–65] at all possible speeds of the sensor movement.

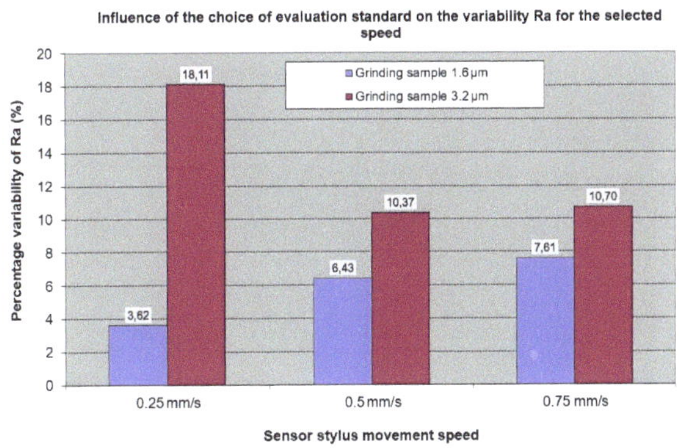

Figure 9. Percentage expression of the variability of the evaluated data *Ra*, if the measurements were carried out for specific sensor movement speeds and only the evaluation norm, was changed during the experiments.

3.2. *The Influence of the Cut-Off Setting on the Evaluation of the Arithmetic Mean Height of the Roughness Ra*

According to the cut-off value λc (or the text equivalent *Lc*), the recommendation is also for the choice of the second parameter cut-off λs (or the text equivalent *Ls*) according to Table 2 [ISO 11562; ISO 3274] [35,40]. Both values must be set before measuring the assessed surface of the component. This applies except for the VDA [63–65], JIS1982 [50], and JIS1994 [51] standards, where this parameter is not used.

Table 2. Setting the cut-off parameter λs (ISO 3274, ISO 11562); [35,40]. ISO standards are used to report the nominal values of roughness standards.

Cut-Off λc (mm)	Cut-Off λs (μm)
0.08	0.4
0.25	1.25
0.8	4
2.5	12.5

When choosing the parameters of the cut-off filter, the parameters of the stylus tip must also be taken into account. The recommended combinations of λc, λs, and stylus tip radius are shown in ISO 3274:1996 [35]. The used roughness gauge has a stylus tip that has a radius of 2 μm and an angle of 60° and is pressed with a measuring force of 0.75 mN (Table 3). Some types of surface testers allow you to select only the *Lc* parameter, and the second parameter of the cut-off filter is automatically set. However, there are also surface testers that allow you to independently set both parameters of the cut-off filter, so the mentioned standards provide a recommendation in the form of a table (Table 3) for choosing these settings. In some cases, the surface tester also allows the replacement of the stylus and thereby changing its geometry, where the standards bring certain recommendations.

Table 3. Recommended relationship between the cut-off parameters λc, λs according to the used sensing probe ISO 3274, ISO 11562; [35,40].

Cut-Off λc (mm)	Cut-Off λs (μm)	Roughness Cut-Off Wavelength Ratio $\lambda c/\lambda s$	r_{tip} Max (μm)	Maximum Sampling Spacing (μm)
0.08	2.5	30	2	0.5
0.25	2.5	100	2	0.5
0.8	2.5	300	2	0.5
2.5	8	300	5	1.5
8	25	300	10	5

Similarly, the newer standard ISO 21920-3 [47] introduces default settings for functional parameters (*Rk*, *Rpk*, *Rvk*, etc.) and R- and W- parameters without defined defaults (Table 4). The so-called setting class, while the difference is only in the values of the parameter λs.

Table 4. Setting classes—recommended settings (ISO 21920-3) [39].

Setting Class Sc	Sc1	Sc2	Sc3	Sc4	Sc5
Profile L-filter nesting index *Nic* (cut-off λc (*Lc*) for *R*-parameters) (mm)	0.08	0.25	0.8	2.5	8
Evaluation length *lm* (mm)	0.4	1.25	4	12.5	40
Profile S-filter nesting index *Nis* (cut-off λs) (μm)	0.8	0.8	2.5	8	25
Maximum sampling interval *dx* (μm)	0.15	0.15	0.5	1.5	5
Maximum tip radius r_{tip} (μm)	2	2	2	5	10
Section length *lsc* (mm)	0.08	0.25	0.8	2.5	8
Number of sections *nsc*	5	5	5	5	5

The problem is that if we determine the *Ra* of the assessed surface, we choose L-filter *Lc*. If, for example, we choose a too-small value of L-filter *Lc* for the sample, then the result is *Ra* with an almost zero size. However, if we choose a value of *Lc* that is too large, then the value of *Ra* may also include the effect of waviness and will distort the value of *Ra*. The mistakes we make when choosing a bad cut-off can be huge and can have a fatal impact on the result of evaluating the roughness of the surface of the component under consideration. We will try to solve this problem in this work.

For the practical measurement of surface roughness parameter values, it is important to choose the L-filter *Lc* (the cut-off value for the stylus-based surface roughness tester) in such a way that it includes a sufficient number of profile irregularities characterizing its roughness and at the same time excludes the influence of irregularities with a larger spacing (with longer wavelength) of inappropriate roughness (profile waviness and profile shape). For the comparability of the surface roughness measurement results, the values of the basic length *Lc* (cut-off) are determined by the standard. The nominal values of the limit lengths of the profile filter *Lc* (cut-off) are selected from the following series: 0.08 mm, 0.25 mm, 0.8 mm, 2.5 mm, and 8 mm. To obtain sufficiently accurate values of the surface roughness parameters, it is recommended to choose the evaluated length as several times the basic length (usually five times) (Table 4).

For sensor feed speed rates, the manufacturer's recommendation is to use selected values of the measuring feed speed. The assessment of the influence of *Lc* on *Ra* values will therefore be carried out at a feed speed of 0.5 mm/s because this speed is recommended for all *Lc* values, and at the same time, the results of our experiments in the previous chapter confirm the correctness of this choice (Table 5).

Table 5. Recommended measuring speed values for *Lc* (ISO 11562; ISO 3274) [35,40].

Cut-Off λc (mm)	Sensor Feed Speed (mm/s)
0.08	0.25; 0.5
0.25	0.25; 0.5
0.8	0.25; 0.5
2.5	0.25; 0.5; 0.75

According to the standards for evaluating the surface roughness, the cut-off parameter λc (also referred to as *Lc*) determines the basic length of the evaluated profile *lm*. The standard STN EN ISO 4288 [34] provides recommended values (see Table 6) for the choice of cut-off *Lc* and section and evaluation of lengths according to the values of the roughness parameters of periodic and non-periodic profiles. In the ISO 21920-3 [39] standard, the Setting class parameters (Table 4) are similarly designed but already uniformly for all types of surfaces and without justification. In our opinion, it is also necessary to take into account the technology used to evaluate the surface of the component. The origin and conditions of the mentioned recommendations are not clear, and therefore, in the next part of this work, experiments will be carried out on samples produced by different technologies to verify these recommendations.

Table 6. Roughness length section for measurement and curves and corresponding parameters for non-periodic profiles (EN ISO 4288, ISO 21920-3) [34,39].

Arithmetical Mean Height of the Roughness Profile *Ra* (mm)	Section Length of Profile *lsc* (μm)	Evaluation Length of Profile *lm* (mm)
$Ra < 0.02$	0.08	0.4
$0.02 < Ra \leq 0.1$	0.25	1.25
$0.1 < Ra \leq 2$	0.8	4
$2 < Ra \leq 10$	2.5	12.5
$10 < Ra \leq 80$	8	40

In the next part, we will distinguish the so-called periodic and non-periodic surface roughness profiles (See Appendix A).

The ISO 21920-3 [39] standard introduces no distinction between periodic and non-periodic profiles compared to the edition of ISO 4288:1996 [34]. Nevertheless, the following part is focused on the analysis in the course of the choice of cut-off filter for individual technologies, which were used to produce individual samples.

3.2.1. Effect of Cut-Off Filter Setting for Non-Periodic Surface Roughness Profiles

The goal is to identify the effect of setting the cut-off filter *Lc* on *Ra* values for individual types of reference samples processed by different technologies.

The standard (EN ISO 4288 and ISO 21920-3) [34,39] recommends setting the cut-off parameter *Lc* or section length *lsc* and evaluation length *lm* according to the *Ra* value. Evaluation length is recommended as five times the section length *lsc*.

Here the aforementioned problem arises concerning how to set the L-filter or section length and evaluation length according to an unknown *Ra* value. This problem does not make sense and creates a challenge with the measurement procedure defined in this way because it is the choice of the basic length that has a significant impact on the result of the *Ra* evaluation process. This paradoxical recommendation can significantly influence the result of the surface roughness evaluation.

Some standards for the evaluation of surface roughness quantities differ in these recommendations, stating different values of the parameter *Lc* for the values of *Ra*. These differences are shown in the following Figure 10. The JIS B0601-1982 [50] standard, for example, recommends using a cut-off filter with a value of *Lc* = 0.8 mm for values of *Ra* < 12.5 μm. For all larger *Ra* values, it is recommended to use *Lc* = 2.5 mm. Other standards for evaluation use more finely graded cut-off filters, which complicate the situation when setting the cut-off *Lc* filter during measurement.

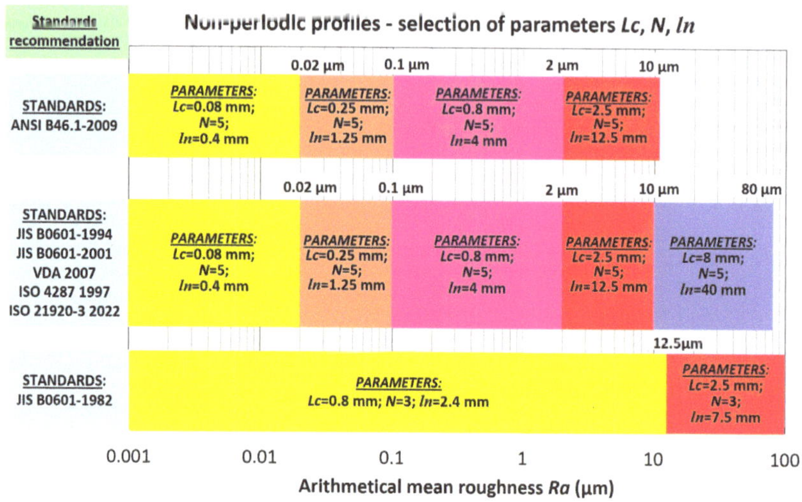

Figure 10. Recommendations of the standards [33–40,50–65] for the selection of the cut-off *Lc* parameter for non-periodic profiles.

For the experimental investigation, selected samples of non-periodic surface roughness profiles with different *Ra* values were processed using the technology (Figure 11), (Table 7):

- Lapping operation samples.
- Peripheral grinding operation samples.
- Face grinding operation samples.

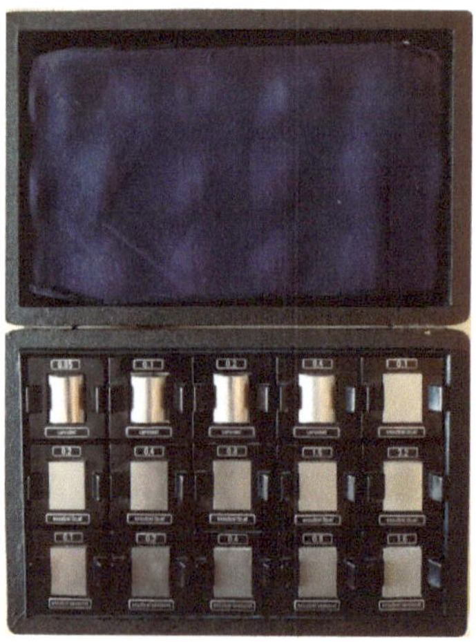

Figure 11. Etalon samples of surface roughness—Lapping and grinding operation.

Table 7. Selected evaluated non-periodic profile samples.

Samples: Lapping Sample Set: 591182	Sample 1	Sample 2	Sample 3	Sample 4	Sample 5	Sample 6
Nominal value *Ran* (μm)	0.05	0.1	0.2	0.4	-	-
Nominal corrected value *Ra* (μm)	0.0415	0.083	0.166	0.408	-	-
Samples: Peripheral Surface Grinding Sample Set: 591182	**Sample 1**	**Sample 2**	**Sample 3**	**Sample 4**	**Sample 5**	**Sample 6**
Nominal value *Ran* (μm)	0.1	0.2	0.4	0.8	1.6	-
Nominal corrected value *Ra* (μm)	0.103	0.14	0.412	0.784	0.91	-
Samples: Face Surface Grinding Sample Set: 591182	**Sample 1**	**Sample 2**	**Sample 3**	**Sample 4**	**Sample 5**	**Sample 6**
Nominal value *Ran* (μm)	0.1	0.2	0.4	0.8	1.6	3.2
Nominal corrected value *Ra* (μm)	0.11	0.176	0.384	0.72	1.46	3.42

Measurements were carried out for each setting of the cut-off filter *Lc* and for each sample at ten randomly selected locations in the central region of this sample. The average values were shown in graphs (Appendix B), also showing the value of *Ra* for each reference sample.

To assess suitability, it is necessary to observe the deviation of the average value from the *Ra* value of the reference sample. We were looking for a setting of the cut-off filter *Lc* for which the average value has the smallest deviation with respect to the reference nominal value *Ra*.

For experimentally assessed non-periodic surface roughness profiles (lapping, peripheral grinding, and face grinding), the optimal setting of the cut-off filter *Lc* was observed for some reference samples, which is recommended in the ISO standard [33–40]. However, for some samples, other recommended settings of the cut-off filter *Lc* (for samples with larger *Ra* values) were obtained.

For the mentioned non-periodic samples with values up to *Ra* = 0.1 μm, it is appropriate to use *Lc* = 0.25 mm. Then, for values from *Ra* = 0.1 μm to *Ra* = 0.4 μm, it is appropriate to use *Lc* = 0.8 mm. From a value of *Ra* = 0.4 μm, it is advisable to use *Lc* = 2.5 mm. It is shown more vividly in the following Figure 12, while for comparison, the recommended settings of the cut-off *Lc* filter according to the ISO standard [33–40] are also shown.

The shift of the boundary between *Lc* = 0.8 mm and *Lc* = 2.5 mm may be due to measurement uncertainties. The area with the cut-off value of the filter *Lc* = 0.25 mm starts only at the value *Ra* = 0.04 μm because reference samples with a smaller value of *Ra* were not available for the experiments.

On the basis of the above, it can be concluded that the guidelines according to the standard for the settings of the cut-off filter *Lc* are applicable for obtaining relevant results for the evaluation of the surface roughness quantity *Ra* (Figure 12).

3.2.2. Effect of Cut-Off Filter Setting for Periodic Surface Roughness Profiles

For the experimental investigation, standard samples of periodic surface roughness profiles (Figure 13) with different *Ra* values were selected and processed using the technology (Table 8):

- Milling operation with a slab mill cutter
- Milling operation with a face mill cutter
- Turning operation

- Planing operation

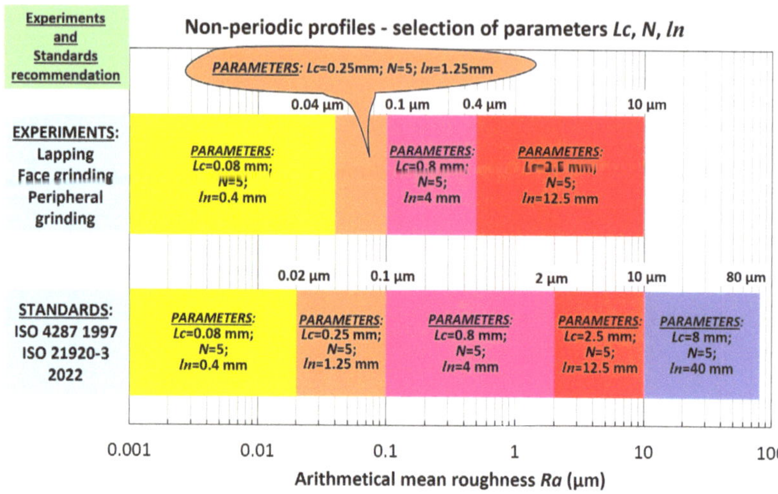

Figure 12. Optimal setting of the cut-off filter *Lc* for tested reference samples of non-periodic profiles and setting recommended by standards [33–40,50–65].

Figure 13. Etalon samples of surface roughness—milling operation, turning operation, and planing operation.

For periodic profiles (turning, milling, planing), the ISO 4288:1996 standard [34] recommends monitoring the mean spacing of the profile elements *RSm* (Appendix A) (ISO 21920-2) [38] as a parameter for choosing the cut-off filter *Lc*. Here again, the problem arises in that the setting of the cut-off filter *Lc* is dependent on the value of the parameter, in this case, *RSm*, which, however, will only be available as a result of the evaluation of the surface roughness measurement. Standard ISO 4288:1996 presents a table of recommended values of section length or filter cut-off according to *RSm* values (Table 9).

The results for individual technologies for periodic profiles (Appendix C) were subjected to the above analyses, and the optimal setting of the cut-off filter *Lc* for each technology was determined. Figure 15 shows a comparison of the selection of the filter cut-off value *Lc* for periodic profiles with the recommendations for the selection of the filter cut-off value *Lc* according to ISO 4288/1996 [34].

Table 8. Selected evaluated periodic profile samples.

Samples: Turning Sample Set: 591182	Sample 1	Sample 2	Sample 3	Sample 4	Sample 5
Nominal value *Ran* (μm)	0.8	1.6	3.2	6.3	12.5
Nominal corrected value *Ra* (μm)	0.784	1.55	2.78	6.55	12.63
Samples: Slab Mill Cutter Sample Set: 2580663	**Sample 1**	**Sample 2**	**Sample 3**	**Sample 4**	**Sample 5**
Nominal value *Ran* (μm)	1.6	3.2	6.3	12.5	25
Nominal corrected value *Ra* (μm)	1.52	3.57	6.05	10.25	16.25
Samples: Face Mill Cutter Sample Set: 2580663	**Sample 1**	**Sample 2**	**Sample 3**	**Sample 4**	**Sample 5**
Nominal value *Ran* (μm)	1.6	3.2	6.3	12.5	25
Nominal corrected value *Ra* (μm)	1.71	3.46	6.17	11	19.25
Samples: Planing Sample Set: 580845	**Sample 1**	**Sample 2**	**Sample 3**	**Sample 4**	**Sample 5**
Nominal value *Ran* (μm)	1.6	3.2	6.3	12.5	25
Nominal corrected value *Ra* (μm)	1.5	3.04	6.57	11.5	20.25

Table 9. Recommended cut-off values of the *Lc* parameter or section lengths *lr* and evaluated lengths ln for periodic profiles according to ISO 4288:1996 [34].

Mean Spacing of the Profile Elements *RSm* (mm)	Roughness Sampling Length *lr* (mm)	Roughness Evaluation Length *ln* (mm)
$0.013 < RSm \leq 0.04$	0.08	0.4
$0.04 < RSm \leq 0.13$	0.25	1.25
$0.13 < RSm \leq 0.4$	0.8	4
$0.4 < RSm \leq 1.3$	2.5	12.5
$1.3 < RSm \leq 4$	8	40

Note: in the experimental measurements in this work, where the *RSm* value is not given, it means that it could not be evaluated because there were not enough peaks and valleys on the measured surface profile. This condition occurs if the selected cut-off parameter is inappropriate, for example, the cut-off value is too small) or it does not allow the shape of the surface profile of the evaluated surface of the sample, or *Ra* has a large value, and the device does not provide a sufficient cut-off value. The recommendation for the selection of the cut-off filter is also summarized graphically in the following Figure 14. In this part of the article, an attempt is made to verify these recommendations for different types of machining samples. For experimental testing, selected samples were machined by turning, planing, milling with a cylindrical milling cutter, and milling with a face milling cutter. The parameters *Ra* and *RSm* were observed.

Samples produced by both methods of milling (Figure 15) are in accordance with the recommendations of the mentioned ISO standard. For the turned and planed samples, slight disproportions are visible in the intervals suitable for the free cut-off values of the *Lc* filter; however, these interval shifts may be related to measurement uncertainties.

The analysis of surface roughness profiles, *RSm* values, and *Ra* values reveals the interesting fact that for all samples of periodic roughness surfaces, it is advisable to proceed in such a way that to determine the parameter *RSm*, it is advisable to carry out measurements at all settings of the cut-off filter *Lc* (similarly to what was carried out in the case of individual technologies of periodic profiles) and according to the detected surface roughness profiles, it is then possible to determine the value of *RSm*. When evaluating *RSm*, the character of the course of the surface roughness profile must be observed and whether there are signs of periodicity typical for the given type of machining, and whether the influence of the waviness of the surface profile is no longer visible. For the *RSm* value determined in this way, it is then possible to select the cut-off filter parameter *Lc* for further repeated measurements according to the recommendations in the ISO 4288/1996 standard [42], and with this procedure, it is then possible to obtain the correct value of *Ra*.

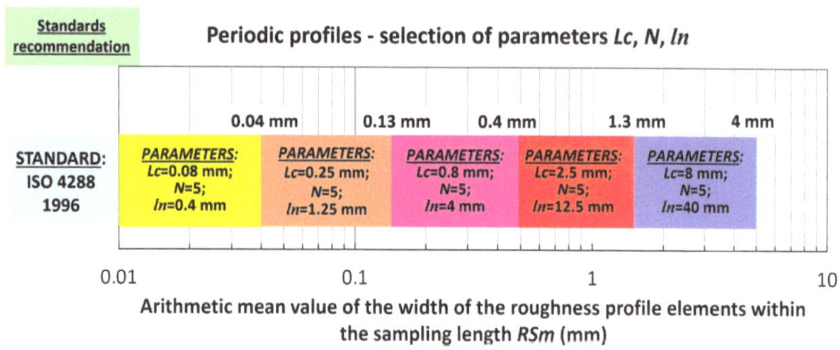

Figure 14. Recommendation for the selection of parameters of the cut-off filter *Lc* for evaluating the surface roughness of samples according to the ISO 4288:1996 standard.

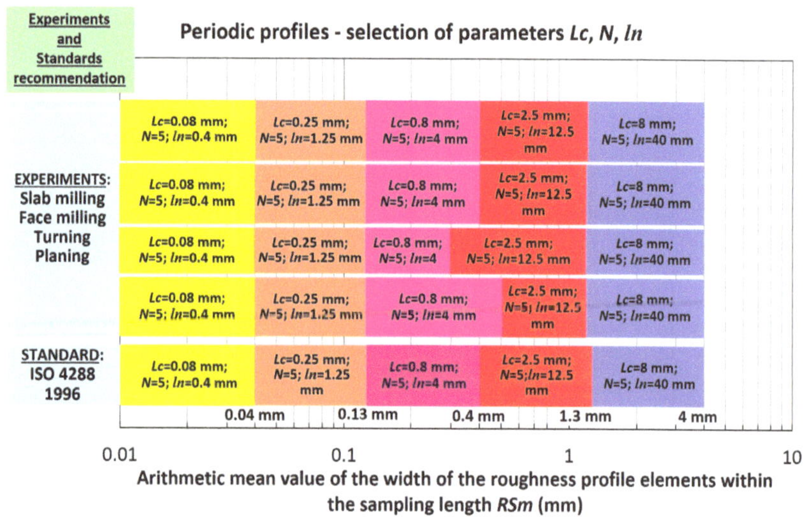

Figure 15. Obtained recommendations for the choice of cut-off filter *Lc* and recommended according to ISO 4288/1996.

In order to verify the correct determination of the *Ra* value, it is also possible to perform comparative measurements using the Comparex comparative microscope, similar to what was done with non-periodic samples. In this way, it is possible to confirm the correct selection of the cut-off filter value *Lc*.

3.2.3. No Discrimination Periodic and Non-Periodic Profiles for Procedure of *Ra* Estimation

It was complicated for users working with unknown surfaces to identify whether it was a periodic profile or a non-periodic one. In some cases, even with the same evaluated surface, it could happen that a profile that was periodically changed to a non-periodic one. This made the strategy of measuring and evaluating surface roughness quantities difficult. Subjective assessment of component surfaces can thus cause errors caused by the operator himself. ISO 21920-1, ISO 21920-2, and ISO 21920-3 standards [37–39] eliminate these problems. The ISO 21920-3 standard specifies the non-discrimination procedure for periodic and non-periodic profiles for the roughness quantities estimation procedure. This edition of ISO 21920 cancels and replaces the edition of ISO 4287:1997 [33] and ISO 4288:1996 [40], which have been technically revised. ISO 21920-3 introduces no distinction between periodic and non-periodic profiles. The ISO 21920-3 standard introduces a unified strategy for setting measurement parameters using setting classes—Sc1, Sc2, Sc3, Sc4, and Sc5.

In cases of missing part surface specifications, one of the options for determining the surface roughness of the examined part is to compare the surface roughness profile with the surface roughness profile of the standard sample whose nominal *Ra* value is known. For this purpose, the Comparex monocular comparison microscope (Figure 16) can be used.

It is a device that enables the determination of surface roughness on different types of surfaces obtained by all available technologies. The eyepiece is divided into two halves, and both surfaces of the surface roughness are displayed in it, namely the surface of the part and the surface of the reference sample, so it is possible to optically compare both surface profiles. According to these displayed profiles, it is possible to make a fairly reliable estimate of the *Ra* of the examined sample.

Figure 16. Comparex comparison microscope for determining surface roughness.

By successively optically comparing the surfaces of the reference samples with the examined surface of the component, it is possible to determine the desired value of *Ra* for the surface of the component (Figure 17). In Figure 17 is a comparison with the standard sample *Ra* = 1.6 μm, and from the image of the eyepiece, it can be seen that the standard has a greater surface fragmentation than the examined surface of the part. Then the surface of the component is compared with the reference sample *Ra* = 0.8 μm, and a match is visible (Figure 18), based on which we can estimate the value of *Ra* and then subsequently set the cut-off parameter *Lc* for the subsequent measurement and evaluation of *Ra* using stylus-based surface roughness tester.

The ISO 21920-3 standard [39] also provides recommendations for the selection of Setting class settings according to the estimate of the unknown quantity of surface roughness *Ra* (Table 10). The graph (Figure 19) shows a comparison of the recommendations of both ISO 4287 [33] and ISO 21920-3 [39] standards and the results of experiments for non-periodic part surface profiles. It results in only slight differences, which can be considered measurement errors and uncertainties.

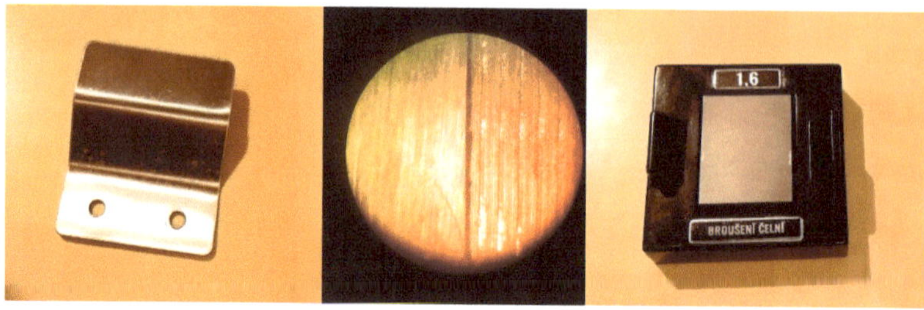

Figure 17. Comparison of the surface roughness profile of the part and the reference sample with a nominal value of *Ra* = 1.6 µm.

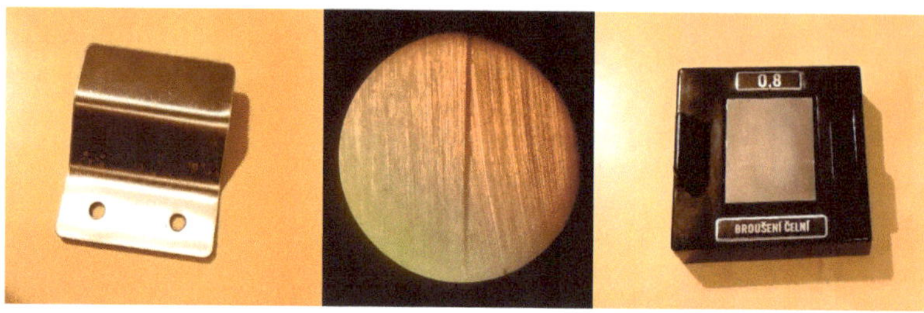

Figure 18. Comparison of the surface roughness profile of the part and the reference sample with a nominal value of *Ra* = 0.8 µm.

Table 10. Setting classes—recommended settings for cases of missing specification (ISO 21920-3) [39].

Setting Class Sc	Sc1	Sc2	Sc3	Sc4	Sc5
Parameter *Ra* (µm)	$Ra \leq 0.012$	$0.012 < Ra \leq 0.006$	$0.06 < Ra \leq 1.2$	$1.2 < Ra \leq 6$	$Ra > 6$
Profile L-filter nesting index *Nic* (cut-off λc (*Lc*) for *R*-parameters) (mm)	0.08	0.25	0.8	2.5	8
Evaluation length *lm* (mm)	0.4	1.25	4	12.5	40
Profile S-filter nesting index *Nis* (cut-off λs) (µm)	0.8	0.8	2.5	8	25
Maximum sampling interval *dx* (µm)	0.15	0.15	0.5	1.5	5
Maximum tip radius r_{tip} (µm)	2	2	2	5	10
Section length *lsc* (mm)	0.08	0.25	0.8	2.5	8
Number of sections *nsc*	5	5	5	5	5

For periodic profiles, preferred values resulting from experiments and recommended values according to the ISO 4288 standard and according to the ISO 21920-3 standard are shown. Except for planing technology, all values for periodic profiles are in accordance with the recommendations of the ISO 21920-3 standard. Planing technology does not agree with these recommendations, and the reasons for this are probably found in the essence of the technology itself. It is important to note that reference samples were not available for ranges with small *Ra* values, which is related to the nature of these technologies, which are used to achieve larger *Ra* values, and thus it was not possible to assess the recommendations of the standards for smaller *Ra* values. Interestingly, however, the experimental optimum settings for the periodic profiles are in agreement with the recommendations of the previous ISO 4288 standard.

With the exception of planing technology, however, we can state that the recommendations of the ISO 21920-3 standard can be applied to periodic and non-periodic profiles of the surface of components (Figure 20), and therefore the investigation of planing technology will be the subject of further research.

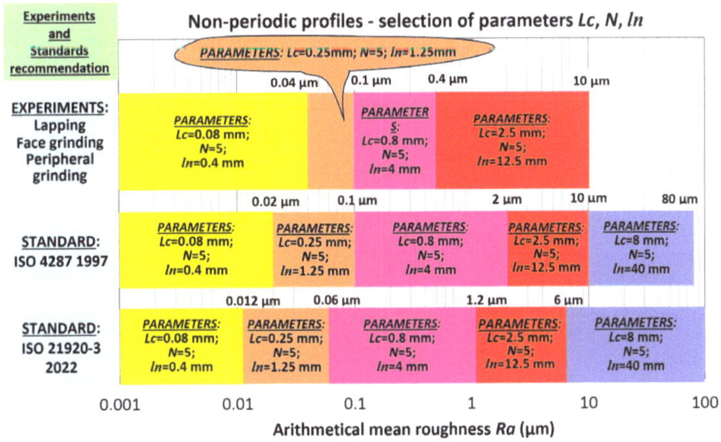

Figure 19. Optimal setting of the cut-off filter Lc for tested reference samples of non-periodic profiles and recommended settings according to ISO 4287 and ISO 21920-3 standards.

A comparison of the validity of using *RSm* to select the cut-off filter *Lc* according to ISO 4288 and using *Ra* according to ISO 21920 can be realized from the evaluation of the previous graphs (Figure 20). According to this evaluation, it can be concluded that better compliance results were achieved using the *RSm* quantity according to ISO 4288. At the same time, however, it can be seen that there were not enough samples available for a reliable evaluation, mainly with low *Ra* values, which follows from the nature of these technologies, as they are designed to achieve surfaces with larger *Ra* values. We will further verify these results in future research.

3.3. Identification of the Minimum Number of Measurements for Assessing the Surface Roughness of the Part

To define the problem, it is necessary to determine the minimum number of measurements needed to assess the surface roughness. It is known from common metrological practice that one measurement is insufficient and, especially for quantities with a high degree of variability, it is necessary to carry out repeated measurements and an estimate of the measured quantity together with an expression of the uncertainty of the determination of this estimate of the measured quantity.

To solve this problem, a method of successive evaluation of the standard deviation of repeated measurements and evaluation of the quantity *Ra* is proposed. After each additional measurement is added, the standard deviation will be recalculated with the addition of a new value to the set of measured data. We called it the cumulative standard deviation, which will capture the effect of the number of performed measurements on the value of the standard deviation of the quantity *Ra*.

First, 100 measurements were carried out at different measurement points on selected reference samples from each technology for which samples were available while the *Ra* quantity was evaluated. Thus, the influence of the variability of the sample implementation on the resulting variability of the measured data will be clear. Figure 21 shows the cumulative standard deviations for non-periodic profiles, and it follows that the first values of the cumulative standard deviations are unstable, which indicates a high degree of variability of the evaluated *Ra* data. Thus, it is clear that the number of measurements is insufficient. Only after 30 measurements is the cumulative standard deviation stabilized, and therefore, it is the optimal number for determining a reliable value of the surface roughness when measuring at different places of the examined surface of the component. Similarly, for samples with periodic profiles, the cumulative standard deviations are evaluated (Figure 22). It also follows from them that the optimal number of measurements for determining a reliable value of the quantity *Ra* is 30. Further increasing the number of measurements does not significantly improve the cumulative standard deviation of the measurement.

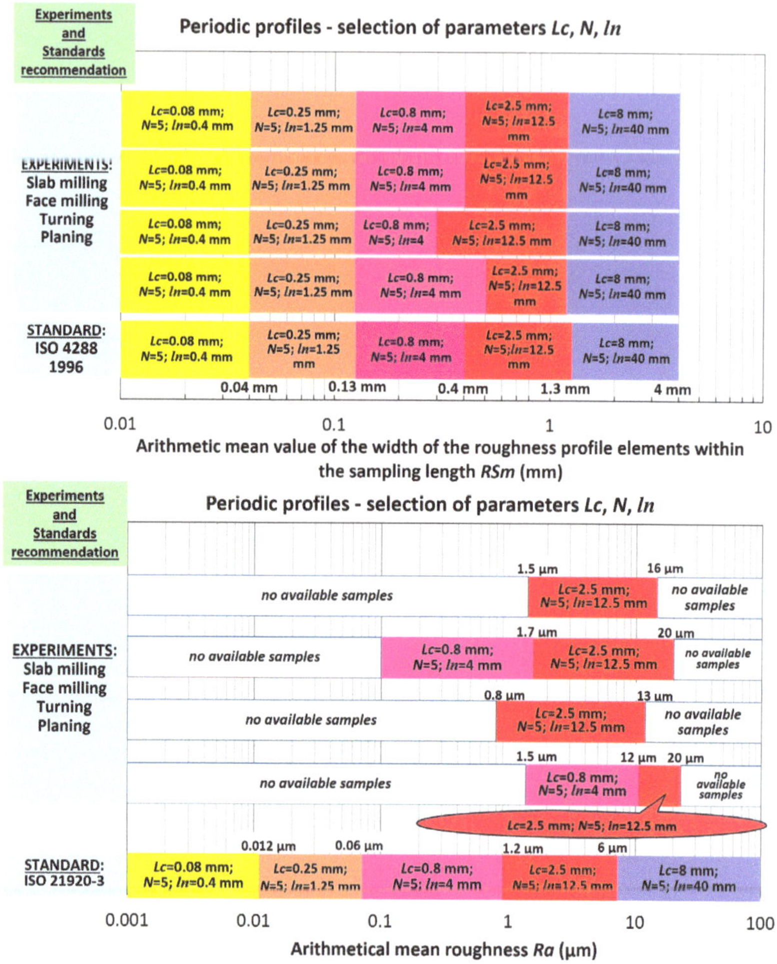

Figure 20. Comparison of optimal cut-off filter settings for individual technologies for tested samples of pe-riodic profiles and recommended settings according to ISO 4287 and ISO 21920-3 standards.

In the next step, repeated measurements and evaluations of *Ra* were again carried out for the same reference samples from each technology that was used for deposition. However, the measurements were carried out 30 times at the same place under the same measurement conditions. These measurements and evaluations should provide information on the variability of *Ra* values, excluding the influence of the technology itself on the uniformity of the surface roughness of the examined sample.

In Figure 23, cumulative standard deviations for non-periodic profiles are evaluated. In the first values, instability is visible, which indicates a high variability of the *Ra* values, but after 15 measurements, the value of the cumulative standard deviation stabilizes, which indicates that the number of measurements of 15 is sufficient for all technologies and further measurements will no longer bring a rapid improvement in the variability and uncertainty of the value of the quantity *Ra*.

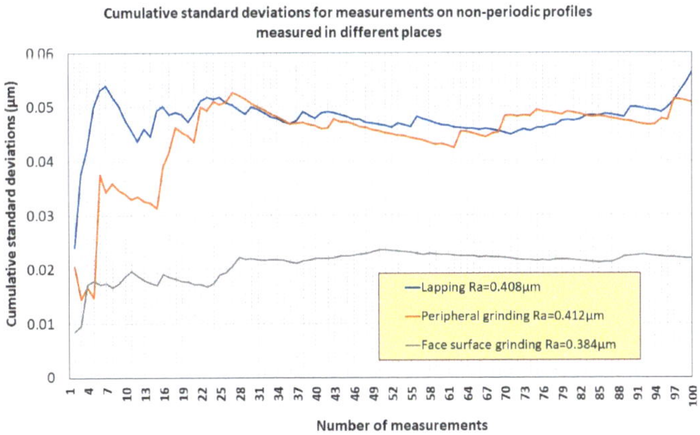

Figure 21. Cumulative standard deviations for the evaluation of the quantity *Ra* when measured at different locations of samples with a non-periodic surface profile.

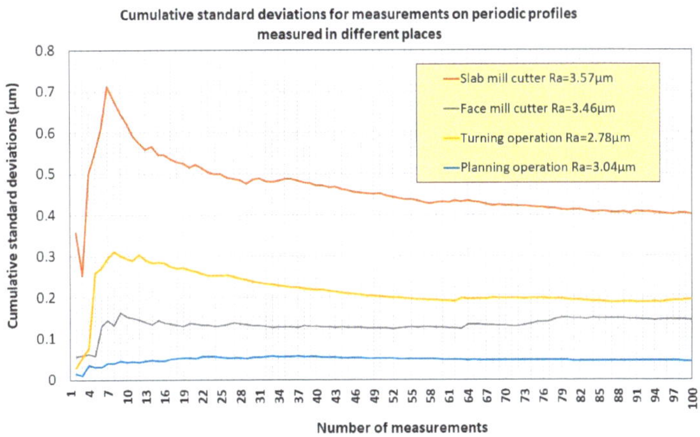

Figure 22. Cumulative standard deviations for the evaluation of the quantity *Ra* when measured at different locations of samples with a periodic surface profile.

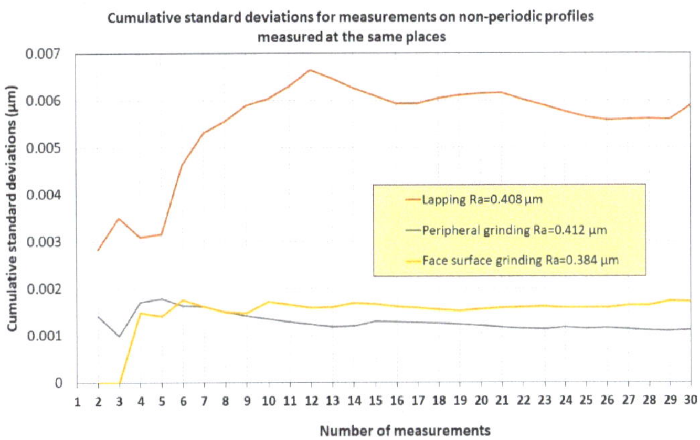

Figure 23. Cumulative standard deviations for the evaluation of the quantity *Ra* when measured at the same place on the surface of samples with a non-periodic surface profile.

In Figure 24, cumulative standard deviations for periodic profiles are similarly evaluated. Additionally, in this case, with these technologies, 15 measurements are sufficient to obtain the value of the quantity *Ra*.

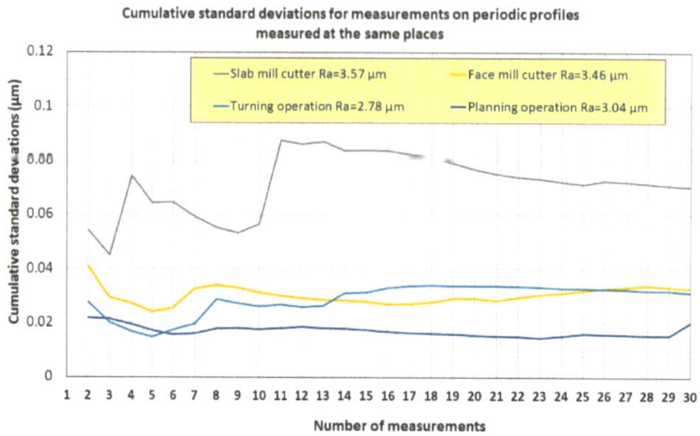

Figure 24. Cumulative standard deviations for the evaluation of the quantity *Ra* when measured at the same place on the surface of samples with a periodic surface profile.

4. Discussion

The roughness of the surface is defined by the micro-geometry of the surface of the components, and to define this micro-geometry, a range of quantities are used, which are determined in this contact type of measurement from the identified surface profile. Since these are micro-dimensions, high demands are placed on the resolution and precision of the stylus and evaluation unit of the surface roughness tester.

The surface roughness and its quantities are characterized by a high degree of variability, and this is also related to the fact that it is quite difficult to implement the individual surfaces of the components so that their surface roughness is the same in every place of this surface.

The variability of the measured and evaluated data in this type of measurement is also related to the fact that the measurements are carried out indirectly, i.e., the profile of the surface of the examined surface of the contact method is measured using a stylus, and a very small displacement is seen in the vertical and horizontal direction.

The preparation and setting of the measurement parameters require a lot of attention, and poor implementation will cause incorrect measurement and evaluation results to be obtained. This can introduce additional errors and uncertainties into the measurement process; therefore, it is necessary to solve this problem with research such as that in this article, which can provide guidance on how to correctly implement the measurement process and evaluation of the quantity *Ra*.

Before carrying out the measurement, it is necessary to check the surface tester using a reference specimen and, if necessary, to carry out the calibration of the meter, and we recommend carrying out this check continuously during a large number of measurements or at least at the end of a set of measurements in order to eliminate measurement errors related to the detuning of the surface tester. The internal structure of the stylus contains fine mechanics and sensor and actuator systems that are extremely sensitive to improper handling and unsuitable environmental conditions; therefore, measurements and calibrations must be approached very carefully and sensitively, as rough handling often results in damage to the surface tester or reference specimen.

Determining the uncertainty for the roughness surface tester is, in this case, a very difficult task that exceeds the scope of this work, but it is possible to make at least an estimate of the measurement uncertainty using the resolution values of the displacement measurement for both axes X, Z. Evaluation of the uncertainty *Ra* based on data on errors and uncertainties of the sensor and positioning of the stylus system is, therefore, a task that may be addressed in further future research.

In order to determine at least a rough estimate of the uncertainty of the gauge, 100 measurements were made at the same place under unchanged measurement conditions, and the values of the quantity *Ra* were evaluated. The cumulative standard deviation method was used to monitor the influence of the number of measurements as well as the statistical model of the distribution of the measured data. From these measurements and evaluations, it can be concluded that the used surface tester is in excellent condition, and the maximum standard deviation did not exceed 0.005 μm. The distribution of the measured values did not match the Gaussian normal law of the distribution of random values, so for the further use of the device and the evaluation of the surface roughness values, a uniform law of the distribution of the measured values will be considered.

The actual measurement of the surface structure usually takes only a few seconds and is determined by the feed speed and the scanned length, and other settings, but the measurement preparation time is considerably longer, on which the success of the measurement depends. Before the measurement and evaluation process, it is necessary to set several input parameters for the measurement, which can have a very significant impact on the actual measurement and evaluation of individual surface roughness quantities. The effort was to solve this problem, and thus the influence of individual settings of the surface tester on the process of measurement and evaluation was investigated; however, due to the difficulty of this issue, this article focuses mainly on the quantity *Ra*, which is evaluated by the used surface tester. The effect of the selection of the evaluation standard, the effect of the speed of the stylus movement, and the effect of the cut-off filter value on the resulting *Ra* value were monitored for different types of machined parts surfaces, where standard samples with a known *Ra* value were used.

To investigate the influence of the stylus speed and the choice of standard, experimental measurements were carried out on a standard sample of surface roughness for grinding with nominal values of *Ra* = 1.6 μm and 3.2 μm, where for each combination of the selected speed of the sensor and the evaluation standard, ten measurements were carried out on it in a certain place to exclude the variability of the standard sample. The influence of the choice of the evaluation standard and the choice of the sensor movement speed can have a significant impact on the resulting *Ra* value, and it can be concluded that for some standards, this influence is negligible, but for some standards, it can be seen that increasing the sensor movement speed tends to reduce the resulting *Ra* value and this influence it is mainly visible in samples with a larger *Ra* value. A good choice seems to be to use a speed of movement of the sensor with a value of 0.5 mm/s, and the evaluation standard can be selected according to the country where the evaluation is carried out. For our conditions, it was preferred in the experiments and then also the evaluation according to the ISO 1997 standard. To assess the impact of the choice of the evaluation standard and the choice of speed of movement of the sensor, the variability of the measured and evaluated *Ra* data obtained under the same conditions is also an interesting indicator.

The variability of the data, depending on the choice of the evaluation standard and the speed of the sensor movement, is expressed as a percentage with respect to the average evaluated *Ra* values, and from its evaluation, it follows that the variability of the *Ra* data is a maximum of 18% in percentage terms.

Another monitored and problematic setting is the setting of the filter cut-off value *Lc*, and this can be seen from the conducted experiments, where it is clear that if, for example, we choose a too-small *Lc* value for the sample, then the result is *Ra* with an almost zero size. However, if we choose a value of *Lc* that is too large, then the value of *Ra* may also include the influence of waviness, and this may distort the value of *Ra*, so the errors that we make with a bad free cut-off may be huge and may have a fatal impact on the result of evaluating the roughness of the surface being assessed parts.

The ISO 3274:1996 [35] standard contains guidelines on how to set the cut-off value of the *Lc* filter for non-periodic surface roughness profiles, while the EN ISO 4288 [34] standard recommends setting the cut-off parameter *Lc* or of the basic length *lr* and the evaluated length ln according to the *Ra* value, while the evaluated length *ln* is recommended as five times the basic length *lr*. Here the problem arises of how to determine the basic length and the evaluated length according to the unknown *Ra* value, which creates a problem with the measurement procedure defined in this way because the choice of the basic length has a significant impact on the result of the process of evaluating the average arithmetic deviation of the considered profile. This paradoxical recommendation can influence the result of the surface roughness evaluation to a significant extent.

The origin and conditions of the mentioned recommendations of these standards are not clear, and therefore in this article, experiments were carried out on samples produced by different technologies to verify these recommendations. It is, therefore, necessary to take into account the technology used to evaluate the surface of the component, and therefore the goal was to identify the effect of the cut-off filter setting *Lc* on the *Ra* values for individual types of reference samples processed by different technologies.

For experimentally assessed non-periodic surface roughness profiles (lapping, peripheral grinding, and face grinding), the optimal setting of the cut-off filter *Lc* was observed in some reference samples as the one recommended in the ISO standard, but for some samples, other recommended cut-off settings were found off filter *Lc* (for samples with larger *Ra* values). However, on the basis of the above results, it can be concluded that the guidelines according to the standard for the settings of the cut-off filter *Lc* are generally applicable for obtaining relevant results for the evaluation of the surface roughness quantity *Ra*; however, the company that produces machined parts must verify these settings for its technology in order to achieve reliable results of measurement and evaluation of surface roughness values.

In the article, a methodology was proposed as one of the options for determining the surface roughness of the examined part by comparing the surface roughness profile with the surface roughness profile of the standard sample whose nominal *Ra* value is known. For this purpose, it is possible to use the Comparex comparative monocular microscope, with which experimental measurements were carried out in the work, and then based on these measurements, it is possible to estimate the value of *Ra* and then subsequently set the cut-off parameter *Lc* for the subsequent measurement and evaluation of *Ra* using Stylus-based surface roughness tester.

For periodic profiles (turning, milling, planing) the ISO 4288:1996 standard [34] recommends monitoring the quantity—the average distance of the profile elements *RSm* as a parameter for choosing the cut-off filter *Lc*, and thus the problem arises again in that the cut-off setting of the filter *Lc* depends on the value of the parameter in this case *RSm*, which, however, will be available only as a result of the evaluation of the surface roughness measurement. The experimental results of measurements and evaluation of *Ra* for individual technologies for periodic profiles were subjected to the above analyses, and the optimal setting of the cut-off filter *Lc* for each technology was determined. For some samples, slight disproportions are visible in the intervals suitable for the free cut-off values of the *Lc* filter; however, these interval shifts may be related to measurement uncertainties.

The analysis of surface roughness profiles, *RSm* values, and *Ra* values shows the interesting fact that for all samples of periodic surface irregularities, it is advisable to proceed in such a way that to determine the parameter *RSm*, it is advisable to carry out measurements at all settings of the cut-off filter *Lc* (similarly to what was done in the case of individual technologies of periodic profiles) and according to the detected surface roughness profiles, it is then possible to determine the value of *RSm*. When evaluating *RSm*, the character of the course of the surface roughness profile must be monitored and whether there are signs of periodicity typical for the given type of machining, and whether the influence of the waviness of the surface profile is no longer visible.

For the *RSm* value determined in this way, it is then possible to select the filter cut-off parameter *Lc* for further repeated measurements according to the recommendations in the ISO 4288/1996 standard. With this procedure, it is then possible to obtain the correct

Ra value and, in addition, it is also possible to carry out comparative measurements using the Comparex comparison microscope for the verification of the correct determination of the value of *Ra*, similar to what was done with non-periodic samples, and thus it is possible to confirm the correct selection of the cut-off filter value *Lc*.

The existence of newer standards of the ISO 21920-1, ISO 21920-2, and ISO 21920-3 series [37–39] brings the methodology of non-discrimination of periodic and non-periodic profiles and unifies the recommended gauge settings for measuring surface roughness quantities. For non-periodic profiles, the recommended settings are slightly shifted to smaller *Ra* values, and it can be stated that the new methodology, according to the ISO 21920-3 standard [39], is in agreement with the experimental values from the previous standards ISO 4287 [33], ISO 4288 [34]. For periodic profiles, the situation is more complicated, as the previous standard ISO 4288 [34] used the *RSm* parameter to determine the optimal settings, but by comparing the experimentally determined optimal settings, it was found that, except for planing technology, all values for periodic profiles are in accordance with the recommendations of the ISO 21920-3 standard [39] and therefore the investigation of planning technology will be the subject of further research.

Another problem addressed in this article is how to determine the necessary number of repeated measurements needed to determine the value of the quantity *Ra*, and to solve this problem, a method of successive evaluation of the standard deviation of repeated measurements and evaluation of the quantity *Ra* is proposed. After each additional measurement, the standard deviation will be recalculated with the addition of each new value to the set of measured data. That is why this method was called cumulative standard deviation, which will thus capture the impact of the increasing number of performed measurements on the value of the standard deviation of the quantity *Ra*.

One hundred measurements were carried out at different measurement points on selected reference samples from each technology for which samples were available, while the quantity *Ra* was evaluated, and from these experiments, it follows that the optimal number of measurements is 30 for all used samples from the available technologies. In the next stage, repeated measurements and evaluations of the quantity *Ra* were carried out again for the same reference samples, but the measurements took place at the same place of each sample under the same measurement conditions, and from the calculated values of the cumulative standard deviations, it follows that the optimal number of measurements for this case on all samples is 15 measurements.

5. Conclusions

In this work, specific problems related to the process of measurement and evaluation of surface roughness quantities were solved, and samples of non-periodic surface roughness profiles with different *Ra* values processed using technology were selected for experimental investigation, including lapping operation samples, peripheral grinding operation samples, and face grinding operation samples.

Additionally, samples of periodic surface roughness profiles with different *Ra* values were processed using the following technologies: milling operation with slab mill cutter, milling operation with face mill cutter, turning operation, and planing operation.

The novelty and contribution of the article is:

(a) Assessment of the condition of the used surface tester by the proposed methodology for determining the variability of the obtained results.

(b) Identification of the influence of individual surface tester parameter settings. A methodology was proposed for assessing the influence of parameters on the resulting value of *Ra*.

(c) Verification of the standard recommended settings of the cut-off filter for the determination of the quantity *Ra* and determination of the methodology for the process of measurement and evaluation for periodic and non-periodic profiles.

(d) Experimental determination of the optimal number of repeated measurements and proposal of a methodology for determining these data.

The results of the work confirmed the validity of assessing the condition of the gauge, and the proposed method is a relatively quick method to implement it.

In conclusion, it should be said that the experiments have shown that for individual technologies, it is appropriate to experimentally verify the methodology for the selection of measurement parameters and the evaluation of surface roughness quantities in order to avoid potential errors in the evaluation of surface quality.

It should be said that all experiments were carried out on a specific surface tester, so it is possible that another device may behave differently. Consequently, it is in the interest of each laboratory to know the status of its surface tester in this way. Additionally, it is also necessary to emphasize that this condition must be regularly checked, and, if necessary, a service intervention must be undertaken.

So far, we have not found studies of a similar type, so it was not possible to compare our results to the results of other studies. The scarcity of similar studies is why this research was conducted. The goal was to solve problems that we encountered in common tasks of evaluating surface roughness.

It should be emphasized that these results are tied to specific reference samples and a specific surface tester. Further experiments on other surface testers and other samples will be necessary to generalize these results. For practical use, it is essential that users identify their surface tester and the manufacturing capabilities of their products in terms of process stability and surface roughness variability. Therefore, in this article, we have tried to solve some practical and scientific problems associated with the assessment of surface roughness.

Many research works solve this problem; however, the influence of surface roughness and frictional forces tend to be neglected, and problems are solved without these important factors. Thus, the real situation can differ very dramatically from mathematical models and simulations where contact surfaces without surface roughness and frictionless were considered forces, and this can result in serious problems in real systems. As a result, many authors have focused on identifying surface roughness and planning technologies to achieve it and also take into account the effect on frictional forces. These are mainly applied in the field of mechanical engineering, mechatronics, robotics, production engineering, machining, biomedical engineering, automotive engineering, tribology, microsystems, precision mechanics, etc. [66–85].

In future research, we plan to develop a method for a more precise evaluation of measurement uncertainties for surface roughness quantities because, for practical use, a suitable methodology is needed to effectively determine the measurement uncertainty.

Author Contributions: Conceptualization, K.P. and T.K.; methodology, M.K.; software, M.K.; validation, K.P., T.K. and M.K.; formal analysis, T.K.; investigation, K.P.; resources, M.K.; data curation, T.K.; writing—original draft preparation, T.K.; writing—review and editing, M.K.; visualization, M.K.; supervision, T.K.; project administration, M.K. All authors have read and agreed to the published version of the manuscript.

Funding: This research received no external funding.

Institutional Review Board Statement: Not applicable.

Informed Consent Statement: Not applicable.

Data Availability Statement: Data are unavailable due to privacy or ethical restrictions.

Acknowledgments: The authors would like to thank the Slovak Grant Agency-project VEGA 1/0168/21 and this publication is the result of the project implementation Research and development of intelligent traumatological external fixation systems manufactured by digitalization methods and additive manufacturing technology (Acronym: SMARTfix), ITMS2014+: 313011BWQ1 supported by the Operational Programme Integrated Infrastructure funded by the European Regional Development Fund.

Conflicts of Interest: The authors declare no conflict of interest.

Appendix A

Appendix A.1. Surface Roughness Assessment Methodology According to the Standards and Used Methods of Measurement and Data Processing

Every technology used in the realization of technical surfaces leaves inequalities that are of fundamental importance in the function of these surfaces. It is very difficult to assess the irregularities on the surface, which represent a certain spatial formation. This problem of assessing irregularities (surface structure) is solved by reducing the plane of the section to a plane perpendicular to the surface of the part (Figure A1). The profile of the surface of the component, which is obtained in the plane of the section, is the basic source of information for assessing the structure of the surface of the component.

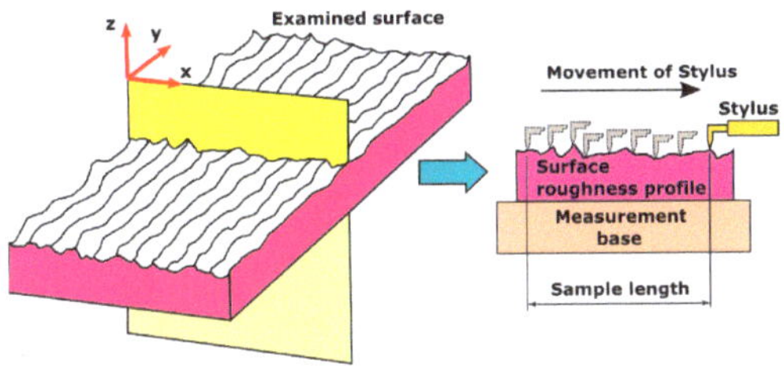

Figure A1. Part surface profile—as the intersection of the actual surface and the given plane.

Surface texture—geometrical irregularities contained in a scale-limited profile, or the basic surface profile, is divided into two components according to the spacing of the respective irregularities, namely the component with the smallest spacing forming the surface roughness profile and the component with the largest spacing of irregularities determined by the basic profile, namely the waviness profile. The basic profile is the complete profile before the application of the short-wave filter λc (the filter defining the interface between roughness and waviness present on the examined surface). The basic profile represents the basis for the digital processing of the profile using profile filters and for the calculation of profile parameters (Figure A2). Waviness profile—as a profile derived by sequentially applying filters (a filter defining the interface between waviness and longer wave components present on the surface) and profile filters to the base profile (ISO 4287:1997, ISO 4288:1996) [33,34].

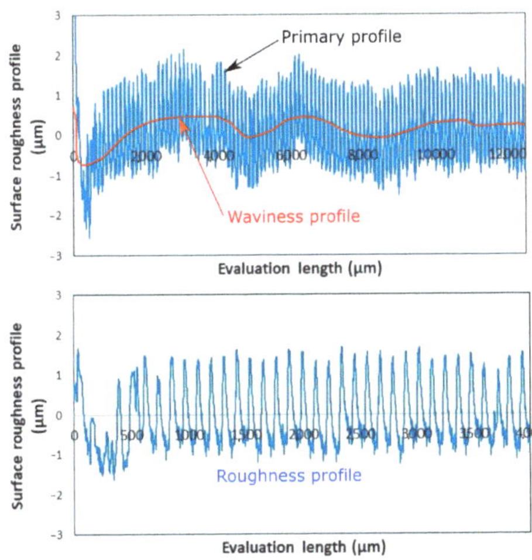

Figure A2. Profile of waviness and profile of surface roughness.

The cut-off system is created to filter the influence of surface waviness on the process of evaluating surface roughness values (Figure A3). The evaluated profile length ln is divided into several basic lengths (e.g., 5), and these are evaluated separately. The cut-off system is specified by two parameters (Figure 4):

- Cut-off λc (also referred to as *Lc* in the text), a profile filter that defines the intersection between roughness and even shorter wave components on the surface. We also call it the shortest cut-off.
- Cut-off λs (also referred to as *Ls* in the text), a profile filter that defines the intersection between roughness and waviness of the profile of the surface under consideration. We also call it the longest cut-off.

Filtering consists of evaluating the wavelengths between these two limits in order to correctly determine the roughness values of the surface under consideration.

Some devices for measuring surface roughness do not allow setting the *Ls* value for some evaluation standards. This can have an impact on the *Ra* evaluation results, which was also shown when assessing the impact of evaluation standards on *Ra* values (see the previous subsection).

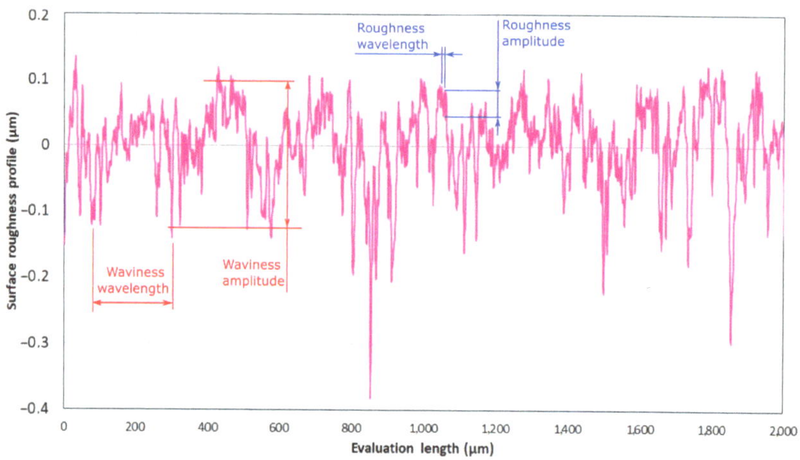

Figure A3. Amplitudes and wavelengths of the roughness profile and the waviness profile.

For filtering the waviness and shape of the surface of the component, it is necessary to select the values of the profile filters of the wavelength λc, λs, and λf (Figure A4), so that this filtering is possible.

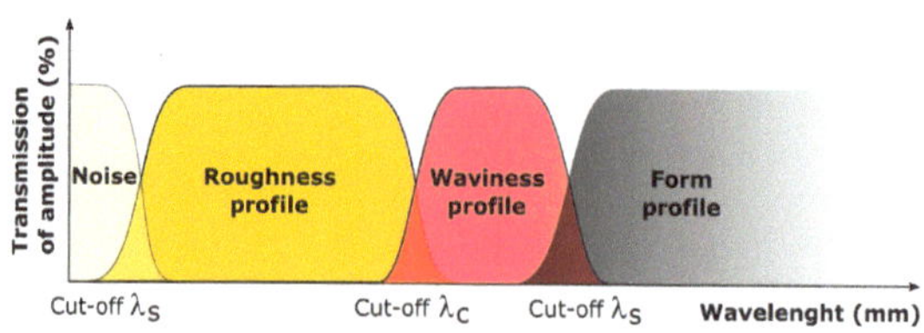

Figure A4. Wavelengths of surface roughness, waviness, and surface shape of the component.

After applying profile filters for waviness and shape, it is possible to separate the course of surface roughness, course of waviness, and course of surface shape from the obtained surface profile (Figure A5). For standard surface profile measurements, the surface roughness is filtered into two parts, namely the surface roughness profile and the waviness profile, with the shape of the surface of the part.

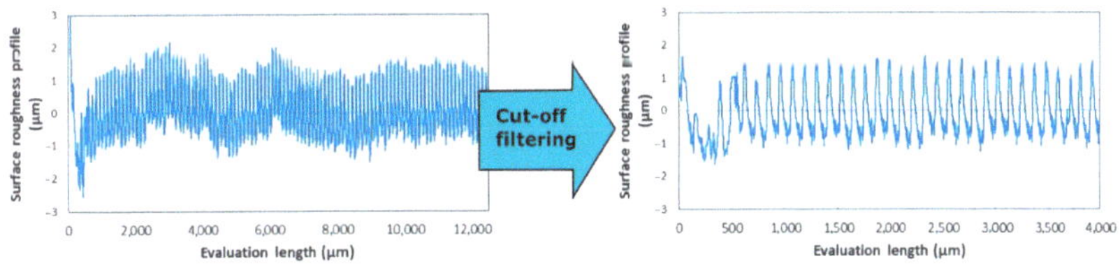

Figure A5. After applying a cut-off filter to separate the waviness of the surface profile and the surface roughness of the part.

The newer standard ISO 21920-1 [37] defines these filters with newer designations—profile **S-filter type and profile L-filter type for R-parameter or profile S-filter type for W-parameter**. The indication of the profile L-filter nesting is a value to specify the large-scale lateral components which are removed for the R-parameter. The indication of the profile S-filter nesting index is a value to specify the small-scale lateral components which are removed for P- and W-parameter (ISO 21920-2) [38]. Profile S-filter is a profile filter that removes small lateral scale components from a profile. Profile L-filter is a profile filter that removes large lateral scale components from a profile.

The setting of measurement parameters is defined in ISO 21920-1 [37] as Symbol 'Scn' **to specify a setting class number**. The symbol Scn indicates what setting class number for the determination of the default settings shall be applied. The setting class number or a default profile L- or S-filter nesting index value shall be specified for the parameter without defined defaults (ISO 21920-3) [39].

Primary surface profile—*P*: surface profile trace obtained when a surface profile trace is represented as a specified primary mathematical model with a specified nesting index *Nis*. Primary profile—*p* profile is scale limited profile at any position *x* derived from the primary surface profile by removing the form using a profile F-operation with nesting index *Nif*. The primary profile is the basis for the evaluation of the primary profile parameters.

Waviness profile—*W* profile is scale limited profile at any position *x*; derived from the primary profile by removing small-scale lateral components by a profile S-filter with a specified nesting index *Nic* and with a specified type of filter. The waviness profile is the basis for the evaluation of the waviness profile parameters.

Roughness profile—*R* profile is scale limited profile at any position *x* derived from the primary profile by removing large-scale lateral components by a profile L-filter and with the same nesting index *Nic* and the same type of filter specified for the profile S-filter to obtain the waviness profile. The roughness profile is the basis for the evaluation of the roughness profile parameters.

For the evaluated quantities, the following are then distinguished:

P-prefix parameter—parameter calculated from the primary profile,

W-prefix parameter—parameter calculated from the waviness profile,

R-prefix parameter—parameter calculated from the roughness profile.

Evaluation length *lm*—The indication of the evaluation length (Figure 6) is a value to specify the length used for identifying the geometric structures characterizing the scale-limited profile (ISO 21920-2) [38]. In the former ISO 4287 [33] profile standard, the evaluation length was given by *ln*. Evaluation length is length in the direction of the *x*-axis used for identifying the geometric structures characterizing the scale-limited profile. The evaluated length can contain one or more section lengths. The profile filter is a tool for dividing the profile into long-wave and short-wave components. The filtration process is carried out in several stages, providing modified profiles.

Section length *lsc*—The indication of the section length is a value to specify the length used to obtain height parameters based on profile peaks and profile valleys (ISO 21920-2) [38]. Section length *lsc* or *lp*, *lr*, *lw* is the length in the direction of the x-axis used to recognize the irregularities characterizing the evaluated profile. Section length for the roughness *lr* and for the waviness profile *lw* are numerically equal to the characteristic wavelength of the profile filter λc and λf. Section length for the basic profile *lp* is equal to the evaluated length.

Number of sections—The indication of the number of sections (Figure A6) is a value to specify the number used to obtain height parameters based on profile peaks and profile valleys (ISO 21920-2) [38]. The Number of sections shall be an integer number.

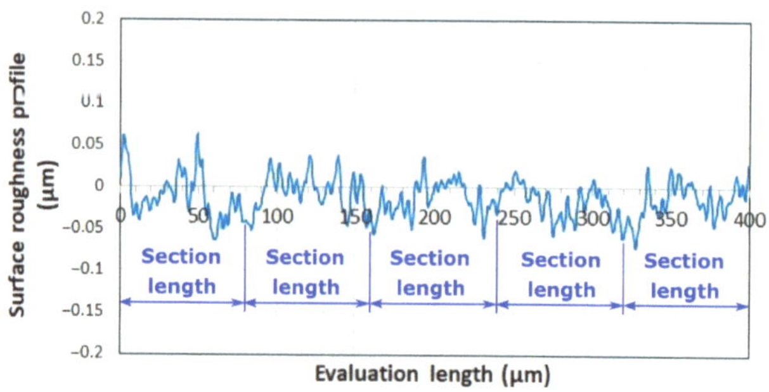

Figure A6. Evaluated and section length of the surface roughness profile.

STN EN ISO 4287 [33] and ISO 21920-2 [38] expands the possibilities of surface evaluation by the fact that almost all parameters defined in it can be applied to the primary profile, roughness profile, and waviness profile. The relations for their calculation are the same, except they are applied to a different profile. When marking, these parameters are distinguished by the first character, e.g., arithmetic mean height of the absolute values *Pa*, of the roughness profile *Ra*, and the waviness profile *Wa*.

In this article, the arithmetic mean height of the roughness profile will be assessed and evaluated. The arithmetical mean height of the roughness profile—*Ra* is the arithmetic mean of the absolute values of the coordinates in the range of the basic length (Figure A7):

$$R_a = \frac{1}{l_m}\int_0^{l_m} z(x)dx. \tag{A1}$$

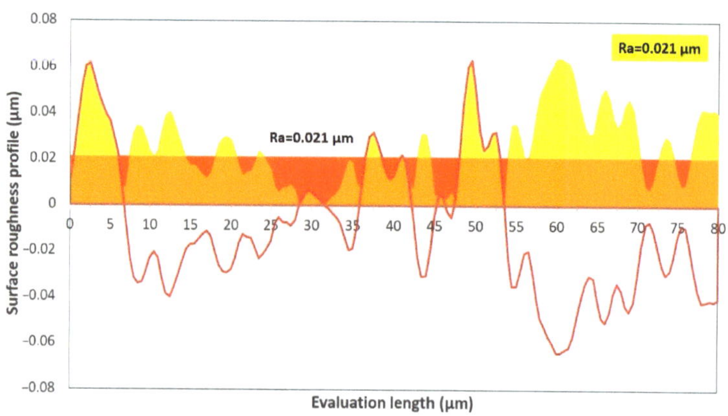

Figure A7. Arithmetical mean deviation of the assessed roughness profile.

Appendix A.2. The Surface Roughness Standards

The basis for measuring surface roughness parameters and characteristics is the unit of length, defined in EN ISO standards and implemented by a set of micro-length etalons (line depth and line spacing), whose conventionally true value of line depth is measured by wavelength (on the device—interference microscopes), these etalons are for the area of surface roughness measurement—Primary Surface Roughness Standards.

Secondary Surface Roughness Standards (Figure A8) are sets of standards that embody the values of roughness parameters—*Ra*, *Rz*, *Ry*, and others. They are used to calibrate working gauges using the direct measurement method, where a measured standard of the same nominal value is directly calibrated using standard gauges.

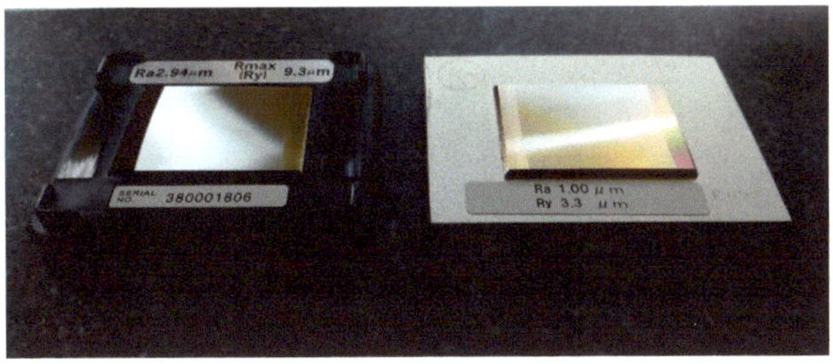

Figure A8. Secondary Surface Roughness Standards.

The average distance of profile elements *RSm* can be determined as the average value of individual widths *Xsi* of profile elements' surface roughness in the range of section length (Figure A9) according to the relationship:

$$R_{sm} = \frac{1}{n_{pe}} \cdot \sum_{i=1}^{n_{pe}} Xs_i \tag{A2}$$

where n_{pe} is the total number of profile elements.

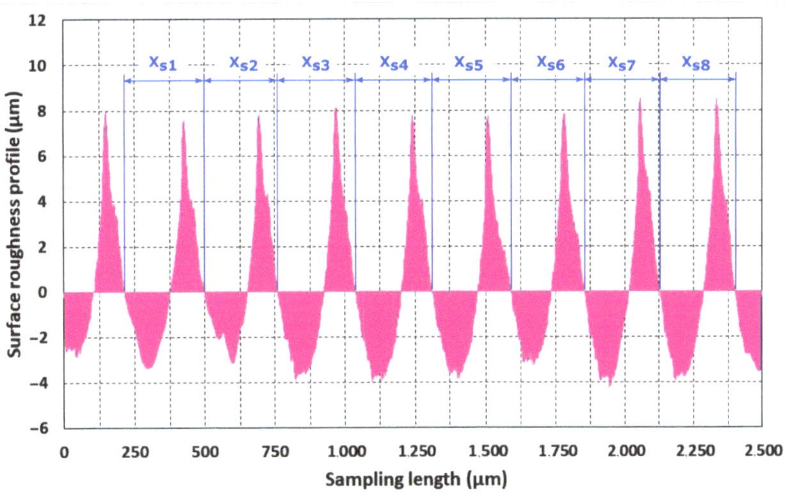

Figure A9. Mean spacing of the profile elements *RSm*.

In the experimental part of this paper, we distinguish the so-called periodic and non-periodic surface roughness profiles.

Non-periodic surface roughness profiles arise mainly during finishing technologies such as grinding, polishing, and lapping, and with these methods, the *Ra* value reaches a maximum of 0.8 μm. Grinding technology is implemented in such a way that the material is removed by geometrically undefinable and randomly arranged cutting wedges on the grains of the abrasive material of the grinding disc, which separate the particles of the surface layers of the workpieces with abrasive grains stored in the binder, on the supporting tool or freely. Lapping is a finishing machining method that is used precisely to refine the surface. In principle, this technology belongs to grinding technologies. Lapping is the sanding of the surface of the product with a fine grinding abrasive material that is freely applied in a lapping liquid (e.g., oil, kerosene) or paste (e.g., paraffin, stearin) with dispersed fine abrasive that is applied to the lapping surface. During lapping, the peaks of unevenness are cut off with abrasive grains. Abrasive grain and fine abrasive can be understood as a cutting knife that has a random geometry and orientation, so the surfaces that arise after these types of machining have a non-periodic and randomly oriented character.

Periodic profiles of surface roughness arise during technological operations such as turning, milling, planing, which reach *Ra* values of more than 0.8 μm. With periodic profiles, there is a periodic movement of the tool along the surface of the component and the gradual removal of material, so the waviness of such a surface is more pronounced than with non-periodic profiles. *Ra* values greater than 25 μm are typical for the surfaces of parts that arise during casting and forging technologies.

When assessing the roughness of the surface, it is also necessary to take into account the direction of the unevenness of the surface, which is related to the machining method used. The unevenness of the profile can be parallel, perpendicular, oblique, or crossed with the contour line of the component. During some machining operations, circular irregularities or irregularities directed from the center of the component or unevenness are random in nature.

The ISO 21920-3 [39] standard introduces no distinction between periodic and non-periodic profiles compared to the edition of ISO 4288:1996 [34]. Nevertheless, the paper is focused on the analysis of the choice of cut-off filter for individual technologies, which were used to produce individual samples.

Appendix B

Non-Periodic Surface Roughness Profiles—Experiments

For the Lapping operation, samples (Figure A10) with nominal values *Ra* = 0.0415 μm, *Ra* = 0.083 μm, *Ra* = 0.166 μm, *Ra* = 0.408 μm were selected from the etalon samples set of surface roughness. On the course of the surface roughness profile for one of the samples, it can be seen how the number of profile unevenness increases according to the selected cut-off filter value *Lc*, where at the smallest value, too small an amount of unevenness of the surface roughness profile can be seen for *Ra* evaluation.

Similar evaluations of *Ra* were carried out for other samples produced by the Lapping operation technology, and the obtained values are shown in Figure A11, where the [ISO] symbol indicates the cut-off value of the L-filter *Lc* recommended according to the ISO standard [33–40], and the red triangle with the OK (▼ ✅) symbol shows the value of *Lc*, which is optimal according to the performed experiments when the determined value of *Ra* is the closest nominal value of sample *Ra*. The displayed results show that the recommendations of the ISO standard are not in agreement with all experimental observations. These are mainly samples with larger *Ra* values.

For Face grinding operation, samples (Figure A12) with nominal values *Ra* = 0.11 μm, *Ra* = 0.176 μm, *Ra* = 0.384 μm, *Ra* = 0.72 μm, *Ra* = 1.46 μm, *Ra* = 3.42 μm were selected from the surface roughness sample book. The *Ra* = 0.384 μm sample is shown when measured for different values of the cut-off filter setting (Table 2). Even in this case, it can be seen with small values of the cut-off filter *Lc* that there is not a sufficient amount of unevenness of the surface roughness profile available for evaluation.

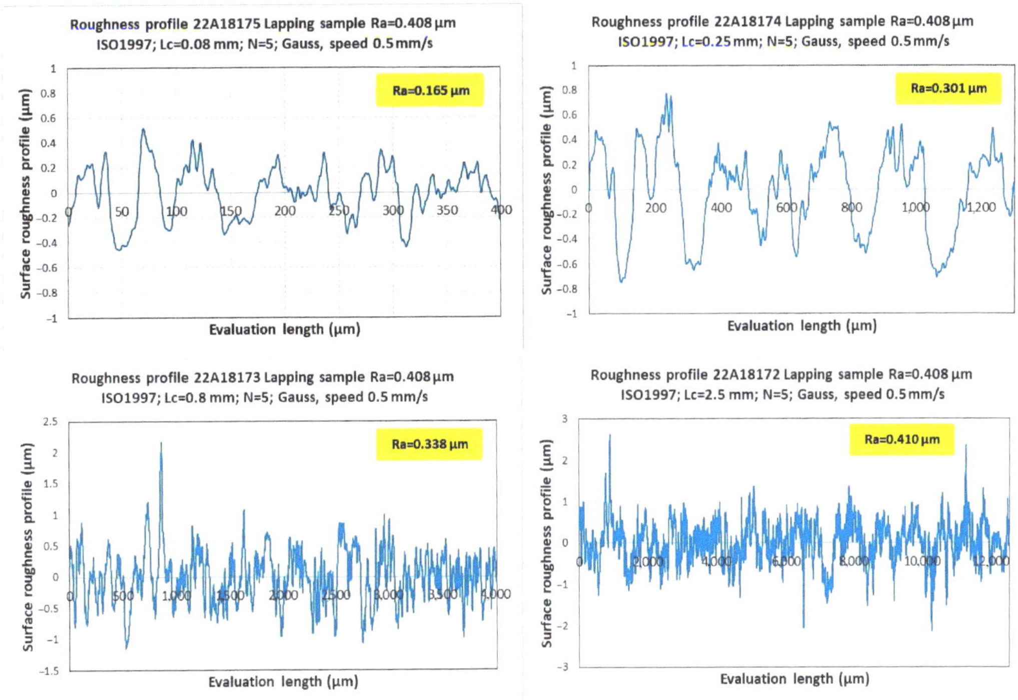

Figure A10. Profile of surface roughness measured using different settings of the cut-off filter *Lc* for lapping—sample with value *Ra* = 0.408 µm.

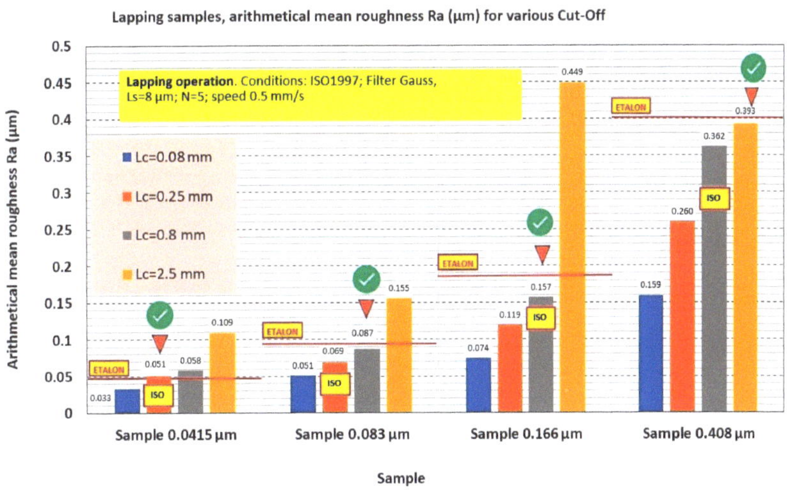

Figure A11. Lapping—average value of *Ra* for different *Lc* cut-off filter settings.

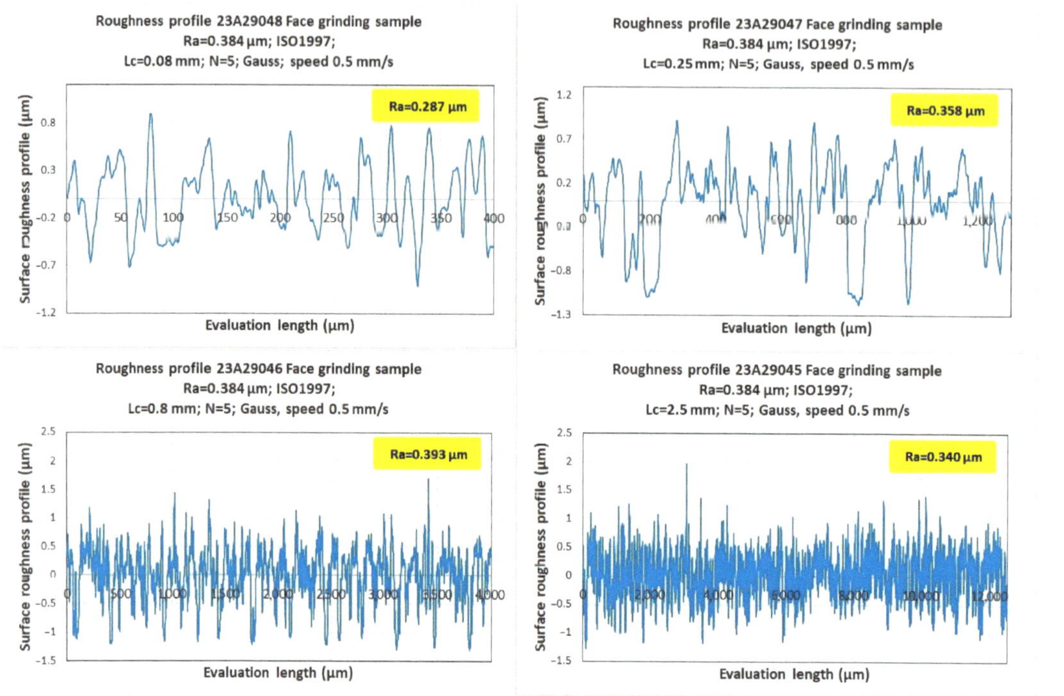

Figure A12. Surface roughness profile measured using different settings of the cut-off filter *Lc* for face grinding for a sample with a nominal value of *Ra* = 0.384 µm.

Evaluations were carried out for other Face grinding operation samples and are shown in Figure A13, where it is also indicated which cut-off values of the *Lc* filter are recommended and which are optimal according to our experimental observations, where the cut-off value is marked using the symbol [ISO] of the filter *Lc* recommended according to the ISO standard [33–40], and the red triangle with the symbol OK (▼✅) shows the cut-off value of the filter *Lc*, which is optimal according to the conducted experiments, when the determined *Ra* value is closest to the nominal *Ra* value of the sample. The displayed results show that the recommendations of the ISO standard are also not in agreement with all experimental observations. These are mainly samples with larger nominal *Ra* values (Figure A13).

For the Peripheral grinding operation, samples (Figure A14) with nominal values *Ra* = 0.103 µm, *Ra* = 0.14 µm, *Ra* = 0.412 µm, *Ra* = 0.784 µm, *Ra* = 0.91 µm were selected from the etalon samples set of surface roughness. For the *Ra* = 0.392 µm sample, measurements at different cut-off filter values were evaluated (Figure A14).

Even in this case, it can be seen with small values of the cut-off filter *Lc* that there is not a sufficient amount of unevenness of the surface roughness profile available for evaluation (Figure A14).

Evaluations were carried out for other peripheral grinding operation samples and are shown in Figure A15, where it is also indicated which cut-off values of the filter *Lc* are recommended and which are optimal according to our experimental observations, where the cut-off value is marked using the symbol [ISO] of the filter *Lc* recommended according to the ISO standard [33–40], and the red triangle with the symbol OK (▼✅) shows the cut-off value of the filter *Lc*, which is optimal according to the conducted experiments, when the determined value of *Ra* is closest to the nominal value of the sample. The displayed results show that the recommendations of the ISO standard are also not in agreement with all experimental observations. These are mainly samples with larger *Ra* values (Figure A15).

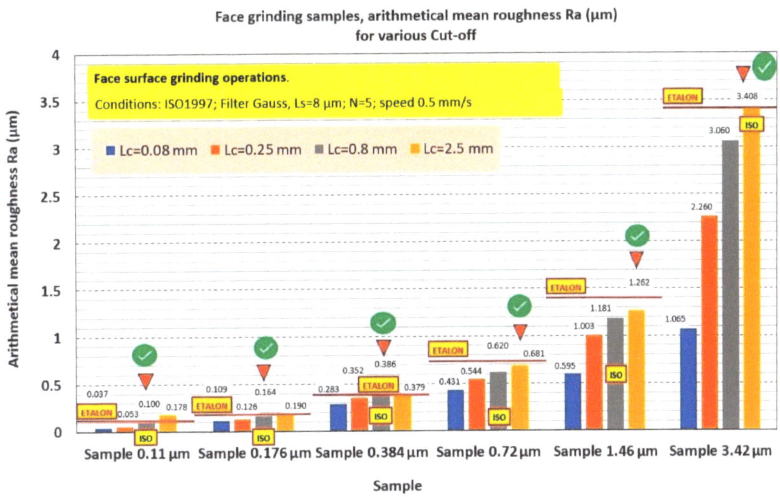

Figure A13. Face grinding—average values of *Ra* for different settings of the cut-off filter *Lc*.

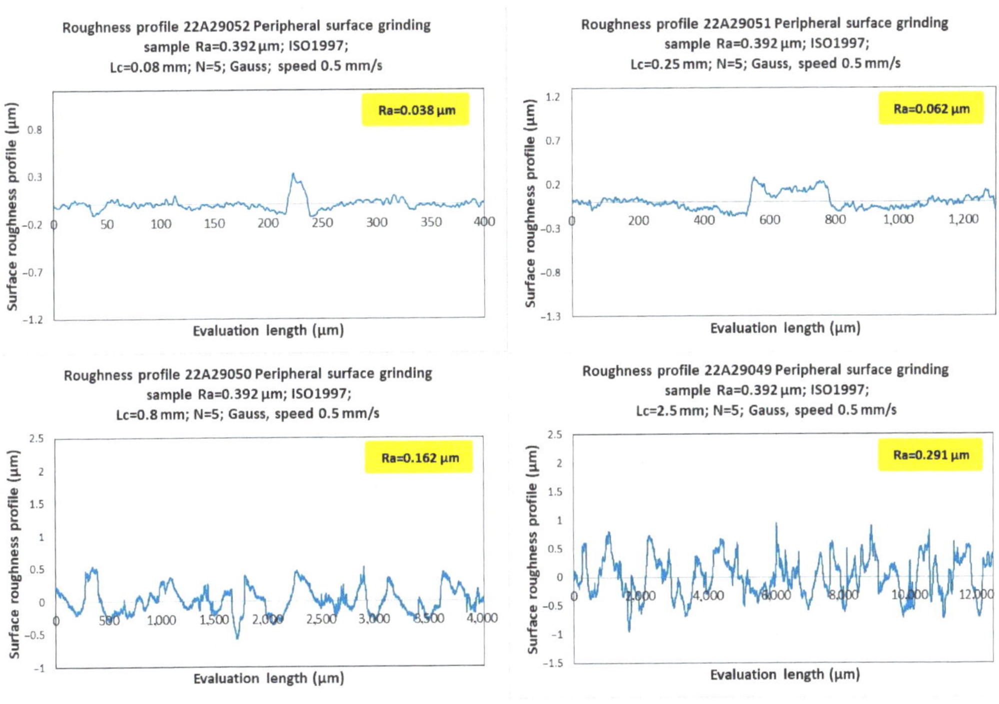

Figure A14. Surface roughness profile measured using different settings of the cut-off filter *Lc* for peripheral grinding for a sample with a nominal value of *Ra* = 0.392 µm.

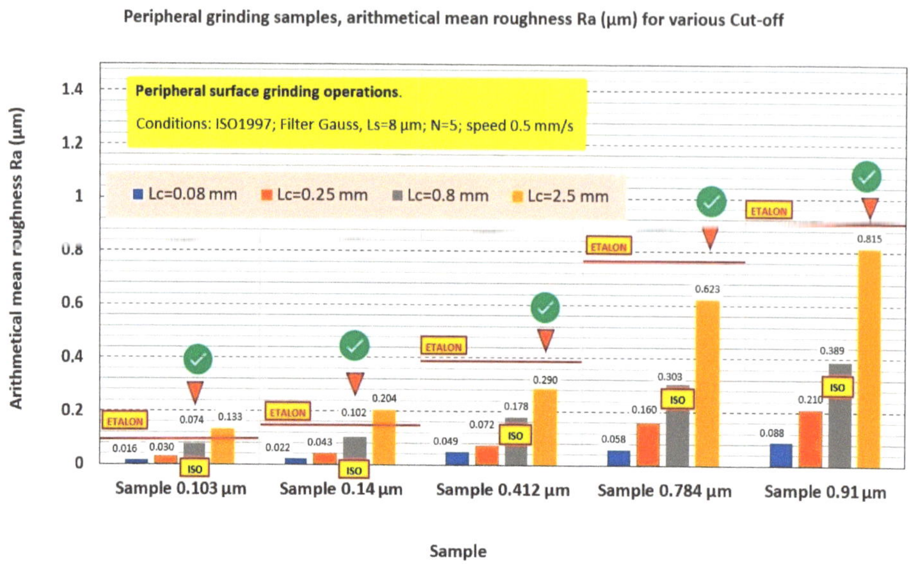

Figure A15. Peripheral grinding—average values of *Ra* for different settings of the cut-off filter *Lc*.

Appendix C

Periodic Surface Roughness Profiles–Experiments

For milling with a slab mill cutter (Milling operation with slab mill cutter), a selected sample, measured at the same place but with different values of the cut-off filter *Lc*, is shown (Figure A16). It can be seen from the given graphic curves that for the cut-off filter *Lc* with the smallest value, its value is too small and therefore does not capture the actual value of the quantity Ra after evaluation. At other values, the periodic pattern characteristic of this technology is already visible. The effect of waviness is only slightly visible at the largest used value of the cut-off filter *Lc* = 2.5 mm.

For milling with a cylindrical cutter (Milling operation with slab mill cutter), samples with nominal values *Ra* = 1.52 µm were experimentally examined; 3.57 µm; 6.05 µm; 10.25 µm; 16.25 µm. The *RSm* values for all these samples measured at different settings of the cut-off filter *Lc* are shown in Figure A17, and it follows from them that the only suitably chosen value of the cut-off filter *Lc* is 2.5 mm. At the other values of the cut-off filter *Lc*, the values of *RSm* were outside the range of the instrument or could not be evaluated.

The *Ra* values for milling with a slab mill cutter (Milling operation with slab mill cutter) were determined for different values of the cut-off filter *Lc* and are in accordance with ISO recommendations (Figure A18).

For milling operation with a face mill cutter, a selected sample with a nominal value of *Ra* = 3.46 µm measured at the same place but with different values of the cut-off filter *Lc* is shown. From the contours of the surface roughness profile of the sample, it can be seen that the first two contours have a poorly chosen cut-off filter *Lc* and the contours do not capture a representative sample of the course characteristic of this type of technology. The other two surface roughness profile records already provide a surface roughness profile that can already be analyzed. In the last surface roughness profile, a slight influence of the waviness of the surface profile is already visible (Figure A19).

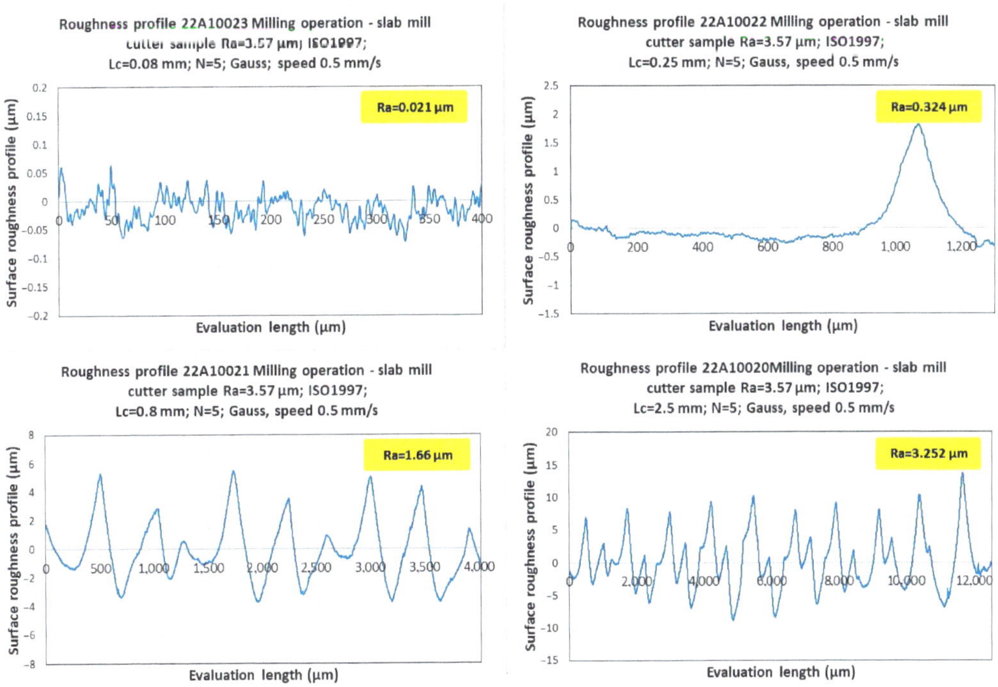

Figure A16. Surface roughness profile measured using different settings of the cut-off filter *Lc* for milling with a slab mill cutter (Milling operation with slab mill cutter) for a sample with a nominal value of *Ra* = 3.57 µm.

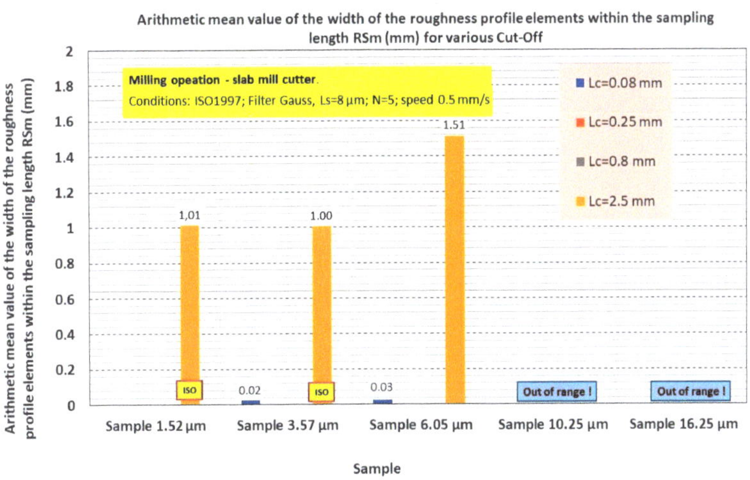

Figure A17. Milling with a cylindrical cutter—average *RSm* values for different settings of the cut-off filter *Lc*.

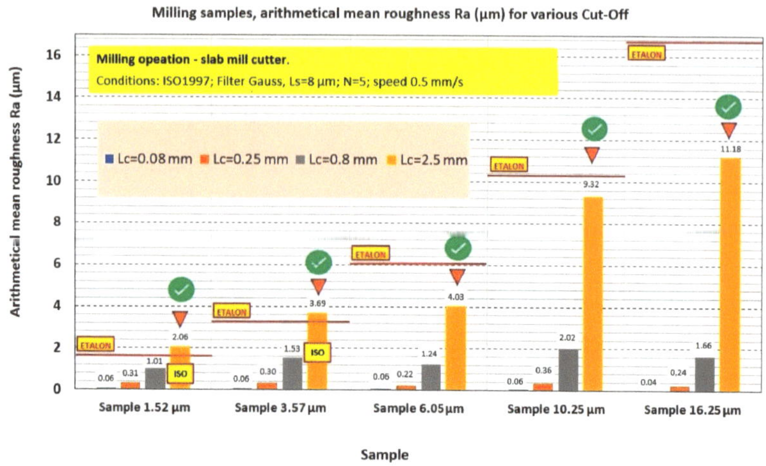

Figure A18. Milling with a slab mill cutter—average values of *Ra* for different settings of the cut-off filter *Lc*.

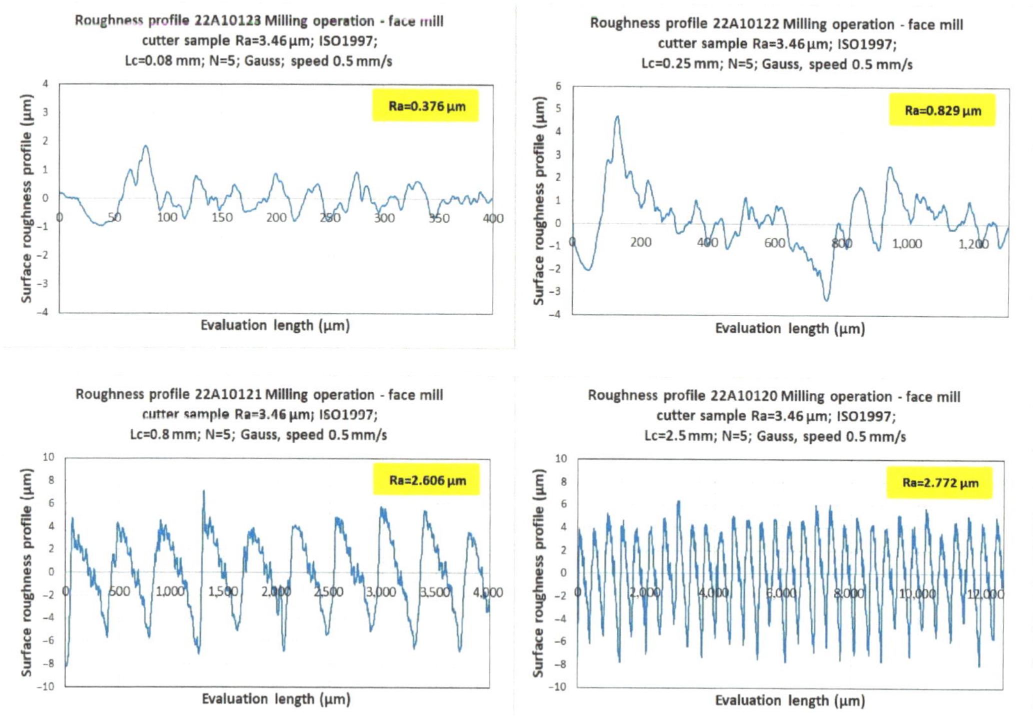

Figure A19. Surface roughness profile measured using different settings of the cut-off filter *Lc* for face milling for a sample with a nominal value of *Ra* = 3.46 µm.

For milling operation with a face mill cutter, samples with nominal values *Ra* = 1.71 µm 3.46 µm; 6.17 µm; 11 µm; 19.25 µm were experimentally examined. The *RSm* values for all these samples determined at different settings of the cut-off filter *Lc* are shown in Figure A20, and from them, it follows that for samples *Ra* = 1.71 µm; 3.46 µm, *RSm* values are similar at Lc = 0.8 mm and at Lc = 0.25 mm. For other samples, the only appropriately chosen cut-off value of the Lc filter is 2.5 mm, and at other cut-off values of the Lc filter, the RSm values were very low or could not be evaluated (Figure A20).

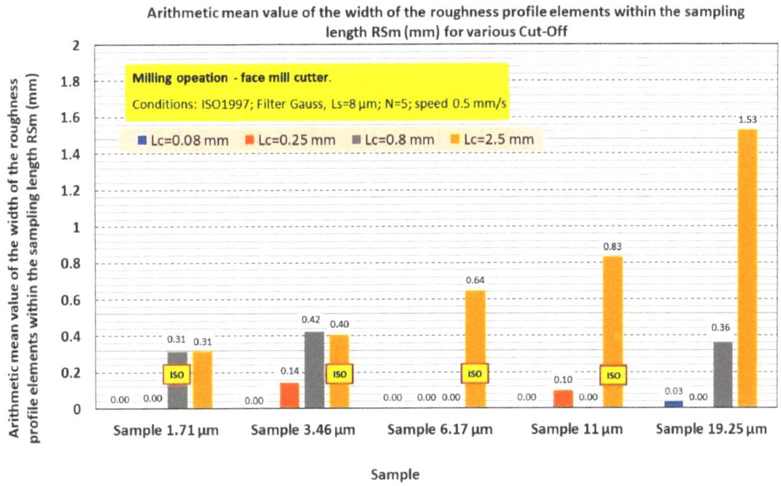

Figure A20. Milling operation with face mill cutter—average RSm values for different settings of the cut-off filter Lc.

The Ra values for face milling were determined for different values of the cut-off filter Lc and are in accordance with ISO recommendations (Figure A21). For the 19.25 µm sample, the recommended filter cut-off value is Lc = 8 mm, which is not shown on the graph because the used surface tester does not have this setting available (Figure A21).

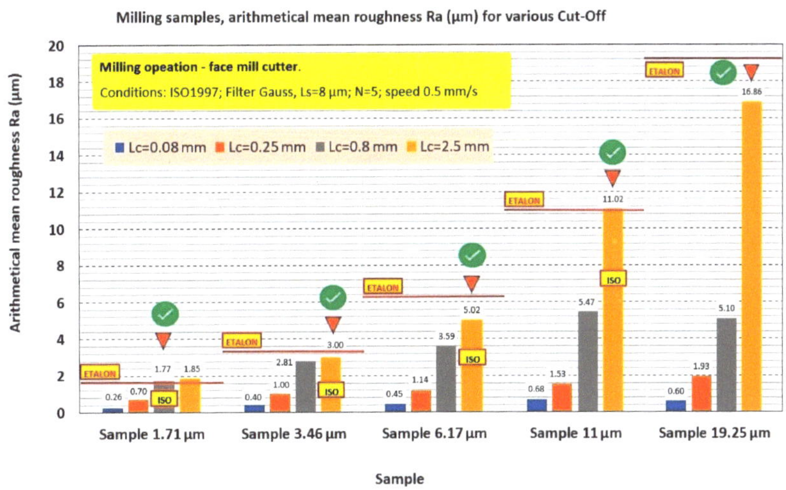

Figure A21. Milling operation with face mill cutter—average Ra values for different settings of the cut-off filter Lc.

For turning, a sample with a nominal value of Ra = 2.78 µm is shown (Figure A22). With the first two values of the cut-off filter Lc, it is again visible that the sample does not correspond to the nature of the turning technology, and the cut-off filter parameter Lc is poorly chosen. In the case of a sample with a filter cut-off parameter Lc = 0.8 mm, the course of the surface roughness profile already corresponds to the technology of this sample, and the filter cut-off parameter *Lc* is appropriately chosen. In the last sample, the effect of waviness is already visible, so it is obvious that the cut-off filter parameter *Lc* is no longer appropriately chosen (Figure A22).

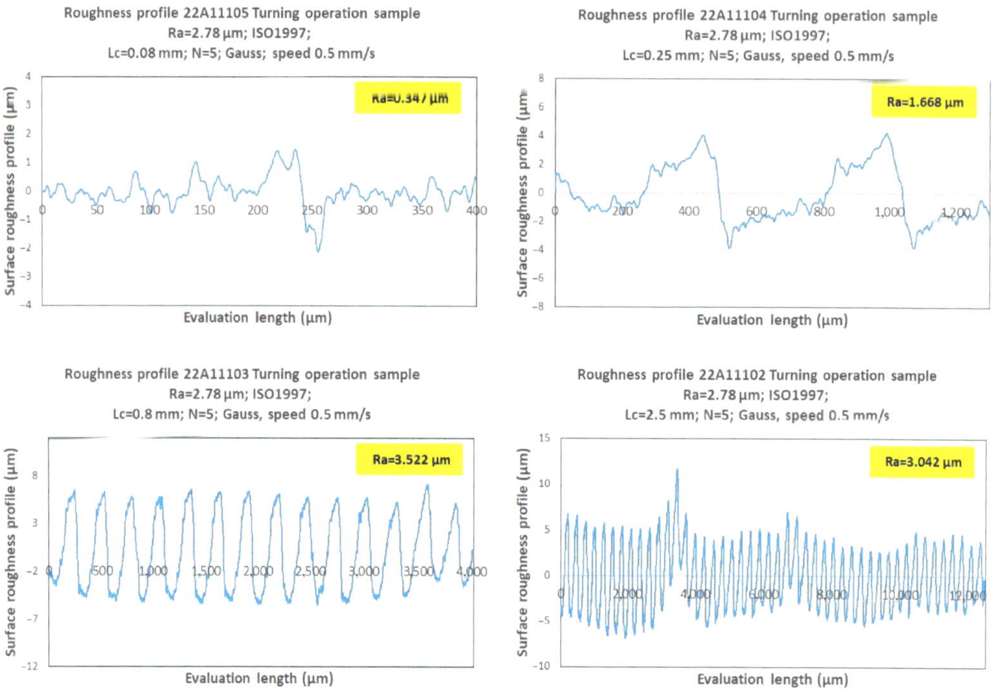

Figure A22. Surface roughness profile measured using different settings of the cut-off filter *Lc* for the turning operation for a sample with a nominal value of *Ra* = 2.78 µm.

For the turning operation, samples (C8) with nominal values of *Ra* = 0.784 µm; 1.55 µm; 2.78 µm; 6.55 µm; 12.63 µm were experimentally examined. The *RSm* values for all these samples determined at different settings of the cut-off filter *Lc* are shown in Figure A23, and it follows from them that, apart from one sample, the *RSm* values are evaluated only for the cut-off filter values *Lc* = 0.8 mm and 2.5 mm and show approximately the same values.

The values of *Ra* for turning operation were determined for different values of the cut-off filter *Lc* and, except for one sample, they are not in accordance with ISO recommendations. However, this may be due to the uncertainty of the measurement because there are only small differences between the values that are optimal and the values at the settings recommended by the standard (Figure A24).

For the planing operation, a sample with a nominal value of *Ra* = 3.04 µm is shown (Figure A25). With the first two values of the cut-off filter *Lc*, it is again visible that the sample does not correspond to the nature of the planing technology, and the cut-off filter parameter *Lc* is poorly chosen. In the case of a sample with a filter cut-off parameter *Lc* = 0.8 mm, the course of the surface roughness profile already corresponds to the technology of this sample, and the filter cut-off parameter *Lc* is appropriately chosen. In the last sample, the effect of waviness is already visible, so it is obvious that the cut-off filter parameter *Lc* is no longer appropriately chosen (Figure A25).

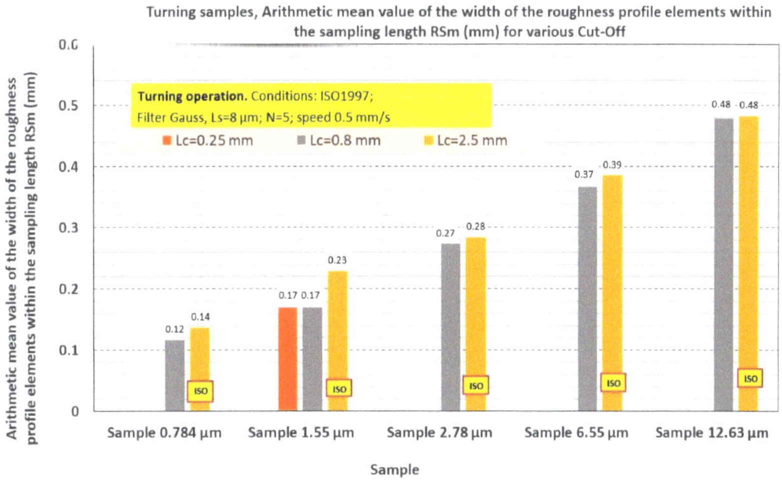

Figure A23. Turning operation—average *RSm* values for different settings of the cut-off filter *Lc*.

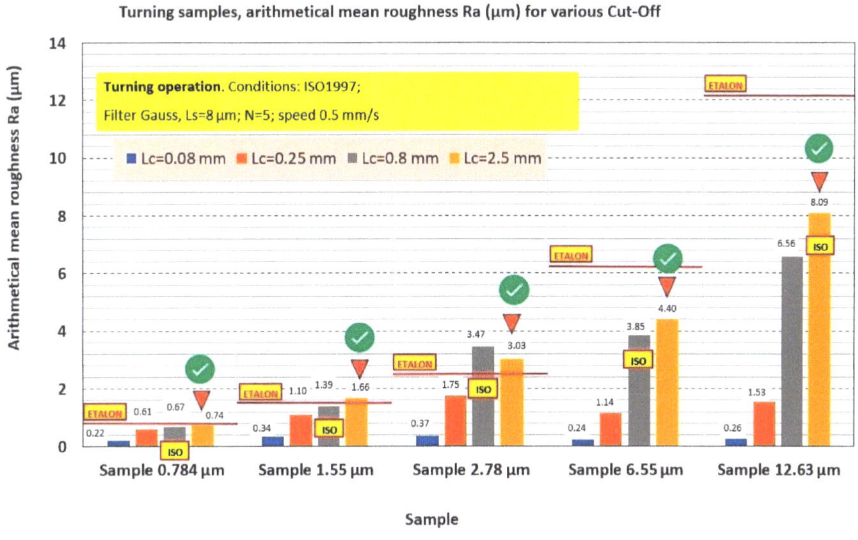

Figure A24. Turning operation—average *Ra* values for different settings of the cut-off filter *Lc*.

Samples with nominal values of *Ra* = 1.5 µm; 3.04 µm; 6.57 µm; 11.5 µm; 20.25 µm were experimentally examined for planing operation. The *RSm* values for all these samples determined at different settings of the cut-off filter *Lc* are shown in Figure A26, and it follows from them that, except for one sample, the *RSm* values are evaluated only for the cut-off filter values *Lc* = 0.8 mm and 2.5 mm and show approximately the same *RSm* values. For the sample with the largest value of *Ra*, the value *RSm* is determined only for cut-off filter values *Lc* = 2.5 mm; with other settings of the cut-off filter *Lc*, it was not possible to determine *RSm* (Figure A26).

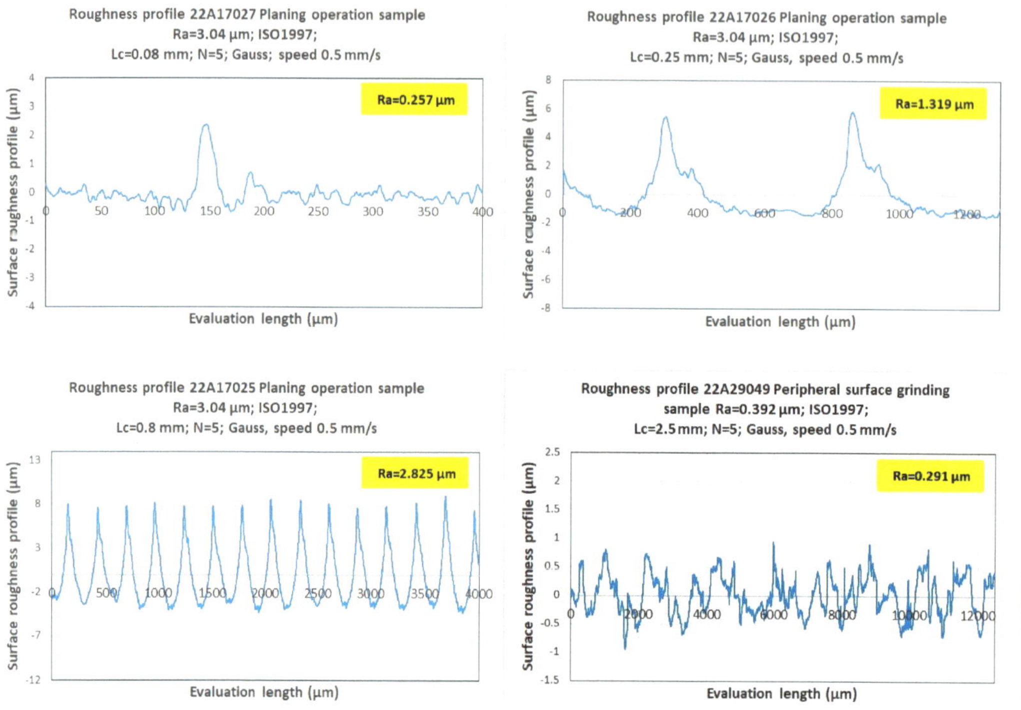

Figure A25. Surface roughness profile measured using different settings of the cut-off filter *Lc* for the planing operation for a sample with a nominal value of Ra = 3.04 µm.

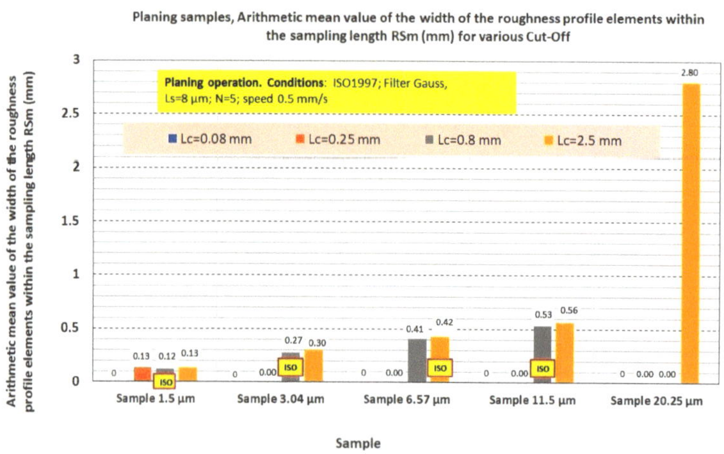

Figure A26. Planing operation—average *RSm* values for different settings of the cut-off filter *Lc*.

The Ra values for the planing operation were determined for different values of the cut-off filter *Lc* and the first two samples are in accordance with ISO recommendations (Figure A27). The *Ra* values of other samples do not match the values determined for the cut-off filter parameters *Lc* recommended by the standard. However, this may be due to the uncertainty of the measurement because there are only small differences between the values that are optimal and the values at the settings recommended by the standard (Figure A27).

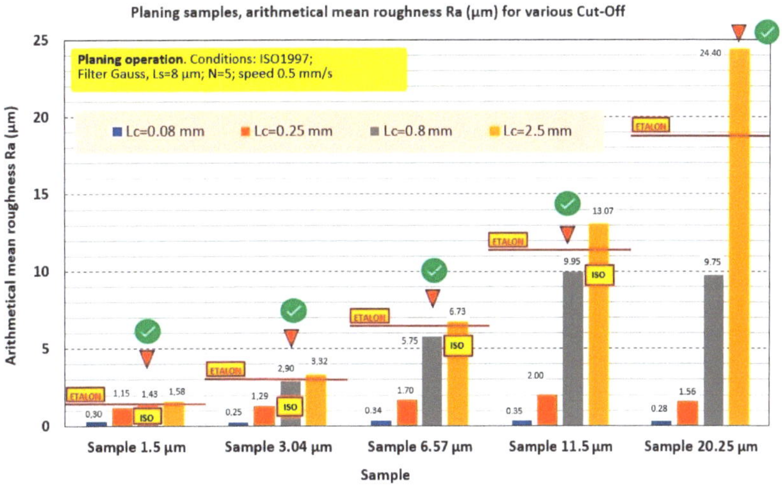

Figure A27. Planing operation—average values of *Ra* for different settings of the cut-off filter *Lc*.

References

1. Whitehouse, D.J. *Surfaces and Their Measurement*, 1st ed.; Butterworth-Heinemann: Oxford, UK, 2004; 432p, ISBN 9781903996607.
2. Vorburger, T.V.; Rhee, H.G.; Renegar, T.B.; Song, J.F.; Zheng, A. Comparison of optical and stylus methods for measurement of surface texture. *Int. J. Adv. Manuf. Technol.* **2007**, *33*, 110–118. [CrossRef]
3. Kumar, S.P.L. Measurement and uncertainty analysis of surface roughness and material removal rate in micro turning operation and process parameters optimization. *Measurement* **2019**, *140*, 538–547. [CrossRef]
4. Beňo, J.; Maňková, I.; Vrabel, M.; Kottfer, D. Roughness measurement methodology for selection of tool inserts. *Measurement* **2013**, *46*, 582–592. [CrossRef]
5. Davinci, M.A.; Parthasarathi, N.L.; Borah, U.; Albert, S.K. Effect of the tracing speed and span on roughness parameters determined by stylus type equipment. *Measurement* **2014**, *48*, 368–377. [CrossRef]
6. Misumi, I.; Naoi, K.; Sugawara, K.; Gonda, S. Profile surface roughness measurement using metrological atomic force microscope and uncertainty evaluation. *Measurement* **2015**, *73*, 295–303. [CrossRef]
7. Molnár, V. Minimization Method for 3D Surface Roughness Evaluation Area. *Machines* **2021**, *9*, 192. [CrossRef]
8. Bazan, A.; Turek, P.; Przeszłowski, Ł. Assessment of InfiniteFocus system measurement errors in testing the accuracy of crown and tooth body model. *J. Mech. Sci. Technol.* **2021**, *35*, 1167–1176. [CrossRef]
9. Bazan, A.; Turek, P.; Przeszłowski, Ł. Comparison of the contact and focus variation measurement methods in the process of surface topography evaluation of additively manufactured models with different geometry complexity. *Surf. Topogr. Metrol. Prop.* **2022**, *10*, 035021. [CrossRef]
10. Płodzień, M.; Żyłka, Ł.; Sułkowicz, P.; Żak, K.; Wojciechowski, S. High-Performance Face Milling of 42CrMo4 Steel: Influence of Entering Angle on the Measured Surface Roughness, Cutting Force and Vibration Amplitude. *Materials* **2021**, *14*, 2196. [CrossRef]
11. Bazan, A.; Turek, P.; Sułkowicz, P.; Przeszłowski, Ł.; Zakręcki, A. Influence of the Size of Measurement Area Determined by Smooth-Rough Crossover Scale and Mean Profile Element Spacing on Topography Parameters of Samples Produced with Additive Methods. *Machines* **2023**, *11*, 615. [CrossRef]
12. Bhushan, B. Chapter: Surface Roughness Analysis and Measurement Techniques. In *Modern Tribology Handbook, Two Volume Set*, 1st ed.; CRC Press: Boca Raton, FL, USA, 2000; 1760p, ISBN 9780429126727. [CrossRef]
13. Kreith, F. *The Mechanical Engineering Handbook Series*, 2nd ed.; CRC Press: New York, NY, USA, 2005; 2508p, ISBN 0-8493-0866-6.
14. Webster, J.G.; Eren, H. *Measurement, Instrumentation, and Sensors Handbook*, 2nd ed.; CRC Press: Boca Raton, FL, USA, 2014; 1640p, ISBN 9781315217451. [CrossRef]
15. Leach, R. *The Measurement of Surface Texture Using Stylus Instruments*; NPL National Physical Laboratory, HMSO: Teddington, UK, 2001.
16. Poon, C.Y.; Bhushan, B. Comparison of surface roughness measurements by stylus profiler, AFM and non-contact optical profiler. *Wear* **1995**, *190*, 76–88. [CrossRef]
17. Jester, W.; Klik, A. Soil surface roughness measurement—Methods, applicability, and surface representation. *Catena* **2005**, *64*, 174–192. [CrossRef]

18. Stout, K.J. Surface roughness ~ measurement, interpretation and significance of data. *Mater. Des.* **1981**, *2*, 260–265. [CrossRef]
19. Verheijen, E. A survey on roughness measurements. *J. Sound Vib.* **2006**, *293*, 784–794. [CrossRef]
20. Chang, W.-R.; Kim, I.-J.; Manning, D.P.; Bunterngchit, Y. The role of surface roughness in the measurement of slipperiness. *Ergonomics* **2001**, *44*, 1200–1216. [CrossRef]
21. Whitehouse, D.J. Surface Characterization and Roughness Measurement in Engineering. In *Photomechanics*; Rastogi, P.K., Ed.; Topics in Applied Physics; Springer: Berlin/Heidelberg, Germany, 2000; Volume 77. [CrossRef]
22. Wieczorowski, M.; Cellary, A.; Majchrowski, R. The analysis of credibility and reproducibility of surface roughness measurement results. *Wear* **2010**, *269*, 480–484. [CrossRef]
23. Farkas, G.; Drégelyi-Kiss, Á. Measurement uncertainty of surface roughness measurement. In Proceedings of the IOP Conference Series: Materials Science and Engineering, Volume 448, XXIII International Conference on Manufacturing (Manufacturing 2018), Kecskemét, Hungary, 7–8 June 2018; pp. 1–10. [CrossRef]
24. Rodriguez, F.V.; Sukumaran, J.; Ando, M.; De Baets, P. Roughness measurement problems in tribological testing. In Proceedings of the Sustainable Construction and Design 2011 (SCAD), Ghent, Belgium, 16–17 February 2011; Laboratory Soete, Ghent University: Ghent, Belgium, 2011; Volume 2, pp. 115–121.
25. Fedotkin, A.Y.; Akimchenko, I.O.; Tran, T.-H.; Shugurov, A.R.; Shesterikov, E.V.; Kozelskaya, A.I.; Rutkowski, S.; Tverdokhlebov, S.I. Improvement of the Surface Properties of Polyether Ether Ketone via Arc Evaporation for Biomedical Applications. *Materials* **2023**, *16*, 2990. [CrossRef]
26. La, S.; Wang, C. Experimental and Numerical Calculation of the Friction Performance of a Concrete Surface. *Materials* **2023**, *16*, 2989. [CrossRef]
27. Płodzień, M.; Żyłka, Ł.; Stoić, A. Modelling of the Face-Milling Process by Toroidal Cutter. *Materials* **2023**, *16*, 2829. [CrossRef]
28. Každukauskas, E.; Butkus, S.; Jukna, V.; Paipulas, D.; Sirutkaitis, V. Scanning Algorithm Optimization for Achieving Low-Roughness Surfaces Using Ultrashort Laser Pulses: A Comparative Study. *Materials* **2023**, *16*, 2788. [CrossRef]
29. Xia, Y.; Lu, Y.; Yang, G.; Chen, C.; Hu, X.; Song, H.; Deng, L.; Wang, Y.; Yi, J.; Wang, B. Application of Nano-Crystalline Diamond in Tribology. *Materials* **2023**, *16*, 2710. [CrossRef] [PubMed]
30. Podulka, P.; Macek, W.; Branco, R.; Nejad, R.M. Reduction in Errors in Roughness Evaluation with an Accurate Definition of the S-L Surface. *Materials* **2023**, *16*, 1865. [CrossRef] [PubMed]
31. Smak, K.; Szablewski, P.; Legutko, S.; Krawczyk, B.; Miko, E. Investigation of the Influence of Anti-Wear Coatings on the Surface Quality and Dimensional Accuracy during Finish Turning of the Inconel 718 Alloy. *Materials* **2023**, *16*, 715. [CrossRef]
32. Carek, A.; Slokar Benić, L.; Komar, D.; Krebelj, E. Roughness of the Surface of Zirconia Reinforced Lithium Disilicate Ceramic Treated by Different Procedures. *Materials* **2023**, *16*, 265. [CrossRef]
33. *ISO 4287:1997*; Geometrical Product Specifications (GPS)—Surface Texture: Profile Method—Terms, Definitions and Surface Texture Parameters. 1st ed. International Organization for Standardization: Geneve, Switzerland, 1 April 1997.
34. *ISO 4288: 1996*; Geometrical Product Specifications (GPS)—Surface Texture: Profile Method—Rules and Procedures for the Assessment of Surface Texture. Edition DIN EN ISO 4288; European Committee for Standardization: Brussels, Belgium, April 1998.
35. *ISO 3274: 1996*; Geometrical Product Specifications (GPS)—Surface Texture: Profile Method—Nominal Characteristic of Contact (Stylus) Instruments. European Committee for Standardization: Brussels, Belgium, December 1996.
36. *ISO 5436-1:2000*; Geometrical Product Specifications (GPS)—Surface Texture: Profile Method; Measurement Standards—Part 1: Material Measures. European Committee for Standardization: Brussels, Belgium, March 2000.
37. *ISO 21920-1:2021*; Geometrical Product Specifications (GPS)—Surface Texture: Profile—Part 1: Indication of Surface Texture. European Committee for Standardization: Brussels, Belgium, December 2021.
38. *ISO 21920-2:2021*; Geometrical Product Specifications (GPS)—Surface Texture: Profile—Part 2: Terms, Definitions and Surface Texture Parameters. European Committee for Standardization: Brussels, Belgium, December 2021.
39. *ISO 21920-3:2021*; Geometrical Product Specifications (GPS)—Surface Texture: Profile—Part 3: Specification Operators. European Committee for Standardization: Brussels, Belgium, December 2021.
40. *ISO 11562:1996*; Geometrical Product Specifications (GPS)—Surface Texture: Profile Method—Metrological Characteristics of Phase Correct Filters. European Committee for Standardization: Brussels, Belgium, December 1996.
41. Semyon, G. *Rabinovich: Measurement Errors and Uncertainties. Theory and Practice*, 3rd ed.; Springer: New York, NY, USA, 2005; p. XII. 308p, ISBN 978-0-387-25358-9/978-0-387-29143-7.
42. Semyon, G. Rabinovich: Accuracy of single measurements. *Accredit. Qual. Assur.* **2007**, *12*, 419–424.
43. Taylor, B.N.; Kuyatt, C.E. *Guidelines for Evaluating and Expressing the Uncertainty of NIST Measurement Results*; NIST Technical Note 1297; US Department of Commerce, Technology Administration, National Institute of Standards and Technology: Gaithersburg, MD, USA, 1994.
44. Taylor, J.R. *An Introduction to Error Analysis: The Study of Uncertainties in Physical Measurements*, 2nd ed.; University Science Books: Herndon, VA, USA, 1 August 1996; 327p, ISBN 9780935702750.
45. Mohajan, H. Two Criteria for Good Measurements in Research: Validity and Reliability. *Ann. Spiru Haret Univ.* **2017**, *17*, 58–82. Available online: https://mpra.ub.uni-muenchen.de/83458/MPRA (accessed on 8 April 2023). [CrossRef]
46. *EA-4/02 1999*; Expression of the Uncertainty of Measurement in Calibration. European Co-Operation Accreditation Publication Reference: Brussels, Belgium, December 1999.

47. *JCGM 100 2008*; JCGM 100—Evaluation of Measurement Data—Guide to the Expression of Uncertainty in Measurement (ISO/IEC Guide 98-3). 1st ed. ISO: Geneva, Switzerland, September 2008. Available online: http://www.iso.org/sites/JCGM/GUM-JCGM100.htm (accessed on 8 April 2023).
48. *JCGM 104 2009*; Evaluation of Measurement Data—An introduction to the "Guide to the Expression of Uncertainty in Measurement" (ISO/IEC Guide 98-1). 1st ed. ISO: Geneva, Switzerland, July 2009. Available online: http://www.bipm.org/en/publications/guides/gum_print.html (accessed on 8 April 2023).
49. JCGM 200—International Vocabulary of Metrology—Basic and General Concepts and Associated Terms (VIM) 3rd Edition (2008 Version with Minor Corrections). © JCGM 2012. Available online: http://www.iso.org/sites/JCGM/VIM-JCGM200.htm (accessed on 8 April 2023).
50. *JIS B 0601-1982*; Geometrical Product Specifications (GPS)—Surface Texture: Profile Method—Terms, Definitions and Surface Texture Parameters. 1982 Edition; Japanese Industrial Standards Committee Standards Board, Japanese Standards Association (JSA): Tokyo, Japan, 15 June 1982.
51. *JIS B 0601: 1994*; Geometrical Product Specifications (GPS)—Surface Texture: Profile Method—Terms, Definitions and Surface Texture Parameters. 1994 Edition; Japanese Industrial Standards Committee Standards Board, Japanese Standards Association (JSA): Tokyo, Japan, 1 January 1994.
52. *JIS B 0601: 2001*; Geometrical Product Specifications (GPS)—Surface Texture: Profile Method—Terms, Definitions and Surface Texture Parameters. Japanese Industrial Standards Committee Standards Board, Japanese Standards Association (JSA): Tokyo, Japan, 1 January 2001.
53. *JIS B 0601: 2010*; Geometrical Product Specifications (GPS)—Surface Texture: Profile Method—Terms, Definitions and Surface Texture Parameters. 2010 Edition; Japanese Industrial Standards Committee Standards Board, Japanese Standards Association (JSA): Tokyo, Japan, 1 October 2010.
54. *JIS B 0601:2013*; Geometrical Product Specifications (GPS)—Surface Texture: Profile Method—Terms, Definitions and Surface Texture Parameters. 2013 Edition; Japanese Industrial Standards Committee Standards Board, Japanese Standards Association (JSA): Tokyo, Japan, 1 January 2013.
55. *JIS B 0601: 2022*; Geometrical Product Specifications (GPS)—Surface Texture: Profile Method—Terms, Definitions and Surface Texture Parameters. 2022 Edition; Japanese Industrial Standards Committee Standards Board, Japanese Standards Association (JSA): Tokyo, Japan, 20 October 2022.
56. *JIS B0632: 2001*; Geometrical Product Specification (GPS)—Surface Texture: Profile Method—Metrological Characteristics of Phase Correct Filters. 2001 Edition; Japanese Industrial Standards Committee Standards Board, Japanese Standards Association (JSA): Tokyo, Japan, 1 January 2001.
57. *JIS B 0651:2001*; Geometrical Product Specification (GPS)—Surface Texture: Profile Method—Nominal Characteristics of Contact (Stylus) Instruments. 2001 Edition; Japanese Industrial Standards Committee Standards Board, Japanese Standards Association (JSA): Tokyo, Japan, 1 January 2001.
58. *JIS B 0610:2001*; Geometrical Product Specification (GPS)—Surface Texture: Profile Method—Definitions and Designation of Rolling Circle Waviness. 2001 Edition; Japanese Industrial Standards Committee Standards Board, Japanese Standards Association (JSA): Tokyo, Japan, 22 January 2001.
59. *ANSI B46.1-1995*; Surface Texture (Roughness, Waviness, Lay). ASME International (ASME): New York, NY, USA, January 1995.
60. *ANSI B46.1-2002*; Surface Texture (Roughness, Waviness, Lay). ASME International (ASME): New York, NY, USA, January 2002.
61. *ANSI B46.1-2009*; Surface Texture (Roughness, Waviness, Lay). ASME International (ASME): New York, NY, USA, January 2009.
62. *ASME B46.1-2019*; Surface Texture (Roughness, Waviness, Lay). ASME International (ASME): New York, NY, USA, January 2019.
63. *VDA 2005*; Geometrical Product Specifications Technical Drawings—Specification of Surface Texture. Verband Der Automobilindustrie E.V. (VDA): Berlin, Germany, February 2007.
64. *VDA 2006*; Geometrical Product Specifications—Surface Texture—Rules and Procedures for The Assessment of Surface Texture. Verband Der Automobilindustrie E.V. (VDA): Berlin, Germany, July 2003.
65. *VDA 2007*; Geometrical Product Specifications—Surface Texture—Definitions and Characteristic Values of The Dominant Waviness. Verband Der Automobilindustrie E.V. (VDA): Berlin, Germany, 1 February 2007.
66. Álló, Š.; Kročko, V.; Korenko, M.; Andrássyová, Z.; Földešiová, D. Effect of chemical degreasing on corrosion stability of components in automobile industry. *Adv. Mater. Res.* **2013**, *801*, 19–23. [CrossRef]
67. Bujna, M.; Prístavka, M.; Dostál, P.; Korenko, M.; Kadnár, M. Influence of Roughness on Quality Molybdenum Deposit Layer by Thermal Spraying. *Acta Univ. Agric. Silvic. Mendel. Brun.* **2016**, *64*, 9–14. [CrossRef]
68. Duplak, J.; Duplakova, D.; Zajac, J. Research on Roughness and Microhardness of C45 Material Using High-Speed Machining. *Appl. Sci.* **2023**, *13*, 7851. [CrossRef]
69. Krenicky, T.; Nikitin, Y.; Božek, P. Model-Based Design of Induction Motor Control System in MATLAB. *Appl. Sci.* **2022**, *12*, 11957. [CrossRef]
70. Obaton, A.-F.; Fain, J.; Meinel, D.; Tsamos, A.; Léonard, F.; Lécuelle, B.; Djemaï, M. In Vivo Bone Progression in and around Lattice Implants Additively Manufactured with a New Titanium Alloy. *Appl. Sci.* **2023**, *13*, 7282. [CrossRef]
71. Song, X.; Ke, F.; Zhu, K.; Ren, Y.; Zhou, J.; Li, W. Study on Preparation and Processing Properties of Mechano-Chemical Micro-Grinding Tools. *Appl. Sci.* **2023**, *13*, 6599. [CrossRef]

72. Gradov, O.M. Optimization of Ultrasonic Powder Coatings on the Surface of Treated Materials. *Appl. Sci.* **2023**, *13*, 6034. [CrossRef]
73. Khan, A.A.; Abdullah Alkhureif, A.; Bautista, L.S.J.; Alsunbul, H.; Vellappally, S. Peroxide-Free Bleaching Gel: Effect on the Surface and Mechanical Properties of Nano- and Micro-Hybrid Restorative Composite Materials. *Appl. Sci.* **2023**, *13*, 5935. [CrossRef]
74. Ružbarský, J. The Difficulty of Measuring the Roughness of Glossy Surfaces Using the Triangulation Principle. *Appl. Sci.* **2023**, *13*, 5155. [CrossRef]
75. Zhao, W.; Gao, Y.; Ren, X.; Lu, K.; Yang, Y. Dynamic Characteristics Analysis of a Rod Fastening Rotor System Considering Contact Roughness. *Appl. Sci.* **2023**, *13*, 5009. [CrossRef]
76. Balonji, S.; Tartibu, L.K.; Okokpujie, I.P. Prediction Analysis of Surface Roughness of Aluminum Al6061 in End Milling CNC Machine Using Soft Computing Techniques. *Appl. Sci.* **2023**, *13*, 4147. [CrossRef]
77. Kong, C.; Xu, Y.; Li, Z.; Zhang, C.; Li, T.; Macleod, I.; Jiang, X.; Tang, D.; Lu, J. A Multivariate Local Descriptor Registration Method for Surface Topography Evaluation. *Appl. Sci.* **2023**, *13*, 3311. [CrossRef]
78. Ružbarský, J.; Gašpár, Š. Analysis of Selected Production Parameters for the Quality of Pressure Castings as a Tool to Increase Competitiveness. *Appl. Sci.* **2023**, *13*, 8098. [CrossRef]
79. Krenicky, T.; Olejarova, S.; Servatka, M. Assessment of the Influence of Selected Technological Parameters on the Morphology Parameters of the Cutting Surfaces of the Hardox 500 Material Cut by Abrasive Water Jet Technology. *Materials* **2022**, *15*, 1381. [CrossRef]
80. Gulyaev, P.; Krenicky, T.; Shelkovnikov, E.; Korshunov, A. Particle and Particle Agglomerate Size Monitoring by Scanning Probe Microscope. *Appl. Sci.* **2022**, *12*, 2183. [CrossRef]
81. Sinčák, P.J.; Virgala, I.; Kelemen, M.; Prada, E.; Bobovský, Z.; Kot, T. Chimney Sweeping Robot Based on a Pneumatic Actuator. *Appl. Sci.* **2021**, *11*, 4872. [CrossRef]
82. Kelemenová, T.; Dovica, M.; Božek, P.; Koláriková, I.; Benedik, O.; Virgala, I.; Prada, E.; Miková, L.; Kot, T.; Kelemen, M. Specific Problems in Measurement of Coefficient of Friction Using Variable Incidence Tribometer. *Symmetry* **2020**, *12*, 1235. [CrossRef]
83. Miková, L'.; Prada, E.; Kelemen, M.; Krys, V.; Mykhailyshyn, R.; Sinčák, P.J.; Merva, T.; Leštach, L. Upgrade of Biaxial Mechatronic Testing Machine for Cruciform Specimens and Verification by FEM Analysis. *Machines* **2022**, *10*, 916. [CrossRef]
84. Hroncova, D.; Sincak, P.J.; Merva, T.; Mykhailyshyn, R. Robot Trajectory Planning. *MM Sci. J.* **2022**, *2022*, 6098–6108. [CrossRef]
85. Mykhailyshyn, R.; Duchoň, F.; Virgala, I.; Sinčák, P.J.; Fey, A.M. Optimization of Outer Diameter Bernoulli Gripper with Cylindrical Nozzle. *Machines* **2023**, *11*, 667. [CrossRef]

applied sciences

MDPI

Article

New Calculation Methodology of the Operations Number of Cold Rolling Rolls Fine Grinding

Sergey Bratan [1], Zuzana Ságová [2,*], Milan Sága [3], Boris Yakimovich [4] and Ivan Kuric [2]

[1] Polytechnical Institute, Sevastopol State University, 33 Universitetskaya St., 299053 Sevastopol, Russia
[2] Department of Automation and Production Systems, Faculty of Mechanical Engineering, University of Žilina, Univerzitná 8215/1, 010 26 Žilina, Slovakia
[3] Department of Applied Mechanics, Faculty of Mechanical Engineering, University of Žilina, Univerzitná 8215/1, 010 26 Žilina, Slovakia
[4] Institute of Nuclear Energy and Industry, Sevastopol State University, 7 Kurchatova St., 299015 Sevastopol, Russia
* Correspondence: zuzana.sagova@fstroj.uniza.sk

Abstract: This article considers the methods of calculating the number of operations for the fine grinding of cold rolling rolls which allows one to ensure the least labor-intensive processing of the part. Based on the analysis of the patterns in the operation of single abrasive grains, the dependence of the productivity of the fine-grinding process on the grain size of the wheel, the grinding allowance, the grinding mode, the size of the workpiece, and the wheel, is proposed. To analyze the productivity of the finishing grinding processes, a productivity indicator, independent of the size of the wheel and the workpiece, is analyzed, and the impact on the productivity indicator of the number of finishing grinding operations, as well as the following sequence of their optimization, is established. To simplify the calculation of the optimal number of operations, graphs and a nomogram are constructed. According to the proposed methodology, the calculation of the number of finishing operations of cold rolling rolls of various sizes was carried out. Experimental verification of the proposed theoretical dependencies and methods for calculating the optimal number of grinding operations was also carried out.

Keywords: fine-grained wheels; grinding; operation; performance indicator; the least labor intensity of processing; grinding mode

Citation: Bratan, S.; Ságová, Z.; Sága, M.; Yakimovich, B.; Kuric, I. New Calculation Methodology of the Operations Number of Cold Rolling Rolls Fine Grinding. *Appl. Sci.* **2023**, *13*, 3484. https://doi.org/10.3390/app13063484

Academic Editors: Tibor Krenicky, Maros Korenko and Juraj Ruzbarsky

Received: 7 February 2023
Revised: 27 February 2023
Accepted: 3 March 2023
Published: 9 March 2023

1. Introduction

One of the high-performance methods of finishing parts is grinding them with fine-grained wheels. Circular external grinding with fine-grained wheels can provide parts with a processing accuracy of five accuracy standards, with a surface roughness of $Ra = 0.02 - 0.01$ microns [1]. The productivity of the process is higher than the productivity of such widespread finishing methods as superfinishing and semi-mechanical finishing [2–7].

Grinding of parts with fine-grained wheels is used in the ball bearing, machine tool, and instrument-making industries, as well as in the processing of cold rolling rolls in the metallurgical industry [8]. In recent years, models of high- and especially high-precision circular grinding machines have been developed and mastered, and manufacturing technology and formulations of fine-grained wheels have also been developed. All of these are prerequisites for the wider introduction of grinding with fine-grained wheels into the industry as a progressive finishing method.

The main issues in the development of technological processes for grinding parts to obtain a surface roughness of $Ra = 0.02$ and surface accuracy according to the fifth quality are: determining the number of grinding operations of parts, choosing the characteristics of grinding wheels, and assigning grinding modes [9–12].

Recommendations on these issues are diverse [12]. For example, for the case of grinding the rolls of cold rolling machines to obtain $Ra = 0.02 - 0.01$ microns, various recommendations suggest carrying out finishing in one, two, three, four, or more operations [10]. The variety of recommendations for grinding with fine-grained wheels is explained not only by the multiplicity of factors affecting the process, but also by the lack of methods for optimizing the process.

The use of fine-grained wheels makes it possible to provide a given roughness of $Ra = 0.04 - 0.02$ microns for the surface of the rolls; however, the complexity of their processing increases sharply, and intensive wear of the abrasive tool is observed [13]. Attempts to intensify the process lead to a deterioration in the output quality indicators; that is, it is not possible to obtain a part with a given surface roughness. It is possible to solve this problem by optimizing the number of technological operations. Similarly to other process parameters, the optimal number of operations when grinding parts with fine-grained wheels should be determined on the basis of a technical and economic analysis. The complexity of processing the part can be taken as a parameter reflecting a variable share of the cost of the process. With an increase in the number of finishing operations, the auxiliary time increases, due to the additional costs of rearranging the part and the wheel, and the machine processing time of the part decreases, since larger-grained wheels are used for intermediate operations, providing greater productivity [12]. When constructing technological processes of grinding in one, two, three, or more operations, each option must ensure that the part is obtained with a given surface roughness, a given accuracy, and the required quality of surface layers. The optimal one can be taken as the one that provides the least complexity in processing the part [13].

Thus, the object of research is the process of the fine grinding of cold rolling rolls.

The subjects of the study are the regularities of determining the optimal number of operations that ensures the roughness of the surface of cold rolling rolls is $Ra = 0.04 - 0.02$ microns with the least complexity of processing the part.

The present work aims, based on theoretical and experimental study of the grinding process with fine-grained wheels, to develop a methodology for calculating the number of operations for the fine grinding of cold rolling rolls in order to obtain a surface roughness of $Ra = 0.04 - 0.02$ microns, which allows for the least labor-intensive processing of the part.

2. Mathematical Derivation of Applied Relationships

The complexity of finishing the part T_K during the grinding process in one, two, three, or more operations can be calculated according to the generally accepted formula [14]:

$$T_K = T_M + T_B + T_{eq} + T_{nn} + \frac{T_{nc}}{n_{pp}} \tag{1}$$

where T_M is machine part processing time, min; T_B is auxiliary time, min; T_{eq} is time for organizational and technical maintenance of the workplace, min; T_{nn} is time for rest and natural needs, min; T_{nc} is preparatory and final time, min; n_{pp} is the number of parts in the package. Values T_B, T_{eq}, T_{nn}, and T_{nc} can be found according to the normative data, and the machine grinding time of the part can be determined by the equation:

$$T_M = f/Q_\Omega \tag{2}$$

where f is the surface area to be processed, in mm^2; Q_Ω is the grinding performance by processing area, mm^2 $\cdot$min^{-1}.

Grinding performance with fine-grained wheels on the area processed per unit of time Q_Ω, can be expressed in terms of the grinding performance with regard to the volume of metal removed per unit of time, Q_M, by the formula:

$$Q_\Omega = Q_M/\Omega \tag{3}$$

where Ω is the radial allowance for grinding, in mm.

During grinding, the metal is removed from the workpiece by single abrasive grains, so the grinding performance, Q_M, depends on the number of slices and the volume of metal removed during a single slice, which are determined by the shape and size of the slice. The most typical cuts in finishing grinding processes are segmented cuts [14].

Figure 1 shows a diagram of a single mark applied to a polished surface, where S_1 and S_2 are areas of bulk formed on the surface of the material due to plastic deformation by its abrasive grain, in μm^2. At the accepted coordinates, the elementary volume of one-quarter of the segmented mark is determined by the equation:

$$dV_p' = \phi(y,z)dydz \tag{4}$$

where $\phi(y,z)$ is the equation of the surface limiting the volume of the mark.

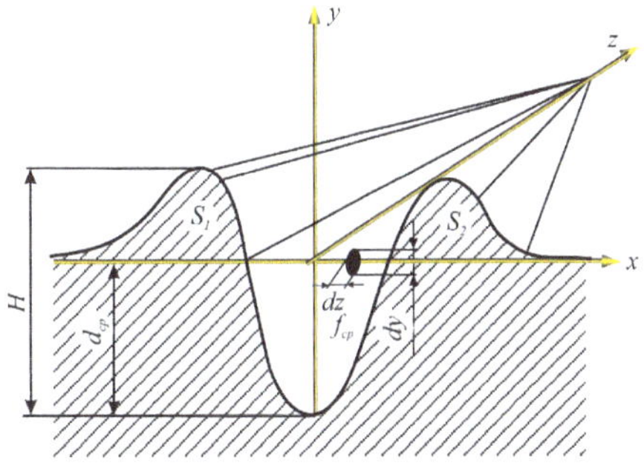

Figure 1. Diagram of a single mark applied by a diamond cone to the polished surface of a steel sample.

Integrating Equation (4) and multiplying the result by four, we obtain the full volume of the segmented mark, dV_p:

$$dV_p = \int_0^y \int_0^z \phi(y,z)dydz \tag{5}$$

The number of single slices per minute N can be determined by the equation:

$$N = \Psi \cdot B \cdot V_K \cdot n_g \tag{6}$$

where n_g is the number of grains located on a single site in 1 mm^2 of the wheel surface, Ψ is the scale factor (during processing with fine-grained wheels, $\Psi = 6 \times 10^4$), B is wheel height, and V_K is the circumferential speed of the wheel.

It is shown in [15–17] that the abrasive grains on the surface of the wheel are located at different levels, and therefore are embedded in the processed metal at different depths. On a single site perpendicular to the radius of the wheel, the distribution of vertices can be expressed by the dependence:

$$dn_g = f(y)dy \tag{7}$$

where dn_g is the number of grains enclosed between the surface of the wheel in depth y and $y + dy$ in the direction of the radius; and $f(y)$ is the function of the distribution of the vertices of abrasive grains in depth.

The volume of metal removed per minute by grains enclosed in a layer dy, taking into account Equations (5)–(7), is determined by:

$$dQ_M = 2.4 \cdot 10^5 \cdot B \cdot V_K \left[f(y) \int_0^y \int_0^z \phi(y,z)dydz \right] dy \tag{8}$$

Integrating and substituting the value (8) into Equation (3) we obtain:

$$Q_{\Omega} = \frac{2.4 \cdot 10^5 \cdot B \cdot V_K \left[f(y) \int_0^y \int_0^z \phi(y,z) dy dz \right] dy}{\Omega} \tag{9}$$

To use Equation (9), it is necessary to know the function of the distribution of abrasive grains in depth and the equation of the surface that limits the volume of the mark.

According to the data of [1], in the first approximation, it is quite acceptable to consider a segmented structure as two equal pyramids with a common base. In this case, the volume of a single mark can be determined by the equation:

$$V_p = \frac{2 \cdot 10^{-6}}{3} \cdot f_{cp} \cdot L \tag{10}$$

where f_{cp} is the median cross-sectional area of the mark, in μm^2; L is half of the contact area with a segmented cut, equal to the contact arc with a comma cut, in mm.

According to Maslov [14], the length of the contact area, L, during circular external grinding can be determined by the equation:

$$L = 3.16 \cdot 10^{-2} \cdot \left(1 \pm \frac{V_W}{60 \cdot V_K} \right) \cdot \sqrt{\frac{D \cdot d \cdot a_{cp}}{D + d}} \tag{11}$$

where D is the diameter of the grinding wheel, in mm; a_{cp} is the average depth of embedding of the abrasive grain into the processed metal, in mm; V_W is the circumferential speed of the workpiece; and $m \cdot s^{-1}$; d is the diameter of the workpiece, in mm.

The middle area of the mark, neglecting the elastic deformations of the metal, can be determined by the geometry of the vertex of the abrasive grain [1].

For the case when the thickness of the slice lies within $a_{cp} \backslash \rho \leq 0.29$ the median cross-sectional area can be calculated using an approximate dependence:

$$f_{cp} = 1.8 \cdot a_{cp} \cdot \sqrt{\rho \cdot a_{cp}} \tag{12}$$

The dependence (12) is obtained by processing mathematical tables of the segment area at different values of the arrow height. The radius of an abrasive grain, ρ, depends on the grain size and grain material [12].

In the works of Bogomolov [1], Vakser [18], and Novoselov [12], it is shown that the volume of marks formed is always greater than the volume of metal actually removed from the part. Part of the metal is displaced from the volume of the mark, forming bulk along its edges. The ratio of the volume of metal removed, V_M, during the formation of marks and the volume of the mark, V_P, can be considered as a first approximation by the ratio of bulk areas to the area of the mark in the middle section (Figure 1):

$$\frac{V_M}{V_p} = 1 - K_b = 1 - \frac{S_1 + S_2}{f_{cp}} \tag{13}$$

where K_b—extrusion ratio.

Substituting the value V_P into Equation (13) from Equation (10), we determine the volume of metal removed by a single abrasive grain:

$$V_M = 2.1 \cdot 10^{-8} \cdot (1 - K_b) \cdot f_{cp} \cdot \left(1 \pm \frac{V_W}{60 \cdot V_K} \right) \cdot \sqrt{\frac{D \cdot d \cdot a_{cp}}{D + d}} \tag{14}$$

As a first approximation, it can be assumed that, when grinding, all the abrasive grains involved in the cutting process are located on the surface of the wheel at the same level. In this case, the number of sections of the processed metal with abrasive grains is determined by the equation:

$$N = \frac{6 \cdot 10^4 \cdot B \cdot V_K}{l_\varphi^2} \tag{15}$$

where l_φ^2 is the distance between the abrasive grains located on the surface of the wheel and involved in the cutting process, in mm.

According to the results of the work [12]:

$$l_\varphi \approx 3.5 \cdot 10^{-3} \cdot l_0 \tag{16}$$

where l_0 is the diameter of the average and most likely grain size, in µm.

The volume of metal removed per minute N by abrasive grains will be determined by:

$$Q_\Omega = \frac{103 \cdot (1 - K_b) \cdot f_{cp} \cdot B \cdot V_K \cdot \left(1 \pm \frac{V_W}{60 \cdot V_K}\right) \cdot \sqrt{\frac{D \cdot d \cdot a_{cp}}{D + d}}}{l_0^2} \tag{17}$$

In [8], it is shown that the surface roughness during grinding is determined by a set of single drawings. The correspondence of the surface roughness to the depth of individual drawings allows one to determine the average depth of penetration of single abrasive grains into the processed metal by roughness. The data in Figure 1 show that the surface roughness formed during cutting-scratching, $Ra = H$, is greater than the depth of embedding by the amount of bulk at the edges of the mark.

$$a_{cp} = K_R \cdot Ra_1 \tag{18}$$

where $K_R = a_{cp}\ / H$ is the coefficient of excess surface roughness.

Substituting the value of grinding performance by the volume of metal, Q_M, from Equation (17) into Equation (3) and taking into account the possibility of increasing productivity in the initial period of time by changing the mode by the coefficient K_Q, we obtain:

$$Q_\Omega = \frac{103 \cdot K_Q \cdot (1 - K_b) \cdot f_{cp} \cdot B \cdot V_K \cdot \left(1 \pm \frac{V_W}{60 \cdot V_K}\right) \cdot \sqrt{\frac{K_R \cdot Rz_1 \cdot D \cdot d}{D + d}}}{l_0^2 \cdot \Omega} \tag{19}$$

Formula (19) takes into account the impact of the size of the abrasive grain on the grinding performance, the grinding mode, the grinding allowance, the size of the workpiece, and the wheel. The formula can be divided into two functional parts:

$$Q_\Omega = G \cdot S \tag{20}$$

where

$$G = \frac{103 \cdot K_Q \cdot (1 - K_b) \cdot f_{cp} \cdot V_K \cdot \left(1 \pm \frac{V_W}{60 \cdot V_K}\right) \cdot \sqrt{K_R \cdot Rz_1}}{l_0^2 \cdot \Omega} \tag{21}$$

does not depend on the size of the part and the wheel;

$$S = B \cdot D_e \tag{22}$$

where $D_e = \sqrt{(D \cdot d)/(D + d)}$—equivalent diameter, a parameter that takes into account the dimensions of the workpiece and the wheel.

The functional part of the formula (19), which does not depend on the size of the workpiece and the size of the wheel, can be taken as an indicator of the performance of finishing grinding processes.

To analyze the influence of grinding conditions on the grinding performance indicator, it is necessary to experimentally determine the coefficients included in Equation (21), K_b and K_R.

3. Experimental Determination of Coefficients K_b and K_R

To experimentally determine the values of the coefficients K_b and K_R, a series of experiments on cutting and scratching polished steel samples L3 (USA standard) HRC = 55. . .60, with a diamond pyramid with an angle of 136° at the top, a radius of 60 microns at the top, and a series of experiments on cutting and scratching samples with grains of an abrasive bar of green silicon carbide GC 150 N7 V220 with a height of 10 mm and an angle of 45° at the top, was carried out.

To conduct these experiments, the diamond pyramid (1) was fixed in a metal disk (2) mounted on the faceplate of a circular grinding machine (3), Figure 2. To obtain drawings with different depths when cutting-scratching with a diamond pyramid, the axis of the part (4) was deflected by turning the table from its line of movement by 1–2 degrees. Cutting-scratching with a diamond pyramid was carried out at four speeds: 6, 10, 20 and 38 $m \cdot s^{-1}$. Cutting-scratching with abrasive bar grains was carried out at a bar speed of 35 $m \cdot s^{-1}$, a part speed of 40 $m \cdot min^{-1}$, $S = 1.0$ $m \cdot min^{-1}$, and a cross feed $t = 0.05$ $mm \cdot min^{-1}$. A 3% solution of Na_2CO_3 was used as a cutting fluid.

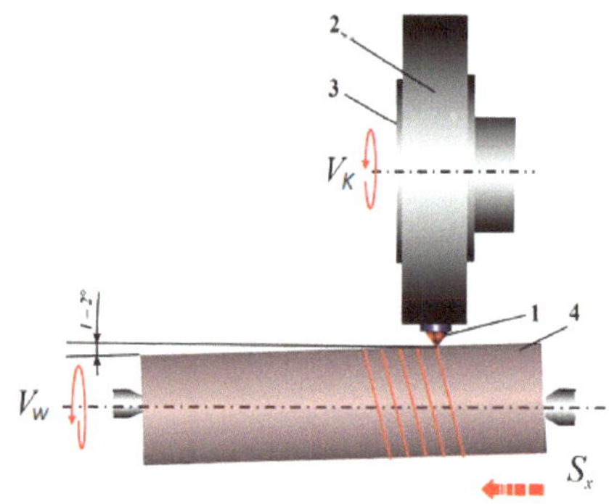

Figure 2. Scheme of cutting-scratching of a part with a diamond cone and an abrasive bar on a circular grinding machine.

The median profile of a single mark was recorded on a profiler model 33319-13. The measurement of the areas $S1$, $S2$, and fcp was carried out with a planimeter of the Planix 5 model.

The results of processing profilograms were calculated using the least squares method.

Figure 3 shows an example of the dependence of the change in the extrusion coefficient on the speed and depth of cutting-scratching. As the figure data show, the extrusion coefficient decreases with the increasing depth of cutting-scratching and the speed of the wheel, while the influence of the speed of the wheel on the extrusion coefficient is significantly less than the influence of the depth of cutting-scratching.

Mathematical processing of experimental data from 120 experiments established the following dependence of the extrusion coefficient on the depth and speed of cutting-scratching:

$$K_b = 1 - 0.34 \cdot a_{cp}^{0.56} \cdot V^{0.06}, r_{ka} = 0.58, r_{kV} = 0.22, \tag{23}$$

where r_{ka} and r_{kV} are correlation coefficients of the coefficient K_b on the depth and speed of cutting-scratching.

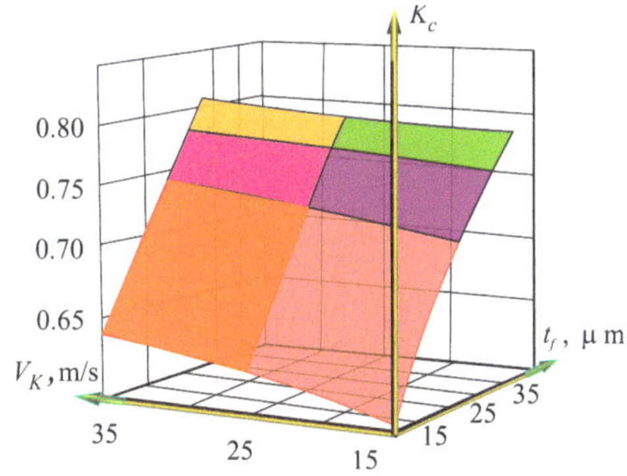

Figure 3. The effect of the speed and depth of cutting-scratching with a diamond cone on the extrusion coefficient.

Analysis of Equation (23) shows that the exponent at the cutting-scratching speed is seven times less than the exponent at the cutting-scratching depth, Figure 3. Figure 4 shows the dependence of the coefficients K_b and K_R on the depth of cutting-scratching by the grains of the abrasive bar. The coefficient of extrusion, as well as cutting-scratching with a diamond pyramid, decreases with increasing depth of cutting-scratching. So, by increasing a_{cp} from 0.2 to 1 μm, coefficient K_b decreases by 1.85 times. The coefficient of excess surface roughness increases with increasing cutting-scratching depth. So, by increasing a_{cp} from 0.2 to 1 μm, coefficient K_R increases from 0.42 to 0.8.

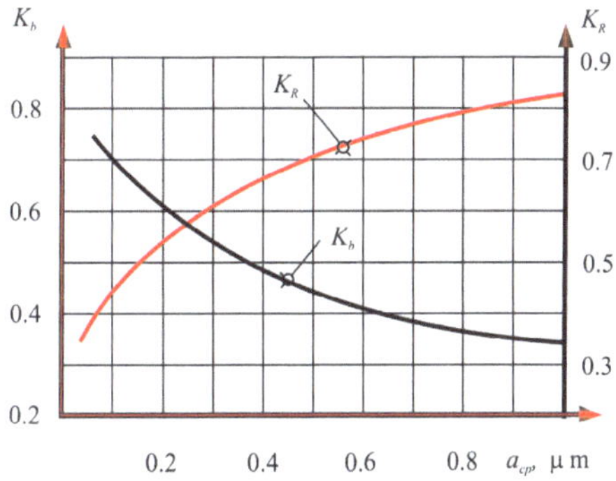

Figure 4. The effect of cutting depth: scratching by abrasive bar grains by 36 CM BBΠ on the extrusion coefficient and the coefficient of excess surface roughness: $V_K = 35$ m·min^{-1}; $V_w = 20$ m·min^{-1}; $S = 1.0$ m·min^{-1}; $t_x = 0.05$ mm·unit^{-1}.

By processing the experimental data from 80 experiments, the following dependences of the extrusion coefficient and the coefficient of excess surface roughness on the depth of cutting-scratching were established:

$$K_b = 1 - 0.66 \cdot a_{cp}^{0.38},\ r_{kb \cdot a} = 0.44 \quad (24)$$

$$K_R = 0.8 \cdot a_{cp}^{0.27},\ r_{kR \cdot a} = 0.57 \quad (25)$$

where $r_{kb \cdot a}$ and $r_{kR \cdot a}$ are correlation coefficients.

During grinding in wheels GC 150 N7 V220, the surface roughness, depending on the mode, is within $Ra_0 = 0.32$–0.08 μm, with an average value: $Ra = 0.16$ μm. For the value of the depth of cutting-scratching, $a_{cp} = 1.0$ μm. The coefficients can be assumed to be equal: $K_b = 35$; $K_R = 0.8$. The considered dependences of the coefficients K_b and K_R on the cutting-scratching depth indicate an improvement in chip formation conditions with an increase in the depth of penetration of single grains and metal. With a decrease in the depth of cutting-scratching, the proportion of metal removed during the formation of a single mark decreases. This phenomenon is explained by Maslov [14] and Bogomolov [1] by the fact that the vertices of abrasive grains have radii of rounding. When the cutting-scratching depth decreases, the front cutting angle increases, and the folding resistance of the deformed layer increases. With a decrease in the depth of cutting-scratching, the volume of metal removed by a single grain becomes so insignificant that the metal of the treated surface will practically not be sanded, but instead deformed.

Thus, when analyzing the shaping process of a polished surface, the plastic deformation of the metal by single abrasive grains cannot be neglected. The extrusion coefficient varies depending on the depth of cutting-scratching in the range from 0.2 to 0.7, and the coefficient of excess surface roughness in the range from 0.4 to 0.8.

4. Performance Indicator Analysis

To analyze the impact of the number of finishing operations on the performance indicator G, it is necessary to express the components in the Equation (21).

To ensure that there are no traces of previous processing on the grinding surface, it is necessary to remove a layer of metal from the part Ω (Figure 5):

$$\Omega = 10^{-3} \cdot (C_0 \cdot Ra_0 - C_1 \cdot Ra_1) \tag{26}$$

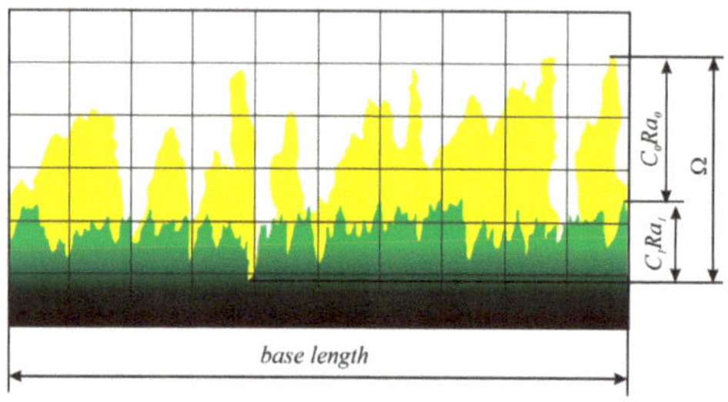

Figure 5. Scheme of surface roughness before and after grinding with a fine-grained wheel.

Abrasive grain size l_0 is not the only factor determining the roughness of the surface of the part after grinding, so the relationship between them cannot be presented in an unambiguous form.

Reducing the roughness of the surface of the part is possible:

(a) Through reduction of the depth of penetration of single grains into the metal at the same granularity as that of the wheel, $l_0 = const$;
(b) By changing the grain size of the wheel, the bundle, and/or the grinding mode, and changing the composition of the coolant. At the same time, the dependence l_0 from Ra can be expressed by the equation:

$$l_0 = a_0 \cdot R_a^{a_1} \tag{27}$$

where a_0 and a_1 are regression coefficients.

Consider the effect of surface roughness after grinding on the performance indicator when $l_0 = const$ and $l_0 = 25.5 \cdot R_a^{0.72}$ (Figure 6). In the calculations in accordance with the experimental data, the values of the coefficients were assumed to be equal: $K_Q = 2$; $K_b = 0.65$; $K_R = 0.8$; and $C_0 = C_1 = 2$. The radius of rounding of abrasive grains was assumed to be equal to: $\rho = 0.077 \cdot l_0$ µm [12].

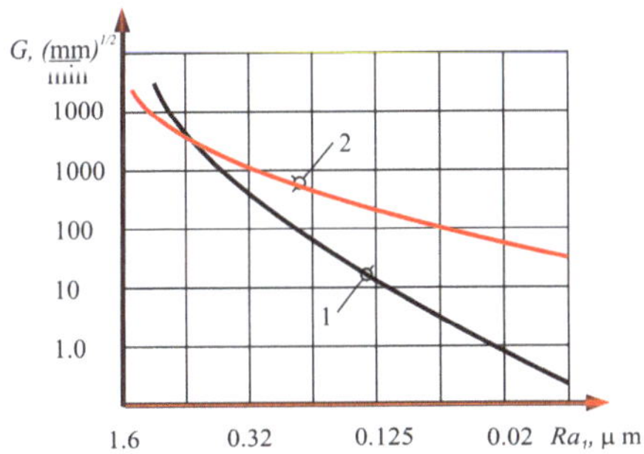

Figure 6. The dependence of the performance indicator on surface roughness after grinding: $V_K = 35\ \text{m}\cdot\text{s}^{-1}$; $1:\ l_0 = const$; $2:\ l_0 = 25.5\cdot R_a^{0.72}$.

The analysis of Figure 6 shows that the productivity indicator decreases with a decrease in surface roughness after grinding. Higher values of the performance indicator correspond to the case when the reduction of roughness is achieved by reducing the grain size of the wheel, the grinding mode, and the composition of the coolant. Therefore, the use of fine-grained wheels is a more effective method of reducing roughness than nursing. Considered dependencies Ω and l_0 on the preliminary and required roughness make it possible to calculate the performance indicator during grinding of parts, not only with one finishing operation, but also with two, three, or more operations. When grinding in several operations, the performance indicator can be determined by the equation:

$$G = \frac{1}{\frac{1}{G_1} + \frac{1}{G_2} + \ldots + \frac{1}{G_n}} \tag{28}$$

where n is number of finishing grinding operations; $G_1 \ldots G_n$ is the performance indicator of each individual operation.

Expressing the performance indicator of individual operations by the Formula (21) we obtain:

$$G = \frac{103}{\frac{\Omega_{01}\cdot l_{01}^2}{K_Q\cdot(1-K_b)\cdot f_{cp}\cdot V_K\cdot\left(1\pm\frac{V_W}{60\cdot V_K}\right)\cdot\sqrt{K_R\cdot Rz_1}} + \ldots + \frac{\Omega_{0n}\cdot l_{0n}^2}{K_Q\cdot(1-K_b)\cdot f_{cp}\cdot V_K\cdot\left(1\pm\frac{V_W}{60\cdot V_K}\right)\cdot\sqrt{K_R\cdot Rz_1}}} \tag{29}$$

When substituting the values into Equation (29)—f_{cp} from Equation (12), Ω from Equation (26), and l_0 from Equation (27)—we obtain:

$$G = \frac{187.5}{\frac{Ra_0 - Ra_{01}}{Ra_{01}^{0.87}} + \frac{Ra_{01} - Ra_{02}}{Ra_{02}^{0.87}} + \ldots + \frac{Ra_{0n-1} - Ra_1}{Ra_1^{0.87}}} \tag{30}$$

where $Ra_{01}, Ra_{02}, \ldots, Ra_{0n}$—surface roughness after the first, second, etc. operations.

Consider the example of grinding a part to obtain surface roughness ($Ra_1 = 0.025$ μm; $Ra_0 = 0.50$ μm), the effect on the performance indicator of the number of operations and surface roughness after each operation.

During the grinding of the part in three operations, the dependence is obtained:

$$G = \frac{187.5}{\frac{0.50 - Ra_{01}}{Ra_{01}^{0.87}} + \frac{Ra_{01} - Ra_{02}}{Ra_{02}^{0.87}} + \ldots + \frac{Ra_{0n-1} - Ra_1}{0.04}} \tag{31}$$

Figure 7 shows a diagram of the levels of values of the productivity indicator from the surface roughness after the first and second finishing operations, constructed according to Equation (31).

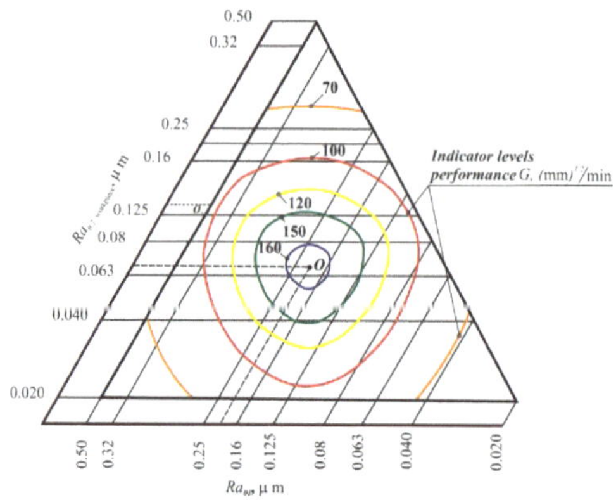

Figure 7. The dependence of the productivity indicator on the surface roughness after the first and second finishing operations when grinding parts in three operations to obtain: $Ra_0 = 0.50$ µm; $Ra_1 = 0.025$ µm; $V_K = 35$ m·s^{-1}.

The surface roughness after the first finishing operation is postponed along the axis X; the surface roughness after the second finishing operation is postponed along the axis Y. The vertices of the triangle ABC correspond to the efforts:

$$\begin{aligned} A - Ra_{01} &= Ra_0 \text{ and } Ra_{02} = Ra_1 \\ B - Rz_{01} &= Rz_{02} = Rz_0 \\ C - Ra_{01} &= Ra_{02} = Ra_1 \end{aligned} \quad (32)$$

The sides of the triangle ABC correspond to the efforts:

$$\begin{aligned} AB - Ra_{01} &= Ra_0 \\ BC - Ra_{01} &= Ra_{02} \\ AC - Ra_{02} &= Ra_1 \end{aligned} \quad (33)$$

For all sides, one of the finishing operations falls out, and grinding is carried out in two operations.

The analysis of Figure 7 shows that the performance indicator significantly depends on the surface roughness of each intermediate operation. The highest value of the performance indicator for grinding in three operations corresponds to the roughness after the first finishing operation. Any other combination of roughness values after the first and second finishing operations provides a lower value of the performance indicator. The roughness values determined by the point O, under the accepted conditions, should be considered optimal.

During grinding of a part in two operations, the maximum value of the performance indicator is determined by the point O_1, which corresponds to the roughness after the first finishing operation. Analytically optimal values of surface roughness after intermediate operations can be determined by studying the maximum of Equation (30).

Studies on the maximum of Equation (30) and the calculation of the process performance indicator were carried out by the approximation method. The dependence of the performance indicator on the number of grinding operations, and on the specified and preliminary surface roughness, are shown in Figure 8. The performance indicator increases with the introduction of additional grinding operations, and with a decrease in the preliminary surface roughness. With a decrease in surface roughness after grinding, the performance indicator decreases.

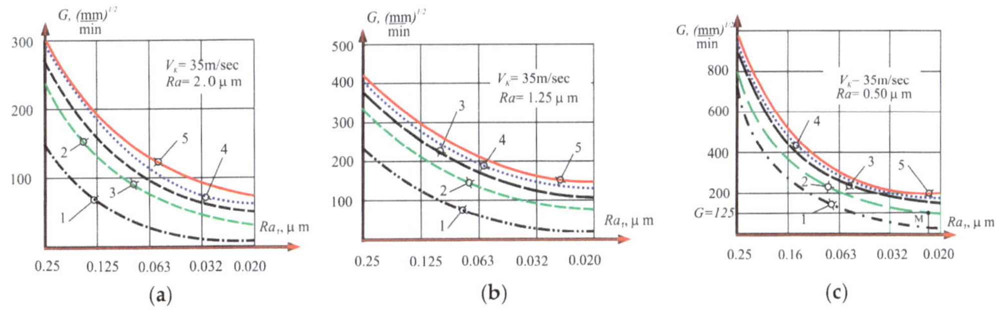

Figure 8. The dependence of the performance indicator on the surface roughness after grinding, and the number of finishing operations with preliminary roughness: (**a**) $Ra = 0.63$ μm, (**b**) $Ra = 0.32$ μm, (**c**) $Ra = 0.16$ μm: $1 : n = 1$; 2: $n = 2$; $3 : n = 3$; $4 : n = 5$; 5: $n = 10$; $V_K = 35$ m/min).

The analysis of the influence of grinding conditions on the performance indicator shows that, to determine it when grinding a part in several operations, it is enough to know the values of the preliminary and required surface roughness.

At a certain indicator G, the productivity of the grinding process is determined by Equation (20), and the complexity of processing the part is determined according to Equation (1). Analysis of the complexity of grinding parts in one, two, or more operations, allows one to establish the optimal variant of the technological process.

Thus, the optimal number of finishing grinding operations, with the requirement of obtaining a surface roughness of $Ra = 0.02 - 0.04$, is determined in the following sequence:

(1) according to Equation (30), the productivity indicator is determined for different variants of the technological process (with one, two, or more finishing operations);
(2) according to Equation (20), the productivity of the grinding process is determined for each variant;
(3) according to Equation (1), the complexity of processing the part is calculated for all options and the option that provides the least complexity is selected.

As an example, let's use Figures 8–10. Using the dependencies shown in Figure 8, the value of performance indicator G is calculated. According to the graphs shown in Figure 9, the value of the equivalent diameter is determined $\sqrt{D{\cdot}d/D+d}$; according to the nomogram, Figure 10, the productivity of the grinding process is determined.

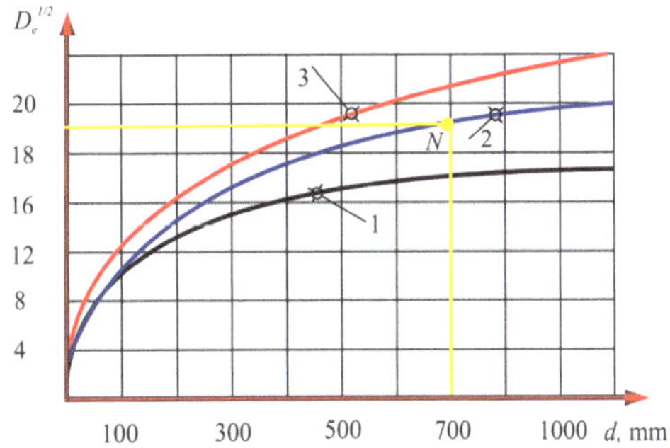

Figure 9. The effect of the diameter of the part and the diameter of the wheel on the value $D_e{}^{1/2} = \sqrt{D{\cdot}d/D+d}$; $1 : D = 300$ mm; $2 : D = 600$ mm; $3 : n = 1000$ mm.

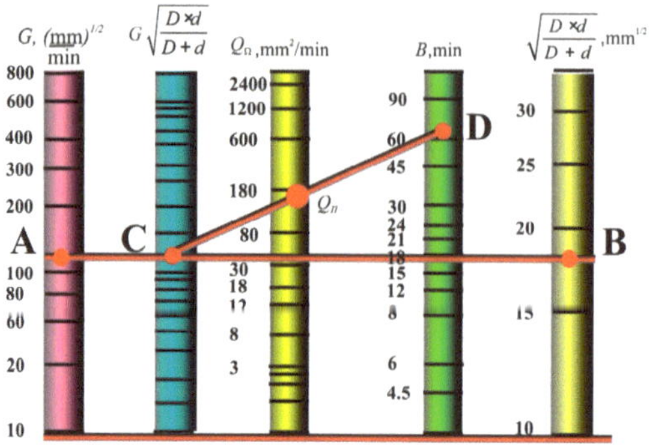

Figure 10. Nomogram for determining the performance of the grinding process with fine-grained wheels.

5. Experimental Verification of the Methodology

Consider the calculation of the optimal number of finishing operations of grinding cold rolling rolls to obtain roughness $Ra_0 = 0.50$ µm; $Ra_1 = 0.025$ µm. The dimensions of the grinding wheel are 1600 × 75 × 305 mm, and the roll dimensions are $d \times l = 690 \times 1680$ mm. During grinding in two operations for the example in question, $G = 125\ \text{mm}^{1/2}{\cdot}\text{min}^{-1}$ (Figure 9). According to Figure 10, we determine the values of the square root with a wheel diameter of 600 mm and a roll diameter of 690 mm, $\sqrt{D{\cdot}d/D+d} = 18$ (point N). On the nomogram (Figure 10), we draw from point A the scale of the performance indicator corresponding to the value $G = 125$, through the point B of the scale of the square root line to the intersection with the auxiliary line $G\sqrt{D{\cdot}d/D+d}$ in the point C. From point C of the auxiliary line through the point of the wheel height scale corresponding to the value $B = 75$ mm, draw the line to the intersection with the performance scale. The resulting point will determine the estimated productivity of the roll finishing process $Q_\Omega = 1.6 \times 10^5\ \text{mm}^2{\cdot}\text{min}^{-1}$.

According to Equation (1), we calculate the complexity of finishing the roll.

According to the timekeeping conducted at the Kamensk-Ural Metallurgical Plant, the sum of the auxiliary time and the time for organized maintenance of the working time for finishing the roll is: with one finishing operation, 48 min; with two finishing operations, 72 min; and with three finishing operations, 94 min. The time for rest and natural needs according to general technical standards with the accuracy of processing according to the fifth quality is 6% of the operational time, and the preparatory and final time for processing the roll is 8 min. With two finishing operations, the complexity of finishing the roll will be determined:

$$T_K = \frac{3.14{\cdot}690{\cdot}1680}{1.6{\cdot}10^5} + 72 + (22+72){\cdot}\frac{6}{100} + 8 = 108.3\ \text{min} \quad (34)$$

During the grinding of the roll in one operation, the estimated processing time will be 113 min; when grinding in three operations, it will be 124 min. The estimated processing time of the roll in two operations will be 103 min. Since 103 min < 113 min and 103 < 124 min, the optimal variant of the technological process will be the variant with two finishing operations.

Table 1 shows the calculated data on the machine time and labor intensity of finishing grinding of rolls of five standard sizes to obtain surface a roughness of $Ra = 0.04 - 0.02$ in one, two, three, and four operations. The preliminary roughness of the surface is assumed to be equal to $Ra = 0.63$ microns, the dimensions of the wheel are $B = 600$ mm and $B = 75$ mm. Calculations are given under the condition of grinding the roll on the same machine with the rearrangement of grinding wheels. The method of grinding cold rolling rolls with the rearrangement of grinding wheels, compared with the method of grinding with the rearrangement of rolls from machine to machine, provides greater roll accuracy, since the installation bases do not change during the roll processing period.

Table 1. Calculated data on the machine time and labor intensity of finishing grinding of rolls.

Rolls Size $d \times l$	Machine Grinding Time of the Roll T_M, Min during n Finishing Operations				Machine Grinding Time of the Roll T_K, Min during n Finishing Operations			
	$n = 1$	$n = 2$	$n = 3$	$n = 4$	$n = 1$	$n = 2$	$n = 3$	$n = 4$
500 × 740	18.6	7.6	5.8	5.1	73.2	87.5	111.9	137.6
500 × 1200	30.2	12.3	9.4	8.2	86.0	97.7	115.9	141
480 × 1680	40.2	16.4	12.5	11.0	97.0	97.2	119.3	144
690 × 1680	55	22.1	16.8	14.8	113	108.3	124.0	148.2
900 × 2800	113	45	34.5	30.5	177	129	143	165

For all the considered roll sizes (Table 1) with an increase in the number of finishing operations, the machine processing time of the roll decreases. The labor intensity of the roll processing is the lowest for rolls with dimensions $d \times l = 500 \times 740$ mm, $d \times l = 500 \times 1200$ mm and $d \times l = 480 \times 1680$ mm with one finishing operation; it is lowest for rolls with dimensions $d \times l = 690 \times 1680$ mm and $d \times l = 900 \times 2800$ mm with two finishing operations.

The method of grinding cold rolling rolls during the rotation of the grinding wheels, compared with the grinding method with the rearrangement of the rolls from machine to machine, provides greater accuracy of the roll, since the installation bases do not change during the processing of the roll.

The conducted research on the grinding of rolls has shown that, with an increase in the number of finishing operations, the machine processing time of the roll decreases. The labor intensity of the roll processing is the lowest for rolls of size $d \times l = 500 \times 740$ mm, $d \times l = 500 \times 1200$ mm, and $d \times l = 480 \times 1680$ mm with one finishing operation, and for rolls of sizes $d \times l = 690 \times 1680$ mm and $d \times l = 900 \times 2800$ mm, with two finishing operations.

Based on the above, it can be concluded that the complexity of processing rolls and the optimal number of finishing grinding operations largely depends on the size of the roll.

Figure 11 shows the data for calculating the optimal number of operations when grinding rolls to obtain a surface roughness of $Ra_0 = 0.50$ μm; $Ra_1 = 0.025$ μm, depending on the diameter and length of the roll barrel.

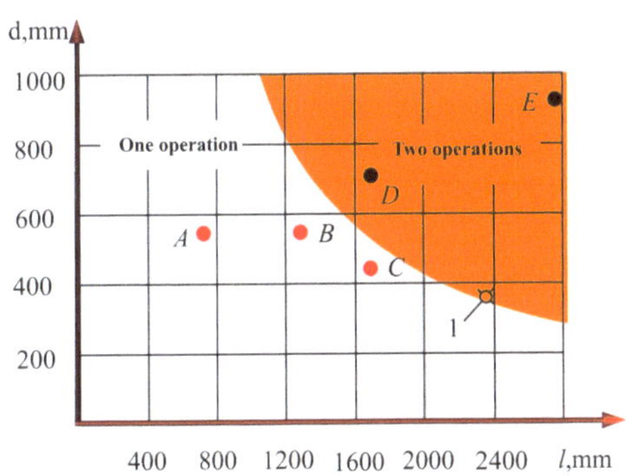

Figure 11. Dependence of the optimal number of finishing operations of grinding rolls to obtain a surface roughness of $Ra = 0.04$–0.02 μm.

With an increase in the allowance for processing, the surface roughness after pre-grinding increases for all cases.

It is advisable to process all rolls whose dimensions correspond to points lying below line 1 in one operation, and rolls whose dimensions correspond to points lying above line 1 in two operations. The points *A*, *B*, *C*, *D*, and *E* are marked by the standard sizes of the rolls, according to which the calculations were summarized. Most of the rolls obtaining a surface roughness of $Ra_0 = 0.04–0.02$ μm can be processed with a single finishing operation. However, rolls of sizes $d \times l = 690 \times 1680$ mm and $d \times l = 90 \times 2800$ mm should be processed with two finishing operations.

Figure 12 shows the dependence of the optimal values of the surface roughness after pre-grinding on the processing allowance when grinding cold rolling rolls $d \times l = 500 \times 1200$ mm to obtain a surface roughness of $Ra = 0.02 - 0.01$ μm when conducting the process in one, two, and three operations.

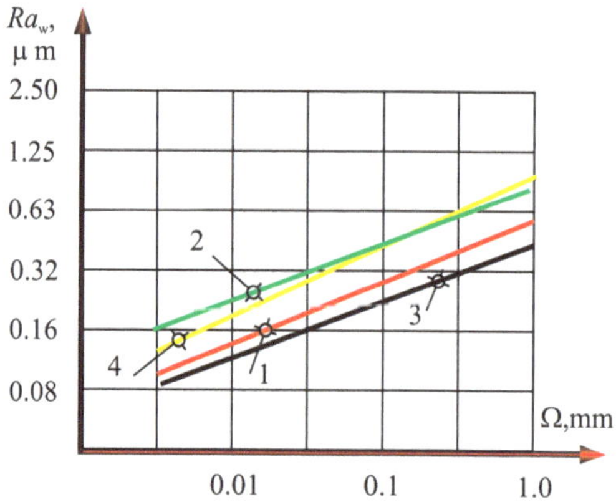

Figure 12. The dependence of the optimal values of the preliminary roughness of the surface on the allowance for processing when grinding rolls with dimensions $d \times l = 500 \times 1200$ mm to obtain surface roughness: $1 : Ra = 0.04–0.02$ μm (in a single operation); $2 : Ra = 0.04–0.02$ μm (in two operations); $3 : Ra = 0.02–0.01$ μm (in two operations); $4 : Ra = 0.02–0.01$ μm (in three operations).

6. Conclusions

The proposed method makes it possible to calculate not only the optimal number of finishing grinding operations, but also the optimal value of the preliminary surface roughness.

Provided that the characteristics of the abrasive tool for the preliminary grinding operation are not set, and the surface roughness after it is not predetermined, the optimal value of the latter can also be calculated from the condition of the lowest labor intensity of the technological process. To determine the optimal surface roughness after preliminary grinding, it is sufficient to supplement the denominator of Equation (28) with the summand $1/G_0$, where G_0 is the performance indicator of the preliminary operation as determined by Equation (21).

According to the results of production tests, technological processes for grinding rolls with the number of finishing operations corresponding to the calculated ones are proposed (Table 1). Grinding of cold rolling rolls with dimensions $d \times l = 500 \times 740$ mm, $d \times l = 500 \times 1200$ mm, and $d \times l = 480 \times 1680$ mm, obtaining a surface roughness of $Ra = 0.02 - 0.01$ μm is recommended conduct with one finishing operation, while grinding of rolls with dimensions $d \times l = 690 \times 1680$ mm should be processed with two finishing operations.

In the future, it is possible to consider automating the whole process of decision-making about and optimization of the number of operations (one, two, three, etc.) via CA systems. The production process can also be controlled and diagnosed in an automated or automatic mode [19]. It is possible to use CNC (Computer Numerical Control) systems, artificial intelligence, fuzzy logic, mechatronic nodes, or neural networks. In the future, the authors plan to support the methodology presented in this paper with automated computer modeling and simulation systems.

Author Contributions: Conceptualization, S.B. and B.Y.; methodology, S.B. and B.Y.; formal analysis, S.B. and B.Y.; writing—original draft preparation, S.B. and B.Y.; investigation, S.B. and B.Y.; writing—review and editing, Z.S., M.S. and I.K.; supervision, M.S.; project administration, I.K.; funding acquisition, Z.S. and M.S. All authors have read and agreed to the published version of the manuscript.

Funding: This article was funded by the University of Žilina, project no. 313011ASY4: "Strategic implementation of additive technologies to strengthen the intervention capacities of emergencies caused by the COVID-19 pandemic".

Institutional Review Board Statement: Not applicable.

Informed Consent Statement: Not applicable.

Data Availability Statement: The data presented in this paper are available upon request from the corresponding author.

Conflicts of Interest: The authors declare no conflict of interest.

References

1. Bogomolov, N.I. The Main Processes in the Interaction of an Abrasive with Metal. Ph.D. Thesis, Kyiv Polytechnic Institute, Kyiv, Ukraine, 1977; p. 46. (In Ukrainian)
2. Görög, A. Simulation of Superfinished Surface Formation. *Adv. Sci. Technol. Res. J.* **2021**, *15*, 219–227. [CrossRef]
3. Vopát, T.; Peterka, J.; Šimna, V.; Kuruc, M. The Influence of Different Types of Copy Milling on the Surface Roughness and Tool Life of End Mills. *Procedia Eng.* **2015**, *100*, 868–876. [CrossRef]
4. Jurko, J.; Džupon, M.; Panda, A.; Zajac, J. Study Influence of Plastic Deformation a New Extra Low Carbon Stainless Steels XCr17Ni7MoTiN under the Surface Finish When Drilling. *Adv. Mater. Res.* **2012**, *538–541*, 1312–1315. [CrossRef]
5. Panda, A.; Duplák, J.; Jurko, J.; Behún, M. New Experimental Expression of Durability Dependence for Ceramic Cutting Tool. *Appl. Mech. Mater.* **2013**, *275–277*, 2230–2236. [CrossRef]
6. Glazyrin, V.; Ružbarský, J.; Nikitin, Y.; Božek, P.; De Silva, W.T. Study of Dynamic Processes during the Finishing of Spherical Parts Made of Difficult-to-Machine Materials. *MM Sci. J.* **2022**, *2022*, 5937–5942. [CrossRef]
7. Beňo, M.; Zvončan, M.; Kováč, M.; Peterka, J. Circular Interpolation and Positioning Accuracy Deviation Measurement on Five Axis Machine Tools with Different Structures. *Teh. Vjesn. Tech. Gaz.* **2013**, *20*, 479–484.
8. Polakovič, M.; Buranský, I.; Peterka, J. Simulation Concept for Machined Surface Roughness and Shape Deviations Prediction. In *Proceedings of the Annals of DAAAM for 2008 & Proceedings of the 19th International DAAAM Symposium, Trnava, Slovakia, 22–25 October 2008*; Annals of DAAAM and Proceedings; DAAAM International: Vienna, Austria, 2008; pp. 1089–1090.
9. Lee, C.W. A Control-Oriented Model for the Cylindrical Grinding Process. *Int. J. Adv. Manuf. Technol.* **2009**, *44*, 657–666. [CrossRef]
10. Shen, N.; He, Y.; Li, J.; Fang, M. An Improved Differential Evolution (IDE) Based on Double Populations for Cylindrical Grinding Optimization. In Proceedings of the 2009 International Conference on Measuring Technology and Mechatronics Automation, Hunan, China, 11–12 April 2009; pp. 724–727.
11. Hung, L.X.; Ky, L.H.; Hong, T.T.; Van Cuong, N.; Trung, D.D.; Phan, N.H.; Tung, L.A.; Vu, N.P. Optimization of Manufacturing Time in Internal Grinding. In Proceedings of the International Conference on Engineering Research and Applications, ICERA 2019, Thai Nguyen, Vietnam, 1–2 December 2020; pp. 557–565.
12. Novoselov, Y.K. *Dynamics of Surface Formation during Abrasive Machining*; SevNTU: Sevastopol, Russia, 2012; p. 304. ISBN 978-617-612-051-3. (In Russian)
13. Bratan, S.M.; Roshchupkin, S.I.; Kharchenko, A.O.; Belousov, S.V. Quality Improvement of Manufacturing Rolling Mill Rolls. *CIS Iron Steel Rev.* **2021**, *22*, 26–31. [CrossRef]
14. Maslov, E.N. *Theory of Grinding Materials*; Mashinostroenie: Moscow, Russia, 1974; p. 320. (In Russian)
15. Bratan, S.; Roshchupkin, S.; Chasovitina, A.; Gupta, K. The Effect of the Relative Vibrations of the Abrasive Tool and the Workpiece on the Probability of Material Removing during Finishing Grinding. *Met. Work. Mater. Sci.* **2022**, *24*, 33–47. [CrossRef]
16. Bratan, S.; Roshchupkin, S.; Kharchenko, A.; Chasovitina, A. Simulation of the Stock Removal in the Contact Zone during Internal Grinding of Brittle Non-Metallic Materials. *Met. Work. Mater. Sci.* **2021**, *23*, 31–39. [CrossRef]
17. Kharchenko, A.; Chasovitina, A.; Bratan, S. Modeling of Regularities of Change in Profile Sizes and Wear Areas of Abrasive Wheel Grains during Grinding. *Mater. Today Proc.* **2021**, *38*, 2088–2091. [CrossRef]

18. Vakser, D.B. *Ways of Increasing the Productivity of Abrasive Tools in Grinding*; Mashinostroenie: Moscow-Leningrad, Russia, 1964; p. 123. (In Russian)
19. Peterka, J.; Nikitin, Y.R.; Božek, P. Diagnostics of Automated Technological Devices. *MM Sci. J.* **2020**, *2020*, 4027–4034. [CrossRef]

applied sciences

MDPI

Article

Advancements in Roundness Measurement Parts for Industrial Automation Using Internet of Things Architecture-Based Computer Vision and Image Processing Techniques

Yazid Saif [1], Anika Zafiah M. Rus [1,*], Yusri Yusof [1], Maznah Lliyas Ahmed [2,*], Sami Al-Alimi [1], Djamal Hissein Didane [1], Anbia Adam [3], Yeong Hyeon Gu [4,*], Mohammed A. Al-masni [4] and Hakim Qaid Abdullah Abdulrab [5]

1 Faculty of Mechanical and Manufacturing Engineering, Universiti Tun Hussein Onn Malaysia, Parit Raja 86400, Johor, Malaysia; yazeedalkosa@gmail.com (Y.S.); yusri@uthm.edu.my (Y.Y.); samialalimi@gmail.com (S.A.-A.); djamal@uthm.edu.my (D.H.D.)
2 Politeknik Sultan Azlan Shah, Behrang Stesion, Behrang 35950, Perak, Malaysia
3 Faculty of Mechanical and Manufacturing Engineering Technology, Universiti Teknikal Malaysia Meleka (UTeM), Durian Tunggal 76100, Melaka, Malaysia; anbia@utem.edu.my
4 Department of Artificial Intelligence, College of Software and Convergence Technology, Sejong University, Seoul 05006, Republic of Korea; m.almasani@sejong.ac.kr
5 Department of Electrical and Electronics Engineering, University Teknologi PETRONAS, Seri Iskandar 32610, Perak, Malaysia; hakim_19001004@utp.edu.my
* Correspondence: zafiah@uthm.edu.my (A.Z.M.R.); maznah.iliyas@psas.edu.my (M.L.A.); yhgu@sejong.ac.kr (Y.H.G.)

Abstract: In the era of Industry 4.0, the digital capture of products has become a critical aspect, which prompts the need for reliable inspection methods. In the current technological landscape, the Internet of Things (IoT) holds significant value, especially for industrial devices that require seamless communication with local and cloud computing servers. This research focuses on the advancements made in roundness measurement techniques for industrial automation by leveraging an IoT architecture, computer vision, and image processing. The interconnectedness enables the efficient collection of feedback information, meeting the demands of closed-loop manufacturing. The accuracy and performance of assemblies heavily rely on the roundness of specific workpiece components. In order to address this problem, automated inspection methods are needed. A new method of computer vision for measuring and inspecting roundness is proposed in this paper. This method uses a non-contact method that takes into account all points on the contours of measured objects, making it more accurate and practical than conventional methods. The system developed by AMMC Laboratory captures Delrin work images and analyzes them using a specially designed 3SMVI system based on Open CV with Python script language. The system can measure and inspect several rounded components in the same part, including external frames and internal holes. It is calibrated to accommodate various units of measurement and has been tested using sample holes within the surface feature of the workpiece. According to the results of both techniques, there is a noticeable difference ranging from 2.9 µm to 11.6 µm. However, the accuracy of the measurements can be enhanced by utilizing a high-resolution camera with proper lighting. The results were compared to those obtained using a computer measurement machine (CMM), with a maximum difference of 8.7%.

Keywords: computer vision; image processing; CMM; 3SMVI; inspection; IoT; roundness

Citation: Saif, Y.; Rus, A.Z.M.; Yusof, Y.; Ahmed, M.L.; Al-Alimi, S.; Didane, D.H.; Adam, A.; Gu, Y.H.; Al-masni, M.A.; Abdulrab, H.Q.A. Advancements in Roundness Measurement Parts for Industrial Automation Using Internet of Things Architecture-Based Computer Vision and Image Processing Techniques. *Appl. Sci.* **2023**, *13*, 11419. https://doi.org/10.3390/app132011419

Academic Editors: Tibor Krenicky, Juraj Ruzbarsky and Maros Korenko

Received: 18 July 2023
Revised: 15 October 2023
Accepted: 16 October 2023
Published: 18 October 2023

1. Introduction

In the present fast-changing industrial world driven by globalization, product customization, and automation, the manufacturing industry plays a crucial role. With advanced technologies and techniques that enable rapid and efficient changes in products, processes, and supply chains, the manufacturing industry is at the forefront of Industry 4.0. In the realm of advanced manufacturing systems, manufacturing metrology holds significant importance as it pertains to the measurement and inspection of nearly all machined objects. The advent of Industry 4.0 brings forth novel methodologies for the development of next-generation manufacturing metrology systems that possess characteristics such as intelligence, autonomy, flexibility, interoperability, high productivity, and self-adaptability [1]. Furthermore, a significant shift is generally known as the fourth industrial revolution (IR 4.0) [2]. Emerging technologies such as machine learning, artificial intelligence, big data, 3D printing, and robotics are constantly evolving within the realm of IoT (Internet of Things). The rapid progress of these technologies brings about substantial transformations, as IoT enables devices to perceive and interact with the surrounding environment, imbuing them with a sense of vitality. This industrial decentralization has paved the way for the efficient and remote control and monitoring of industrial operations. Given that IoT represents the future of computing, it requires the synergistic support of pioneering technologies to facilitate its growth and development [3,4]. IoT provides a wide variety of digital and physical resources for I 4.0. Thus, the established network allows for decentralized decision-making and reacting to cyber–physical systems (CPS) in real time [5]. In addition, a computer vision system contains smart Internet of Things (IoT) cameras looking directly at the production line that capture images, which are then algorithmically compared to a predefined image to detect defective objects [6]. The IoT and digital manufacturing industry significantly accelerate product development, producing products with complexity and precision that previously could not be created by cycles and manufacturers [7].

Circular features are among the fundamental geometric elements of mechanical parts. The performance and lifespan of mechanical products depend on how accurately and efficiently circularity errors can be evaluated [8]. To determine the degree of roundness of rounded parts, the entire profile should be measured in accordance with the geometric roundness tolerance standard [9]. The mission is to use machine vision to inspect an automotive camshaft for roundness errors in real time. Further measurement studies are required to improve accuracy and efficiency in roundness error measurement in order to achieve successful roundness evaluation of the part using image processing procedures and mathematical models [10–13]. Coordinate metrology is an essential inspection tool for ensuring quality in machined workpieces. It can analyze the impact of parameters on milled pocket diameter and roundness accuracy using CNC and CMM specimens, address measurement deviations, and estimate uncertainty using statistical repeatability analysis [14]. The measurement of circular sector elements requires appropriate control, which is driven by the increasing demand for complete control of specific geometrical features [15–17]. Conventional instruments have limitations in their sampling points, making it difficult or impossible to meet current roundness measurement standards. Furthermore, using such instruments to measure the entire profile of rounded parts increases the process's complexity and duration. Coordinate measuring machines (CMMs) are becoming more common in automated inspections of manufactured components, both online and offline, to meet geometric roundness tolerance standards. In the measurement of roundness errors, coordinate measuring machines (CMMs) rely on capturing data from various sampling points along the profile of curved parts. However, the use of CMMs for this purpose can be both resource-intensive and time-consuming, particularly when a large number of points need to be measured [18–20].

The availability of inexpensive digital computers has made it possible to incorporate machine vision systems into automated manufacturing systems without a large financial burden. The progress in machine vision, image processing, computational intelligence, and similar areas has significantly improved the potential of visual inspection methods [21–23]. A stereo vision and image analysis system for automating workpiece referencing in three-axis machining centers is presented. The system captures images with two cameras, which are then processed in custom software to return workpiece coordinates to machining workers [24]. A cloud-based system uses stereo vision and image analysis to automate workpiece referencing in machining companies. The system processes images from two cameras in the spindle in custom software and returns the Workpiece Coordinate System (WCS) position to the CNC machine controller. Experiment results validate the application in actual machining processes [25]. In comparison to coordinate measuring machines (CMMs), a machine vision system equipped with a standard 512 × 512 image processor can gather a large number of boundary points in a short period. Furthermore, machine vision can conduct non-contact inspections of products, whereas most instruments used to measure roundness require physical contact. As a consequence, there has been significant interest in developing machine vision systems for inspection in the industrial sector [2,26–31]. A machine vision-based inspection system has been addressed for real-time geometric inspection of A-grade nuts to ensure their quality. The algorithm was optimized for accuracy and usability, resulting in faster and more appropriate nut detection, making it suitable for all testing processes [32]. Machine vision systems provide benefits such as faster measurement times, improved accuracy, and increased flexibility compared to traditional methods. However, since machine vision is a precise and efficient data collector, specialized assessment algorithms are required for different inspection tasks. The effective development of algorithms that are tailored to manufactured parts and error analysis holds significant importance in the context of vision-based inspection. This research project aims to establish a comprehensive collection of computer vision algorithms capable of analyzing images of circular parts to measure and inspect roundness errors and publish the data cloudily through VNC. The proposed methodology integrates an Internet of Things (IoT) architecture and leverages MQTT protocols to enable seamless data communication.

This article is organized as follows: In Section 2, relevant studies on roundness measurement and the MZC algorithm used in determining the shape of a circular form are discussed. Section 3 covers the methodology and the experimental setup and process proposed. Section 4 presents the development of 3SMVI software and the implementation algorithms for image processing. Section 5 outlines the development of a vision system calibration, and Section 6 explains the CMM approach. Section 7 provides a discussion of the results and their interpretation. Finally, in Section 8, the authors' conclusions are summarized.

2. Previous Studies in Auto-Vision of Inspection System

The two types of measurements are diametral and radial roundness [33]. Diametral measurements are made with two-point measuring tools such as calipers or micrometers, whereas radial measurements require precision spindle instruments that are both expensive and time-consuming. These studies introduced a method for evaluating roundness and cylindricity tolerances, addressing difficulties encountered by technicians and engineers, focusing on optimizing manual procedures, and improving skills [34]. Furthermore, roundness profiles with pseudo and actual diameters were evaluated, and relationship models with neural network regression based on coordinate measuring machines were developed [35]. The "roundness tolerance band" is defined by the International Organization for Standardization (ISO) as the area between two concentric circles at the same cross-section where the difference in their radii equals the roundness tolerance value. This value emphasizes the geometric roundness tolerance standard by representing the allowable range of deviation from perfect roundness in a circular object. It specifies the acceptable margin of error for the roundness of a part [13,36,37]. This study addressed four ISO 14405-1:2016 modifiers, providing simple algorithms for evaluating circularity errors and investigating the impact of measurement system strategies on these new specifications [38].

In engineering processes, the surface structure can be defined by three primary features: form errors, roughness, and waviness. Errors encompass three vital criteria essential for the assembly and connection of different parts, namely straightness, flatness, and circularity, commonly referred to as roundness. Out-of-round (OOR) indicates the disparity between the actual and measured radius at a particular point. Roundness measurement is usually based on four globally recognized reference circles discussed in ISO-GPA and ASME Y14.5(1994), such as Maximum Inscription Radial Circle (MIC), Minimum Circumscribing Circle (MCC), Minimum Zone for Circularity (MZC), and least squares circle (LSC), as addressed by [10,39–41]. The comparison of the benchmark algorithms for calculating round errors based on MCC, MIC, and MZC was shown by [42].

Likewise, this research focused on two robust cylinder fitting alignment algorithms that combined PCA, regression, and algebraic methods, as well as a PCD cylinder alignment algorithm that can handle high-percentage deviations and improve shape recognition [43,44]. This study also conducted an iterative procedure for robust circle fitting that employs Taubin's method to obtain the center and radius, as well as geometric distances to the circle, to identify and remove outliers, thereby reducing the corrupted effect on circle parameter estimates [45]. A new roundness measurement algorithm and least squares circle fit arithmetic were employed in a rapid quality detection system for electrochemically fabricated micro-holes. The findings demonstrated reliable results and high precision, meeting the requirements for quality inspection of electrochemical machining [46]. The optimal wire spark erosion machining (WSET) turning condition minimized roundness deviation and maximized material erosion rate. It produced a suitable surface for a variety of applications, with a roundness deviation of 15.51 m and a material erosion rate of 9.35 mm^3/min. These findings will be useful in the advanced fabrication of miniature products from difficult-to-machine materials such as titanium [47]. The hybrid approach evaluates circularity error using the least squares method (LSM) and probabilistic global search Lausanne (PGSL) techniques, which have been proven to be efficient and suitable computer-aided circularity measuring instruments [48]. A circular profile surface has a maximum point that is connected by a wide circle diameter, as shown in Figure 1a. The center of the circle is represented by C, and its radius can be calculated for all points. The maximum error is determined by measuring the values for the outer and inner boundaries. The inner boundary, outer boundary, and circular center are represented by xi, xo, and c, respectively, as described by [49]. The distance is measured by establishing a reference circle from the measurement data. In the case of shafts, the largest circle that can be contained within the workpiece's profile is used, known as the MIC. Conversely, the smallest circle that can just contain the profile is used for holes, referred to as the MCC. To find the roundness error, it is necessary to find a pair of concentric circles between which the narrowest annular zone, known as the MZC, can be established. The circle that minimizes the sum of the squares of the radial distances of the data points is selected as the LSC. Accurate determination of the center position and radius of the reference circle is crucial in assembly operations. The methods MIC, MCC, and MZC, along with their respective algorithms and mathematical calculations, have been previously discussed [20]. Therefore, the primary objective of this paper is to concentrate on the mathematical calculations with algorithms of the MZC technique for measuring roundness error through the aid of computer vision. Additionally, the study aims to integrate the data collected using MQTT protocols with the Internet of Things (IoT) based on Virtual Network Computing (VNC).

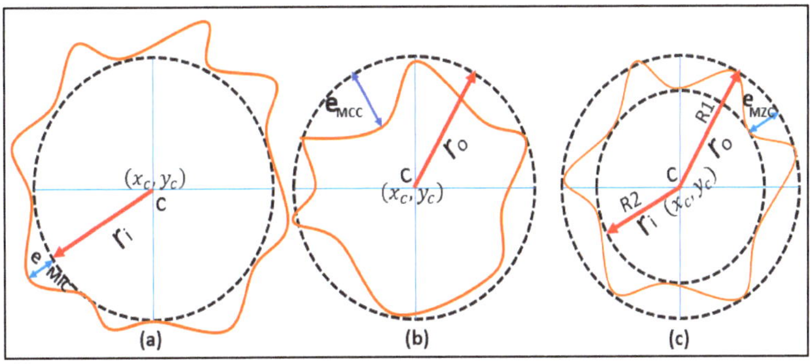

Figure 1. The fundamentals of calculating roundness using three different algorithms: (**a**) MIC, (**b**) MCC, and (**c**) MZC [50].

2.1. The Minimum Zone Circles

The MZC method determines roundness error by using two reference circles, which provide a quantitative measure by comparing the actual contour with two reference circles: the Maximum Inscribed Circle (MIC) and the Minimum Circumscribed Circle (MCC). In this technique, a circle is drawn around the circularity profile, while another circle is drawn inside the roundness profile to fit it as described by [51]. In the measurement of roundness, two circles, namely the circumscribed circle and the inscribed circle, play a crucial role. By identifying the center of these circles, which forms the minimal zone circle, the center of the measured shape is determined. This center is used as a reference point to assess the roundness profile. The roundness error is then computed by measuring the radial distance between the minimum inscribed circle and the maximum circumcised circle. The difference in radii between these circles indicates the magnitude of the roundness error present in the measured object. The roundness error is established using the MZC technique by measuring the radial separation between two circles, specifically the minimum inscribed circle and the maximum circumscribed circle. The difference in radii between these two circles indicates the roundness error of the object being measured.

$$\left\{\begin{array}{l} R_1 - R_2 \\ \sqrt{(xi - xc)^2 + (yi - yc)^2} \leq R_1 \\ \sqrt{(xi - xc)^2 + (yi - yc)^2} \geq R_2, i = 1,2,3,\ldots n \\ R_1 - R_2 \geq 0 \\ R_L \leq R \leq R_U; \\ x_L \leq x_c \leq x_U; \\ y_L \leq y_c \leq y_U \end{array}\right\} \quad (1)$$

where the R_1 and R_2 are the radii of $c1$ and $c2$, respectively, and thus share a center at (xc, yc). The minimum radial separation is defined as the roundness error. It is mathematically formulated as

$$e_{MZC} = R_{max} - R_{min} \quad (2)$$

In recent years, there has been a growing interest in implementing computer vision systems in industrial workplaces to detect defects in products. These systems use various image processing techniques to analyze the features and surface of the model, providing valuable information to the experts. Advanced industrial systems necessarily involve ever-improving product performance and improved product quality during the production process [52–54]. Defects, such as scratches, imperfections, or holes on the surface of the product, on the other hand, have an adverse influence mainly on the product's aesthetics and user comfort but also on its performance [55–58]. Therefore, defect detection is an efficient strategy for minimizing the environmental effects of product defects [59–61]. Machine vision in industrial processes can help with a wide range of industrial tasks [62–64]. A typical industrial visual inspection system is categorized into three components: optical illumination, image acquisition, image processing, and defect detection [29,65]. The development of an automated smart system-based interpreter for detecting and measuring the surface feature of Example 1, Part 21, of ISO 14649 standard using machine vision inspection (3SMVI) is proposed in this study [66]. The 3SMVI system will be developed in the milling machine using a camera system and lighting system [50,66]. The 3SMVI system is widely adopted as a standard tool for measuring and inspecting surfaces. This article explores the increasing use of camera systems to validate the accuracy of surface features in design and measurement. It also discusses the challenges of measuring roundness in circles using vision and validates the system based on IoT.

3. Materials and Methodology

3.1. Materials

The CAD model displayed in Figure 2 is utilized to determine the surface area of the side view layout. The model illustrates the presence of five holes that require inspection, and the inspection circles comprise the largest portion of the model's surface area that can measure the outer edge surface. Machining parameters for a mild Delrin workpiece that has round holes with a 30 mm diameter, a spindle speed of 2000 revolutions per minute (rpm), a feed rate of 0.1 mm per revolution (mm/rev), a cutting depth of 20 mm, and a cutting speed of 250 m per second (m/s) are given. In this study, the inspection hardware is developed with the aid of IoT support devices, including a Raspberry Pi, multiple subsystem cameras, a lighting system, and a Wi-Fi router. The hardware component specifications used in the IoT configuration are listed in Table 1.

Table 1. Characteristics of parameters, hardware, and software components.

Machine Characteristics	Parameter	Specification
Machine Part	Workpiece Size	Delrin 120 × 100 × 50 mm
Machining	Feed rate Spindle speed Depth of cut One Circle diameter Cutting speed	0.1 mm 2500 rpm 22 mm 30 mm 250 m/s
Cutting tool	Material type Diameter Number of flutes Tool type	High-speed steel 0.6 mm 2 New tool
	Number of axes	Three axes, X, Y, and Z
CMM measurement	Machine type Resolution High accuracy Versatility	MITUTOYO QM-353 0.0005 mm (0.00002 in.) Accuracy of min 0.0017 mm Wide range of probe systems are available
Software		
Open CV		Open Vision Library2011, windows 10
Python		3.8.3
Pycharm editor		IDE used in computer programming for Python
CAD design		CATIA v5, R21 2020
Operating System		The Raspbian Debian Buster
MQTT protocol		Messaging protocol designed for low-bandwidth

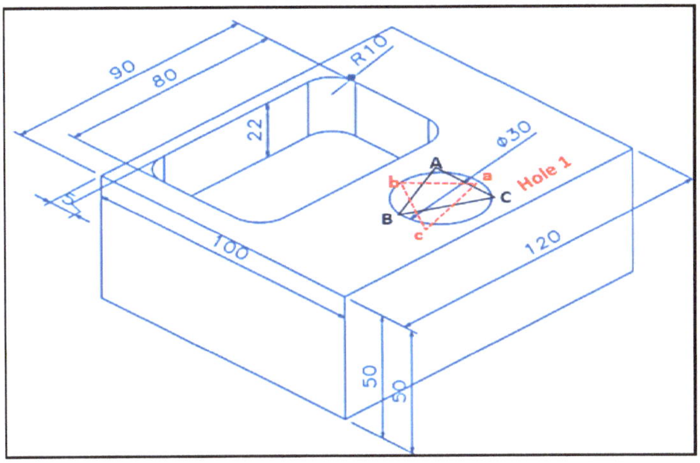

Figure 2. The CAD Design of STEP-NC file 14649 Part 21.

3.2. The Experimental Setup and Procedure

The setup for the experiment includes a combination of hardware and software for carrying out non-contact roundness tests through a vision system. Proper lighting is essential for acquiring good-quality images of the test objects. The workpiece is rotated through the use of a servo motor, while the encoder response on the fixed axis measures the angle of rotation. Simultaneously, a COMS camera captures visual data of the targeted region. The experimental setup for the Smart System-based STEP-NC file-21 for the Machine Vision Inspection (3SMVI) prototype, which incorporates cloud-based MQTT connectivity, is depicted in Figure 3. This system was implemented on an INTELITEK PROLIGHT three-axis milling machine, designed in accordance with Example 1, Part 21, of the ISO 14649 standard.

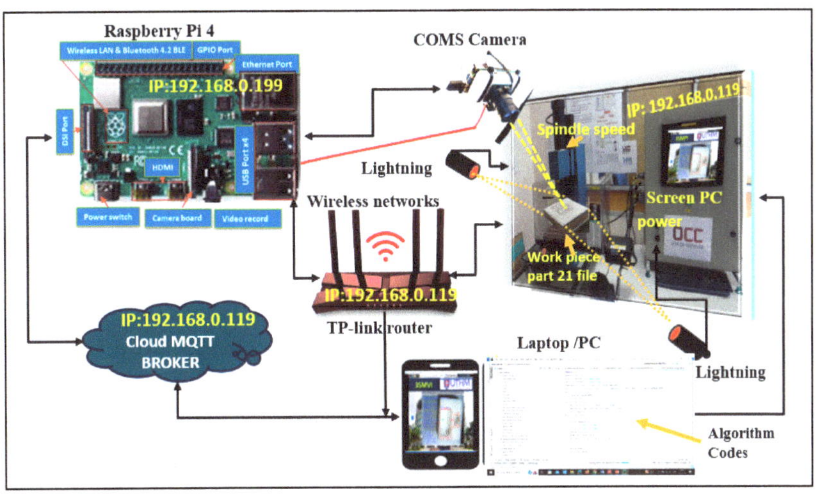

Figure 3. The 3SMVI prototype based on MQTT.

Case Study

Based on the ISO 14649 standard's Example 1, Part 21, a system design was formulated specifically for the INTELITEK PROLIGHT 3-axis milling machine. The CAD model was utilized to identify the surface region of the side view layout, with a specific focus on one hole and one pocket that necessitate inspection. The selected circle designated for inspection represents the most critical section of the model area, ensuring precise measurement of the outer surface of the edge.

This case study provides comprehensive information regarding the machining parameters employed for a Delrin prismatic mill workpiece. Notably, the workpiece has a roundness hole with a diameter of 30 mm. The specified parameters consist of a spindle speed of 2000 rpm, a feed rate of 0.1 mm/rev, a depth of cut of 20 mm, and a cutting speed of 250 m/s. To further elaborate, this case study encompasses an expanded version of Example 1 ISO 14649 Part 21 file, incorporating four holes and one pocket, as illustrated in Figure 2. Additionally, Table 1 presents an overview of the characteristic parameters, hardware, and software components involved in the study.

4. The Development of 3SMVI Software and Integration

The successful integration of machining and inspection capabilities within a single CNC machine marked a significant advancement in the establishment of the 3SMVI (Smart System-based STEP-NC file-21 for Machine Vision Inspection) system. Its primary goal is to create 3SMVI applications that combine software and hardware, as noted in [66]. The 3SMVI system is intended to operate in an IoT environment based on the STEP-NC protocol, with a heavy emphasis on hardware and software integration. The OpenCV library and Python programming language were employed to construct the machine vision system, while the 3SMVI application consisted of four major components: data diffusion mechanism, operational modules, data model, and computational intelligence algorithm. The 3SMVI system provides a comprehensive solution for machining and inspection tasks by integrating these components into a single computer. The software for the 3SMVI system was written in Python 3.8.3 with the PyCharm editor, which supported 64-bit Windows applications. It includes several computer vision and image processing algorithms for accurately measuring and inspecting roundness errors in circular components.

Figure 3 depicts the 3SMVI system based on MQTT protocols with virtual network computing. The object is placed on the backlighting table to begin the measurement or inspection process, and the capturing software, which is based on Open Computer Vision, captures and saves an image in BMP format. The vision inspection system then opens the captured image and performs the necessary inspection and measurement procedures.

4.1. The 3SMVI Hardware and Software Integration

The advancement of a smart system with interpreted files of STEP-NC for machine vision inspection based on an IoT environment. This innovation successfully integrated the machining and inspection of the roundness of the milled workpiece into a PC CNC machine cloud platform under an Open CV environment depending on a new technology approach known as IoT. Every device and machine vision were connected to the internet via the same IP address and communicated wirelessly. Therefore, the IoT application approach enables inspection integration and communication through the server broker. In addition, the information from interpreted files of the STEP-NC-based CNC machine was sensed through the camera device and finally sent back the inspection dataset into the server cloud to the STEP-NC-based IoT application. Figure 4 illustrates the smart system for interpreted files of STEP-NC for machine vision inspection based on IoT or the 3SMVI systems with the integration of both 3SMVI hardware and 3SMVI software. Additionally, the developed system provides an advanced human–machine interface that uses visual network computing (VNC) on any smart device, such as a mobile phone, tablet, and PC, through a single IP address.

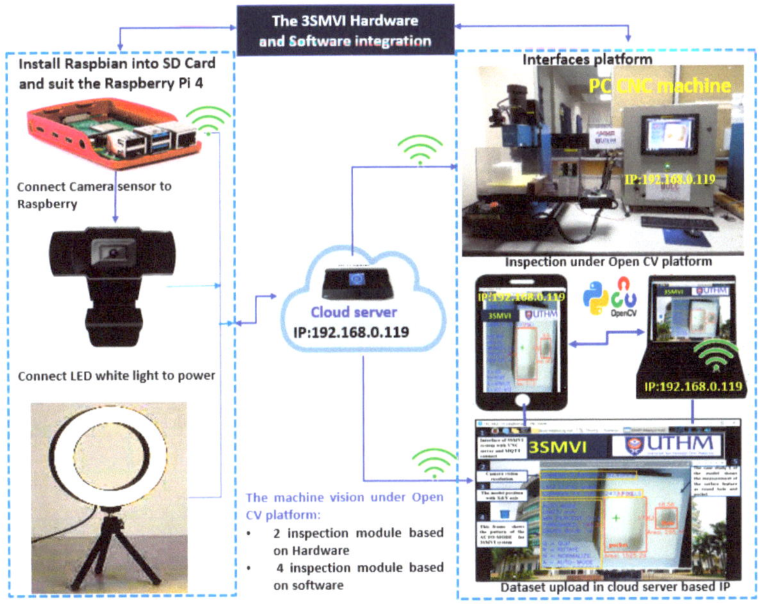

Figure 4. The integration of 3SMVI hardware- and software-based IoT.

4.2. Software Utilization and Algorithms

The data was implemented and published in cloud server-based MQTT on IoT applications. Hence, the captures of the images were data acquisition, facilitating data integration and communication during machining between inspection systems and VNC cloud servers through IoT-based technology. The main interface, as shown in Figure 5, requires two inputs from the user to carry out the measurement and inspection procedures. The 3SMVI system takes two inputs for measuring and inspecting circular objects. The diameter of the object is the first input, which is used to calibrate the system. The second input, which is used for inspection, is the maximum allowable roundness error value. To perform measurement and inspection procedures, the system employs a variety of algorithms based on computer vision and image processing.

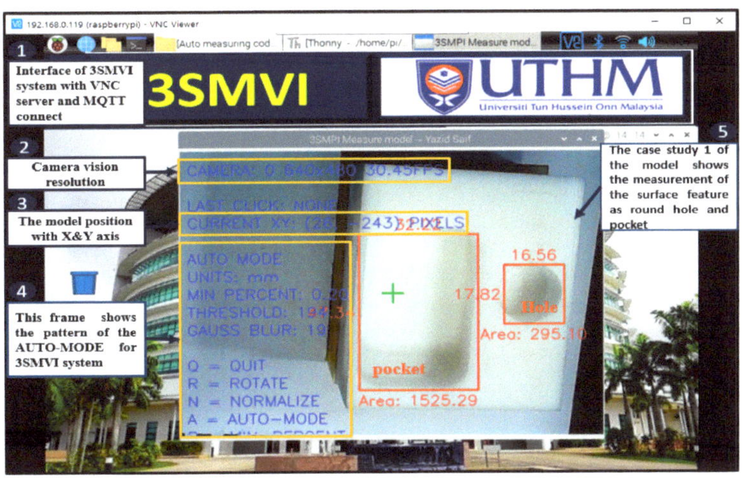

Figure 5. The 3SMVI main interface system.

4.3. Image Analysis and Segmentation

Image analysis is the initial stage of the process of image segmentation in the computer vision process. The image is fed into the system, and features are then extracted from it. Pixels that have similar characteristics are grouped together into an area, and regions with linear structures and 2D shapes are separated based on the success or failure of the computer vision-based process that relies on efficient segmentation. Industrial machine vision is one example of a typical application where the assembly of gadgets is monitored, and objects are automatically recognized and tracked. Apart from that, image segmentation has been used in traffic control, content-based image retrieval, video investigation, and sports arenas. The process of extracting meaningful information and insights from images is referred to as image analysis.

In the segmentation process, the system uses either of two features of an image separation procedure that relies on controlling the direction of illumination as part of the discontinuity. There are defined mechanisms like thresholding, region splitting and merging, and region growth used in the similarity approach to partition the image into comparable areas. Thresholding is used to separate the foreground from the background by selecting a threshold value. A binary image is generated from the gray-level image, with the threshold value being the average value of all pixels. Binary images have the advantage of minimizing data complexity and simplifying the process of image recognition and segmentation, but setting a threshold value can be difficult due to its sensitivity to noise. Clustering is the process of categorizing objects based on their characteristics. Following segmentation, models are extracted using edge detection algorithms such as Canny edges, as illustrated in Figure 10.

The process of measuring the roundness error of a Delrin workpiece involves capturing images, isolating the edge pixels of the object, and applying an edge detection algorithm. Typically, the frame grabber captures color images. However, for this particular application, binary images featuring only two colors, black and white, are required. To achieve this, a gray image must first be generated from the color image using the following formula:

$$Gl = 0.299Rc + 0.587Gc + 0.114Bc \tag{3}$$

where Gl describes the pixel's grey level and Rc, Gc, and Bc are the red, green, and blue components of the pixel color, respectively. Then, the histogram of each gray level is determined to compute a threshold value using the following:

$$T = Abs((Gl_{max}1 - Gl_{max}2)/2) \tag{4}$$

where $Gl_{max}1$ and $Gl_{max}2$ are the two grey levels with the highest frequencies, and T is the threshold value. All pixels with grey levels above the threshold are converted to black pixels to generate the binary image, and all other pixels are converted to white pixels.

4.4. An Algorithm for Detecting Edges

The edge detection process entails detecting the pixels that form the edges in a binary image. This is achieved through the use of an edge detection algorithm that compares the color of each pixel in the binary image with that of its eight neighboring pixels. If a pixel's color is black (0), and any of its neighboring pixels are white (255), it is classified as an edge pixel. However, if the pixel's color is not black or its neighboring pixels are not white, it is considered a background pixel. A process flowchart depicting these algorithms is shown in Figure 6, and further descriptions of these algorithms are provided in the subsequent sections. The following code can be used to implement this algorithm, as shown in Supplementary File.

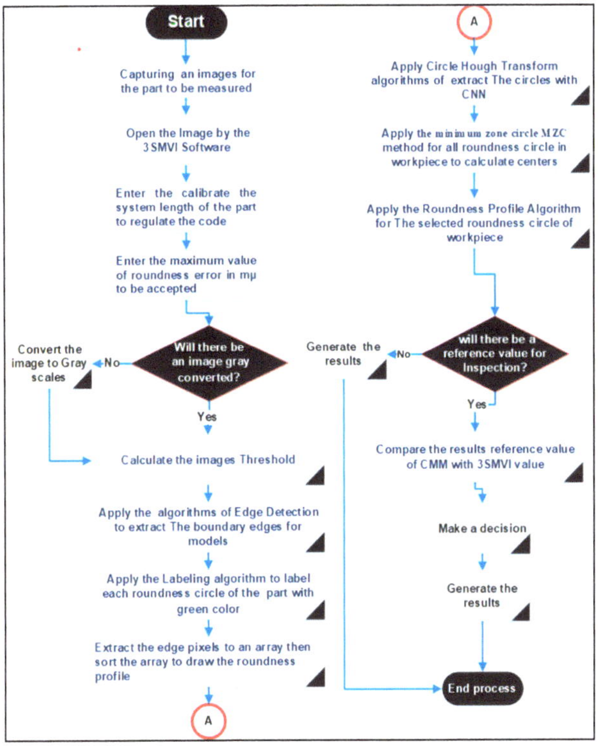

Figure 6. The 3SMVI algorithms process diagram.

Objects or Models Detection

An outline of object detection is seen in Figure 7, which includes a set of two images. Image 1 contains an object's reference, and Image 2 transforms them to grey, adds a bit of filtering to blur the image, and then subtracts the images to a binary or black-and-white image. Finally, the outcome was dilated to merge some of the near pixels or blobs, detecting the largest circle and rectangular blob around it. That was one method of detecting changes in the pixel value of images. Hence, the principle of whether to adjust a bit more behind motion or object detection to improve the final output to suit the circumstances of the image, etc. Firstly, users would need to create a video stream. Once the users have created a video feed, the first image is captured in the same way as Image 1 and used as a reference image, and then the frames are compared with the reference frame.

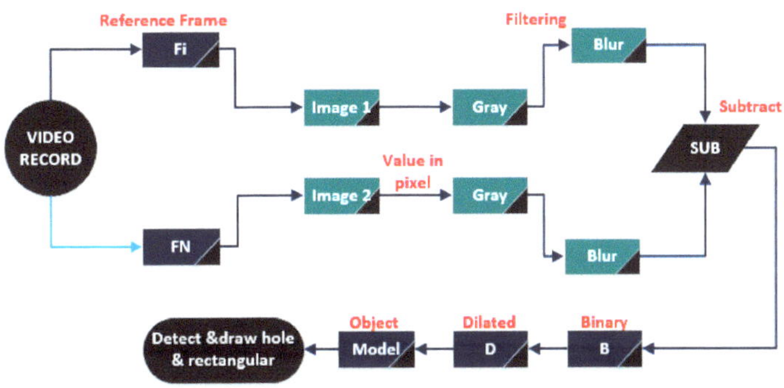

Figure 7. The framework of models with object detection.

4.5. Data Acquisition

This non-contact roundness measurement method automatically generates data acquisition by rotating the workpiece at a specific point using a machine vision system and capturing images with a high-end webcam camera placed directly in front of the object with 90 degrees of freedom. For one cycle of the measurement part, 330 images of the target region were captured, with each measurement performed after rotating the workpiece by 1.8 degrees using an Arduino-controlled two-phase stepper motor, which provides higher resolution due to its smaller step angle and usable torque. The application connects to MQTT clients over the internet by specifying the client's IP address and port number through visual network computing. Once connected, the VNC module displays the image file and serves as the primary interface for the machine vision platform. A sample image obtained from the vision system is shown in Figure 6.

Table 2 presents the surface feature of a hole pocket with four circle hole diameter values, namely 22 mm, 22 mm, 22 mm, and 22 mm, generated at different angles ranging from 0 degrees to 360 degrees in the 3SMVI vision system. The average circular error in micrometers for each hole was also calculated.

Table 2. The diameter values from the vision system for mild Delrin workpiece.

Image No.	Angle in Degree	Mild Delrin Dia. 30 mm
1	0°	29.9383
2	31°	29.9397
3	62°	29.9378
4	93°	29.9292
5	124°	29.9364
6	155°	29.9281
7	186°	29.9368
8	217°	29.9311
9	248°	29.9315
10	279°	29.9299
11	310°	29.9324
12	360°	29.9308
Roundness Circle Hole	*Error in* μm $= Rc_{max} - Rc_{min}$	**2.9 μm**

4.6. The Algorithm for Edge Labelling

To compute the roundness error of each circular part individually, it is necessary to first identify the edge pixels of each part from the captured images. This can be achieved through the use of an edge labeling algorithm, which generates a labeled image from the edge image and assigns a unique color label to the pixels of each circular part. The algorithm follows a general process described in the reference and comprises several steps [43], including the following steps:

1. The edge-pixel image is encoded using the run-length algorithm;
2. After scanning the runs, each run is given a preliminary label. The label equivalent is then entered into a local equivalent table;
3. The classes that are resolved have equivalence;
4. Finally, based on the resolved equivalence classes, the runs are given labels.

4.7. The Error Algorithm for Roundness Holes

The roundness error algorithm, which is explained in Section 2.1, uses the minimum zone circles technique and includes the following steps:

1. The labeled image is scanned from left to right and top to bottom. During this process, the (x, y) coordinates of the edge pixels of a selected part (color) are extracted and stored in an array known as Edge Pixels;
2. The minimum zone method is then applied to the pixels in the Edge Pixels array to determine the center and radius of the minimum zone circle;
3. The center of the minimum zone circle is calculated, and the distances between this center and all pixels in the Edge Pixels array are computed. The minimum and maximum distances are identified as R_{min} and R_{max}.

5. System Calibration

To calibrate the system, it is necessary to compute the pixel sizes in both the x and y directions based on the actual size of the holes in the workpiece being measured. Camera calibration involves determining internal parameters like focal length and correcting lens distortions, as well as establishing the camera's position and orientation. This process, aided by specialized software and calibration targets, enhances measurement accuracy and computer vision applications, especially in robotics, augmented reality, and industrial quality control. The following steps are involved in this process:

1. The user provides the 3SMVI platform with the actual diameter (in millimeters) of the object being measured. If the object is comprised of multiple circular parts, the outer part's size (maximum diameter) should be utilized.
2. The software searches for the two edge pixels on the outer contour with the minimum and maximum x coordinates to determine the maximum diameter of the captured image in the x direction ($D_{max}x$) using the formula

$$D_{max}x = Abs(X_{max} - X_{min}) \tag{5}$$

   where X_{max} and X_{min} are the maximum and minimum x coordinates of the edge pixels of the outer contour, respectively.
3. The calibration factor in the x direction (CFx) is calculated as follows:

$$CFx = Dactual / D_{max}x \tag{6}$$

4. In the same manner, the calibration process also involves determining the calibration factor in the y direction (CFy) by identifying the two pixels with the minimum and maximum y coordinates, which can be obtained by using the formula

$$CFy = Dactual / D_{max}y \tag{7}$$

5. After calculating CFx and CFy, all x and y coordinates of the edge pixels are multiplied by these calibration factors.
6. The primary roundness error $(RE\ primary)$ is then calculated using the least squares circle technique on the edge pixels.
7. Furthermore, using a formula derived from Excel, the fixed error (FE) in pixels is determined based on the user-provided value of the smoothing factor (SF).
8. The fixed error is then calibrated and expressed in millimeters $(FE\ mm)$ using the following equation:

$$FE\ mm = FE\ pixel\ CFx \times CFy \tag{8}$$

9. The final roundness error $(RE\ final)$ is calculated as follows:

$$RE\ final = RE\ primary - FE\ mm \tag{9}$$

Camera-to-Workpiece Distance Estimation

After calibration, the distance to the object was calculated using a single photograph. Thus, the concept of the given solution was the image analysis resulting from the object moving away from the point for which it defined the focus, how to use a single camera to figure out the distance to an object, where the focal distance (f) and height (h) and distance the distance between the object and the lens known as (d). b was the height, while m was the distance between the camera and a and b, which was named the reflection of height. Figure 8 below shows the points between the camera distance and the object

$$\frac{a}{f} = \tan\theta_1 = \frac{h}{d} \tag{10}$$

$$\frac{b}{f} = \tan\theta_2 = \frac{h}{d-m} \tag{11}$$

By dividing Equations (10) and (11), they may represent the distance between the object and lens and calculate the pixel distance to figure out the length.

$$\begin{aligned} &\Rightarrow \tfrac{a}{b} = \tfrac{\tan\theta_1}{\tan\theta_2} = \tfrac{h}{d} \times \tfrac{d-m}{h} = 1 - \tfrac{m}{d}; \\ &\Rightarrow \tfrac{m}{d} = 1 - \tfrac{a}{b}; \\ &\Rightarrow d = \tfrac{m}{1-\frac{a}{b}} \end{aligned} \tag{12}$$

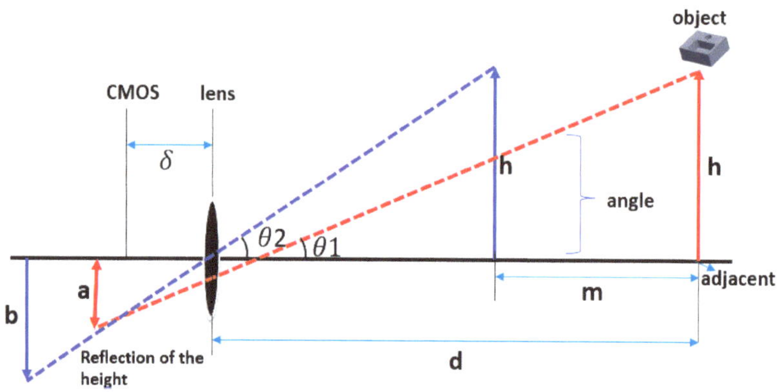

Figure 8. The camera-to-workpiece distance estimation.

6. CMM Inspection of Contact Measurement

In this validation section, a case study was used to inspect the roundness of a circle hole known as an offline coordinate measuring machine. Figure 9 shows the illustration of a model that measures the surface feature during the inspection process using the CMM technique, as described in this section. Furthermore, the research employed an offline measuring tool to evaluate the Delrin components utilizing Example 1, Part 21, file based on the STEP-NC environment. The QM-353 manual CMM equipment at UTHM Metrology was used to examine samples of the same workpiece holes. The CMM measurements were conducted in a particular sequence to determine the roundness of the holes and surface characteristics of mild Delrin workpieces.

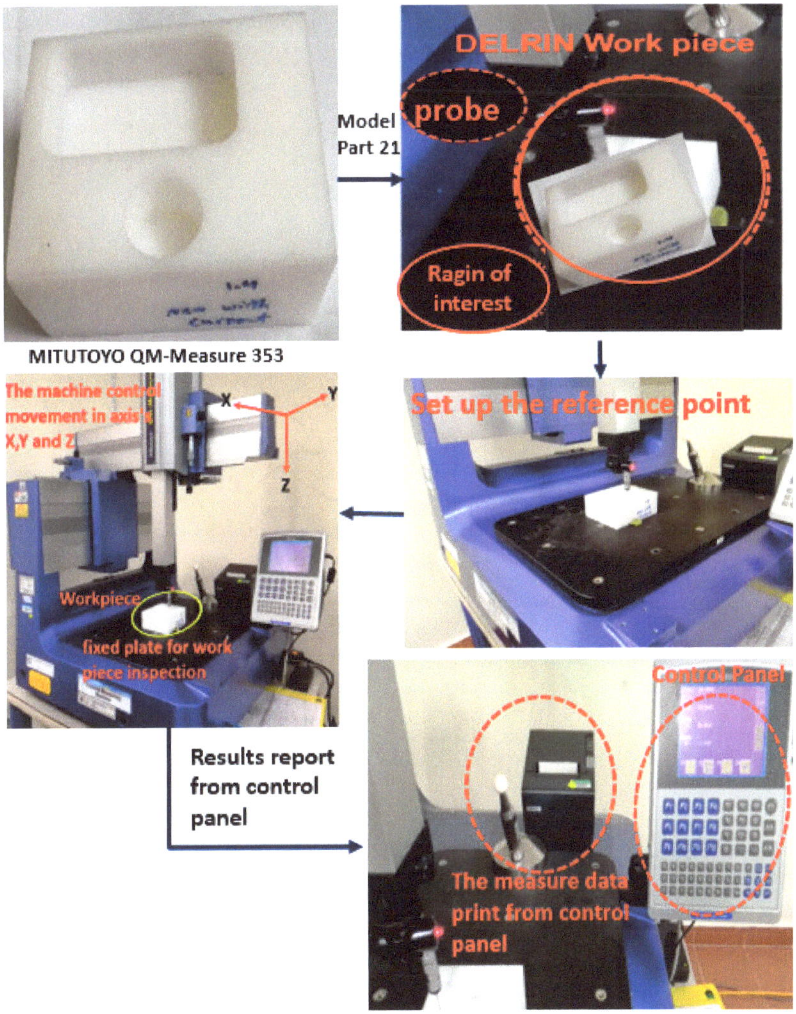

Figure 9. The CMM implementation for Case Study 1.

7. Result and Discussion

The study utilized a new cutting tool to produce Delrin parts with the Example 1, Part 21, file based on the STEP-NC environment using a PROLIGHT CNC milling machine in this validation section. However, the inspection processes utilizing image processing techniques remained the same as described in the previous section on the 3SMVI system implementation. Figure 10 depicts a graphical representation of the inspection procedure, illustrating the steps involving image processing, image analysis, and computational intelligence for measurement identification. Images were pre-processed to identify their contents for further analysis, such as counting or measuring roundness circles and plane surface features. When identifying objects in an image, edges play a crucial role. Boundaries were the lines that differentiate groups of identical pixels in the image from groups of distinct pixels. The edge considers image pixels that delineate the boundary of a target, such as whether the image's background finishes and the image continues.

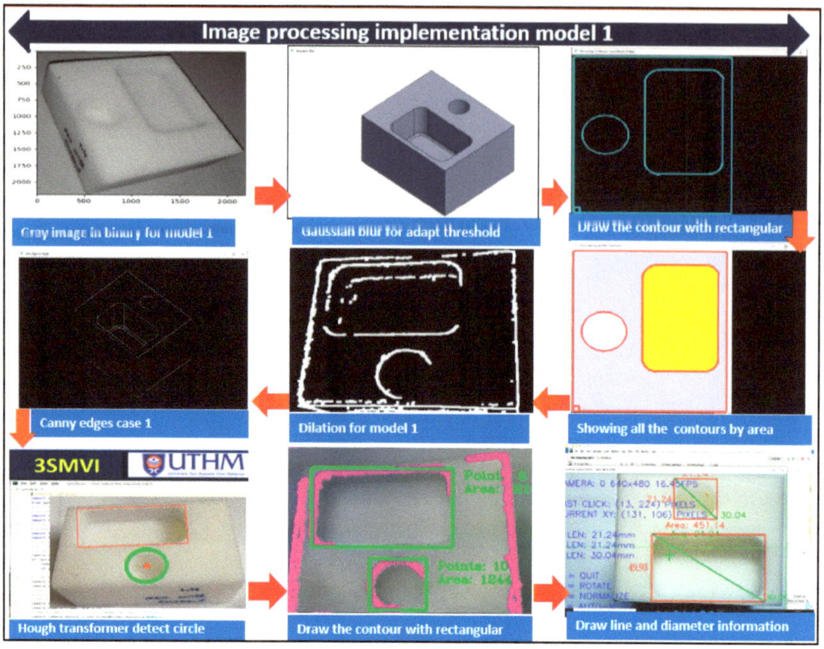

Figure 10. Implementation of image processing technique for Case Study 1.

Table 2 illustrates the surface feature diameter of the roundness circle values of Example 1, Part 21 file based on ISO 14649, developed using the 3SMVI vision system in twelve points. Consequently, Figure 11 displays consistent and predictable diameter data at each point, covering various angles ranging from zero degrees to 360 degrees of position. In addition, it depicted the data in red color decrease as the minimum value in three measurements.

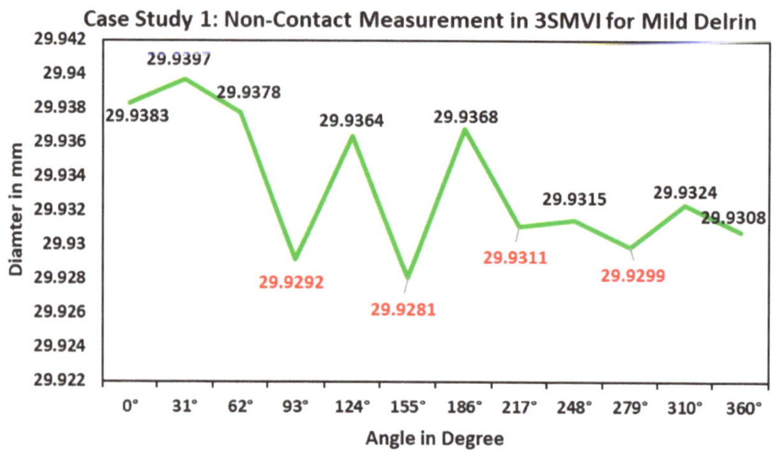

Figure 11. Case study with non-contact measurement in 3SMVI for mild Delrin.

Table 3 presents the results of the measurements taken with the MITUTOYO QM-Measure CMM of the surface feature of a roundness hole of the workpiece. Figure 12 shows that the diameter data obtained at each point were uniform and predictable, with positions ranging from 0° to 360°. The red data points represented the minimum value measurements, which were considered the optimal values for the vision system. However, while the contact approach provided accurate results, it was more time-consuming compared to the vision system. The main issue with the contact method was the wear and tear that occurred between the measuring probe and the workpiece.

Table 3. Dimensional information from the CMM for a mild Delrin workpiece.

No. of Point	Points	Mild Delrin Dia. 30 mm
3	A,B,C	30.0121
3	A,C,B	30.0099
3	B,A,C	30.0112
3	B,C,A	30.0123
3	C,A,B	30.0133
3	C,B,A	30.0113
3	a,b,c	30.0104
3	a,c,b	30.0109
3	b,a,c	30.0124
3	b,c,a	30.0116
3	c,a,b	30.0128
3	c,b,a	30.0103

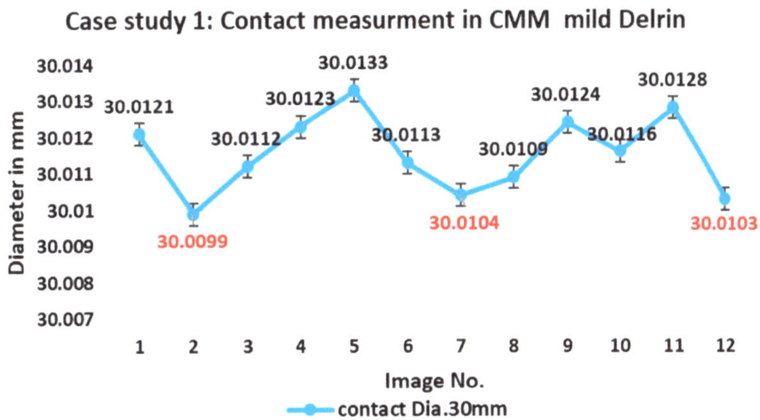

Figure 12. CMM of contact measurement for mild Delrin workpiece in Case Study 1.

Table 4 illustrates the average value of the difference errors in microns between contact and non-contact measurements of the dimensions in each parameter for the case study. Therefore, Figure 13 shows the decline of deference errors in microns between the 3SMVI and CMM.

Table 4. Comparison of the difference of average circle hole error between 3SMVI and CMM measurements for mild Delrin Case Study 1.

Diameter in mm	Hole Circle Error from 3SMVI System in μm	Hole Circle Error from CMM in μm	Difference in μm
30	11.6 μm	2.9 μm	8.7

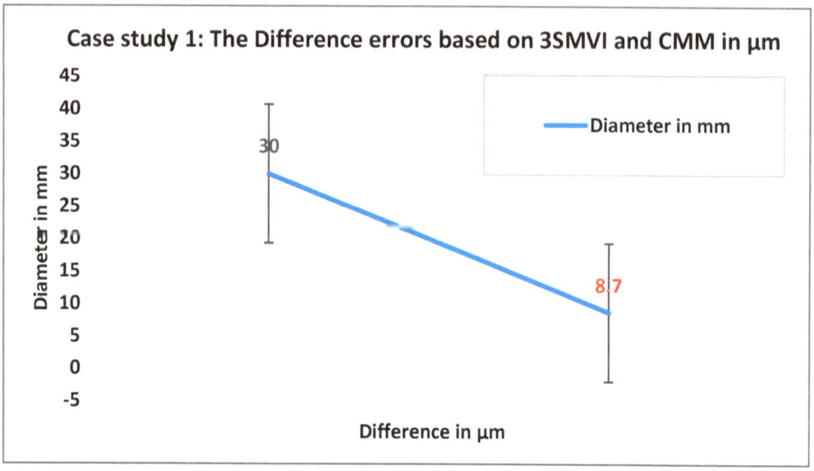

Figure 13. The difference error based on 3SMVI and CMM in µm.

8. Conclusions and Future Research

The fusion of IoT architecture, computer vision, and image processing methods has driven progress in measuring roundness within industrial automation. This integration enhances the accuracy and effectiveness of assessing roundness in industrial applications. IoT architecture facilitates real-time data exchange among devices, enabling instant monitoring and analysis of roundness measurements. This interconnectedness enables automated systems to gather and process image data from diverse sensors and cameras, offering a holistic perspective on an object's roundness.

This paper discusses the development of an intelligent system utilizing machine vision inspection within an IoT framework and ISO 14649 interpreted files. A case study, including Example 1, Part 21, of STEP-NC, validated the system's functionality and confirmed its high accuracy compared to previous models. The article stresses the significance of assessing uncertainty from the perspective of the MVIS system rather than solely focusing on algorithm development. The 3SMVI system, built on IoT architecture, has been fully developed, and the next step involves integrating advanced intelligent algorithms to enhance its capabilities. The primary contribution of this research is the examination of workpiece roundness using the 3SMVI system within an IoT and CMM framework. The IoT architecture facilitates seamless communication between the roundness measurement system and connected devices, enabling efficient data exchange, remote monitoring, and data-driven decision-making.

Computer vision algorithms play a crucial role in extracting meaningful information from the captured images. These algorithms can identify and analyze geometric features, enabling accurate roundness measurements. Through the use of edge detection, contour analysis, and shape-fitting algorithms, the system can precisely determine the roundness of objects with minimal human intervention. A vision system was introduced to measure and inspect circular parts' roundness errors using a non-contact method that adheres to geometric roundness tolerance standards. The 3SMVI system was developed to assist measurement and inspection processes by analyzing captured images using image processing and computer vision algorithms based on Open Computer Vision. Therefore, to ensure the system's accuracy, the roundness error of holes in workpiece parts was measured using both a coordinate measuring machine (CMM) and the machine vision inspection system (MVIS). The results were compared, and the maximum difference between the outcomes of the two systems was found to be 8.7%.

In a future study, we aim to enhance the 3SMVI system's performance by upgrading hardware and software algorithms, improving image quality, and using more protocols for information transmission via cloud computing servers. It will focus on enhancing surface feature detection and manufacturing optimization using deep learning techniques and incorporating machine learning and artificial intelligence for continuous improvement and intelligent decision-making in industrial automation.

Supplementary Materials: The following supporting information can be downloaded at: https://www.mdpi.com/article/10.3390/app132011419/s1.

Author Contributions: Conceptualization, Y.S., A.Z.M.R., A.A. and H.Q.A.A.; Methodology, Y.S., A.Z.M.R., Y.Y., M.L.A., S.A.-A. and A.A.; Software, Y.S.; Validation, M.L.A., S.A.-A., D.H.D., M.A.A.-m. and H.Q.A.A.; Formal analysis, M.L.A., S.A.-A., D.H.D. and M.A.A.-m.; Investigation, Y.Y., S.A.-A., D.H.D., A.A., Y.H.G. and H.Q.A.A.; Resources, A.Z.M.R., A.A. and M.A.A.-m.; Data curation, Y.S. and M.A.A.-m.; Writing—original draft, Y.S.; Writing—review & editing, A.Z.M.R., Y.Y., M.L.A. and Y.H.G.; Supervision, A.Z.M.R.; Funding acquisition, Y.H.G. and H.Q.A.A. All authors have read and agreed to the published version of the manuscript.

Funding: The authors would like to express their sincere gratitude to the Ministry of Higher Education (MOHE) in Malaysia for their invaluable support through the Fundamental Grant Scheme (FRGS/1/2020/STG01/UTHM/02/2). Furthermore, the authors would like to acknowledge the generous provision of facilities by the Sustainable Polymer Engineering, Advanced Manufacturing, and Material Centre (SPEN-AMMC). Moreover, this work was supported by Institute of Information & communications Technology Planning & Evaluation (IITP) grant funded by the Korea government (MSIT) (2021-0-00755), Dark data analysis technology for data scale and accuracy improvement.

Institutional Review Board Statement: Not applicable.

Informed Consent Statement: Not applicable.

Data Availability Statement: All the relevant data are within the manuscript and supporting information file as Supplementary File.

Conflicts of Interest: The authors declare no conflict of interest.

Abbreviations

ASME	American Society of Mechanical Engineers
ACO	Colony Optimization
CNC	Computer Numerical Control
CMM	Coordinate measuring machine
C	Center of this circle
CMOS	Complementary metal–oxide–semiconductor
EO	Engineering Ontology
e_{MZC}	Error-based Minimum Zone radial circles
CLIM	Closed-loop inspection manufacturing
CAIP	Computer-aided Inspection Planning
FTP	File Transfer Protocol
GA	General Assembly Simulation
GD&T	Geometric Dimensioning and Tolerance
I 4.0	The Fourth Industrial Revolution
IoT	Internet of Things
WSET	Wire spark erosion machining
ISO	International Standard Organization
LED	Light-emitting Diodes
MQTT	Message Queuing Telemetry Transport
MZC	Minimum Zone for Radial Circles
MTD	Metrology for the Digitalization
Open CV	Open Computer Vision
OLP	Offline Programming System
OOR	Out-of-the-Round
QM	Quality MITUTOYO Measure
RGB	Three channels of color: Red, Green, and Blue

R_i	Radius inner
R_U	Upper radius
R_L	Lower radius
STEP-NC	The Standard for the Exchange of Product Model Data for Numerical Control
VNC	Visual network center
3SMVI	Smart System based on interpreted STEP-NC for Machine Vision Inspection
3D	Three-dimension model

References

1. Stojadinovic, S.M.; Majstorovic, V.D.; Durakbasa, N.M. Toward a cyber-physical manufacturing metrology model for industry 4.0. *Artif. Intell. Eng. Des. Anal. Manuf.* **2020**, *35*, 20–36. [CrossRef]
2. Ahmad, M.I.; Yusof, Y.; Daud, E.; Latiff, K.; Kadir, A.Z.A.; Saif, Y. Machine monitoring system: A decade in review. *Int. J. Adv. Manuf. Technol.* **2020**, *108*, 3645–3659. [CrossRef]
3. Ashima, R.; Haleem, A.; Bahl, S.; Javaid, M.; Mahla, S.K.; Singh, S. Automation and manufacturing of smart materials in additive manufacturing technologies using Internet of Things towards the adoption of industry 4.0. *Mater. Today Proc.* **2021**, *45*, 5081–5088. [CrossRef]
4. Nate, K.; Tentzeris, M.M. A novel 3-D printed loop antenna using flexible NinjaFlex material for wearable and IoT applications. In Proceedings of the 2015 IEEE 24th Electrical Performance of Electronic Packaging and Systems (EPEPS), San Jose, CA, USA, 25–28 October 2015; pp. 171–174. [CrossRef]
5. Qin, W.; Chen, S.; Peng, M. Recent advances in Industrial Internet: Insights and challenges. *Digit. Commun. Netw.* **2020**, *6*, 1–13. [CrossRef]
6. Benbarrad, T.; Salhaoui, M.; Kenitar, S.B.; Arioua, M. Intelligent Machine Vision Model for Defective Product Inspection Based on Machine Learning. *J. Sens. Actuator Netw.* **2021**, *10*, 7. [CrossRef]
7. Pan, Y.; White, J.; Schmidt, D.C.; Elhabashy, A.; Sturm, L.; Camelio, J.; Williams, C. Taxonomies for Reasoning About Cyber-physical Attacks in IoT-based Manufacturing Systems. *Int. J. Interact. Multimedia Artif. Intell.* **2017**, *4*, 45. [CrossRef]
8. Wen, X.; Xia, Q.; Zhao, Y. An effective genetic algorithm for circularity error unified evaluation. *Int. J. Mach. Tools Manuf.* **2006**, *46*, 1770–1777. [CrossRef]
9. Renzi, C.; Ceruti, A.; Leali, F. Integrated geometrical and dimensional tolerances stack-up analysis for the design of mechanical assemblies: An application on marine engineering. *Comput.-Aided. Des. Appl.* **2018**, *15*, 631–642. [CrossRef]
10. Guu, S.M.; Tsai, D.M. Measurement of roundness: A nonlinear approach. *Proc. Natl. Sci. Counc. Repub. China Part A Phys. Sci. Eng.* **1999**, *23*, 348–352.
11. Mohamed, A.; Esa, A.H.; Ayub, M.A. Roundness measurement of cylindrical part by machine vision. In Proceedings of the International Conference on Electrical, Control and Computer Engineering 2011 (InECCE), Kuantan, Malaysia, 21–22 June 2011; pp. 486–490. [CrossRef]
12. Ayub, M.A.; Mohamed, A.B.; Esa, A.H. In-line Inspection of Roundness Using Machine Vision. *Procedia Technol.* **2014**, *15*, 807–816. [CrossRef]
13. Gadelmawla, E.; Khalifa, W.; Elewa, I. Measurement and Inspection of Roundness Using Computer Vision. *MEJ Mansoura Eng. J.* **2008**, *33*, 20–32. [CrossRef]
14. Ali, S.H.R.; Mohamd, O.M. Dimensional and Geometrical Form Accuracy of Circular Pockets Manufactured for Aluminum, Copper and Steel Materials on CNC Milling Machine Using CMM. *Int. J. Eng. Res. Afr.* **2015**, *17*, 64–73. [CrossRef]
15. Gapinski, B.; Kołodziej, A. Measurement of diameter and roundness deviation for circle with incomplete contour. In Proceedings of the 11th International Symposium on Measurement and Quality Control 2013, Cracow-Kielce, Poland, 11–13 September 2013; pp. 142–145.
16. Janusiewicz, A.; Adamczak, S.; Makieła, W.; Stpień, K. Determining the theoretical method error during an on-machine roundness measurement. *Meas. J. Int. Meas. Confed.* **2011**, *44*, 1761–1767. [CrossRef]
17. Lei, X.; Song, H.; Xue, Y.; Li, J.; Zhou, J.; Duan, M. Method for cylindricity error evaluation using Geometry Optimization Searching Algorithm. *Meas. J. Int. Meas. Confed.* **2011**, *44*, 1556–1563. [CrossRef]
18. Mears, L.; Roth, J.T.; Djurdjanovic, D.; Yang, X.; Kurfess, T. Quality and Inspection of Machining Operations: CMM Integration to the Machine Tool. *J. Manuf. Sci. Eng.* **2009**, *131*, 051006. [CrossRef]
19. Stojadinovic, S.M.; Majstorovic, V.D.; Durakbasa, N.M.; Sibalija, T.V. Towards an intelligent approach for CMM inspection planning of prismatic parts. *Meas. J. Int. Meas. Confed.* **2016**, *92*, 326–339. [CrossRef]
20. Gadelmawla, E.S. Computer Aided Measurement software for roundness evaluation from the coordinate measurement data. In Proceedings of the Ninth Cairo University International Conference on Mechanical Design and Production (MDP-9), Cairo, Egypt, 8–10 January 2008; pp. 270–285.
21. Hong, R.; Xiang, C.; Liu, H.; Glowacz, A.; Pan, W. Visualizing the Knowledge Structure and Research Evolution of Infrared Detection Technology Studies. *Information* **2019**, *10*, 227. [CrossRef]
22. Xu, P.; Guan, C.; Zhang, H.; Li, G.; Zhao, D.; Ross, R.J.; Shen, Y. Application of Nondestructive Testing Technologies in Preserving Historic Trees and Ancient Timber Structures in China. *Forests* **2021**, *12*, 318. [CrossRef]

23. Yang, J.; Li, S.; Wang, Z.; Dong, H.; Wang, J.; Tang, S. Using deep learning to detect defects in manufacturing: A comprehensive survey and current challenges. *Materials* **2020**, *13*, 5755. [CrossRef]
24. de Araujo, P.R.M.; Lins, R.G. Computer vision system for workpiece referencing in three-axis machining centers. *Int. J. Adv. Manuf. Technol.* **2020**, *106*, 2007–2020. [CrossRef]
25. de Araujo, P.R.M.; Lins, R.G. Cloud-based approach for automatic CNC workpiece origin localization based on image analysis. *Robot. Comput. Manuf.* **2021**, *68*, 102090. [CrossRef]
26. Kumar, B.M.; Ratnam, M.M. Machine vision method for non-contact measurement of surface roughness of a rotating workpiece. *Sens. Rev.* **2015**, *35*, 10–19. [CrossRef]
27. Jones, J.M.; Foster, W.; Twomey, C.R.; Burdge, J.; Ahmed, O.M.; Pereira, T.D.; A Wojick, J.; Corder, G.; Plotkin, J.B.; Abdus-Saboor, I. A machine-vision approach for automated pain measurement at millisecond timescales. *eLife* **2020**, *9*, 1–22. [CrossRef] [PubMed]
28. McCarthy, C.L.; Hancock, N.H.; Raine, S.R. Applied machine vision of plants: A review with implications for field deployment in automated farming operations. *Intell. Serv. Robot.* **2010**, *3*, 209–217. [CrossRef]
29. Di Leo, G.; Liguori, C.; Pietrosanto, A.; Sommella, P. A vision system for the online quality monitoring of industrial manufacturing. *Opt. Lasers Eng.* **2016**, *89*, 162–168. [CrossRef]
30. Fernández-Robles, L.; Azzopardi, G.; Alegre, E.; Petkov, N. Machine-vision-based identification of broken inserts in edge profile milling heads. *Robot. Comput. Manuf.* **2017**, *44*, 276–283. [CrossRef]
31. Saif, Y.; Yusof, Y.; Latif, K.; Kadir, A.Z.A.; Ahmed, M.L. Systematic review of STEP-NC-based inspection. *Int. J. Adv. Manuf. Technol.* **2020**, *108*, 3619–3644. [CrossRef]
32. Lin, E.-Y.; Tu, C.-T.; Lien, J.-J.J. Nut Geometry Inspection Using Improved Hough Line and Circle Methods. *Sensors* **2023**, *23*, 3961. [CrossRef]
33. Meadows, J.D. *Measurement of Geometric Tolerances in Manufacturing*; Marcell Dekker, Inc.: New York, NY, USA, 1998; Volume 1998.
34. García-Martínez, E.; García-González, N.; Manjabacas, M.C.; Miguel, V. Validation of a Manual Methodology for Measuring Roundness and Cylindricity Tolerances. *Appl. Sci.* **2023**, *13*, 9702. [CrossRef]
35. Zhao, Z.; Xi, J.; Zhao, X.; Zhang, G.; Shang, M. Evaluation of the Calculated Sizes Based on the Neural Network Regression. *Math. Probl. Eng.* **2018**, *2018*, 4078456. [CrossRef]
36. Cioboata, D.; Dontu, O.; Besnea, D.; Ciobanu, R.; Soare, A. Mecatronic equipment for bearing ring surface inspection. *Rom. Rev. Precis. Mech. Opt. Mechatron.* **2015**, *2015*, 262–266.
37. Du, C.L.; Luo, C.X.; Han, Z.T.; Zhu, Y.S. Applying particle swarm optimization algorithm to roundness error evaluation based on minimum zone circle. *Meas. J. Int. Meas. Confed.* **2014**, *52*, 12–21. [CrossRef]
38. Jbira, I.; Tahan, A.; Mahjoub, M.A.; Louhichi, B. Evaluation of the Algorithmic Error of New Specification Tools for an ISO 14405-1:2016 Size. In *International Design Engineering Technical Conferences and Computers and Information in Engineering Conference, Quebec City, QC, Canada, 26–29 August 2018*; American Society of Mechanical Engineers: New York, NY, USA, 2018. [CrossRef]
39. Gosavi, A.; Cudney, E.A. Form Errors in Precision Metrology: A Survey of Measurement Techniques. *Qual. Eng.* **2012**, *24*, 369–380. [CrossRef]
40. Kshaurad, K.; Kiran, M.; Shanmuganatan, S. Minimum zone tolerance algorithm to detect roundness error for machined rods using vision system. *Mater. Today Proc.* **2021**, *46*, 5997–6003. [CrossRef]
41. Sui, W.; Zhang, D. Four Methods for Roundness Evaluation. *Phys. Procedia* **2012**, *24*, 2159–2164. [CrossRef]
42. Rhinithaa, P.T.; Selvakumar, P.; Sudhakaran, N.; Anirudh, V.; Lawrence, K.D.; Mathew, J. Comparative study of roundness evaluation algorithms for coordinate measurement and form data. *Precis. Eng.* **2018**, *51*, 458–467. [CrossRef]
43. Nurunnabi, A.; Sadahiro, Y.; Lindenbergh, R. Robust cylinder fitting in three-dimensional point cloud data. *Int. Arch. Photogramm. Remote Sens. Spat. Inf. Sci.-ISPRS Arch.* **2017**, *42*, 63–70. [CrossRef]
44. Nurunnabi, A.; Sadahiro, Y.; Laefer, D.F. Robust statistical approaches for circle fitting in laser scanning three-dimensional point cloud data. *Pattern Recognit.* **2018**, *81*, 417–431. [CrossRef]
45. Guo, J.; Yang, J. An iterative procedure for robust circle fitting. *Commun. Stat.-Simul. Comput.* **2019**, *48*, 1872–1879. [CrossRef]
46. Liu, Y.; Zhu, L.S.; Xiong, Q.J. Development of Quality Detecting System for Micro Holes by ECM Based on Machine Vision Technology. *Appl. Mech. Mater.* **2013**, *397–400*, 1482–1485. [CrossRef]
47. Chaubey, S.K.; Gupta, K. An experimental study on out-of-roundness and material erosion rate during wire spark erosion turning of titanium cylindrical bars. *J. Mater. Res. Technol.* **2023**, *24*, 7539–7551. [CrossRef]
48. Srinivasu, D.S.; Venkaiah, N. Minimum zone evaluation of roundness using hybrid global search approach. *Int. J. Adv. Manuf. Technol.* **2017**, *92*, 2743–2754. [CrossRef]
49. Liu, Q.M.; Li, X. Machine Vision Detection on Circle with Non-Uniform Points. *Appl. Mech. Mater.* **2014**, *687–691*, 819–822. [CrossRef]
50. Saif, Y.; Yusof, Y.; Latif, K.; Kadir, A.Z.A.; Ahmed, M.B.L.; Adam, A.; Hatem, N.; Memon, D.A. Roundness Holes' Measurement for milled workpiece using machine vision inspection system based on IoT structure: A case study. *Meas. J. Int. Meas. Confed.* **2022**, *195*, 111072. [CrossRef]
51. Sun, T.-H. Applying particle swarm optimization algorithm to roundness measurement. *Expert Syst. Appl.* **2009**, *36*, 3428–3438. [CrossRef]
52. Wang, T.; Chen, Y.; Qiao, M.; Snoussi, H. A fast and robust convolutional neural network-based defect detection model in product quality control. *Int. J. Adv. Manuf. Technol.* **2018**, *94*, 3465–3471. [CrossRef]

53. Liao, Z.; Abdelhafeez, A.; Li, H.; Yang, Y.; Diaz, O.G.; Axinte, D. State-of-the-art of surface integrity in machining of metal matrix composites. *Int. J. Mach. Tools Manuf.* **2019**, *143*, 63–91. [CrossRef]
54. AKim, D.-H.; Kim, T.J.Y.; Wang, X.; Kim, M.; Quan, Y.-J.; Oh, J.W.; Min, S.-H.; Kim, H.; Bhandari, B.; Yang, I.; et al. Smart Machining Process Using Machine Learning: A Review and Perspective on Machining Industry. *Int. J. Precis. Eng. Manuf.-Green Technol.* **2018**, *5*, 555–568. [CrossRef]
55. Bulnes, F.G.; Usamentiaga, R.; Garcia, D.F.; Molleda, J. An efficient method for defect detection during the manufacturing of web materials. *J. Intell. Manuf.* **2016**, *27*, 431–445. [CrossRef]
56. Song, K.; Yan, Y. A noise robust method based on completed local binary patterns for hot-rolled steel strip surface defects. *Appl. Surf. Sci.* **2013**, *285*, 858–864. [CrossRef]
57. Zhang, Y.; Chen, H.; Bo, X.; Li, A.; Zhan, Y.; Gu, H. A multiscale evaluation of the surface integrity in boring trepanning association deep hole drilling. *Int. J. Mach. Tools Manuf.* **2017**, *123*, 48–56. [CrossRef]
58. Rao, X.; Zhang, F.; Lu, Y.; Luo, X.; Chen, F. Surface and subsurface damage of reaction-bonded silicon carbide induced by electrical discharge diamond grinding. *Int. J. Mach. Tools Manuf.* **2020**, *154*, 103564. [CrossRef]
59. Huang, S.-H.; Pan, Y.-C. Automated visual inspection in the semiconductor industry: A survey. *Comput. Ind.* **2015**, *66*, 1–10. [CrossRef]
60. Ravimal, D.; Kim, H.; Koh, D.; Hong, J.H.; Lee, S.-K. Image-Based Inspection Technique of a Machined Metal Surface for an Unmanned Lapping Process. *Int. J. Precis. Eng. Manuf. Technol.* **2020**, *7*, 547–557. [CrossRef]
61. Ren, Z.; Fang, F.; Yan, N.; Wu, Y. State of the Art in Defect Detection Based on Machine Vision. *Int. J. Precis. Eng. Manuf.-Green Technol.* **2021**, *9*, 661–691. [CrossRef]
62. Penumuru, D.P.; Muthuswamy, S.; Karumbu, P. Identification and classification of materials using machine vision and machine learning in the context of industry 4.0. *J. Intell. Manuf.* **2020**, *31*, 1229–1241. [CrossRef]
63. Ali, M.A.H.; Lun, A.K. A cascading fuzzy logic with image processing algorithm–based defect detection for automatic visual inspection of industrial cylindrical object's surface. *Int. J. Adv. Manuf. Technol.* **2019**, *102*, 81–94. [CrossRef]
64. Badmos, O.; Kopp, A.; Bernthaler, T.; Schneider, G. Image-based defect detection in lithium-ion battery electrode using convolutional neural networks. *J. Intell. Manuf.* **2020**, *31*, 885–897. [CrossRef]
65. Elias, M.Z.; Malamas, N.; Petrakis Euripides, G.M. A survey on industrial vision systems, applications and tools. *Image Vis. Comput.* **2003**, *2*, 171–188. [CrossRef]
66. Saif, Y.; Yusof, Y.; Latif, K.; Kadir, A.Z.A.; Ahmad, M.B.I.; Adam, A.; Hatem, N. Development of a smart system based on STEP-NC for machine vision inspection with IoT environmental. *Int. J. Adv. Manuf. Technol.* **2021**, *118*, 4055–4072. [CrossRef]

applied sciences

MDPI

Article

An Analysis of Selected Technological Parameters' Influences on the Tribological Properties of Products Manufactured Using the FFF Technique

Gerhard Mitaľ *, Ivan Gajdoš, Emil Spišák, Janka Majerníková and Tomáš Jezný

Department of Technology, Materials and Computer Supported Production, Faculty of Mechanical Engineering, Technical University of Košice, Mäsiarska 74, 040 01 Košice, Slovakia; ivan.gajdos@tuke.sk (I.G.); emil.spisak@tuke.sk (E.S.); janka.majernikova@tuke.sk (J.M.); tomas.jezny@tuke.sk (T.J.)

* Correspondence: gerhard.mital@tuke.sk; Tel.: +421-55-602-3507

Abstract: This study investigates how layer deposition in printing 3D models made of polyetherimide (PEI) using the fused filament fabrication (FFF) technique affects resistance of these models to abrasive wear. Samples made by additive manufacturing with different setting of build orientation and path generation strategy of extruded fibers were used in the experiment. The experiments were conducted on a tribometer according to the ASTM G65-16 standard. The friction force, normal force, and temperature in the contact area during the tribometer operation were measured using a strain gauge sensor. The tribometer allowed us to perform the tribological experiments using a rubber-coated or a metal disc without a lubricant in the so-called "dry" operation. Following the ASTM G65-16 standard, a rubber-coated disc and Fe_3Al_2 $(SiO_4)_3$ garnet abrasive were used for the wear resistance tests. The analysis of experimental data showed correlations among orientation selection, the strategy of layer deposition by the FFF technique, and the surface life in terms of abrasive wear. The results of this research also showed the suitability of the chosen building orientation and deposition strategy for part production by additive manufacturing, depending on the required tribological properties, such as the coefficient of friction (yield path) and wear behavior. Based on the results of the study, it is concluded that continuity of wear and friction force depends on the path traveled under the model production orientation. The size of wear (material loss) ranged from 0.451–0.809%. It was shown that the weight loss of the sample under loading was greater, on average, with the chosen fiber orientation strategy in the Z direction than in the X direction.

Keywords: tribology; abrasive wear; FFF; polymers; additive technologies

Citation: Mitaľ, G.; Gajdoš, I.; Spišák, E.; Majerníková, J.; Jezný, T. An Analysis of Selected Technological Parameters' Influences on the Tribological Properties of Products Manufactured Using the FFF Technique. *Appl. Sci.* **2022**, *12*, 3853. https://doi.org/10.3390/app12083853

Academic Editors: Tibor Krenicky, Juraj Ruzbarsky and Maros Korenko

Received: 28 February 2022
Accepted: 5 April 2022
Published: 11 April 2022

1. Introduction

One of the technologies for processing plastics is the 3D printing technology namely fused filament fabrication (FFF). In technical practice, it is a widely used technique due to its ease of use, low cost of input material, and minimal waste generated by the process. The FFF technique can reduce production costs of a product while maintaining high reliability of plastic parts [1,2]. In the field of 3D printing, a number of research activities have been carried out in order to examine the mechanical properties of components which have been produced using new technological processes. Strength, stiffness, weight, and other properties of components made by 3D printing depend on various conditions such as layer thickness and height, shell thickness, and fill density. The most common materials used in 3D printing are acrylonitrile butadiene styrene (ABS) and polylactic acid (PLA). Many studies have addressed the mechanical properties of 3D-printed parts; a study by Bellini et al. [3] was among the most notable studies, which investigated the tribology of 3D-printed parts and mainly focused on the mechanical properties of polymeric materials, in particular, PLA material [4–7]. However, research on the ULTEM 9085 material is lacking.

Research by Song et al. aimed to measure the mechanical wear of 3D-printed parts made of PLA [8]. Mechanical properties of colored PLA types were described by Wittbrodt et al. [9]. Research by Es-Said et al. addressed ABS plastic parts, the influence of orientation of the individual printed layers, and their mechanical properties, such as tensile strength and impact resistance [10]. Suitable orientation of a part in 3D printing can affect the structure of its surface (roughness), which affects the tribological properties of components manufactured in such a way [11]. Lack of information about the tribological properties of components produced by 3D printing can in practice lead to their excessive wear and subsequent failure of function. Dawoud et al. investigated tribological properties such as the coefficient of friction and wear behavior in samples produced by 3D printing from ABS material [12]. Panneerselvam et al. described the tribological behavior of ABS composites with added alumina [13]. Gurrala and Regalla et al. reported that tribological properties of 3D-printed polymer components could generally be improved by appropriate settings of the printing parameters [14,15]. These studies were followed by Hanon et al., who took temperature and the color of the fiber into consideration [16].

Gurrala and Regalla conducted research on the friction and wear of ABS polymer produced using the FFF technique. They took note of three parameters: (1) load, (2) speed, and (3) face-centered central composite design (FCCCD) orientation and their influences on the coefficient of friction as well as abrasive wear. It was shown that yield speed was affected by the friction of a firmer load [14,17]. Parts manufactured using the FFF technique show unevenness, but as the surface smoothens over time, the coefficient of friction decreases and becomes stable. Damage to contact surfaces is usually caused by wear and tear such as abrasion, fatigue, and surface undulation, or erosion and cavitation. Abrasive wear often leads to irreversible changes to the surface of a part which leads to a change between the bonds of the material particles. The friction coefficient usually rises sharply from low values to a peak value and stabilizes at slightly lower value following nearly linear curve [18].

Optimization of parameters from the point of view of tribology was addressed by Norani et al., who analyzed the coefficient of friction and wear rate of components made of ABS using the (Box–Behnken Design) methodology [18–20]. The study found that layer height significantly affected the coefficient of friction (COF) and the wear rate; a layer height of 0.10 mm, a temperature of 234 °C, and the use of an oblique print pattern were the most optimal parameters for minimizing the COF (0.2788) and the wear rate (2.1136×10^{-4} mm^3/Nm).

Another study by Garg et al. compared the behavior of an ABS component in the course of wear with a component made of Nylon6 Fe composite prepared using the FFF technique. ABS was found to have a higher COF (0.35 μm) than the Nylon6-Fe composite (0.26 μm). Wear of the ABS (0.15 g) was much greater than wear of the composite (0.01 g). The Nylon6-Fe composite showed much lower weight loss, indicating that it had very high wear resistance. This could be attributed to the presence of Fe in the composite. The wear that occurred was caused by a combination of abrasion and adhesion. Therefore, higher wear occurred at higher feed rates. This was the case for both ABS and composite materials [15,21].

Tahir et al. concluded that the internal cavities between fibers were 100% due to the absence of material, which improved the absorption of fluids, because 3D-printed samples had two to four times higher absorption rate as compared with pressed ABS samples [22]. In a study by Elsner et al., it was suggested that this higher rate of absorption facilitated microelastohydrodynamic lubrication (EHL) and also increased lubricant capacity in the component. Nevertheless, the 3D-printed ABS had a higher COF (0.040 to 0.055) due to fractures or cracks, and thus, increased the actual contact area of the component. This led to an increase in temperature and a decrease in strength of the material, which resulted in increased friction [20,23,24]. The main mechanism causing the wear of polymers and their composites is abrasive wear. As wear increases, the surface roughness smoothens and the effect of wear decreases. At the same time, increasing the load smoothens the yield surfaces through plastic deformation and evens out irregularities. This is likely to increase lubricant wear and to reduce the rate of friction as the yield surfaces become smoother [25,26]. In addition to these studies, other authors have reached similar conclusions based on their research on the wear of 3D-printed polymer components [26–28].

The yield properties of ABS pins were also investigated by Abdullah et al. (2020). In their research, the pins were printed with different internal geometries, and then, were subjected to wear tests under varying normal loads and feed rates. The results showed that under the wear test conditions of 58.68 N and 800 rpm, the pins with an internal triangular structure had a minimum COF value of 0.27 and a wear rate of 2.7×10^{-5} mm^3/Nm. This suggested that wear rates and COF values were relatively dependent on normal load, feed rates, and internal geometry. The internal triangular structure had lower maximum stresses distributed over the contact surface, and there was no statistically significant correlation between tribological and mechanical properties of the ABS pins, which had been 3D-printed with different internal geometries. Abrasion and lower fatigue wear were considered to be the main wear mechanisms that caused mild and severe wear [29].

It has been found that 3D-printed ABS pins with an internal triangular flip structure have the most favorable shortest run-in time and the lowest COF with high wear resistance [30].

The current state of scientific knowledge aimed at determining the tribological properties of componets produced using the FFF technique has mostly focused on the materials most frequently used in technical practice (PLA and ABS) [31]. Knowledge of the tribological properties of materials used in the FFF technique to a lesser extent, such as PEI, is currently lacking. In terms of their prospective use, as well as their use under demanding conditions, the application of high-performance materials in components manufactured using the FFF technique will be unavoidable, and this is associated with the need to know their tribological properties. The present study aims at determining the tribological properties of PEI (polyetherimide) production parts manufactured using the FFF technique.

The summary of research, in Table 1, shows that the polymers and binders have good tribological properties, and their application in technical practice is beginning to be justified [32–35].

Table 1. A summary of studies on the tribological properties of polymers, testing methods, and findings.

Authors	Research Objective	Materials	Methodology	Findings
Dangnan et al., 2020	Effect of varying contact loads on friction and wear	ABS and VeroGray™ (Stratasys®, Eden Prairie, USA) polymers	Loads of 1. 5 and 10 N were applied under dry sliding contact with a 52,100-steel counter face at room temperature	Loads of 1. 5 and 10 N were applied under dry sliding contact with a 52,100-steel counter face at room temperature.
Rathaur et al., 2019	Mechanical and tribological characterization	Epoxy resin blended with graphite/talc micro fillers	Epoxy resin blended with graphite/talc micro fillers for 15 min at a constant speed of 1200 rpm and an applied load of 100 N as per the ASTM D4172 under dry conditions	Graphite (10 wt%)/talc (10 wt%)/epoxy demonstrated a significant reduction by ~63% (from 0.103 to 0.276) in COF and a moderate increment by ~34% in wear resistance as compared with pure epoxy bearing balls under dry conditions. Hardness improved by ~5% in the graphite (10 wt%)/talc (10 wt%)/epoxy composite bearing ball.

Table 1. *Cont.*

Authors	Research Objective	Materials	Methodology	Findings
Marathe et al., 2020	Performance and tribological properties	50 wt% PEEK, 30 wt% 3 mm (GF), and 20 wt% synthetic graphite	Injection moulding (I) and compression moulding (C)	Final fiber length of 200–300 μm and 1.5 mm were observed for I and C composites, respectively, based on thermal degradation of composites at 600 °C. Low specific wear rate (K0) (~10–16 m^3/Nm) and friction coefficient (μ) (~0.03–0.05) were observed in both composites. Similarly, ~10–10 m^3/Nm and ~0.5–0.7, correspondingly, were observed for abrasive wear. For K0 and μ, in both cases, the μ of the I composite was smaller. The C composite was superior to the I composite in adhesive wear efficiency, but not in abrasive wear.
Sable et al., 2020	Yield strength and friction phenomena at high strain rate using uniaxial and oblique impact configurations	PU and epoxy	Impact velocities: 50 to 120 m/s for oblique, up to 1200 m/s for uniaxial	At a high strain rate, the COF values for both polymers were found to be inversely proportional to pressure. For PU and epoxy, respectively, minimum values of 0.11 and 0.26 were measured under maximum pressure. Pressure dependency of shear strength persisted at high strain rate for both PU and epoxy.

2. Materials and Methods

2.1. Method of Sample Preparation

The samples were produced by Fortus 400mc (Stratasys®, Eden Prairie, MN, USA) 3D printer, with default melting temperature of 380 °C. ULTEM9085 (Stratasys®, Eden Prairie, MN, USA) was chosen as the test material, which is a highly effective thermoplastic material suitable for digital production and rapid prototyping. It is ideal for the transport industry due to its high strength-to-weight ratio and flame, smoke, and toxicity (FST) rating [35].

The sample preparation experiment was conducted using two types of thermoplastic layer extrusions, with five deposition strategies in terms of individual fibers' orientations. Three identical samples were made for each fiber type and orientation, for a total of 30 samples. The samples were block shaped, each with a size of 70 × 20 × 6 mm (a × b × c) (Figure.1) and the contact surface in contact with the rotating disc was a 70 × 20 mm wall of the sample that was compatible with the tribometer sample holder. The extruded samples were all 6 mm thick. The samples marked with an X were printed in the horizontal direction, while samples marked with a Z were printed in the vertical direction (Figure 1).

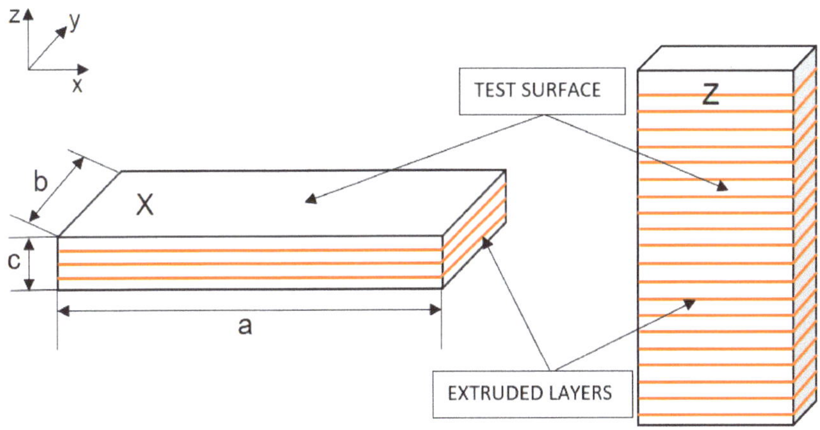

Figure 1. Dimensions and designation of sample types X and Z according to building orientation (xy–reference plane parallel to building platform).

After the sample preparation experiment, samples were cleaned with an ultrasonic cleaner for 30 min in a pure ethanol bath and their surfaces were observed under a Leica S9i microscope. Preprocessing was done using the INSIGHT software under default settings for the material used.

2.2. Measurements

The measurements were performed at the Technical University in Košice in the Tribological laboratory under constant ambient conditions.

The degree of wear of surface layers was monitored during the testing, depending on the yield path and load. At the same time, the different effects of abrasive particles on the wear of the samples were investigated. Depending on the requirements, it was possible to change the size of the load and the speed of the disc rotation (so-called yield speed). The magnitude of the load was based on the load unit [36]. The device could test samples with different physical and mechanical properties. The advantage of this technical solution was that it had the ability to record the friction force, normal force, and temperature of the sample during the testing process (Figure 2). The parameter settings during the testing process are shown in Table 2.

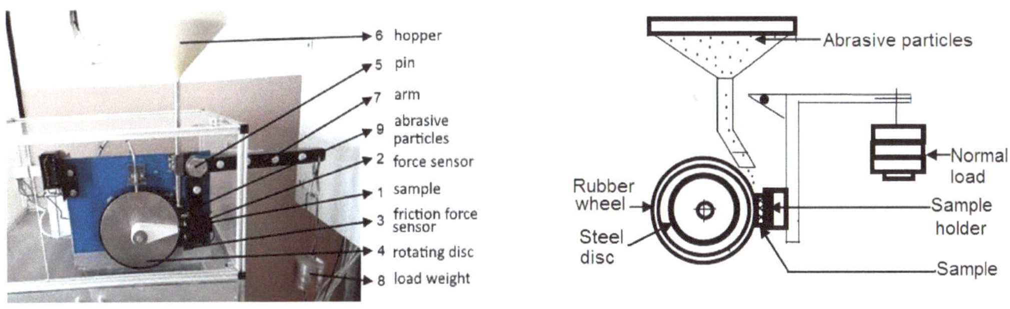

Figure 2. Real tribological laboratory equipment (**left**) and schematic representation (**right**).

Table 2. ASTM G65-16 parameter settings during the testing process.

Disc diameter	(mm)	229
Rotations per minute	(RPM)	278
Sample size	(mm)	70 × 20 × 6
Load	(N)	30
Path	(m)	200

The tribometer ensured the creation of conditions for abrasive wear by the particles (Figure 2, 9) poured in from the hopper (Figure 2, 6). Prior to the testing process, it was possible to adjust the magnitude of the normal force by fixing and positioning the weight on the lever; the normal force was measured by means of a force sensor (Figure 2, 2) located in the arm (Figure 2, 7). The arm with the test sample holder was mounted with a pin enabling its rotation (Figure 2, 5). To measure the friction force between the sample (Figure 2, 1) and the rotating disc (Figure 2, 4), a friction force sensor (Figure 2, 3) was placed in the load (Figure 2, 8). Yield speed (rotation) was controlled to produce the required value of the time course of the motion or the number of cycles. The electric motor used in the experiment ensured stable operation and friction conditions.

The wheel consisted of a steel disk with an outer layer of neoprene rubber molded to its periphery The rubber was made of Elastomer-Neoprene GW. The durometer hardness of the rubber was A-60.

2.3. Experimental Parameters

The samples were weighed at 0.0001 g before and after testing to detect volume loss. Fifteen ASTM G65-16 tests with Fe_3Al_2 $(SiO_4)_3$ garnet abrasive for the orientation in the X direction (X-orientation) and 15 tests for the orientation in the Z direction (Z-orientation) were conducted to determine the weight loss arithmetic mean and the standard deviation.

The three-point abrasive process involved free particles that can move freely (the abrasive). The testing procedure used garnet abrasive (AFS 50–70) with a particle size of 200 micron, as specified in the norm ASTM G65-16. The composition of the abrasive used is given in Table 3. Figure 3 shows a snapshot of the structure of the abrasive used.

Table 3. Chemical composition of the garnet abrasive [36].

Garnet abrasive	**SiO_2** (%)	**FeO** (%)	**Fe_2O_3** (%)	**Al_2O_3** (%)	**CaO** (%)	**MgO** (%)	**MnO** (%)
$Fe_3Al_2(SiO_4)_3$	41.34	9.72	12.55	20.36	2.97	12.35	0.85

Figure 3. Garnet abrasive.

The samples used in the experiment were printed using ten different production strategies in which the orientation of the production changed (Figure 1) and, therefore, the angle of application of the fibers in the layers (Figure 4). Three samples were produced from each strategy, and the weight loss was determined as the average of the weight loss of individual samples of the same production strategy. The same rubber disc was used in all tests to eliminate any variations in rubber chemical composition or hardness.

2.4. Experimental Samples

The toolpath generated for sample production was done under default parameters (0.254 mm layer height and 0.5 mm fiber width) in the Stratasys INSIGHT 9.0™ software (Stratasys®, Eden Prairie, MN, USA). The fiber deposition strategies in the X- and Z-orientations are shown in Figure 4 and Table 3, respectively.

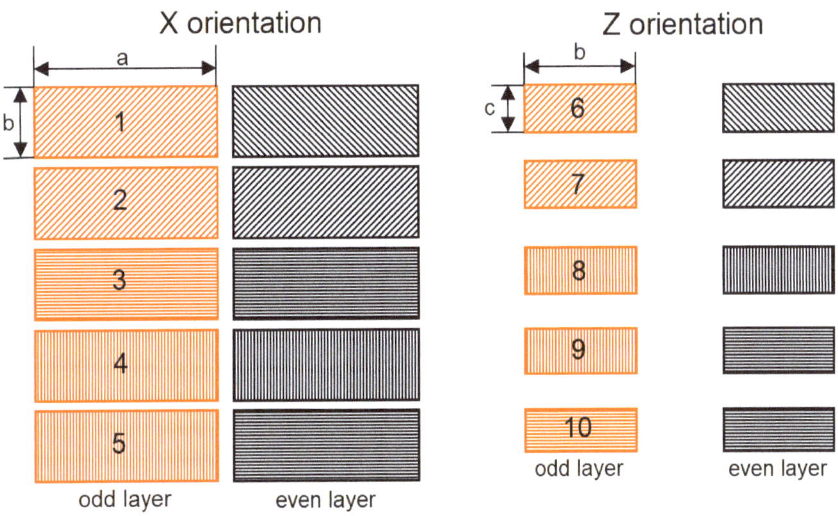

Figure 4. Samples 1–5 diagram and Samples 6–10 diagrams (building orientation). Dimension a, b and c according to Figure 1.

The first column of the two strategies (marked in red) corresponds to the odd layer deposited by the FFF technique, and the printing process also finished with this layer. Thus, the column on the right shows the evenly printed layers at the beginning of the printing process. As we can see, Samples 2, 3, 4, 7, 8, and 10 have the same fiber orientation in all deposited layers, whereas the fiber orientation of the layers in Samples 1, 5, 6, and 9 is alternating (Table 4). The samples used in the experiment are shown in Figure 5.

Table 4. Description of the strategies used with respect to the angle of rotation toward the tested sample wall.

Sample Type	Strategy	Sample Type	Strategy
1	45° R, 45° L	**6**	45° R, 45° L
2	45°	**7**	45°
3	0°	**8**	90°
4	90°	**9**	90°, 0°
5	90°,0°	**10**	0°

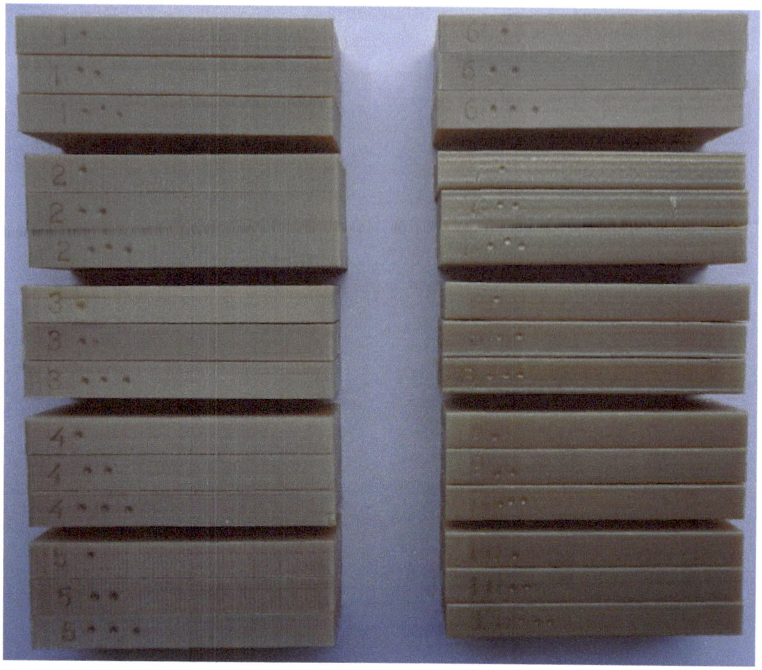

Figure 5. The series of samples used in the experiment.

Subsequently, the prepared samples were used in the abrasive test, where their resistance to abrasive wear was evaluated based on the weight loss of the material and the coefficient of friction and wear rate.

Table 5 shows the values of the initial weight of the sample before the abrasive test, the values of the weight loss of the material, and the percentage of this loss. Weight loss ranged from 0.967% (Samples 3) to 0.451% (Sample 5) (Table 5). This number shows that more material was consumed for the Z-orientation samples and that there was also a greater loss of material because of the abrasive wear.

Table 5. Summary of samples' initial weights and weight losses.

Building Orientation	Sample Type	Initial Weight (g)	Weight Loss (g)	Weight Loss (%)
X	1	13.9903	0.0729	0.520
	2	14.0028	0.0847	0.605
	3	13.8603	0.134	0.967
	4	13.8683	0.0641	0.462
	5	13.9822	0.0631	0.451
Z	6	13.9439	0.0933	0.669
	7	14.5401	0.1143	0.786
	8	14.5558	0.1177	0.809
	9	14.4343	0.1074	0.744
	10	14.5100	0.0988	0.681

3. Evaluation of the Measured Values

Experiments were carried out to determine the influence of the samples building orientation and deposition strategy on the coefficient of friction course and wear behavior. Three samples were made with each type of fiber orientation. A total of 30 samples were tested with a tribometer according to ASTM G65-16 (dry sand).

3.1. Comparisons of Samples 1–5

From the measured values of the resistance of the material to abrasive wear, dependences of the weight of the material on the type of fiber deposition strategy in the FFF process were recorded. Table 6 shows the weight losses of the individual samples and the average weight losses under each strategy.

Table 6. Weight loss results for Samples 1–5 (g).

Sample Type	1	2	3	Average Loss (g)
1	0.0632	0.0797	0.0759	0.0729
2	0.0845	0.0835	0.0860	0.0847
3	0.1308	0.1398	0.1314	0.1340
4	0.0656	0.0543	0.0724	0.0641
5	0.0705	0.0599	0.0558	0.0631

The plotted dependence of the samples with fiber deposition in the X-orientation (Figure 6) following the ASTM G65-16 tests shows the weight losses of the materials tested. Among these, Sample 3 showed the lowest resistance to abrasive wear and, thus, the greatest material loss (material loss was below 0.14 g), where the fiber deposition strategy was identical to the direction of wear (0°). The second largest weight loss was shown in Sample 2 (material loss was below 0.09 g), where the fiber deposition strategy was the same in both layers under a 45° angle and Sample 1 (material loss was below 0.08 g), with the fiber deposition strategy under a 45° angle (left) in the even layer and under a 45° angle (right) tilted in the opposite direction in the odd layer. Sample 5 showed the greatest resistance to abrasive wear according to weight loss, with the orientation alternating between 0° and 90° as well as Sample 4 with a 90° strategy in both layers with respect to direction of wear. The material loss in Samples 4 and 5 was below 0.06 g.

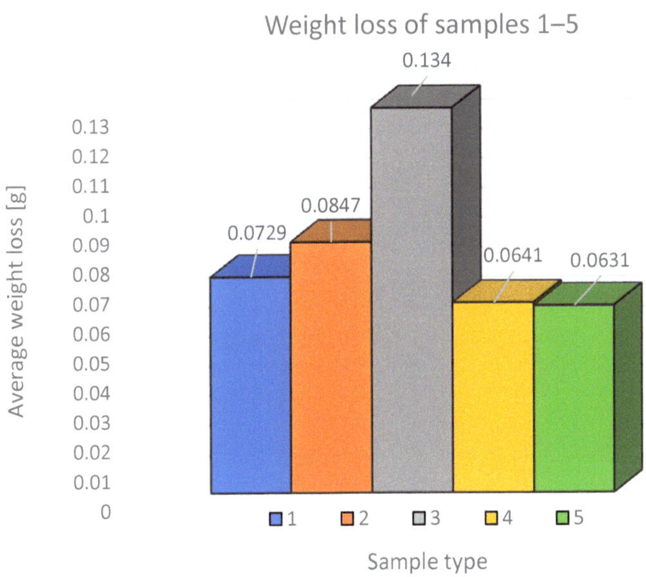

Figure 6. Weight loss in individual Samples 1–5 (X-orientation).

To determine the effect of the layer deposition strategy on the tribological properties of the surface, we subtracted weight losses between the samples similar in fiber orientation and different only by individual layers. This loss is expressed numerically in Table 7 and is represented graphically in Figure 7. Comparing the layers, it can be confirmed that the layers parallel to the direction of wear show a greater surface resistance to abrasion.

Table 7. Weight differences among similar Samples 1–5 (g).

Compared Samples	Same Fiber Orientation	Different Fiber Orientation	Compared Samples
2 vs. 1	0.0817	0.0729	0.0117
3 vs. 5	0.1340	0.0631	0.0709
4 vs. 5	0.0641	0.0631	0.001

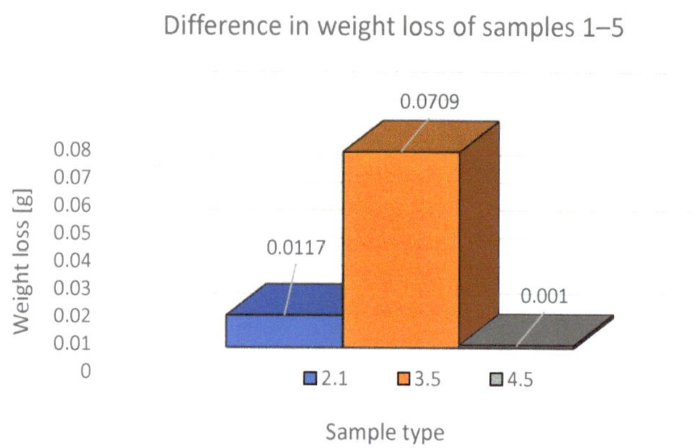

Figure 7. The differences in weight losses for Samples 1–6, represented graphically.

The graphs in Figures 6 and 7 for the X-orientation clearly show that fiber orientation has a significant effect on the abrasive wear of Samples 1–5. If the fibers are laid in the same direction in all layers, this results in a greater loss of material, as shown in the graph in Figure 6. Sample 3, with the fiber deposition strategy under 0° and, thus, parallel to direction of wear in both the even and the odd layer, with a material loss of 0.134 g, showed the most significant wear. Due to a change in the fiber deposition strategy, a smaller weight loss was recorded in Sample 2 under a 45° fiber deposition strategy with the same tilt in the even and the odd layer with a material loss of 0.0847 g, and in Sample 1 under a 45° angle of fiber deposition strategy with opposite slope in the even and the odd layers, with a material loss of 0.0729 g. Sample 5 (with the strategy of fiber deposition parallel to the direction of wear in even layers and perpendicular to the direction of wear in odd layers) with a material loss of 0.0631 g and Sample 4 (with the strategy of fiber deposition perpendicular to direction of the fibers) with a material loss of 0.0641 g showed the highest resistance of the material to abrasive wear in terms of orientation of fibers in even and odd layers.

The difference in weight between the largest and smallest weight loss, i.e., between Samples 3 and 5, is 0.0709 g.

3.2. Comparisons of Samples 6–10

Table 8 shows the weight loss of the individual samples and the average weight loss under each strategy.

Table 8. Weight losses for Samples 6–10 (g).

Sample Type	1	2	3	Average Loss (g)
6	0.0981	0.0906	0.0911	0.0933
7	0.1102	0.1200	0.1127	0.1143
8	0.1003	0.1258	0.1269	0.1177
9	0.1087	0.1114	0.1022	0.1074
10	0.0931	0.0998	0.1045	0.0988

The plotted dependence of the samples with Z-orientation fiber deposition (Figure 8) after the ASTM G65-16 tests shows the weight loss of the materials tested. Among these, Sample 8 (material loss was more than 0.11 g) showed the lowest resistance to abrasive wear and, thus, the largest weight loss of the material, where the fiber deposition strategy was in the horizontal plane (90°) in both layers with respect to direction of wear. The second largest weight loss was shown in Sample 7 (material loss was just over 0.11 g) with the fiber deposition strategy under 45° in both the even and the odd layers in the same direction. Sample 9 followed (material loss exceeded 0.10 g), where the fiber deposition strategy in the even layer was under the 90° angle in the direction of wear and, in the odd layer, the orientation was perpendicular to direction of wear. This was followed by Sample 10 with a strategy of fiber deposition parallel to the direction of wear in both layers with a weight loss just below 0.10 g. Sample 6 showed the greatest resistance to abrasive wear due to weight loss, with alternating strategies under a 45° angle in the layers. The loss of material in Sample 6 was just over 0.09 g.

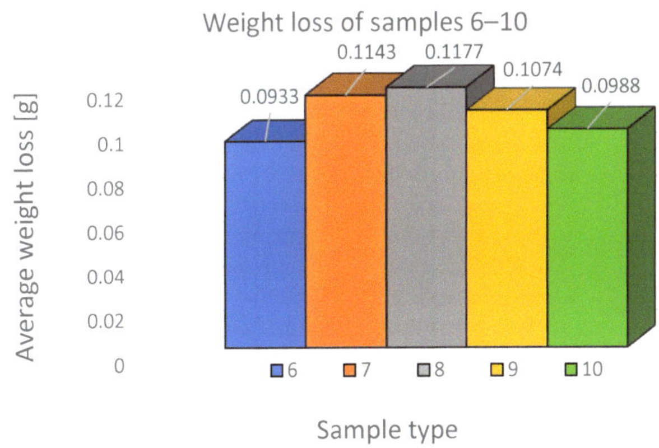

Figure 8. Weight differences among similar Samples 6–10 (g).

To determine the effect of the layer deposition strategy on the tribological properties of the surface, we subtracted weight losses between the samples similar in fiber orientation and different only by individual layers. This loss is expressed numerically in Table 9 and represented graphically in Figure 9.

Table 9. Weight differences in samples 6–11 (g).

Compared Samples	Same Fiber Orientation	Different Fiber Orientation	Compared Samples
7 vs. 6	0.1143	0.0933	0.0210
8 vs. 9	0.1177	0.1074	0.0102
10 vs. 9	0.0988	0.1074	0.0086

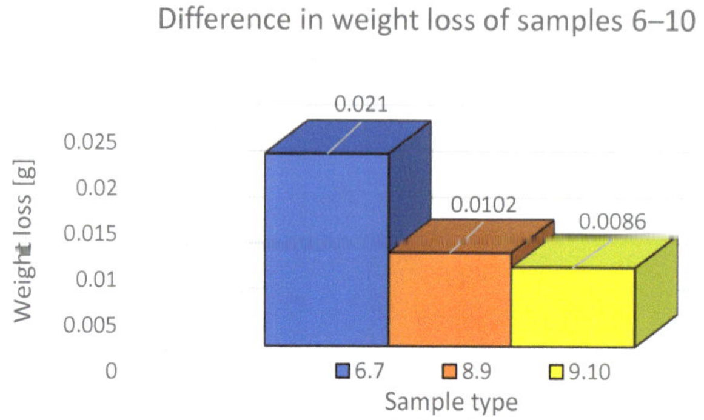

Figure 9. Difference in weight loss in Samples 6–10, represented graphically.

From the comparison of the layers, we can see higher abrasion resistance of the surface layers perpendicular to the direction of wear. However, the greatest abrasion resistance, in this case, is shown in Sample 6, with the fiber orientation under a 45° angle, but with the opposite tilt in the individual layers.

In samples extruded vertically (with Z-orientation of fiber extrusion, Samples 6–10), the values of the variance of weight loss are not as large as in the samples with the X-orientation (Figure 9).

In Samples 6–10, the layering strategy is not significant in nature. When the fibers are laid in the same direction in all layers, this results in a greater loss of material, as shown graphically in Figure 9. Sample 8 (with the fiber deposition strategy under a 90° angle with respect to the direction of wear in both the even and the odd layers) which had a material loss of 0.1177 g, and Sample 7 (with fibers deposition strategy under the same 45° angle in both layers) which had a material loss of 0.1143 g, showed the greatest wear. Due to a change in fiber deposition orientation, smaller weight losses were recorded for Sample 9 (with the fiber deposition strategy under the 90° angle with respect to the direction of wear in the even layer and perpendicular to the wear in the odd layer), with a material loss of 0.1074 g, Sample 10 (with fibers deposition strategy perpendicular to direction of wear in the even and the odd layers), with a material loss of 0.0988 g, and Sample 6 (with the fibers oriented under a 45° angle with respect to the direction of wear in even layers and the opposite strategy under a 45° angle in odd layers) with a material loss of 0.0933 g, and therefore, these samples showed the greatest resistance of the material to abrasive wear in terms of the fiber deposition strategy.

3.3. Monitoring of Friction Force

Another monitored parameter was friction force, which was directly measured and read from the tribometer. The course of friction force depends on many parameters such as the material's mechanical properties, contact pressure value, the coefficient of friction, contact surface roughness, mutual dissolution of materials, contact time, tribosystem elasticity, and the presence of foreign bodies in the contact zone. The results of the friction forces as a function of the path traveled for the X and Z-orientations are shown in Figures 10 and 11.

The graph in Figure 10 shows that the friction force increases from the initial value at the beginning of the experiment and stabilizes over a 50–60 m long path, where it then oscillates around a certain value until the end of the experiment. The friction force was lower at the beginning of the testing procedure than at the end of the testing procedure. The lowest friction force at the beginning of the experiment can be observed in Sample 5, which remains the lowest at the end, together with the friction force in Sample 4.

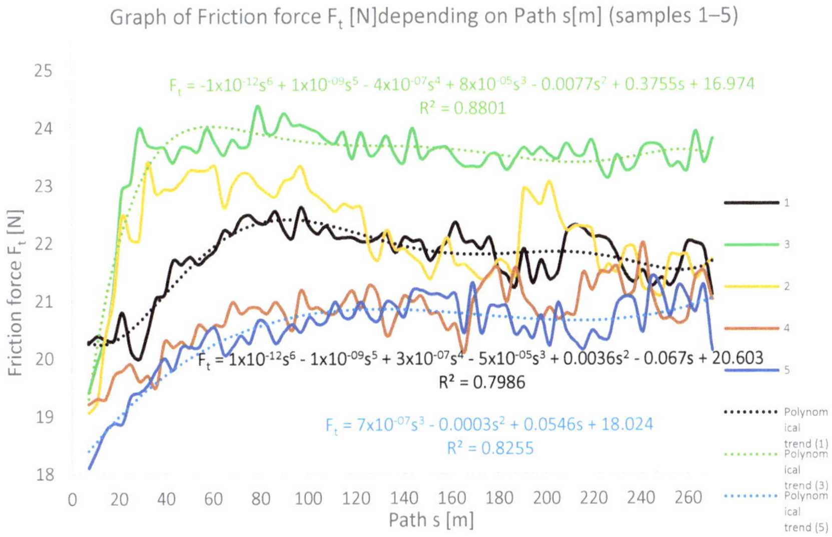

Figure 10. Plotted coefficient of friction depending on the path of Samples 1–5.

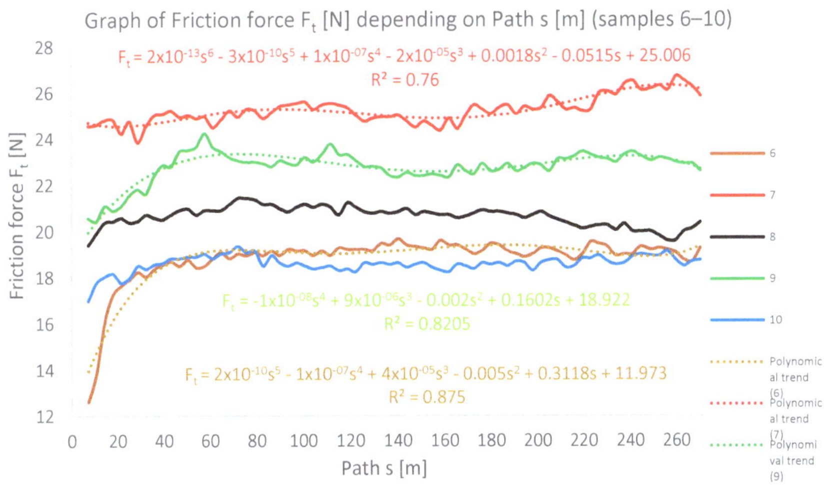

Figure 11. Plotted coefficient of friction depending on the path of Samples 6–10.

The highest friction force in samples with X-orientation (Samples 1–5) was recorded in Sample 3, where the fiber deposition orientation was parallel to the direction of wear (under 0° angle with respect to tested sample wall) and Sample 2, where the fiber orientation was under 45° angle and in both layers. Samples with a lower friction force followed, such as Sample 1, with a 45° orientation but the opposite tilt in the even and the odd layer.

The graph in Figure 11 shows a similar course of friction as in the X-oriented samples, i.e., that the friction force increases from the initial value at the beginning of the experiment and stabilizes along a 40–50 m long path, where it then oscillates around a certain value until the end of the experiment. The lowest friction force at the beginning of the experiment can be observed in Sample 10, which remains the lowest at the end, together with the friction force in Sample 6. The X-oriented samples show greater variance in values, i.e., more pronounced jumping as compared with the Z-oriented samples. In addition, we also found lower coefficients of friction as compared with the Z-oriented samples.

In terms of achieving the optimal parameters for the tribological properties of surfaces of models produced by the FFF technique from ULTEM 9085 material, it is advisable to use the layer deposition strategy, such as in the case of the X-oriented Samples 4 and 5 and the Z-oriented Samples 6 and 10. Using these strategies, a lower friction force (or its course) was found, as well as less weight loss.

3.4. Microscopic Observation of Material Failure

Table 10 shows microscopic images of the sample surfaces after the testing procedure with a transition area between the worn and non-worn surface. In these images, we can see how the surface wore off at different orientations of the fibers laid. The images confirm that the worn surface shows no signs of adhesive or cohesive failure and, thus, the coherence between the individual layers has been maintained in both types of samples produced. With the X- and Z-orientations of the fibers deposition, exclusively abrasive wear (abrasion) of the material was noted.

Table 10. Microscopic images of the sample surfaces of Samples 1–10.

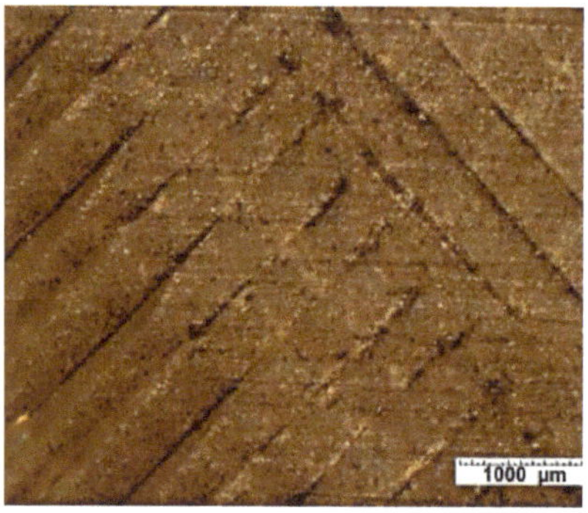	
1	2
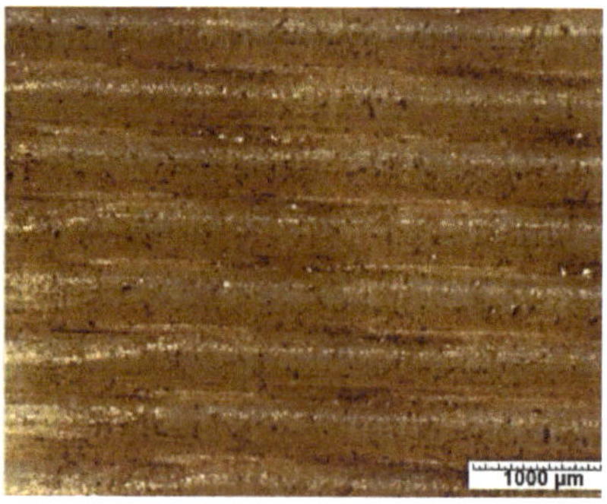	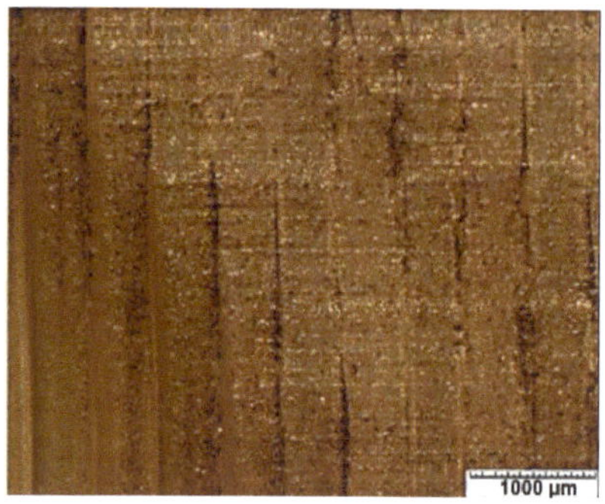
3	4

Table 10. *Cont.*

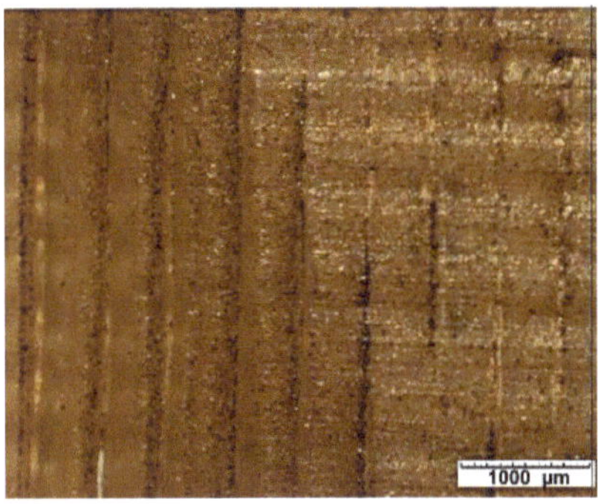	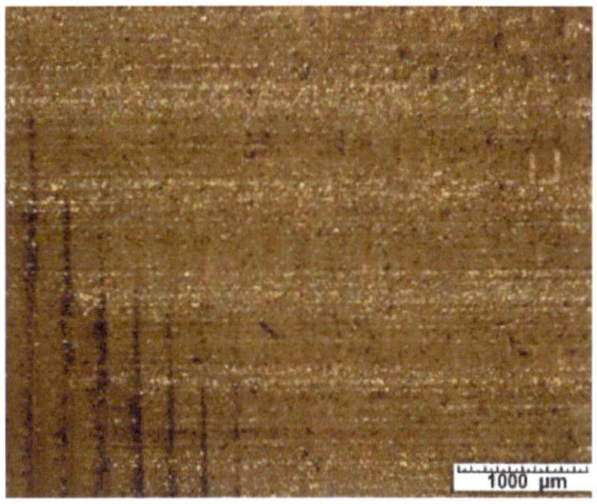
5	6
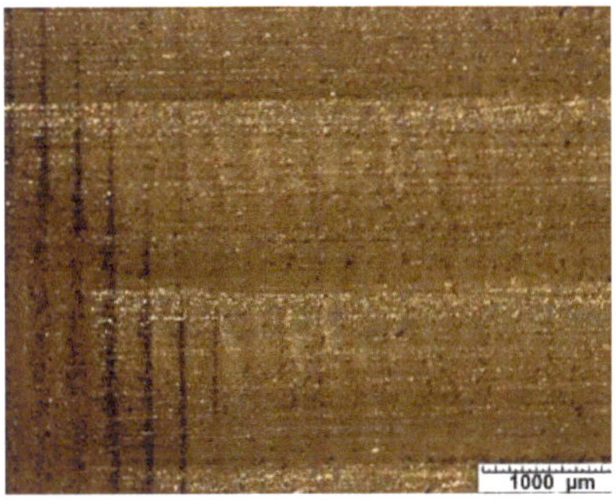	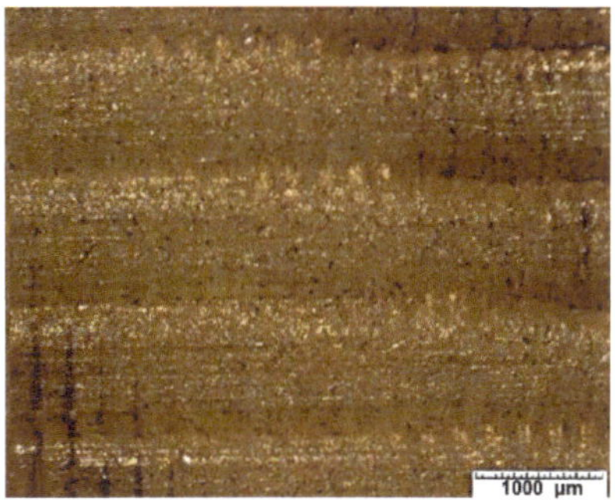
7	8
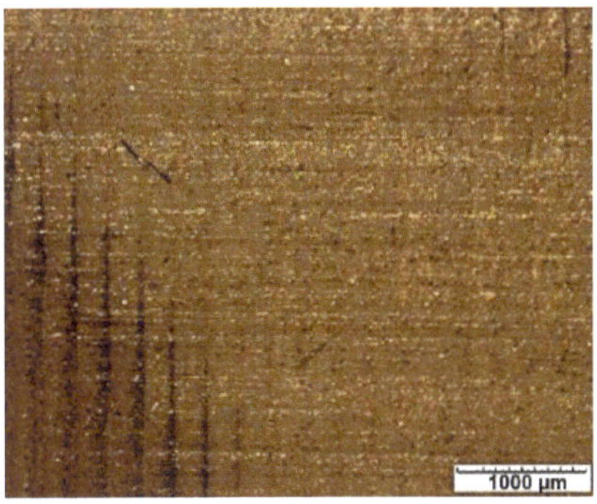	
9	10

4. Evaluation of the Measured Values

The experimental results evaluated weight loss and the course of the coefficient of friction depending on the path traveled.

When evaluating weight loss and its dependence on the chosen fiber-laying strategy, it can be stated that the largest weight loss in the X-oriented samples occurred in Sample 3, where the fibers were oriented parallel (0°) to the tested sample wall in both layers. In Samples 1 and 2, where the fibers formed a 45° angle, the weight loss was 36.79–45.6% lower than in Sample 3. In Sample 4, where the deposition strategy of all layers was the same (90°), the weight loss decreased by 52.16% as compared with Sample 3. Sample 5

(90° and 0°), where the layers with perpendicular fibers and fibers parallel to the tested sample wall alternated, showed the lowest weight loss as compared with Sample 3, by up to 52.91%. It can be concluded from the results of measuring the weight losses that by using the same strategy in both layers (Sample 2, 45°) or a strategy parallel to the direction of wear in both layers (Sample 3) greater weight loss occurred. With the same strategy in both layers (Sample 4, 0°) or with the combined strategy the weight loss in the layers decreases (Sample 5, 0° and 90°) (Figure 12).

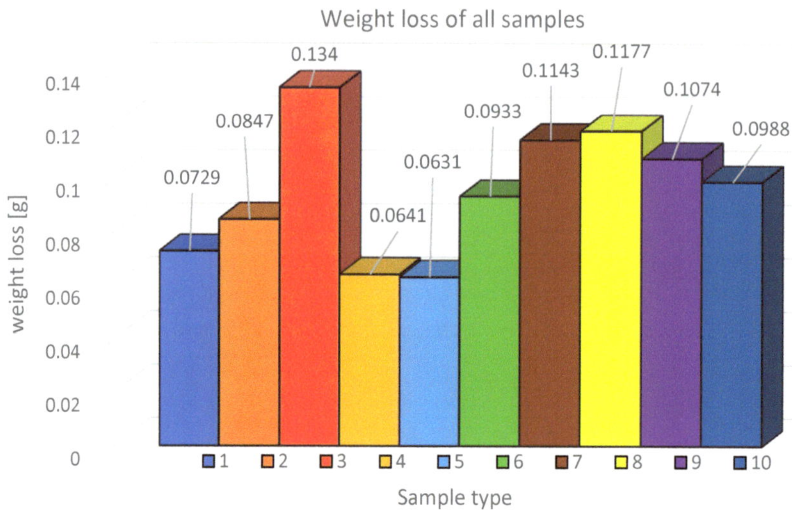

Figure 12. Summary of weight loss in all samples (Samples 1–5 X-orientation and Samples 6–10 Z-orientation).

When evaluating weight loss and its dependence on the chosen fiber-laying strategy, it can be stated that the largest weight loss occurred in the Z-oriented samples of the deposited fibers in Sample 8, where the fibers were oriented perpendicular to the tested sample wall. In Samples 6 and 7, where the fibers formed a 45° angle, the reduction was 2.89–20.73% lower as compared with Sample 8. In Sample 9, where the layers with perpendicular fibers and fibers parallel to the tested wall alternate, the weight loss was 8.75% smaller as compared with Sample 8. In Sample 10, the weight loss decreased by up to 16.06% as compared with Sample 8. The results of the measured weight losses lead to the conclusion that when using the strategy of 45° in the even and the opposite 45° in the odd layer direction, the smallest weight loss occurs. Using the same strategy in both the even and odd layers, the wear increases (Samples 7 and 8) or when using a combined strategy, the weight loss in the layers decreases in the opposite way (Samples 6 and 9) (Figure 12).

Table 11 shows the weight losses in individual samples with an overall indication of the orientation for each type of loss (X- or Z-orientation). This table shows that, although Sample 3 shows the highest wear from among the whole group of samples, this type of fiber deposition orientation is more suitable in terms of abrasive wear because smaller overall weight losses are recorded. The X-orientation shows 21.2% less weight loss than the Z-orientation.

Table 11. Comparison of weight loss in the X- and Z-orientations.

Numbers of Compared Samples of Type X and Z	X (g)	Z (g)	Compared Quantity
1 vs. 6	0.0729	0.0933	
2 vs. 7	0.0847	0.1143	
3 vs. 10	0.1340	0.0988	Weight loss
4 vs. 8	0.0641	0.1177	
5 vs. 9	0.0631	0.1074	
Total loss in all samples	**0.4188**	**0.5315**	Overall difference

5. Conclusions

Nowadays, 3D-printed products are increasingly being used in various areas of industry. One of these technologies is the production technique that uses fused filament fabrication (FFF). With increasing applications and a demand for this technology, there is a need to expand our knowledge about the mechanical properties of materials and about their surface resistance or resistance to various types of wear. Currently, the available studies have addressed the most widespread materials used in experimental investigations of the friction and wear behavior of acrylonitrile butadiene styrene (ABS) specimens. The research on specimens made using the FFF technique has been conducted by considering the normal load, yield speed, and part orientation as parameters in the face-centered central composite design (FCCCD) setup of experiments on a pin-on-disc machine [34] or by studying the friction and wear behavior of parts made of newly developed Nylon6-Fe as compared with the existing acrylonitrile butadiene styrene (ABS) filament of the FFF machine [35].

The PEI material allows the printed products to be used in more demanding applications and this is the reason why we decided to perform experimental measurements of the resistance of this material to abrasive wear.

This study is focused on analysis of building orientation and deposition strategy the influence of the FFF additive manufacturing technique on the course of the coefficient of friction and wear of samples (tribological properties of surface and subsurface layers) of ULTEM9085™ thermoplastic, as demonstrated by the material weight loss, in compliance with the ASTM G65-16 standard.

A total of 10 samples prepared under two different model production orientations were used in the experiment with respect to the orientation of the layers during the testing procedure; five samples were prepared under each orientation, and according to different fiber-laying strategies. The weight loss of the material as a function of the path traveled and the coefficient of friction and wear behavior were observed as the main factors of the experiment. The experimental results show that fiber orientation and the fiber-laying strategy during a model's printing affect its tribological properties.

The results show that the influences of different orientations cause differences in weight loss and fluctuations in the coefficient of friction. With respect to the total material losses, the X-orientation appears to be more suitable, where the total material loss is 0.4188 g; with the Z-orientation, the material loss is 0.5315 g, which represents an increase in weight loss of 26.91%.

The findings from the comparison of individual deposition strategies show that the strategy of Sample 5 (X-orientation) is the most resistant, where it is a question of alternating the deposition strategy of fibers between 0° in the even and 90° in the odd layer with respect to the tested sample wall. Sample 3 (X-orientation) shows the greatest weight loss, where the fiber deposition strategy was parallel to wear direction of 0° in both the even and the odd layers. The study's findings suggest that printing in a horizontal orientation (X) and a 0° and 90° strategy (Sample 5) of layers deposition helps to reduce material wear. In conclusion, it can be stated that if we require higher abrasion resistance from a model made of the ULTEM 9085 material, it is suitable to choose the X-orientation and the strategy of Sample 5.

Author Contributions: Conceptualization, G.M. and I.G.; methodology, I.G.; software, E.S. and T.J.; validation, E.S.; formal analysis, J.M.; investigation, G.M.; resources, T.J.; data curation, G.M.; writing—original draft preparation, J.M. and T.J.; writing—review and editing, G.M. and I.G.; visualization, I.G.; supervision, E.S. and G.M. All authors have read and agreed to the published version of the manuscript.

Funding: This research was funded by VEGA/1/0384/20 and KEGA: 036TUKE-4/2021.

Institutional Review Board Statement: Not applicable.

Informed Consent Statement: Not applicable.

Data Availability Statement: Not applicable.

Acknowledgments: This work was supported by the Slovak Research and Development Agency under the contract no. VEGA/1/0384/20. The autors would like to thank the KEGA grant agency for supporting research work and co-financing the project KEGA: 036TUKE-4/2021.

Conflicts of Interest: The authors declare no conflict of interest.

References

1. Galeta, T.; Raos, P.; Stojši_c, J.; Pakši, I. Influence of structure on mechanical properties of 3Dprinted objects. *Procedia Eng.* **2016**, *149*, 100–104. [CrossRef]
2. Wang, J.; Xie, H.; Weng, Z.; Senthil, T.; Wu, L. A novel approach to improve mechanical properties of parts fabricated by fused deposition modeling. *Mater. Des.* **2016**, *105*, 152–159. [CrossRef]
3. Bellini, A.; Güçeri, S. Mechanical characterization of parts fabricated using fused deposition modeling. *Rapid Prototyp. J.* **2003**, *9*, 252–264. [CrossRef]
4. Cifuentes, S.C.; Frutos, E.; Benavente, R.; Lorenzo, V.; Gonzalez-Carrasco, J.L. Assessment of mechanical behav-ior of PLA composites reinforced with Mg microparticles through depth-sensing indentations analysis. *J. Mech. Behav. Biomed. Mater.* **2017**, *65*, 781–790. [CrossRef] [PubMed]
5. Farah, S.; Anderson, D.G.; Langer, R. Physical and mechanical properties of PLA, and their functions in wide-spread applications–a comprehensive review. *Adv. Drug Deliv. Rev.* **2016**, *107*, 367–392. [CrossRef] [PubMed]
6. Murariu, M.; Dubois, P. PLA composites: From production to properties. *Adv. Drug Deliv. Rev.* **2016**, *107*, 17–46. [CrossRef] [PubMed]
7. Revati, R.; Majid, M.S.A.; Ridzuan, M.J.M.; Basaruddin, K.S.; Rahman Y., M.N.; Cheng, E.; Gibson, A. In vitro degradation of a 3D porous Pennisetum purpureum/PLA biocomposite scaffold. *J. Mech. Behav. Biomed. Mater.* **2017**, *74*, 383–391. [CrossRef]
8. RSong, Y.; Li, Y.; Song, W.; Yee, K.; Lee, K.-Y.; Tagarielli, V. Measurements of the mechanical response of unidirectional 3D-printed PLA. *Mater. Des.* **2017**, *123*, 154–164. [CrossRef]
9. Wittbrodt, B.; Pearce, J.M. The effects of PLA color on material properties of 3-D printed components. *Addit. Manuf.* **2015**, *8*, 110–116. [CrossRef]
10. Es-Said, O.S.; Foyos, J.; Noorani, R.; Mendelson, M.; Marloth, R.; Pregger, B.A. Effect of Layer Orientation on Mechanical Properties of Rapid Prototyped Samples. *Mater. Manuf. Process.* **2000**, *15*, 107–122. [CrossRef]
11. Pham, D.T.; Gault, R.S. A comparison of rapid prototyping technologies. *Int. J. Mach. Tools Manuf.* **1998**, *38*, 1257–1287. [CrossRef]
12. Dawoud, M.; Taha, I.; Ebeid, S.J. Effect of processing parameters and graphite content on the tribological behaviour of 3D printed acrylonitrile butadiene styrene. *Mater. Werkst.* **2015**, *46*, 1185–1195. [CrossRef]
13. Panneerselvam, T.; Kandavel, T.K.; Raghuraman, S. Experimental investigations on tribological behaviour of alumina added Acrylonitrile butadiene styrene (ABS) composites. *Tribol. Ind.* **2016**, *8*, 338–346.
14. Gurrala, P.K.; Regalla, S.P. Friction and wear behavior of abs polymer parts made by fused deposition modeling (FDM). *Technol. Lett.* **2014**, *1*, 13–17.
15. Garg, H.K.; Singh, R. Comparison of wear behavior of ABS and Nylon6—Fe powder composite parts prepared with fused deposition modelling. *J. Central South Univ.* **2015**, *22*, 3705–3711. [CrossRef]
16. Hanon, M.; Kovács, M.; Zsidai, L. Tribology behaviour investigation of 3D printed polymers. *Int. Rev. Appl. Sci. Eng.* **2019**, *10*, 173–181. [CrossRef]
17. Norani, M.N.M.; Abdollah, M.F.B.; Abdullah, M.I.H.C.; Amiruddin, H.; Ramli, F.R.; Tamaldin, N. Correlation of tribo-mechanical properties of internal geometry structures of fused filament fabrication 3D-printed acrylonitrile butadiene styrene. *Ind. Lubr. Tribol.* **2020**, *72*, 1259–1265. [CrossRef]
18. Norani, M.N.M.; Bin Abdollah, M.F.; Abdullah, M.I.H.C.; Amiruddin, H.; Ramli, F.R.; Tamaldin, N. 3D printing parameters of acrylonitrile butadiene styrene polymer for friction and wear analysis using response surface methodology. *Proc. Inst. Mech. Eng. Part J J. Eng. Tribol.* **2021**, *235*, 468–477. [CrossRef]
19. Amiruddin, H.; Bin Abdollah, M.F.; Norashid, N.A. Comparative study of the tribological behaviour of 3D-printed and moulded ABS under lubricated condition. *Mater. Res. Express* **2019**, *6*, 085328. [CrossRef]

20. Hesselbach, J.; Raatz, A.; Wrege, J.; Herrmann, H.; Weule, H.; Buchholz, C.; Tritschler, H.; Knoll, M.; Elsner, J.; Klocke, F.; et al. mikroPro-Untersuchungen zum internationalen Stand der Mikroproduktionstechnik. *wt Werkstattstech. Online Bd.* **2003**, *93*, 119–128.
21. Maeda, H.; Okubo, H.; Sasaki, S. Frictional property of a 3D-capillary-structured surface fabricated by selective laser melting. *J. Tribol.* **2019**, *20*, 87–96.
22. Tahir, N.A.M.; Azmi, M.S.; Abdollah, M.F.B.; Ramli, F.R.; Amiruddin, H.; Tokoroyama, T.; Umehara, N. Tribo-logical properties of 3D-printed pin with internal structure formation under dry sliding conditions. *Proc. Mech.-Cal Eng. Res. Day* **2018**, *2018*, 260–261.
23. Murashima, M.; Kawaguchi, M.; Tanaka, M.; Umehara, N. The effect of inner structures created by 3D-additive manufacturing on wear reduction of 3D printed material. In Proceedings of the SAKURA Symposium on Mechanical Science and Engineering, Nagoya, Japan, 12 September 2017; pp. 39–41.
24. Abdollah, M.F.B.; Norani, M.N.M.; Abdullah, M.I.H.C.; Amiruddin, H.; Ramli, F.R.; Tamaldin, N. Synergistic effect of loads and speeds on the dry sliding behaviour of fused filament fabrication 3D-printed acrylonitrile butadiene sty-rene pins with different internal geometries. *Int. J. Adv. Manuf. Technol.* **2020**, *108*, 2525–2539. [CrossRef]
25. Grytsenko, O.; Dulebova, L.; Suberlyak, O.; Skorokhoda, V.; Spišák, E.; Gajdoš, I. Features of Structure and Properties of pHEMA-gr-PVP Block Copolymers, Obtained in the Presence of Fe^{2+}. *Materials* **2020**, *13*, 4580. [CrossRef]
26. Gajdoš, I.; Spišák, E.; Kaščák, L.; Krasinskyi, V. Surface Finish Techniques for FDM Parts. *Mater. Sci. Forum* **2015**, *818*, 45–48. [CrossRef]
27. Pande, A.-D.; Jozef, J.; Marek, P.; Kačalová, I.M. Advantages and effectiveness of the pow-der metallurgy in manufacturing technologies. In *Metalurgija = Metallurgy.–Záhreb (Chorvátsko): Hrvatsko Meta-Lurško Društvo;* Croatia Metallurgical Society: Zagreb, Croatia, 2018; Volume 57, pp. 353–356.
28. Olejarova, S.; Krenicky, T. Water Jet Technology: Experimental Verification of the Input Factors Variation Influence on the Generated Vibration Levels and Frequency Spectra. *Materials* **2021**, *14*, 4281. [CrossRef]
29. Krenicky, T.; Servatka, M.; Gaspar, S.; Mascenik, J. Abrasive Water Jet Cutting of Hardox Steels—Quality Investigation. *Processes* **2020**, *8*, 1652. [CrossRef]
30. Kováč, I.; Žarnovský, J.; Drlička, R.; Ružbarský, J. An improvement of tribological properties of boron alloyed layers. *Manuf. Technol. J. Sci. Reserch Prod.* **2010**, *10*, 78–80. [CrossRef]
31. Konovalenko, I.; Maruschak, P.; Brezinová, J.; Brezina, J. Morphological Characteristics of Dimples of Ductile Fracture of VT23M Titanium Alloy and Identification of Dimples on Fractograms of Different Scale. *Materials* **2019**, *12*, 2051. [CrossRef]
32. Dangnan, F.; Espejo, C.; Liskiewicz, T.; Gester, M.; Neville, A. Friction and wear of additive manufactured polymers in dry contact. *J. Manuf. Process.* **2020**, *59*, 238–247. [CrossRef]
33. Rathaur, A.S.; Patel, V.K.; Katiyar, J.K. Tribo-mechanical properties of graphite/talc modified polymer composite bearing balls. *Mater. Res. Express* **2019**, *7*, 015305. [CrossRef]
34. Marathe, U.N.; Bijwe, J. High performance polymer composites–Influence of processing technique on the fiber length and performance properties. *Wear* **2020**, *446–447*, 203189. [CrossRef]
35. Sable, P.A.; Neel, C.H.; Borg, J.P. Dynamic strength and friction behavior of thermosetting polyurethane and epoxy. Shock compression of condensed matter–2019. In Proceedings of the Conference of the American Physical Society Topical Group on Shock Compression of Condensed Matter, Portland, OR, USA, 16–21 June 2019; Volume 2272, p. 040010. [CrossRef]
36. GAMIN. Available online: https://www.gamin.sk/fileadmin/user_upload/Airblast_Garnet_SK_01.pdf (accessed on 6 April 2022).

applied sciences

MDPI

Article

Options on Tooth Profile Modification by Hob Adjustment

Jozef Mascenik *, Tomas Coranic and Tibor Krenicky

Department of Design and Monitoring of Technical Systems, Faculty of Manufacturing Technologies with a Seat in Presov, Technical University of Kosice, Bayerova 1, 080 01 Presov, Slovakia; tomas.coranic@tuke.sk (T.C.); tibor.krenicky@tuke.sk (T.K.)

* Correspondence: jozef.mascenik@tuke.sk; Tel.: +421-55-602-6337

Featured Application: This article provides an analysis of the influence of the angular setting in a hobbing mill on the modification of the gear profile. This solution provides valuable insights for gear manufacturing professionals and engineers who seek to achieve higher accuracy, efficiency and quality in the tooth manufacturing process.

Abstract: The presented contribution focuses its attention on the research on the influence of the angular setting in a hobbing mill on the adjustment of the tooth profile created by this technology. The aspect of influence from the angular settings in the hobbing mill on the final shape of the gears is investigated. In this research, the effect of substituting the curvilinear profile of the cutting edges with a straight profile is analyzed, observing how this approach affects the final geometry of the gears. An important point of this research is also the comparison of different profiling methods using the hobbing mill, specifically when using a normal cut, which is perpendicular to the helix of the milling cutter, and also when cutting along the axial plane of the hobbing mill.

Keywords: teeth; profile; modification; hobbing mill; gearing accuracy

1. Introduction

Hobbing mills represent a special category of high-performance tools in the field of engineering, with the main focus being on the production of gears and other complex machine parts, such as splined shafts, sprockets and other components with special screw surfaces. Their effectiveness lies in the combination of a number of cutting edges and their continuous, smooth engagement, which significantly minimizes wasted time and maximizes the productivity of the process. These tools are characterized by a high degree of variability in their use, which means that with the same tool it is possible to produce gear segments of different shapes and sizes, with a variable number of teeth and a variable pitch, with or without gear correction, and also in relation to worm wheels. The method of producing teeth using a shank or disc modular milling cutter is more expensive compared to the hobbing mill method, because when using wheels with the same module and a different number of teeth, the profile of the tooth gap changes, which also results in a more complex profile for the modular milling cutter, which complicates the production and grinding. Gears are essential components for the transmission and transformation of mechanical motion and energy in machines and mechanisms. Their curved shapes require specific mathematical tools for the design of production tools. Although they have been around for centuries, they are constantly being improved. Publications dealing with the development of efficient gear manufacturing tools remain relevant. This text deals with the investigation of the basic shape of the tool in terms of kinematic schemes for material separation. Analytical methods for determining the tool shape require knowledge of analytical geometry in space and the differential geometry of curves and surfaces, which are investigated using differential calculus.

Citation: Mascenik, J.; Coranic, T.; Krenicky, T. Options on Tooth Profile Modification by Hob Adjustment. *Appl. Sci.* **2023**, *13*, 10646. https://doi.org/10.3390/app131910646

Academic Editor: Mark J. Jackson

Received: 26 August 2023
Revised: 18 September 2023
Accepted: 19 September 2023
Published: 25 September 2023

Within the framework of this topic, several studies have been published. Bouzakis et al. [1] aimed their research at the experimental verification of the calculation algorithms, taking into account the wear of the milling cutter teeth in individual generating positions.

The replacement of the classical analytical solution for the machining of involute gears with a general mathematical model was dealt with by Jia et al. [2]. In their work, they dealt with the creation of a general mathematical model by generating two parameters using a discrete method to increase the robustness of the method. They designed a numerical algorithm by transforming multiple cutting edges with respect to the workpiece of the gear wheel based on generated motions. Subsequently, they introduced the derivation of instant contact points using the implicit time sequences of the profile points.

In the article by Spitas et al. [3], the possibilities of gear design using the method of discretization of the gear tooth flanks were analyzed, while instead of an analytical approach, they proposed dividing the flank of the tooth into an infinite number of local involute segments. As an advantage of this method, they presented its relative simplicity compared to the analytical solution. The method was implemented and verified when applied to both globoid and involute gears.

Vasilis et al. [4] contributed to the previously mentioned studies on this topic by focusing on creating an algorithm suitable for CAE simulations, as in addition to the produced gearing, they also monitored the shape and volume of the resulting chip. Within the outputs, the authors present the developed algorithm that was embedded into a CAD environment. They proved that the algorithm is suitable for predicting cutting forces, tool stress and wear, as well as optimizing the parameters of the production process.

The production of gears using hobbing mills enables the production of gear segments with a consistent profile, which reduces the difficulty in setting up and contributes to the efficiency of the production process. Gears are an integral part of many mechanical systems and devices. They are key components for the transmission and transformation of mechanical motion and energy in various machines and mechanisms. However, their shape is often complex and curvilinear, which requires a special approach in the design of production tools. The development of these tools involves a deep understanding of mathematical principles, especially in the field of analytical geometry in space and the differential geometry of curves and surfaces [5]. This knowledge is essential for the design of optimal cutting edges that can create complex gear shapes with high precision and repeatability. Gears are among the most demanding engineering components when it comes to the theoretical, design and manufacturing aspects. Despite the fact that their foundations were laid many centuries ago, their development and improvement continue constantly. Publications and studies focused on the design methods for the efficient production of tools for gears remain current and relevant, because the need to optimize the production process is ever present [6]. All in all, hobbing mills are a key tool in modern engineering, enabling the efficient production of gears and similar machine parts. Their high variability, combined with technical and mathematical know-how, makes it possible to contribute to the improvement of production processes and the performance of mechanical systems where gears are an integral part.

Tooth manufacturing in the case of the hob is accompanied by simulation of the meshing of two-toothed wheels in an involute worm (of the hob) with the machined semi-finished product, i.e., with the toothed wheel with straight or beveled teeth. If the hob is to be identical with the involute worm, it must be profiled by the basic rack. In case of the reverse process, the hob can generate the toothed rack with the basic profile. It means that the meshing of the involute hob with the basic rack is feasible, and the hob meets the basic rack along the straight line [7].

The manufacturing of the involute hob is demanding from the perspective of checking and with regards to the option of undercutting. Therefore, approximate profiling is used, i.e., the curve profile of the axial or of a reference cut by means of the involute screw plane is replaced by the straight-line profile, which can be modified if needed. Further, the influence of the hob adjustment on the modification of the tooth profile in a workpiece is focused on.

Given the contemporary need to enhance the precision of functional surfaces in manufacturing, the primary goal of the research presented in this article was to propose an applicable analytical solution for optimizing the production of gear manufacturing tools.

2. Methodology

Firstly, the hob with the screw plane is selected, which is formed by means of hob profiling in the reference plane (Figure 1), i.e., by replacing the curve profile with the straight line in the cut perpendicular to the helix of the hob.

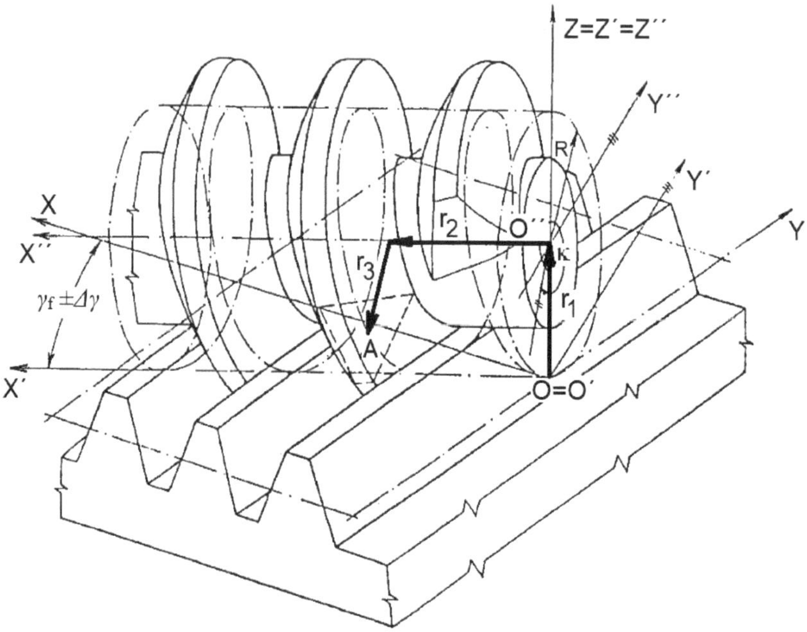

Figure 1. Insertion of the hobbing mill in the basic rack.

The hob is inserted into the basic rack (Figure 1) and turned by an angle of gradient with respect to the hob γ_f. It is evident that the contact with the basic rack did not reach the character of a straight line, which means that the hob profiled in this way does not correspond with the basic rack as to the profile, yet it will accord with any modified toothed rack [8]. The negative shape of the modified rack consequently represents a tool which can approximately replace the profiled hob. Subsequent rolling of such a basic rack (by the tool) results in corresponding teeth profiles, which are identical to the teeth profiles manufactured by means of the approximately profiled hob.

In case the hob is not set in the basic rack precisely, yet correction of its adjustment is allowed (for example, turning by $\pm\Delta\gamma$), the aforementioned procedure helps to detect the influence of the corrections in the modification of the profile of the manufactured workpiece toothing [9].

2.1. Position Vector of the Hob Profile

A description of the screw plane and of the planar areas of the teeth sides in the basic rack is presented as follows: Figure 1 shows vectors $\boldsymbol{r}_1$, $\boldsymbol{r}_2$ and$\boldsymbol{r}_3$. Three coordinate systems are defined in Figure 1, as follows: the fixed coordinate system O (x, y, z) with the coordinate system $O'(x', y', z')$ turned at an angle of γ_f or $\gamma_f \pm \Delta\gamma$ and the coordinate system of the hob $O''(x'', y'', z'')$.

Transformation of the unit vectors of the coordinate axes in a dashed-line coordinate system into a fixed one is as follows:

$$\begin{aligned} \boldsymbol{i}' &= \boldsymbol{i}\cos\gamma_f - \boldsymbol{j}\sin\gamma_f\,, \\ \boldsymbol{j}' &= \boldsymbol{i}\sin\gamma_f + \boldsymbol{j}\cos\gamma_f\,, \\ \boldsymbol{k}' &= \boldsymbol{k}\,. \end{aligned} \tag{1}$$

The end point of vector r_3 is point A in the pitch cylinder screw of the hob. At this point, the moving trihedral A (t, n, b) (Figure 2) is defined in the coordinate system $O''(x'', y'', z'')$ as the following:

$$x'' \parallel x',\ y'' \parallel y',\ z'' \parallel z',\ \text{i.e.,}\ \boldsymbol{i}'' = \boldsymbol{i}',\ \boldsymbol{j}'' = \boldsymbol{j}',\ \boldsymbol{k}'' = \boldsymbol{k}'. \tag{2}$$

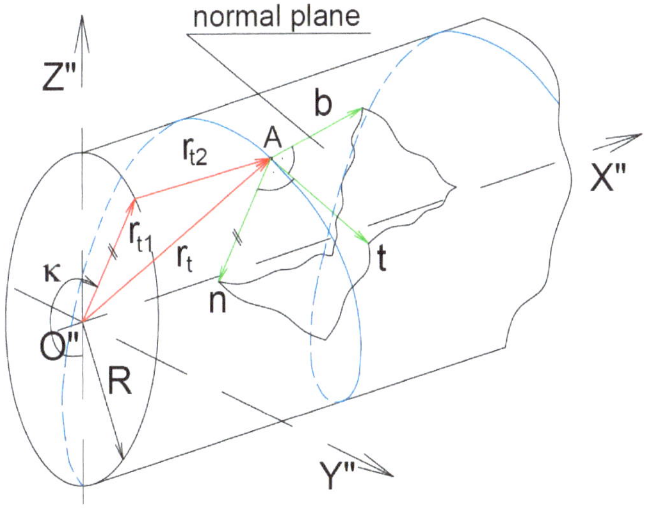

Figure 2. Pitch cylinder screw of the hob and the position vectors.

Based on Figure 2, the following can be expressed:

$$\begin{aligned} \boldsymbol{r}_{t_1} &= -\boldsymbol{j}''R\sin\kappa - \boldsymbol{k}''R\cos\kappa\,, \\ \boldsymbol{r}_{t_2} &= \boldsymbol{i}''R\kappa tg\,\gamma_f\,, \\ \boldsymbol{r}_t &= \boldsymbol{r}_{t_1} + \boldsymbol{r}_{t_2}, \end{aligned} \tag{3}$$

In case of the first derivation of the position vector $\boldsymbol{r}_t$ according to κ, the following is applicable:

$$\dot{\boldsymbol{r}}_t = \frac{d\boldsymbol{r}_t}{d\kappa} = \boldsymbol{i}''Rtg\gamma_f - \boldsymbol{j}''R\cos\kappa + \boldsymbol{k}''R\sin\kappa\,. \tag{3a}$$

The absolute value of $\boldsymbol{r}_t$ is as follows:

$$|\boldsymbol{r}_t| = \sqrt{\boldsymbol{r}_t^2}. \tag{4}$$

Consequently, the unit vector $\boldsymbol{t}$ is expressed as follows:

$$\boldsymbol{t} = \frac{\dot{\boldsymbol{r}}_t}{|\boldsymbol{r}_t|} \tag{5}$$

When Equations (3), (3a) and (4) are substituted into Equation (5), and when Equation (2) is modified and when Equation (2) is used, $\boldsymbol{t}$ is expressed as follows:

$$\boldsymbol{t} = \boldsymbol{i}' \sin\gamma_f - \boldsymbol{j}' \cos\kappa \cos\gamma_f + \boldsymbol{k}' \sin\kappa \cos\gamma_f. \tag{6}$$

The unit vector $\boldsymbol{n}$ is given as follows:

$$\boldsymbol{n} = -\frac{\boldsymbol{r}_{t_1}}{|\boldsymbol{r}_{t_1}|} = -\frac{\boldsymbol{r}_{t_1}}{\sqrt{\boldsymbol{r}_{t_1}^2}}, \tag{7}$$

After substitution and modification of [10] the following is applicable:

$$\boldsymbol{n} = \boldsymbol{j}' \sin\kappa + \boldsymbol{k}' \cos\kappa. \tag{8}$$

The unit vector $\boldsymbol{b}$ is the product of vector multiplication between $\boldsymbol{n}$ and $\boldsymbol{t}$:

$$\boldsymbol{b} = \boldsymbol{n} \times \boldsymbol{t}, \tag{9}$$

and after the substitution of Equations (6) and (8) with Equation (9), and after modification the following is applicable:

$$\boldsymbol{b} = \boldsymbol{i}' \cos\gamma_f + \boldsymbol{j}' \sin\gamma_f \cos\kappa - \boldsymbol{k}' \sin\gamma_f \sin\kappa. \tag{10}$$

The individual partial vectors of the final position vector (Figures 1 and 3) $\boldsymbol{r}$ are as follows:

$$\begin{aligned} \boldsymbol{r}_1 &= \boldsymbol{k}\,R, \\ \boldsymbol{r}_2 &= \boldsymbol{i}'R(2\pi \boldsymbol{i} + \kappa)\,tg\gamma_f, \\ \boldsymbol{r}_3 &= -\boldsymbol{n}\,R, \\ \boldsymbol{r}_4 &= \boldsymbol{n}\,h_n, \\ \boldsymbol{r}_5 &= \boldsymbol{b}\,h_m = \boldsymbol{b}\left(h_n\,tg\alpha_n + \tfrac{p_n}{4}\right). \end{aligned} \tag{11}$$

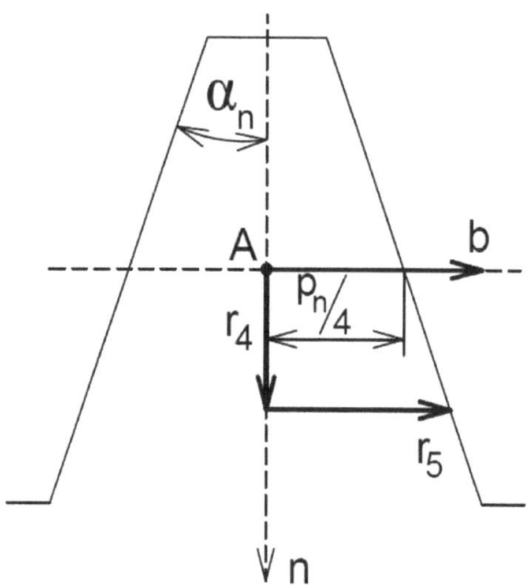

Figure 3. Partial vectors of the position vector.

The final position vector $\boldsymbol{r}$ is given by the total, as follows:

$$\boldsymbol{r} = \boldsymbol{r}_1 + \boldsymbol{r}_2 + \boldsymbol{r}_3 + \boldsymbol{r}_4 + \boldsymbol{r}_5. \tag{12}$$

After transformation and substitution, the following is applicable:

$$\begin{aligned} r &= kR + R(2\pi\, i + \kappa)tg\gamma_f\left(i\,\cos\gamma_f - j\,\sin\gamma_f\right) + \\ &+R\left(-i\,\sin\gamma_f\,\sin\kappa - j\,\cos\gamma_f\,\sin\kappa - k\cos\chi\right) + \\ &+h_n\left(i\,\sin\gamma_f\sin\kappa + j\cos\gamma_f\sin\kappa + k\cos\kappa\right) + \\ &+h_m\left[\begin{array}{l} i\cos^2\gamma_f - j\cos\gamma_f\sin\gamma_f + \\ +\sin\gamma_f\cos\kappa\left(i\sin\gamma_f + j\cos\gamma_f\right) - k\sin\gamma_f\sin\kappa \end{array}\right]. \end{aligned} \tag{13}$$

Position vector Equation (13) is described by the point in the screw plane of the hob in the case its turning corresponds with the angle of gradient of the screw γ_f [11]. In the case of the angular rotation of the hob in the basic rack by the angle of:

$$\gamma_h = \gamma_f \pm \Delta\gamma, \tag{14}$$

the final position vector will be given by the following:

$$r = i\,x + j\,y + k\,z, \tag{15}$$

with

$$\begin{aligned} x = |r_x| &= R(2\pi i + \kappa)\,tg\gamma_f\cos\gamma_h - R\sin\kappa\sin\gamma_h + \\ &+h_n\sin\kappa\sin\gamma_h + h_m\left(\cos\gamma_f\cos\gamma_h + \cos\kappa\sin\gamma_f\sin\gamma_h\right). \end{aligned} \tag{15a}$$

$$\begin{aligned} y = |r_y| &= -R(2\pi i + \kappa)\,tg\gamma_f\sin\gamma_h - R\sin\kappa\cos\gamma_h + \\ &+h_n\sin\kappa\cos\gamma_h + h_m\left(\cos\kappa\sin\gamma_f\cos\gamma_h - \cos\gamma_f\sin\gamma_h\right). \end{aligned} \tag{15b}$$

$$z = |r_z| = R - R\cos\kappa + h_n\cos\kappa - h_m\sin\kappa\sin\gamma_f\,, \tag{15c}$$

In case of which x, y and z refer to the coordinates of the point in the screw plane of the hob.

2.2. Position Vector of the Basic Rack Profile

In order to evaluate the penetration process of the screw plane of the "approximately profiled" hob into the profile of the basic rack it is important to define the planar areas forming the sides of the basic rack teeth [12]. The basic rack has straight teeth, so it is sufficient to define its profile in the plane x–z (Figure 4).

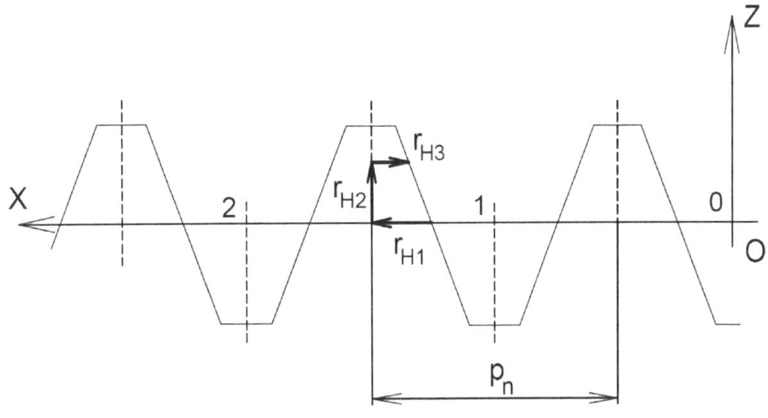

Figure 4. Basic rack profile in plane x–z.

The partial vectors of the final vector $\boldsymbol{r}_{\mathrm{H}}$ are as follows:

$$\boldsymbol{r}_{\mathrm{H}_1} = \boldsymbol{i}\,(p_n\boldsymbol{i} + \tfrac{p_n}{2}),$$
$$\boldsymbol{r}_{\mathrm{H}_2} = \boldsymbol{k}\,z,$$
$$\boldsymbol{r}_{\mathrm{H}_3} = -\boldsymbol{i}\,(\tfrac{p_n}{4} - z\,tg\alpha_n),$$

and the final vector is:

$$\boldsymbol{r}_{\mathrm{H}} = \boldsymbol{r}_{\mathrm{H}_1} + \boldsymbol{r}_{\mathrm{H}_2} + \boldsymbol{r}_{\mathrm{H}_3}. \tag{16}$$

Then, the coordinate x_{H} of the point in the basic rack profile in the plane z = const. is as follows:

$$x_{\mathrm{H}} = |\boldsymbol{r}_{\mathrm{H}_x}| = p_n\left(\boldsymbol{i} + \frac{1}{4}\right) + z\,tg\alpha_n. \tag{17}$$

2.3. Method for Determination of the Deviations in the Profiles of the Basic Rack and the Hob

The modified profile of the teeth in the toothed rack, which is formed when in the reference plane into which the hob is located, will have the form of an envelope formed by the lines.

These lines are formed when the set of axial cuts enters/runs through the screw plane of the hob (Figures 5 and 6). To evaluate the deviations, it is important to apply the system of cuts through the plane z = const. [13]. The reference plane is diverted by the angle γ_f and, therefore, it is important to enter the value h_n in such a manner so that the point of the axial cut through the screw plane of the hob fits into the respective plane z = const.

The deviations in the profiles Δp_b is given by the following relation:

$$\Delta p = (x_{\mathrm{H}} - x)1000\ [\mu\mathrm{m}], \tag{18}$$

with x_{H} as the coordinate of the planar cut of the basic rack profile in the case of a particular planar cut z = const. (mm), and

x as the coordinate of the axial cut through the screw plane of the hob in the case of an identical planar cut z = const. (mm) [14].

The development of deviations in relation to the angular adjustment of the hob is shown in Figure 7 for the module m = 10 mm, with a tip diameter of the hob of R_{H} = 67.5 mm.

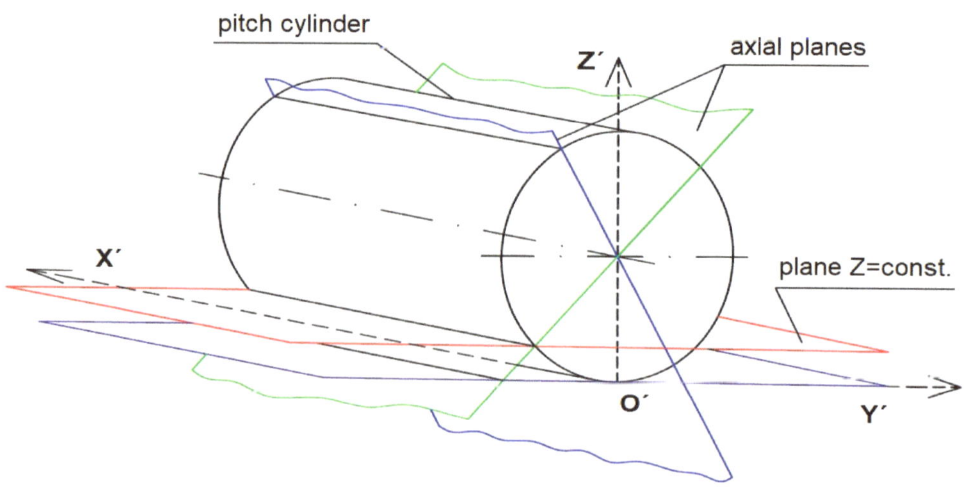

Figure 5. Envelope of lines from the deviations in the profiles.

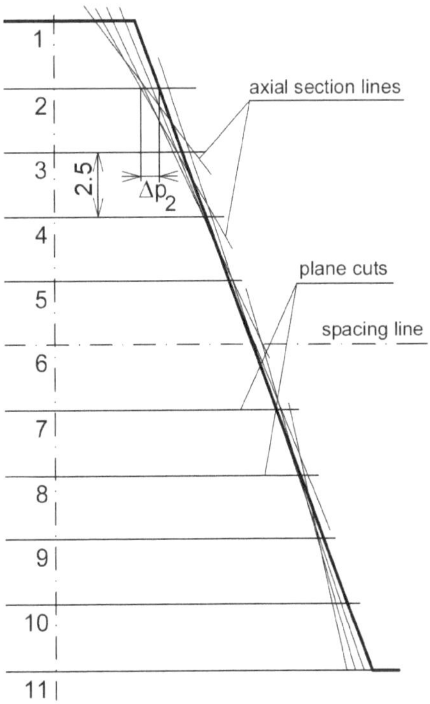

Figure 6. The set of axial cuts enter/run through the screw plane of the hob.

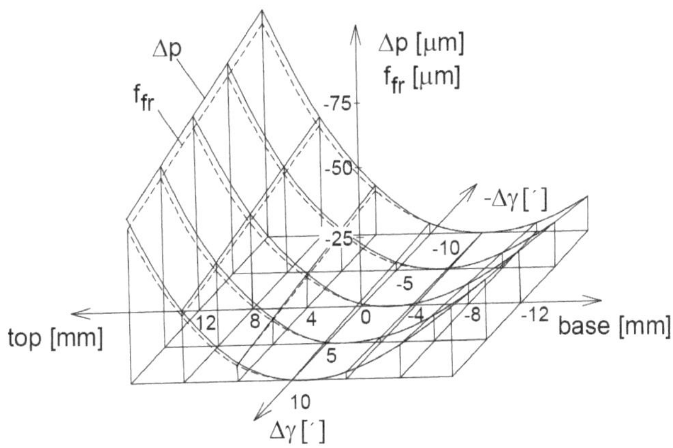

Figure 7. Influence of the setting of the hobbing mill. Module m = 10 mm, R_H = 67.5 mm, profiled in the reference plane.

2.4. Method for Determination of the Profile Deviations in the Workpiece Tooth

To evaluate the deviations, which will be projected in the manufacturing of the tooth by the modified toothed rack, it is important to determine the theoretical profile of the tooth (i.e., the involute profile). It can be achieved by rolling the basic rack [15]. If correction is not taken into consideration, the pitch of the straight line in the rack rolls along the pitch circle of the manufactured wheel (Figure 8).

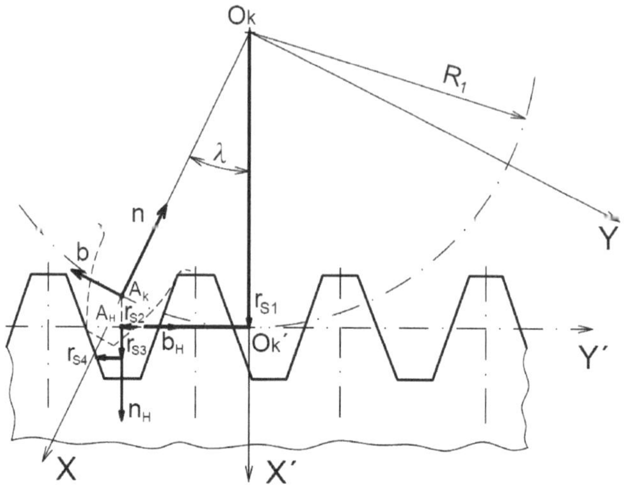

Figure 8. Pitch of the straight line in the rack.

Mathematical notation is a combination of vectors. Figure 8 shows a definition of the coordinate systems O_k (x, y), $O'_k(x', y')$, A_H (n_H, b_H) and A_K (n, b). The transformation equations for the dashed-line coordinate system of the forming rack to transform into the fixed coordinate system of the manufactured toothed wheel are as follows [16]:

$$\begin{aligned} \boldsymbol{i}' &= \boldsymbol{i}\cos\lambda + \boldsymbol{j}\sin\lambda\,, \\ \boldsymbol{j}' &= \bar{\boldsymbol{j}}\cos\lambda - \bar{\boldsymbol{i}}\sin\lambda. \end{aligned} \tag{19}$$

The unit vectors are $n_H = \boldsymbol{i}'$, $b_H = \boldsymbol{j}'$, $n = -\boldsymbol{i}'$ and $b = -\boldsymbol{j}'$.

After transformation, the final position vector of the basic rack $\boldsymbol{r}_s$ is given by the total of the partial vectors (Figure 8):

$$\boldsymbol{r}_s = \boldsymbol{r}_{s_1} + \boldsymbol{r}_{s_2} + \boldsymbol{r}_{s_3} + \boldsymbol{r}_{s_4}, \tag{20}$$

and the individual partial vectors are as follows:

$$\begin{aligned} \boldsymbol{r}_{s_1} &= \boldsymbol{i}\,\boldsymbol{R}_1\cos\lambda + \boldsymbol{j}\,\boldsymbol{R}_1\sin\lambda\,, \\ \boldsymbol{r}_{s_2} &= -\boldsymbol{j}\,\lambda\,\boldsymbol{R}_1\cos\lambda + \boldsymbol{i}\,\lambda\,\boldsymbol{R}_1\sin\lambda\,, \\ \boldsymbol{r}_{s_3} &= \boldsymbol{i}\,h_n\cos\lambda + \boldsymbol{j}\,h_n\sin\lambda\,, \\ \boldsymbol{r}_{s_4} &= -\boldsymbol{j}\,h_b\cos\lambda + \boldsymbol{i}\,h_b\sin\lambda\,, \end{aligned} \tag{20a}$$

and their absolute values are as follows:

$$\begin{aligned} &|\boldsymbol{r}_{s_1}| = \boldsymbol{R}_1\,,\ |\boldsymbol{r}_{s_2}| = \lambda\,\boldsymbol{R}_1\,,\ |\boldsymbol{r}_{s_3}| = h_n\,, \\ &|\boldsymbol{r}_{s_4}| = h_b = \tfrac{p_n}{4} - h_n\,tg\alpha_n. \end{aligned} \tag{20b}$$

Parts of the vector $\boldsymbol{r}_s$ in the direction of the coordinate axes $\boldsymbol{x}$ and $\boldsymbol{y}$ are as follows:

$$\begin{aligned} x_K &= \left|\boldsymbol{r}_{s_x}\right| = \boldsymbol{R}_1\cos\lambda + \lambda\,\boldsymbol{R}_1\sin\lambda + h_n\cos\lambda + h_b\sin\lambda\,\,, \\ y_K &= \left|\boldsymbol{r}_{s_y}\right| = \boldsymbol{R}_1\sin\lambda - \lambda\,\boldsymbol{R}_1\cos\lambda + h_n\sin\lambda - h_b\cos\lambda\,\,. \end{aligned} \tag{21}$$

However, the tooth profile must relate to the coordinates n, b in the coordinate system A_K (n, b). The coordinates of the theoretical profile of the tooth are consequently given by the following:

$$\begin{aligned} n_{\text{teo}} &= \boldsymbol{R}_1 - x_K\,, \\ b_{teo} &= -y_K. \end{aligned} \tag{22}$$

These coordinates define the theoretical or actual profile. However, it is still unknown for which of the values λ or h_n is the point described by the position vector $\boldsymbol{r}_s$, simultaneously by the point in the tooth profile of the manufactured wheel. The toothed wheel profile is formed by the envelope of the profiles in the basic rack. To determine the parameters, i.e., to determine the values λ and h_n, the envelope method is used [17] and, in this case, there is two parametric methods. The rack rolling results in the achievement of a set of lines for the diverse value h_n (Figure 9). A single point lies on each of the lines. At the same time, the point represents the point in the tooth profile of the manufactured wheel. In the case of this point, the determinant Equation (23), which is formed by means of the partial derivations of the position vector sections $\boldsymbol{r}_s$, equals to zero according to parameters λ and h_n.

$$\begin{vmatrix} \dfrac{\partial \boldsymbol{r}_{s_x}}{\partial \lambda} & \dfrac{\partial \boldsymbol{r}_{s_x}}{\partial h_n} \\ \dfrac{\partial \boldsymbol{r}_{s_y}}{\partial \lambda} & \dfrac{\partial \boldsymbol{r}_{s_y}}{\partial h_n} \end{vmatrix} = 0 \tag{23}$$

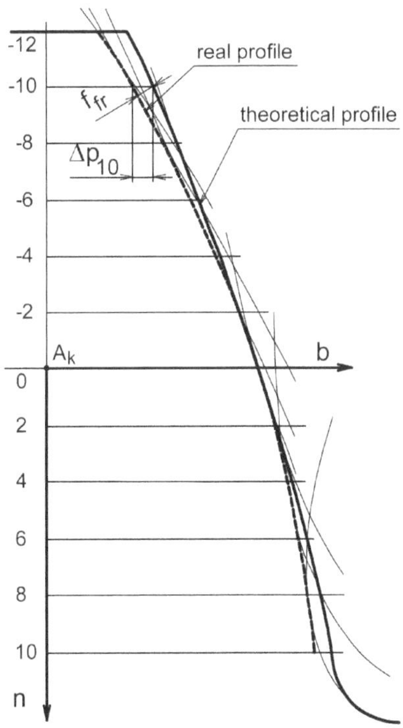

Figure 9. Profile deviation.

To determine the parameters, it is important to use one of the iterative methods (for example Regula Falsi, etc.).

This method is employed in order to determine the theoretical profile of the toothed wheel tooth. The point of the actual profile is achieved by the rolling of the rack, the profile of which is negative, to the modified profile of the rack [18]. To detect the actual tooth profile, which is manufactured after the respective profiling, it is not possible to apply the aforementioned envelope method as the point of the modified rack or its negative shape is too complicated to be expressed implicitly [19].

The profile point in the modified rack is always set to the maximum value of Δp_i, for example, the value of cut 2 is Δp_2 (see Figure 6).

If in the coordinate system A_K (Figure 8) the selected specific is n = const., for which the respective b_{sk} is to be determined, i.e., the point in the actual tooth profile, it is important by means of the envelope method to detect n corresponding to b_{teo}, i.e., the coordinate of the point in the theoretical profile. To determine b_{sk} it is not possible to use the envelope method due to the aforementioned reasons and, therefore, the following procedure is applied [20].

It can be assumed that λ, for which the point in the theoretical profile was detected, will be proximate to λ_s as the point in the actual profile. A particular interval is selected ($\lambda + \Delta\lambda$, $\lambda - \Delta\lambda$) and the rack is rolled within the interval (the rack profile is negative according to the modified profile and represents the hob profiled by the straight line). At the same time, some small step $\Delta\lambda$ will be determined and for each λ from the interval the value b_{sk} will be calculated [21]. This value will always be compared with the previous value and if $b_{ski} > b_{ski+1}$ the rack will be rolled further on. In the case of $b_{ski} < b_{ski+1}$, the rack will not be rolled any further because b_{sk} is the lowest out of the interval and represents the coordinate of the actual profile of the tooth in the manufactured wheel [22]. All the values of b_{ski} are calculated for the same n = const.

The deviation in the manufactured profile from the theoretical one for a particular n = const is then given as follows:

$$\Delta sp = (b_{teo} - b_{sk})1000\ [\mu m]. \tag{24}$$

3. Discussion

The study results on the influence of the angular adjustment of the hob within the range from +10′ up to −10′ upon modification of the tooth profile in the workpiece is a development of the deviations in the manufactured profile of the tooth from the theoretical profile, which is produced during machining by the basic rack [23].

The development of these deviations for the angle of meshing α = 20°, the module m = 10 mm and the tip diameter of the hob R_H = 67.5 mm is shown in Figure 10 for the hob profiling by the straight-line profile in the reference plane, and in Figure 11 for the hob profiling by the straight-line profile in the axial plane. Deviation of the profile f_{fr} (Figure 9) is defined as the perpendicular distance of the nominal profiles [24]. Therefore, when compared with the standard referring to the fitting of toothed wheels, it is recommended to calculate Δs_p in terms of the definition of f_{fr}.

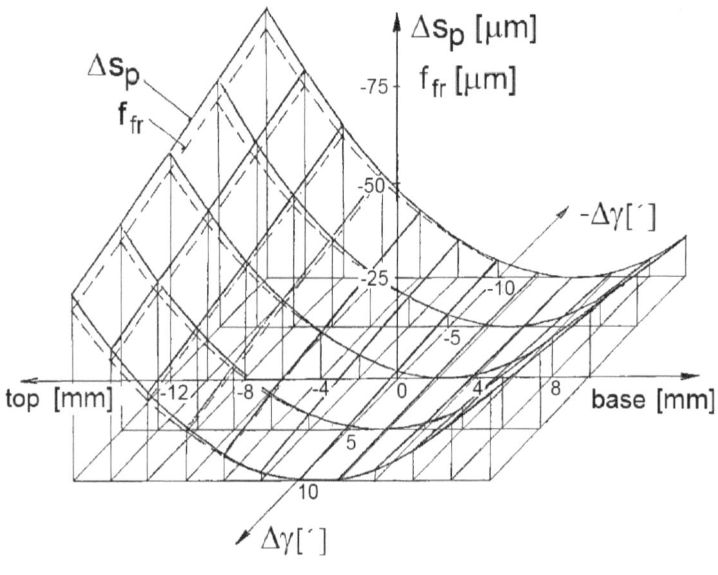

Figure 10. Development of toothing deviations profiled in the reference plane; m = 10 mm, R_H = 67.5 mm.

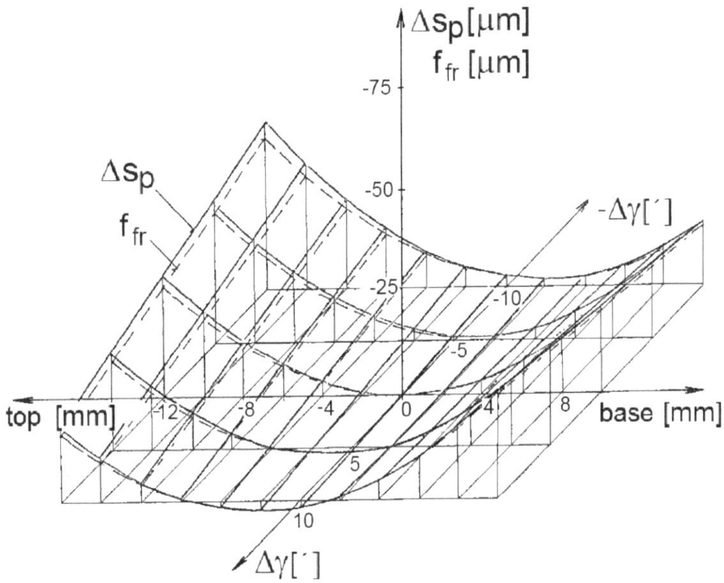

Figure 11. Development of toothing deviations profiled in the axial plane; m = 10 mm, R_H = 67.5 mm.

Based on Figures 10 and 11, it is clear that development of toothing deviations is more favorable for profiling in the axial plane. The values of the deviations are in accordance with [15] or [25]. Adjusting the hob according to $\Delta\gamma = 0$ enables f_{fr} to be reached, which approaches the deviation limit of f_f for the sixth degree of precision [26].

4. Conclusions

This present paper provides a detailed view of the strategy and techniques for describing the tool profile that is optimal for use in the production of gears using a hobbing mill. This highly efficient production process is used very often nowadays and, therefore, measures aimed at increasing the accuracy of this technology are still a current and important topic [27]. The theoretical analysis of the geometry on the formation of connected surfaces provides basic knowledge for a better understanding of this issue and enables the application of this knowledge in a wider context.

The presented procedures provide the designers of hobbing mills with the advantage of working with the initial deviations in the milling cutter profiles from the base ridge profile, which simplifies the analysis [28]. The proposed methods make it possible to simultaneously investigate the influence of various factors that affect production accuracy [29]. In this way, for example, the influence of the diameter of the hobbing mill and the influence of various movement aspects during the process are capable of being investigated, such as throwing or deformations of the components in the production machine, or even torsional and bending vibrations.

These methods make it possible to implement various modifications to tool profiles and make adjustments to the construction of the tool itself. The impact of these changes can be verified and evaluated immediately. The integration of the above processes with computer technology leads to the creation of a complex system for designing these tools intended for the production of gears [30]. The practical implementation of these methods provides space for the creation of more effective tools and the optimization of production processes with an emphasis on the resulting accuracy and quality [31].

The presented approaches are not limited to individual gear types, but can also be applied to different gear geometries. In this way, we obtain a universal methodology that can be adapted to the needs of different production contexts. Overall, this contribution is a significant step in the direction to improve the accuracy and efficiency of gear manufacturing using hobbing mills.

The next step in the research will involve a planned experiment focused on implementing the proposed gear design method in real-world manufacturing conditions. The expected outcome of the experiment will confirm the suitability of the proposed solution and its impact on the qualitative parameters of gearing.

Author Contributions: Conceptualization, J.M. and T.C.; methodology, J.M. and T.K.; software, T.C.; formal analysis, T.K. and T.C.; resources, T.C.; data curation, J.M. and T.K.; writing—original draft preparation, T.C.; writing—review and editing, J.M. and T.K.; visualization, T.C.; project administration and funding acquisition, J.M. All authors have read and agreed to the published version of the manuscript.

Funding: This work was supported by the Slovak Ministry of Education within project KEGA 017TUKE-4/2021 and project VEGA 1/0823/21.

Institutional Review Board Statement: Not applicable.

Informed Consent Statement: Not applicable.

Data Availability Statement: Not applicable.

Conflicts of Interest: The authors declare no conflict of interest.

Nomenclature

$A\ (t, n, b)$	Coordinate system of the rolling mill profile
$A_{\mathrm{H}}\ (n_{\mathrm{H}}, b_{\mathrm{H}})$	Coordinate system of the base ridge profile
$A_{\mathrm{K}}\ (n, b)$	Coordinate system of the tooth profile in the manufactured wheel
$O\ (x, y, z)$	Fixed coordinate system
$O'(x', y', z')$	Coordinate system turned into it
$O''(x'', y'', z'')$	Coordinate system of the hob
$O_k\ (x, y)$	Coordinate systems
$O'_k\ (x', y')$	Coordinate systems
i, j, k, t, n, b	Unit vectors
r_1, r_2, r_3	Components of the position vector
b_{teo}	Coordinate of the theoretical tooth profile in the manufactured wheel (mm)
b_{sk}	Coordinate of the actual tooth profile in the manufactured wheel (mm)
h_n, h_m, h_b	Coordinate of the profile point in the basic ridge, or of the rolling mill in the coordinate system $A_{\mathrm{H}}, A_{\mathrm{K}}$ (mm)
x, y, z	Coordinates of the point in the screw plane of the hob
r_t	Position vector
r_{H}	Final vector
r_s	Final position vector of the basic rack
r	Final position vector
Δp_i	Deviation in the profile of the rolling cutter from the profile of the basic ridge in section i (μm)
Δsp	Deviation in the actual profile of the manufactured tooth (μm)
γ_f	Angle of the gradient of the hob (deg)
$\Delta\gamma$	Inaccuracy in the setting of the rolling mill ($'$),
λ	Base ridge roll angle (deg)
κ	Angle of rotation of the axial cut of the rolling cutter (rad)

References

1. Bouzakis, K.D.; Antoniadis, A. Optimizing of tangential tool shift in gear hobbing. *CIRP Ann.* **1995**, *44*, 75–78. [CrossRef]
2. Jia, K.; Guo, J.; Zheng, S.; Hong, J. A general mathematical model for two-parameter generating machining of involute cylindrical gears. *Appl. Math. Model.* **2019**, *75*, 37–51. [CrossRef]
3. Spitas, V.; Costopoulos, T.; Spitas, C. Fast modeling of conjugate gear tooth profiles using discrete presentation by involute segments. *Mech. Mach. Theory* **2007**, *42*, 751–762. [CrossRef]

4. Vasilis, D.; Nectarios, V.; Aristomenis, A. Advanced computer aided design simulation of gear hobbing by means of three-dimensional kinematics modeling. *J. Manuf. Sci. Eng.* **2007**, *129*, 911–918. [CrossRef]
5. Gupta, K.; Laubscher, R.F.; Davim, J.P.; Jain, N.K. Recent developments in sustainable manufacturing of gears: A review. *J. Clean. Prod.* **2016**, *112*, 3320–3330. [CrossRef]
6. Sałaciński, T.; Przesmycki, A.; Chmielewski, T. Technological aspects in manufacturing of non-circular gears. *Appl. Sci.* **2020**, *10*, 3420. [CrossRef]
7. Maščeník, J.; Pavlenko, S. Determining the exact value of the shape deviations of the experimental measurements. *Appl. Mech. Mater.* **2014**, *624*, 339–343. [CrossRef]
8. Tilavaldiev, B. Methods of manufacturing gears. *Int. J. Adv. Sci. Res.* **2022**, *2*, 127–130. [CrossRef]
9. Lin, C.Y.; Tsay, C.B.; Fong, Z.H. Mathematical model of spiral bevel and hypoid gears manufactured by the modified roll method. *Mech. Mach. Theory* **1997**, *32*, 121–136. [CrossRef]
10. Pavlenko, S.; Maščenik, J.; Krenický, T. *Worm Gears: General Information, Calculations, Dynamics and Reliability*; RAM-Verlag: Lüdenscheid, Germany, 2018; p. 167.
11. Das, H.; Li, X.; Li, L.; Schuessler, B.J.; Overman, N.; Darsell, J.T.; Upadhyay, P.; Soulami, A.; Herling, D.R.; Joshi, V.V.; et al. An innovative and alternative approach toward gear fabrication. *J. Manuf. Process.* **2023**, *102*, 319–329. [CrossRef]
12. Bouzakis, K.D.; Lili, E.; Michailidis, N.; Friderikos, O. Manufacturing of cylindrical gears by generating cutting processes: A critical synthesis of analysis methods. *CIRP Ann.* **2008**, *57*, 676–696. [CrossRef]
13. Čep, R.; Janásek, A.; Čepová, L.; Petrů, J.; Hlavatý, I.; Car, Z.; Hatala, M. Experimental testing of exchangeable cutting inserts cutting ability. *Teh. Vjesn.* **2013**, *20*, 21–26.
14. Fan, Q. Enhanced algorithms of contact simulation for hypoid gear drives produced by face-milling and face-hobbing processes. *J. Mech. Des.* **2007**, *129*, 31–37. [CrossRef]
15. Shih, Y.P.; Fong, Z.H.; Lin, G.C. Mathematical model for a universal face hobbing hypoid gear generator. *J. Mech. Des.* **2007**, *129*, 38–47. [CrossRef]
16. Ivanov, V.; Mital, D.; Karpus, V.; Dehtiarov, I.; Zajac, J.; Pavlenko, I.; Hatala, M. Numerical simulation of the system "fixture–workpiece" for lever machining. *Int. J. Adv. Manuf. Technol.* **2017**, *91*, 79–90. [CrossRef]
17. Boral, P.; Stoić, A.; Kljajin, M. Machining of spur gears using a special milling cutter. *Teh. Vjesn.* **2018**, *25*, 798–802.
18. Durante, S.; Comoglio, M.; Rabezzana, F. *High Performance Gears Hobbing*; Springer: Vienna, Austria, 1999; pp. 207–214.
19. Habibi, M.; Chen, Z.C.; Fang, Z. Tool wear improvement in face-hobbing of bevel gears by re-designing the cutting blades. *IFAC-PapersOnLine* **2015**, *48*, 1888–1893. [CrossRef]
20. Jurko, J.; Miškiv-Pavlík, M.; Husár, J.; Michalik, P. Turned Surface Monitoring Using a Confocal Sensor and the Tool Wear Process Optimization. *Processes* **2022**, *10*, 2599. [CrossRef]
21. Vimercati, M. Mathematical model for tooth surfaces representation of face-hobbed hypoid gears and its application to contact analysis and stress calculation. *Mech. Mach. Theory* **2007**, *42*, 668–690. [CrossRef]
22. Habibi, M.; Chen, Z.C. A new approach to blade design with constant rake and relief angles for face-hobbing of bevel gears. *J. Manuf. Sci. Eng.* **2016**, *138*, 031005. [CrossRef]
23. Golebski, R.; Boral, P. Study of Machining of Gears with Regular and Modified Outline Using CNC Machine Tools. *Materials* **2021**, *14*, 2913. [CrossRef] [PubMed]
24. Fan, Q. Ease-off and application in tooth contact analysis for face-milled and face-hobbed spiral bevel and hypoid gears. In *Theory and Practice of Gearing and Transmissions*; Springer: Berlin/Heidelberg, Germany, 2016; pp. 321–339.
25. Puzyr, R.; Haikova, T.; Majerník, J.; Karkova, M.; Kmec, J. Experimental study of the process of radial rotation profiling of wheel rims resulting in formation and technological flattening of the corrugations. *Manuf. Technol.* **2018**, *18*, 106–111. [CrossRef]
26. Xiang, S.; Li, H.; Deng, M.; Yang, J. Geometric error analysis and compensation for multi-axis spiral bevel gears milling machine. *Mech. Mach. Theory* **2018**, *121*, 59–74. [CrossRef]
27. Ruzbarsky, J.; Mital, G. Diagnostics of selected surface characteristics with laser profilometry. *MM Sci. J.* **2018**, *1*, 2140–2214. [CrossRef]
28. Simon, V.V. Optimal machine-tool settings for the manufacture of face-hobbed spiral bevel gears. *J. Mech. Des.* **2014**, *136*, 081004. [CrossRef]
29. Simon, V.V. Manufacture of optimized face-hobbed spiral bevel gears on computer numerical control hypoid generator. *J. Manuf. Sci. Eng.* **2014**, *136*, 031008. [CrossRef]
30. Gaspar, S.; Pasko, J. Mathematical formulation of the kinematic equations for the control of the robot system with application for the machining conical surfaces. *MM Sci. J.* **2018**, *1*, 2158–2161. [CrossRef]
31. Claudin, C.; Rech, J. Development of a new rapid characterization method of hob's wear resistance in gear manufacturing—Application to the evaluation of various cutting edge preparations in high speed dry gear hobbing. *J. Mater. Process. Technol.* **2009**, *209*, 5152–5160. [CrossRef]

applied sciences

MDPI

Article

Data Exchange with Support for the Neutral Processing of Formats in Computer-Aided Design/Computer-Aided Manufacturing Systems

Piotr Kuryło [1], Peter Frankovský [2,*], Marek Malinowski [1], Tomasz Maciejewski [3], Ján Varga [2], Ján Kostka [2], Łukasz Adrian [4], Szymon Szufa [4] and Svetlana Rusnáková [2]

[1] Faculty of Mechanical Engineering, University of Zielona Góra, 65-516 Zielona Góra, Poland; p.kurylo@iim.uz.zgora.pl (P.K.); m.malinowski@iim.uz.zgora.pl (M.M.)
[2] Faculty of Mechanical Engineering, Technical University of Košice, Letná 9, 042 00 Košice, Slovakia; jan.varga@tuke.sk (J.V.); jan.kostka@tuke.sk (J.K.); svetlana.rusnakova@student.tuke.sk (S.R.)
[3] Independent Researcher, 65-516 Zielona Góra, Poland; tmaciejewski@premiumsolutions.pl
[4] Faculty of Process and Environmental Engineering, Lodz University of Technology, Wolczańka 213, 90-924 Lodz, Poland; lukasz.adrian@eko-look.pl (Ł.A.); szymon.szufa@upwr.edu.pl (S.S.)
* Correspondence: peter.frankovsky@tuke.sk; Tel.: +421-55-602-2623

Abstract: This paper presents an analysis and the research results of system couplers in CAx systems that ensure the correct transfer of product data from the "supplier" system to the "recipient" system. This study presents the results of the compatibility tests between the systems based on the verification of geometric models and their attributes after importing/exporting 2D and 3D objects using neutral data exchange formats. The purpose of the research was to identify neutral formats that do not introduce errors after data transversion in certain types of geometry. Tests and analyses were carried out on selected formats depending on the type of object, such as 2D objects (DXF), solid objects (IGES, STEP, SAT, and PARASOLID), and surface objects (IGES, STEP, SAT, and X_T). One of the results of our research and analysis is the need for continuous development of CAD/CAE systems because current neutral formats are not 100% effective.

Keywords: CAD; data exchange; data format; CAx; DXF; IGES; STEP; SAT; X_T

Citation: Kuryło, P.; Frankovský, P.; Malinowski, M.; Maciejewski, T.; Varga, J.; Kostka, J.; Adrian, Ł.; Szufa, S.; Rusnáková, S. Data Exchange with Support for the Neutral Processing of Formats in Computer-Aided Design/Computer-Aided Manufacturing Systems. *Appl. Sci.* **2023**, *13*, 9811. https://doi.org/10.3390/app13179811

Academic Editor: Jürgen Reichardt

Received: 12 July 2023
Revised: 9 August 2023
Accepted: 16 August 2023
Published: 30 August 2023

1. Introduction

The evolution of manufacturing systems follows the development of technology and their immediate environment in the form of machines, devices, methods, and tools supporting the work on the preparation of technical documentation [1–3]. This documentation describes the product model, processes, and means of production in use. The obvious fact is that computer tools help to meet the production challenges of a rapidly changing world economy, especially in the work of engineers [4,5]. Chen [6] discussed the use of various CAD systems to simplify the product design process. A solution based on commands was proposed to address the problems identified in the development of such a collaborative design system. The authors Chiu and Chlebus designed a representation scheme for multiple material objects in a CAD system [7,8].

In modern production engineering, the phase of design and technical preparation of production is of particular importance, the cost of which comprises 70% of the entire production project [8–10]. Therefore, the designer, regardless of whether they are a constructor, technologist, or planner, must have an efficient tool and technique that enables a comprehensive analysis of the adopted concepts, models, and variants of solutions [11–13].

Previous studies have dealt with the portrayal of the ISO STEP tolerance standard as an enabler of intelligent manufacturing systems [14]. Gargallo [15,16] investigated a method of smoothing and untangling surface meshes independent of CAD parameterization, where he proposed a method to optimize triangular and quadrilateral meshes on parameterized surfaces. Kaščák et al. [17] revealed that data exchange is important in all manufacturing processes, including plastic injection molding, machining simulations, and other technologies [18–21]. Jauhar [22] addressed the heterogeneous environment of CAx systems that have problems in revising and reimporting neutral models, due to the missing required identification information. Jian [23] designed an improved NBA algorithm to optimize the BP neural network algorithm for semantic feature recognition. Xiaoxia et al. [24] revealed that feature-based data exchange (FBDE) between CAD systems is one of the most important issues in collaborative product development and presented a solution based on procedure recovery for FBDE. The result was the functioning of FBDE between different commercial CAD systems and improvements in previous approaches to architecture and procedure searches. Modern CAD systems are used to build models of product shapes based on functions with parameterization and constraints [25].

Integrated computer-aided design/computer-aided manufacturing/computer-aided engineering (CAD/CAM/CAE) were studied by Kirkwood [26,27] in his research, resulting in a set of practices for the effective adaptation of sustained CAD/CAE integration. In their studies, Varga [28] and Kirkwood [29] described the importance of data validation in the CAE process of systems in more detail. Krause [30] highlighted the significance of the rapid transfer of CAD data between different CAD systems to speed up and reduce the number of iterative steps in product development processes. Kreis' research [31] involved the use of a large volume of data in the development process, especially in the applied CAx environment, and an examination of the exchange of data between different areas of design, analysis, and production in the automotive field. The assessment of data formats for layered manufacturing was discussed by Kumar [32] and by Varga in terms of the machining process [33]. Lee [34,35] explained data processing for CAD/CAM/CAE systems. Lipman [36] studied PMI representation in CAD model STEP data exchange files.

Zhang [37] analyzed the corresponding relations of topological entities. He proposed a method for comparing topological entities in the integration of heterogeneous CAD systems. Lu [38] presented data modeling for the integration between specifications and verification for GPS-compliant cylindricity based on category theory. The importance of data transfer was similarly explored by Kumar and Martin [39,40] but in the area of additive manufacturing. In another study [41], the importance of data in assessing the accuracy of machined parts was highlighted. Several other studies involved investigations of data transfer [42–44].

Data exchange between different systems is of particular importance in the process of planning. In the production process, this can be in the field of machining or the case of the production of complex-shaped parts, such as the sintering of powders. In all these cases, data processing and the correct exchange of CAD data are important for an efficient production method. Most often, the team making "strategic" decisions to start production must have a universal data exchange format that ensures the least loss of data from the "giver", i.e., the customer, to the "recipient", i.e., the contractor. They must also have appropriate software that is fully compatible with other systems so that the exchange between them is possible. From this point of view, a full compliance analysis consisting of the verification of geometric models and their attributes after import/export is needed to understand all the issues related to data exchange in CAD/CAE systems [10,21,29,45]. The integration of methods, models, tools, and data is the basis for the comprehensive implementation of computer techniques in a modern manufacturing company.

The overriding goal is to increase awareness of the existence of errors in the transmission of data between CAx systems. Additionally, the existing research aims to highlight the current level of application of neutral formats and assess the compatibility of converted objects [46–48]. Errors at the design stage, which include the transfer and transfer of objects using CAx systems, can significantly increase the cost of implementing the production process [49,50]. This paper draws attention to the possibility of product inconsistency due to incorrect translation between the available programming languages. One of the systems that compares and determines the level of compliance is NIST software 2022.

Measuring compliance includes, among other aspects, geometrical dimensioning and tolerancing, surface specification, annotations regarding finishing, welded joints, and material specification [51].

Generating and optimizing section profile data for rapid prototyping and manufacturing was studied in [52], which presents a slicing algorithm based on a geometric continuity model. Poya [53] analyzed a unified approach to a posteriori generation of arbitrary high-order curvilinear meshes via the solid mechanics analogy, and Pratt [54] described a STEP standard for product data exchange. Different forms of data processing were studied by Adrian et al. [55], who examined the data in heat transfer processes in a heat pipe heat exchanger utilizing a working fluid R134A. Other authors [56–58] examined transfer data in additive processes and its effect on the evaluation of geometrical performance, while Ryou [59] was involved in the development of a data interface for rapid prototyping in STEP-NC. With the use of model-based definitions, most of the data required to design, analyze, and manufacture a product are stored in digital format [60–62]. An effective communication model between CAD, CAPP, and CAM applications in a distributed manufacturing planning environment is considered a key component of CIM [63]. Data handling in NURBS models based on the STEP format was studied by Starly [64]. Staves [65] and Venkiteswaran [66] investigated the associations between CAD models, as well as their geometric dimensions and tolerances.

Other analyses of data exchange formats can be found in [67–69], which examine product data exchange and product model sharing. In [70], the authors described and presented the concept of internal assembly models and the associated set of geometry processing operators. Wu [71] solved the computer modeling of geometric variations in mechanical parts and assemblies. Another data exchange method in additive processing was described in [72–74], which analyzed slice data representation and formats for multi-material objects. Zhao, Zhou, Zhou, and Zhu [75–78] described data processing in the field of geometric tolerancing, which involves the integration of CAD or CAE applications. Stelmach [79] described the important role of data processing in the influence of hydrodynamic changes in a system with a pitched-blade turbine on mixing power. Some studies presented digital induction motor models based on the finite element method or investigated the possibility of modeling parts in different production technologies [80–83], in which the authors analyzed the importance of individual data in these processes. Runde [84] and Kurylo [85] analyzed the data exchange format for the engineering of building automation systems and data work in the design of an automated system for car body measurement.

The importance of data exchange in CAD was discussed by Rappoport [86], who argued that parametric-feature-based data exchange is one of the most important open problems in reliable modeling. In his results, he presented a general outline of the Universal Product Representation (UPR) architecture, providing universal support for all the data levels used by current CAD systems.

The possibility of further data exchange using different methods was discussed by other authors who focused on the FBDE function, the method of topological comparison of entities, or the quantitative optimization of interoperability during data exchange. Data exchange is also encountered with functions (FBDE), which has been one of the key issues in the history of CPD and should be accommodated in the latest CBDM-enabled CPD. In One study, it was concluded that, for the FBDE function, the same geometric shape would be constructed by different modeling procedures, meaning that no unique modeling procedure can be used for the same geometric shape in FBDE. Therefore, for the exchange of the modeling sequence, mapping between source and target elements should be considered, which includes five cases: 1:1 mapping, 1:n mapping, 1:0 mapping, 0:1 mapping, and in the last case, a singular element [87]. Zhang [88] analyzed the sharing of CAD models, which is, in most cases, a complex problem when dealing with heterogeneous CAD systems. The author introduced a new asymptotic strategy to enrich the interoperability theory based on features in singularity solving. The results of his research showed that a properly designed approach can ensure that the exchanged model of the target CAD system can be parametrically adjusted. Varga [89] also focused on the application of CAE systems for machining in the production of shape-complex surfaces.

The analysis of the available literature indicates a lack of studies that specify the percentage level of object compliance between the selected neutral formats. This manuscript is part of the effort to fill the gap in the current state-of-the-art research and is the basis for similar cases of object conversion or testing software that examine compatibility levels.

2. CAx Class Systems

In the modern world, an inherent attribute of an engineer is a computer and the corresponding software. Of the many types of engineering software, the most important are the various CAx systems, which make it possible to significantly accelerate and facilitate (and thus reduce the cost of) the transition from idea to practice [4,10,34,54,77].

Computer-aided systems (CAx) involve the use of computer technology to aid in the design, analysis, and manufacture of products. Advanced CAx tools merge many different aspects of product lifecycle management (PLM), including design, finite element analysis (FEA), manufacturing, production planning, product, and more. CAx systems enable the identification and correction of design defects as early as the virtual model design phase, thereby speeding up the product design and manufacturing process.

There are many CAD/CAE/CAx systems in the engineering software market. Among the many tools used in integrated manufacturing are CAD systems (computer-aided design), which are computer systems supporting the design of structures, details, and tools and machines: SolidWorks, AutoCAD (Mechanical, Electrical, LT), Autodesk (Fusion 360, Inventor, Product Design Suite, and Revit), PTC Cero, CATIA, Siemens NX, Civil3D, MicroStation, Rhino, SolidEdge, Sketchup, Alias, Vectorworks, OpenSCAD, FreeCAD, Draftsight, Spaceclaim, Tekla, Geomagic Design, ArtCam, Inroads, ArchiCAD, BricsCAD, CorelCAD, CADStd Lite, and CorelDRAW Technical. The listed software programs provide applications for all industries including construction, architecture, energy, and automotive.

2.1. Overview of CAx Systems

The following CAEs (computer-aided engineering systems, defined as computer systems supporting calculations and engineering analyses related to strength requirements and the simulation of the complex stress states of products and machines) should be noted: NX Nastran® for, Ansys Design Space, Pro/Engineer® Wildfire, Catia, and Unigraphics® NX.

The many tools in integrated manufacturing belonging to CAx systems include CAD, CAE, and CAM (computer-aided manufacturing), which are used to facilitate resource management (raw materials and semi-finished products) and enable the use of production machines. PDM (product data management) is widely used to support the management of resource planning and engineering processes and, once implemented, to support production management. EDM (engineering data management) and PPC (product planning and control) systems support the data processing facilities in CAx systems and are useful for complex information flow management: production requirements, design data, product versions and engineering changes, product structure, product properties, and development innovations.

2.2. Data Exchange between CAx Systems

In modern industry, it is quite common to encounter a situation where one person creates a model of an element in a CAD system, and another person creates a form for it along with tooling. As a result, the programs in which they work are very likely to be different. This is quite important for the form designer because they need to have a detailed model from another program to perform further modeling. There would be no problem with file exchange if the two systems were the same or if they worked based on an identical data format. The problem with file compatibility only occurs when the systems are different, and worse, when files for one program cannot be directly loaded in another program. In such a situation, it is necessary to translate them into a different format [2,6,12,75,78]. The given potential problem is one of many that require the use of neutral saving formats.

3. Neutral Formats for Data Exchange

Neutral formats (often interchangeable terms such as standards, couplers, converters, norms, or translators are used) of data exchange ensure communication between systems of a similar class, such as CAx.

These systems are often built on the basis of various representation schemes. The best-known translators are DXF, IGES, and STEP [5,34,49].

3.1. DXF Standard

DXF (Drawing Exchange Format) is a format for exchanging drawing (graphic) data and was developed by Autodesk [1,7–9,15,54,70]. Initially, this translator was intended for the AutoCAD program, but over time, it has become a generally available standard used for 2D data exchange.

3.2. IGES Standard

IGES (Initial Graphics Exchange Specification) is a neutral format that was created in 1979 to enable the exchange of information between CAD/CAM systems [49]. Moreover, since 2003, it is also fully compatible with the standard contained in NISTIR 6972 (USA).

The classical IGES converter can be divided into two parts: a preprocessor and a postprocessor. The preprocessor enables the conversion of information stored in the internal CAD database into the IGES format, while the postprocessor enables the conversion of information from the loaded IGES format into the internal CAD format. All CAD programs are now equipped with the IGES post- and preprocessor. However, the use of this format is limited due to the freedom to define component entries, the accuracy of geometry creation, and the type of tool library used [50].

3.3. STEP Standard

STEP (Standard for the Exchange of Product Model Data) is a standard that defines the rules for writing a product model based on geometric, topological, technological, and material data. The product model and its mapping data record are created based on the ISO 10303 Standard, including the STEP standard specification [23–25,27,30,32,41–44,74].

This translator was developed through a program funded by large industrial corporations wishing to bring about standardization in production, mainly to reduce intermediate production costs. When developing the STEP standard, the best features of standards developed in different countries were selected. The IGES standard played the greatest role here [60,82].

The STEP standard defines the rules for creating a product model in a way that allows for the processing and storage of information about the product and its manufacturing processes. As the STEP standard evolves, it becomes possible to flexibly configure CAx software and hardware components while avoiding incompatibility between data and network communication protocols.

All data in this standard are transported (similar to IGES) in the form of ASCII files. This facilitates their transfer between computers and CAx systems with different hardware and software platforms. The application of the STEP standard is based on application protocols, two of which are very important for the typically cooperative automotive industry. These are AP 214, for applications in the mechanical area, and AP 212 for applications in the electrical engineering area, as described in Table 1.

Table 1. The advancement of STEP standardization [26,28,29,59,64].

Protocol No.	Application	Status	Remarks
AP203	Structural design	IS	Structural 3D models in the manufacturing industry. Functional record in AP214
AP212	Description of the structure of electrotechnical devices	CDC	Structure of electrical equipment record in the construction of ships, cars, airplanes, and in the space industry. Compliance with AP214
AP214	Product data structure in the design of manufacturing processes for automotive components	IS	Structure development and record in the machine industry

In 2014, the STEP AP 242 application protocol was introduced, which combines the common features of the above protocols, as shown in Table 1. The main advantages of Step AP 242 are the functionality and capability to support all AP 203 and AP214 protocols, as well as 3D semantic PMI, high-quality 3D shapes, 3D kinematic assemblies, and 3D wire harnesses and pipelines. Due to its complex functionality, the AP242 application protocol is the industry's preferred STEP development protocol.

The main tool (programming language) for modeling the product structure in the STEP application is EXPRESS-M (Modeler) and EXPRESS, which is a text language for describing the data model.

The EXPRESS language is the basic element of the comprehensive, standardized description of a product model in the scope of the ISO 10303 Standard. Apart from the EXPRESS language, the EXPRESS-G graphic module has also been developed. An important benefit resulting from the consistent use of formal language is the unambiguous definition of the data transferred in the application. This makes it possible to test the specification of the data model by means of correct syntax and to generate the so-called adaptation programs, e.g., CAD pre- and postprocessors [14,19,36,70].

3.4. *Formats Compatible with the CAx Kernel*

In addition to neutral data exchange formats, there are also others ensuring proper data exchange between systems of a specific type.

Parasolid Transmit File Format

Parasolid Transmit file: This format enables data exchange between systems built on the Parasolid kernel. It allows for storing topological and geometric information of a created shape [7,11,18,19,21,40].

There are two types of translators:

(1) Parasolid X_T—text (saves information in a text file);
(2) Parasolid X_B—binary (saving data using symbols 0 and 1).

The purpose of the X_B format is usually limited to data exchange between the same type of systems (older or newer versions). The X_T format has a wider application for other systems as well. It does not mean that the file exported in the X_B format cannot be imported to other systems with the above-mentioned kernel.

ACIS Format

ACIS: This is a translator that enables data exchange between systems built on the kernel of the same name. Like Parasolid, it has the ability to save in text (SAT (*Standard ACIS Text*)) and binary (SAB (*Standard ACIS Binary*)). The format commonly used is SAT only, which uses the ASCII code for writing information. The saved file can be viewed using a simple text editor [54,60,62].

4. Translation Errors

Universal formats, apart from their undoubted advantages, also have disadvantages. Whenever the translation process is performed, there is a possibility of incorrect shape interpretation, which may result in geometry errors.

Common mistakes include [16,49,63,72] the following errors:

- Surface shortcomings—a gap is created between the adjacent surfaces, which prevents the connection of both surfaces;
- Holes—areas where, for example, the surface has disappeared and a void is thus generated;
- No surface trim—fragments of the surface protrude from the outline of the solid or the surface that was to be created, the cutting of which was omitted during translation;
- Overlapping geometry elements—when surfaces or edges intersect at some point, creating incorrect geometry;
- Unnecessary "garbage" in the file—when there is other, unnecessary information in the file in addition to the necessary data (e.g., chunks).

Data transfer, interoperability, and unreliability of selected operations

In the article entitled "*Product data exchange and product model sharing*" [13,33,41,55,83], the content of which was presented at one of the lectures at Delft University of Technology in June 2001, Joris Vergeest mentions, among others, the reasons for data transfer between systems, interoperability problems, and symptoms of defective operations.

The main reasons for data exchange between CAx systems are as follows [53,72,76]:

- The need to transfer models from systems that are unable to perform specific operations to systems with the desired functions;
- The continuation of work in another system;
- Communicating with other, often distant colleagues;
- Archival goals.

If the exchange occurs between the same type of systems (when the software modules are the same and of the same version), then the most effective way of communication is direct transfer (Figure 1), which does not cause data loss.

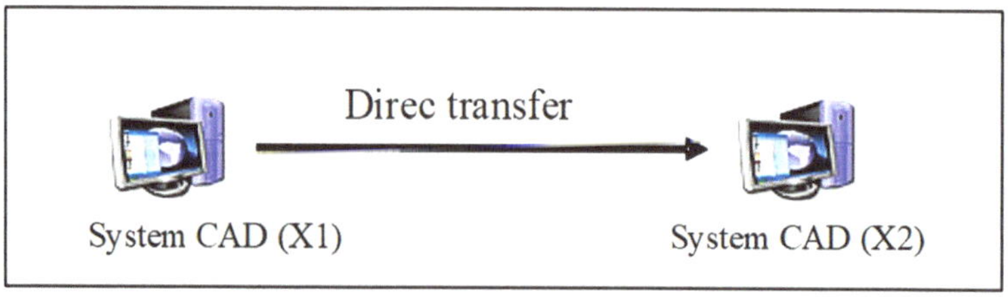

Figure 1. Direct data transfer between systems of the same type.

If the two systems are of a different type, it is usually not possible to transfer data from system A to system B (Figure 2). The file exported by system X is not understood by system Y. So, some kind of processing (conversion) is needed to enable this exchange.

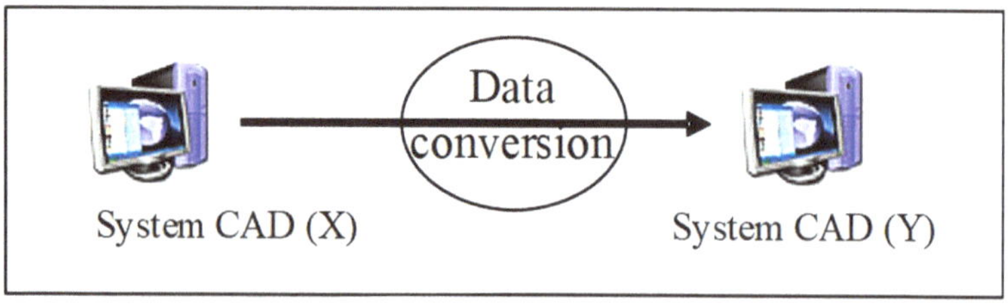

Figure 2. Data sent from system X to Y; the need to process information.

To improve the control of various exchange processes and reduce the number of converters, it is necessary to use CAD transfer standards. If a specific standard is used, the sending system should export the data in a format defined by it. The receiving system must be able to read and correctly interpret the data (Figure 3).

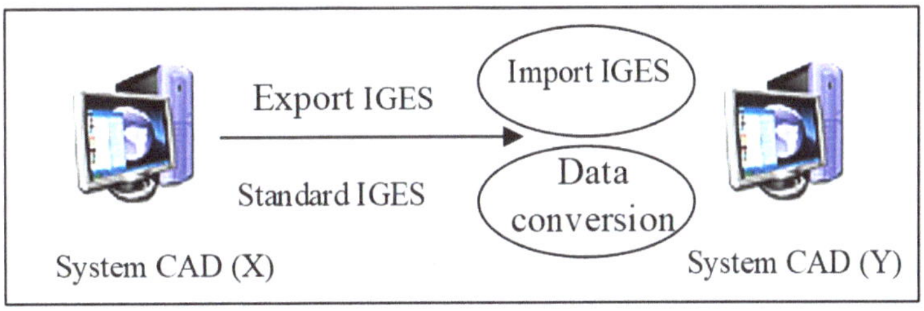

Figure 3. Using IGES data exchange standard.

STEP is the most developed standard. One of the advantages of using STEP is the semi-automatic data export/import resulting from the definition of the standard, which is "computer-readable" (Figure 4).

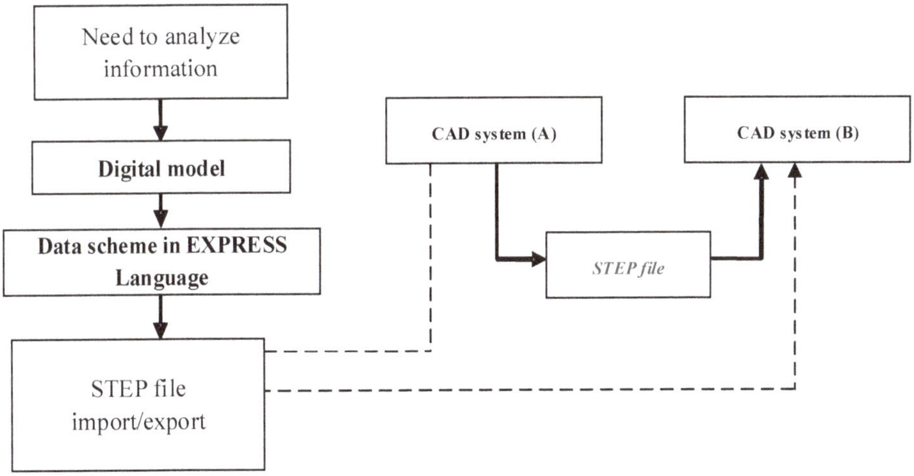

Figure 4. Data sent from system A to B using the STEP standard [14,34,35,66,76,80].

5. Interoperability Problem

With the advent of computer-aided design and quality engineering tools, the speed of software development has changed drastically. The integration of operations, data, models, and results from many disciplines has become the key to effective teamwork in industry. As a result of the high pace of development, the requirements for the information to be transferred and the methods for transferring knowledge, information, and data between designers and engineers have increased. Transfer issues have posed serious obstacles that have only been defined recently [31,67].

In collaborative design and engineering projects, most of the transferred data are reprocessed in CAD and other industrial systems. The increasing amount of information transferred between the CAx family systems is a kind of adoption of data exchange standards, and these practices are successfully supported using the PDM methodology. A significant fraction of transactions between systems fail due to the loss of semantics and/or the poor interoperability of software tools [61].

When formulating the problem of interoperability and operations between users X and Y, we can distinguish three concepts defined for each of the users, languages (Σ^X and Σ^Y) containing the possibility of coding objects, areas of representation (Q^X and Q^Y) containing representations of q^X and q^Y objects, and spaces of reflection (U^X and U^Y) defining the semantics of the objects. A reflection space is a collection of all entities that are relevant to a specific application or context.

Usually, the set of all possible representations of Q is a subset of U. In general, there is a very significant relationship between Q and U. The relation between the concepts is shown in Figure 5. The dashed lines show the process of mapping Q to the subset U. The relationship $a \in \Sigma$ causes the transition from the state q^1 to q^2, the output of y, and the transition from u^1 to u^2 [52,72,85].

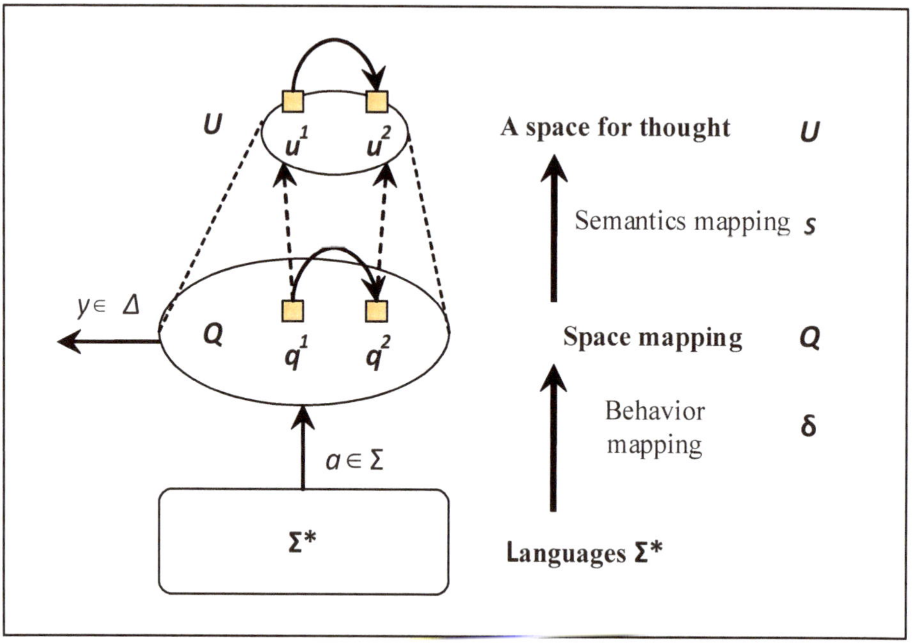

Figure 5. Language specificity Σ* for each user (X, Y, etc.).

In order to treat X and Y users separately, the definition of a reflection space (U) should be known to each side. Let us suppose that a member of team X, while working on a joint project, sends data to team member Y. The transfer initiated by X issuing the command $t \in \Sigma^X$ such as the string $y^X = \lambda^X(t, q^X)$ will be the serial code of the object q^X. Then, the string y^X will serve as an input string for the Y system, which will then be transferred to the q^Y state (Figure 6). The transfer can be considered successful if $u^X = u^Y$.

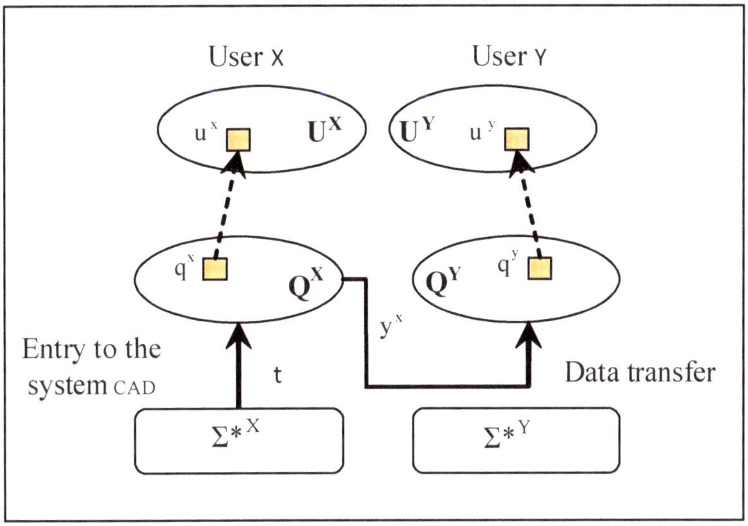

Figure 6. Data transfer from user X to Y.

6. Research

The preparation of this study required an analysis of the purpose, scope, course, and presentation of the results. A series of experiments based on a similar scheme and presentation method was planned.

The experiments consisted of exporting and importing 2D (flat documentation) and 3D (solids and surfaces) objects (Figure 7) to and from various CAD/CAE/CAx systems, in the following data exchange formats:

- For 2D, DXF objects (*.dxf);
- For 3D objects, IGES (*.igs; *.iges); STEP (*.stp; *.step); and ACIS, including SAT (*.sat) for systems with the ACIS kernel and PARASOLID (*.x_t) for systems with the PARASOLID kernel.

The aims of this research were as follows:

(1) To verify the compliance of CAD systems with selected neutral data exchange formats;
(2) To select the translator that is most suitable for the type of transferred entries (offering the lowest number of errors);
(3) To become familiar with the effects of applying certain import/export options.

The research plan included the following steps:

(1) Designing 2D drawings and 3D objects in individual systems;
(2) Saving the document in the default format offered by the program;
(3) Exporting (using the "Save as..." or "Export" command) the file to the specified format;
(4) Planning the course of the experiment;
(5) Defining the evaluation criteria;
(6) Determining the importance of each criterion;
(7) Importing the file into other systems;
(8) Analyzing the results;
(9) The presentation of the results;
(10) The comparison of the tested translators.

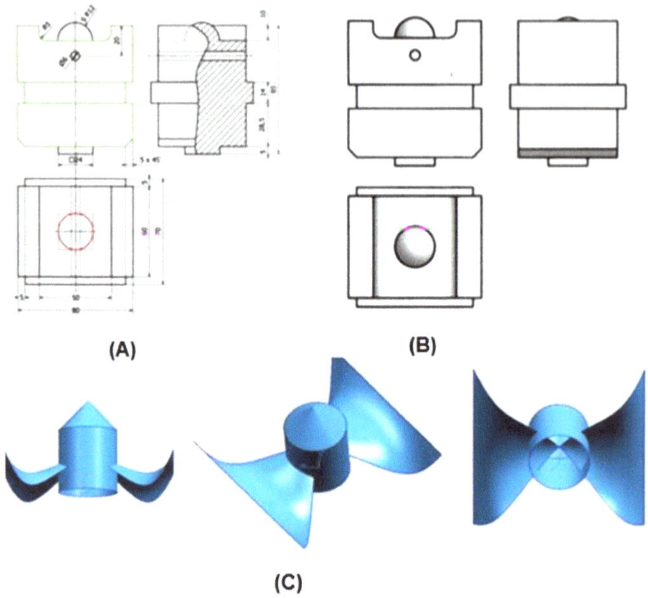

Figure 7. Example of transferred geometry: (**A**) an example of an object in 2D; (**B**) the 3D views of a solid rectangular object; (**C**) an example of a surface object.

The following assumptions were made:

(1) The exchange will be based on the "everyone with everyone" principle;
(2) CAE systems will act only as "recipients" of the file;
(3) Various possible options (if available) will be considered when exporting/importing data.

Ten systems (CATIA, SolidWorks, NX, AutoCAD, CADKey, Autodesk Inventor, Autodesk Fusion, FreeCAD, CADSoftTools, and BRL-CAD) were used for the tests, for which approval from distributors and technical support was obtained. Not every system participating in this study offers the function of saving/reading files in all the tested formats. There were cases in which import/export proved impossible, most often due to the limited license or version incompatibility of individual translators.

A more detailed description of the methodology applied to explain this is shown in Figure 8 for a 2D object, in Figure 9 for a solid object, and in Figure 10 for a surface object.

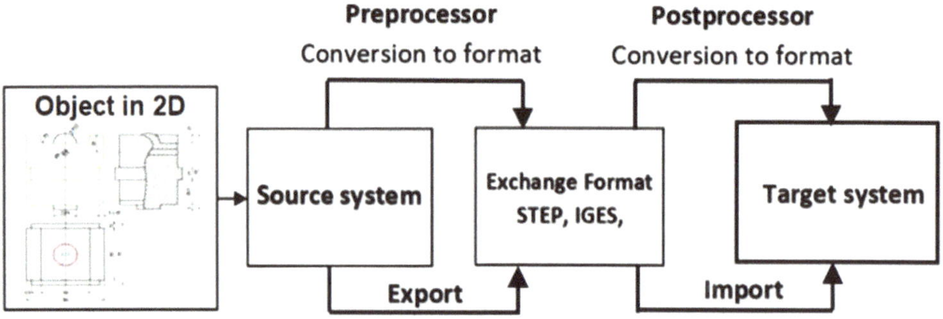

Figure 8. Methodology display example for object in 2D.

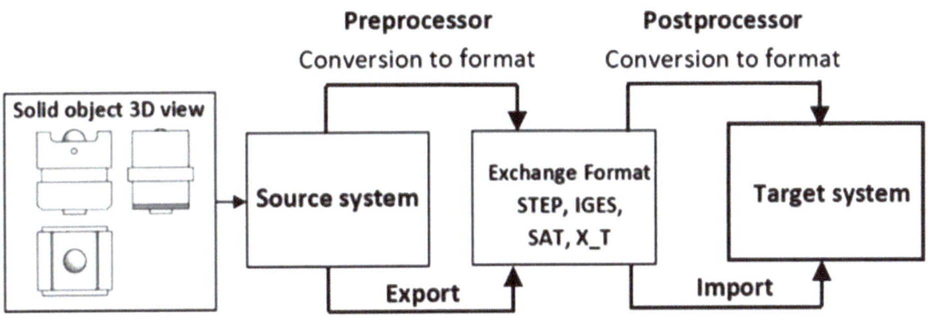

Figure 9. Methodology display example for solid object.

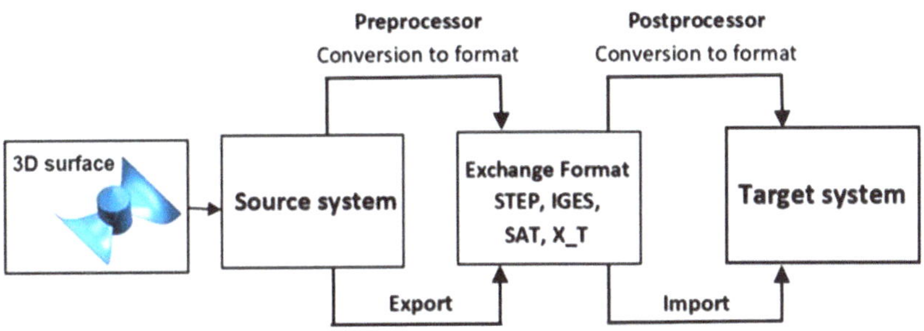

Figure 10. Methodology display example for surface object.

The research plan is schematically presented in Figure 11.

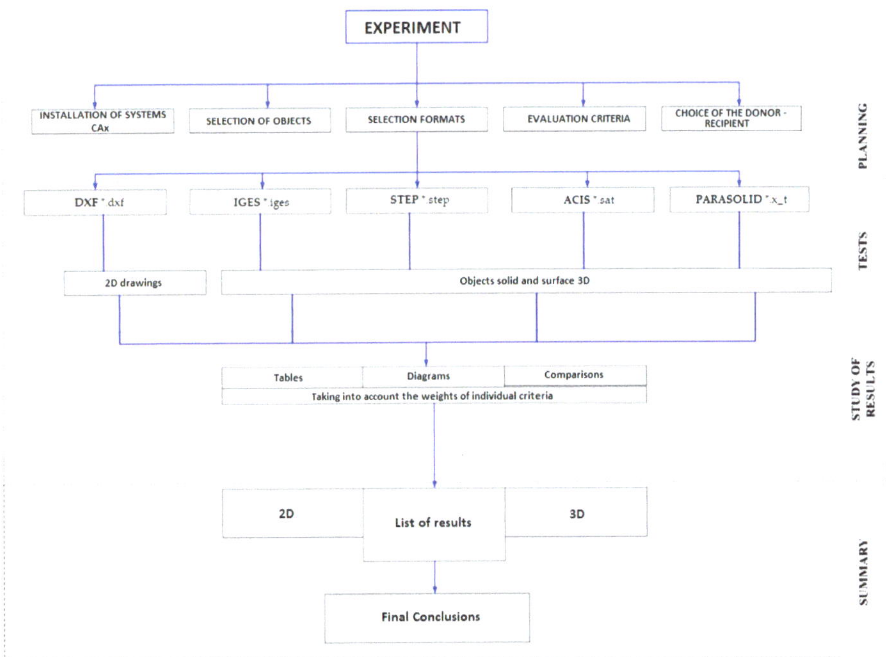

Figure 11. Diagram of the research plan.

7. Analysis of Import/Export of 2D and 3D Objects in Neutral Data Exchange Formats

7.1. Import/Export of 2D Objects in DXF Format

In the manufacturing process, we may encounter parts for which the final design can be implemented as a 2D or 3D model or a model with drawing documentation created in the CAD system. Thus, a model generated in any CAD system can be used for a variety of applications: for machining simulation and toolpath generation in a CAM system, finite element analysis, or documentation applications.

The reason for selecting neutral formats such as DXF, IGES, STEP, SAT, and X-T, which are the focus of this experiment, is their commercial availability. The first reason is that they are standards developed by the ANSI and ISO standards, and the second reason is their wide use in industrial sectors, such as the automotive and aerospace industries. They are also of interest in the field of CNC manufacturing, where data conversion, working with different processing formats, and data exchange between different CAD systems are frequently used. The evaluation criteria were selected and used to solve the problems of data processing between the customer's program and the program used for production, which can be applied to the machining process. In the production of parts, a large part is related to the program in which a given company works and in which a given CAD model is created. However, the range of programs that can be used in the field and the consequent work with them is limited by the format in which the CAD model is saved, sent, and subsequently opened. The selected formats were chosen based on the degree of compatibility between those most used for saving in different CAD systems. The DXF format is the most commonly used data format in current graphics and CAD programs. The choice of comparing DXF formats was based on their use in programming on CNC machines, which makes it possible to open these files and extract points and contours from them. This saves programming and testing time, and the resulting contour is an exact match to the drawing. DXF files usually contain several layers in which the part drawing is arranged.

When producing parts with the required dimensions and shapes, CNC programmers often encounter a problem called data conversion, especially in the case of shaped mold parts, where individual surfaces need to be manually modified using various tools in the CAD system. These errors appear as missing surfaces, gaps, holes, sharp edges, overlapping surfaces, or incorrect visualization of details. The reason for these errors is the format in which the CAD model is saved is different from the format in which it is subsequently opened in another CAD system. This error occurs when different CAD systems are used for different project teams.

Saving CAD models in a certain format, which is the starting point for its further production processing, as well as data processing conversion, has its own justification. Incorrect data processing not only increases the production time but also the time needed to make corrections in the CAD system. It is only when the model is repaired and meets the required properties that it is ready for the next process, e.g., programming toolpaths in the CAM system and generating NC data for a specified CNC machine.

The DXF translator, as a neutral format for the exchange of 2D objects between CAD/CAE/CAx systems, translated the transferred entries correctly in most cases. An imported object with a partial loss of geometry can be quickly repaired in many cases. Before exporting an object, it is always worth considering the "target" system for which it is intended. Some systems, or rather their translators, are not compatible with each other. In these cases, correct data exchange proved impossible. It is possible to experiment with different import options to find the right settings for the type of 2D data being transferred. It is also advisable to export the data several times with different export settings, if available. Sometimes, a change in one parameter (e.g., format version) can greatly improve the result of an import into another system. It is also worth noting which types of drawing entries transfer worse than others. The time spent on drawing them can often be wasted, as these entries are most likely to be lost during replacement [51,57,58,79,87–89].

7.2. Import/Export of 3D Solid Objects

During the analysis of data exchange in the form of a solid object entry in the IGES, STEP, SAT, and X_T formats, we aimed to select the format that was the most suitable for the type of geometry transferred. In order to compare the above translators, it was necessary to divide them into two groups:

(a) IGES and STEP—as neutral formats for data exchange between all CAx systems regardless of the kernel used;
(b) SAT and X_T—as formats enabling the exchange of information between systems using specific spatial modeling kernels.

The IGES translator, in comparison with the STEP translator, was less able to correctly recognize the type of an object, its topological conformity, and its division into integers. However, it was better at transferring sketches. The comparison of the obtained results for a solid object between two formats, IGES and STEP, is shown in Table 2.

Table 2. Comparison of the obtained results for a solid object (IGES and STEP).

Solid Object	IGES			STEP		
	Correct	Possible	Percentage Value	Correct	Possible	Percentage Value
Object type	45.8	56	82%	36	36	100%
Topological compliance	55.2	56	98%	36	36	100%
Division into integers	5	56	9%	4,5	36	13%
Sketches	21	56	38%	6	36	17%

Table 3 shows that the ACIS (*.sat) and PARASOLID translators performed well on the two most important parameters assessed. Unfortunately, they do not allow for the export of the division of an object into integers or a sketch.

Table 3. Comparison of the obtained results for a solid object (SAT and PARASOLID).

Solid Object	SAT			PARASOLID		
	Correct	Possible	Percentage Value	Correct	Possible	Percentage Value
Object type	2	2	100%	12	12	100%
Topological compliance	2	2	100%	12	12	100%
Division into integers	0	2	0%	0	12	0%
Sketches	0	2	0%	0	12	0%

If the CAD system offers object export in all four formats, then the STEP translator should be chosen. In contrast to IGES, it has fewer import/export options; therefore, it is more difficult to make a mistake.

7.3. Import/Export of 3D Surface Objects

By analyzing the results of data exchange in the form of surface object entries in IGES, STEP, SAT, and X_T formats between systems from the CAx family, an attempt was made to select the translator that performed best in the tests conducted.

As can be seen in Table 4, the IGES standard proved to be better than STEP this time. The translation of the type of object, its topological compliance, and a sketch could be performed more accurately using this standard. The data transferred to the Ansys Design Space system did not contain sketches. If this were not the case, the percentage of this type of transferred entries would not be 71% but 83%, as is shown in Table 4.

Table 4. Comparison of the obtained results for the surface object (IGES and STEP).

Surface Object	IGES			STEP		
	Correct	Possible	Percentage Value	Correct	Possible	Percentage Value
Object type	48	48	100%	34	36	94%
Topological compliance	46	48	96%	31,6	36	88%
Division into integers	48	48	100%	36	36	100%
Sketches	34 (40)	48 (42)	71% (83)	24	36	67%

The division of a surface object into integers depends on many factors and is less important here than in the case of solid objects. Unstitched surfaces always retain their division. By using the appropriate stitching sequence, it is possible to achieve the desired effect without major problems. As a result, we often achieve 100% compliance with the assessed parameter. The number of analyses performed in both formats is also worth considering. Typically, the higher the number of tests, the higher the probability of encountering a misinterpreted case, thus lowering the overall number of points obtained. As Table 4 shows, despite the greater number of analyses, the IGES format produced better results than STEP.

An analysis of the research results presented in Table 5 also reveals the relatively low number of tests performed. The systems tested often had limited licenses and did not always allow for this type of replacement. The results obtained with the SAT and X_T translators are very good; they would be even better if these formats also supported sketch import/export. To fully reveal the advantages and disadvantages of the SAT and X_T translators, a much larger number of tests should be carried out.

Table 5. Comparison of the obtained results for the surface object (SAT and X_T).

Surface Object	SAT			X_T		
	Correct	Possible	Percentage Value	Correct	Possible	Percentage Value
Object type	2	2	100%	6	6	100%
Topological compliance	2	2	100%	6	6	100%
Division into integers	2	2	100%	6	6	100%
Sketches	0	2	0%	0	6	0%

8. Conclusions

This paper presents detailed information and research results on computer systems supporting the work of an engineer–designer and provides a general method of solving the potential problems that may arise when working with different formats for 2D, solid, and 3D surfaces, comparing DXF, IGES, STEP, SAT, and X-T formats. In this way, this study allows for a better explanation of and approach to the efficiency of using these formats and makes the further processing of data easier to compare with each other. The models described in this paper can be generally applied to the user's needs only when using the formats outlined in the paper.

We conducted several data exchange studies to identify a neutral format that ensures the least data loss for a specific type of transferred geometry. The analysis was based on a carefully prepared experiment and defined evaluation criteria. All tests were carried out with great care and attention to avoid errors. The obtained results were presented in the form of tables and comparative graphs.

The following propositions can be suggested:

1. For the conversion of data in the form of geometric elements of a CAD model, the STEP file format is the most suitable. It can better recognize the type of object as well as its topological correspondence at a higher level compared with the IGES format. On the other hand, for 2D drawing formats, the IGES format is better.
2. Comparing the ACIS (*.sat) and Parasolid type translators, it can be stated that their main disadvantage is the insufficient export of the object division and the CAD model to integers or sketches. However, they are very good at defining the object type and topological correspondence, which plays a very important role when working with systems in which the main part of the project is meshing the model, i.e., the object. These are, for example, the Moldflow Plastic Insight systems or Simufact Additive, designed for metal additive manufacturing, where the CAD file is imported in Parasolid format, and the imported file is then used to generate a surface mesh of the model, containing data on the number of elements and the number of nodes, which is later important for the creation of the voxel mesh. Therefore, the correct choice of format is also important in this production process.
3. In the case of modeling surfaces, which are most often used in the field of automotive production or mold manufacturing, the IGES format proved to be better than the STEP format, which better handled the parameters of the project in the form of sketches, topological correspondence, object type, etc. Applied research in the form of the evaluation of neutral formats has its justification in the manufacturing process, such as the machining process, where it is necessary to work with different file formats from customers working in different CAE systems. These can be simple or complex surfaces, such as mold cavities, where the way all surfaces are rendered is very important. Therefore, the correct choice of format at the export or import stage of the part plays a crucial role in terms of correctly loading all surfaces, contours, and other geometric elements.
4. Consequently, file transfer and incorrect compatibility between different CAE systems can lead to erroneous data in the form of incorrectly loaded or displayed CAD models, or incorrect geometry elements. This means additional work and time spent on model editing in the form of stitching surfaces, drawing edges, problems with interpolation, the accuracy of drawing geometric elements, etc. All this then has an impact on the productivity of production, as well as on achieving the required accuracy and surface quality of the manufactured parts.

It is hoped that, in the future, design professionals will be able to choose, to some extent, neutral data exchange formats in their work. At the present stage of the development of CAD/CAE systems, neutral data formats do not work 100%; therefore, there is a need for further improvement and thus for the standardization of tools and principles used in design. The development of translators should be aimed at achieving full compatibility with CAx systems. Reaching this level will be largely possible through the acceleration of the introduction of a new product to the market and, consequently, the minimization of the costs associated with this process.

Author Contributions: Conceptualization, P.K., M.M. and T.M; methodology, P.K., P.F., M.M., T.M. and J.V.; investigation, M.M.; P.F., J.K., S.S. and Ł.A.; validation, P.K., P.F., M.M., J.V. and T.M.; writing—original draft preparation, P.K. and J.V.; writing—review and editing, P.K., P.F., S.R. and J.K.; visualization, P.K., M.M., S.S., S.R. and J.K.; funding acquisition, P.K., M.M., P.F. and Ł.A. All authors have read and agreed to the published version of the manuscript.

Funding: P.K. and M.R.D. acknowledge the financial support from the program of the Polish Ministry of Science and Higher Education under the name "Regional Initiative of Excellence" in 2019–2022, project No. 003/RID/2018/19, funding amount 11 936 596.10 PLN, VEGA 1/0500/20, KEGA 036TUKE-4/2021.

Institutional Review Board Statement: Not applicable.

Informed Consent Statement: Not applicable.

Data Availability Statement: Data will be available upon request.

Conflicts of Interest: The authors declare no conflict of interest.

References

1. 3D Interop's Persistent Naming Translation. Available online: http://blog.spatial.com/persistent-naming-translation (accessed on 13 March 2021).
2. 3D Systems. *SLC File Format Information*; 3D Systems, Inc.: Valencia, Spain, 1994.
3. Abdul-Ghafour, S.; Ghodous, P.; Shariat, B.; Perna, E.; Khosrowshahi, F. Semantic Interoperability of Knowledge in Feature-Based CAD Models. *Comput. Aided Des.* **2011**, *56*, 15–57. [CrossRef]
4. Bettig, B.; Shah, J. Derivation of S Standard Set of Geometric Constraints for Parametric Modeling and Data Exchange. *Comput. Aided Des.* **2001**, *33*, 17–33. [CrossRef]
5. Brangé, J.; Delaunay, J.Y.; Moura, E. Whitepaper: Development of STEP AP 242 ed2 Managed Model Based 3D Engineering Version 1.0. 2014. Available online: http://www.asd-ssg.org/step-ap242-ed2 (accessed on 30 March 2014).
6. Chen, J.; Ma, Y.; Wang, C.; Au, C. Collaborative design environment with multiple CAD systems. *Comput. Aided Des. Appl. Des.* **2005**, *2*, 367–376. [CrossRef]
7. Chiu, W.K.; Tan, S.T. Multiple material objects: From CAD representation to data format for rapid prototyping. *Comput. Aided Des.* **2000**, *32*, 707–717. [CrossRef]
8. Edward, C. *Techniki Komputerowe CAx w Inżynierii Produkcji*; Wydawnictwa Naukowo—Techniczne: Warszawa, Poland, 2000.
9. Choi, G.H.; Mun, D.; Han, S. Exchange of CAD Part Models Based on the Macro-Parametric Approach. *Int. J. CAD/CAM* **2002**, *2*, 13–21.
10. Desrochers, A.; Clement, A. A dimensioning and tolerancing assistance model for CAD/CAM systems. *Int. J. Adv. Manuf. Technol.* **1994**, *9*, 352–361. [CrossRef]
11. Dolenc, A.; Malela, I. Leaf: A data exchange format for LMT processes. In *Proceedings of the 3rd International Conference on Rapid Prototyping, Dayton, OH, USA, 23–25 June 1992*; University of Dayton: Dayton, OH, USA, 1992; pp. 4–12.
12. Elber, G.; Patrikalakis, N.; Brunet, P. Integrated feature-based and geometric CAD data exchange. In Proceedings of the Ninth ACM Symposium on Solid Modeling and Applications, Genoa, Italy, 9–11 June 2004.
13. ePLM Interoperability STEP AP242 Project. AP 242 Edition 2 Capabilities for Additive Manufacturing Interoperability. Available online: http://www.ap242.org/additive-manufacturing (accessed on 5 April 2021).
14. Feeney, B.A.; Frechette, S.P.; Srinivasan, V. A Portrait of an ISO STEP Tolerancing Standard as an Enabler of Smart Manufacturing Systems. *J. Comput. Inf. Sci. Eng.* **2014**, *15*, 021001. [CrossRef]
15. Gargallo-Peiró, A.; Roca, X.; Peraire, J.; Sarrate, J. Defining quality measures for validation and generation of high-order tetrahedral meshes. In *Proceedings of the 22nd International Meshing Roundtable, Orlando, FL, USA, 13–16 October 2013*; Springer: Berlin/Heidelberg, Germany, 2014; pp. 109–126.
16. Gargallo-Peiró, A.; Roca, X.; Sarrate, J. A surface mesh smoothing and untangling method independent of the CAD parameterization. *Comput. Mech.* **2014**, *53*, 587–609. [CrossRef]
17. Kaščák, L'.; Varga, J.; Bidulská, J.; Bidulský, R.; Grande, M. Simulation tool for material behaviour prediction in additive manufacturing. *Acta Metall. Slovaca* **2023**, *29*, 113–118. [CrossRef]
18. Handschuh, S.; Dotzauer, R.; Fröhlich, A. Standardised formats for visualisation: Application and development of JT. In *Concurrent Engineering Approaches for Sustainable Product Development in a Multi-Disciplinary Environment*; Stjepandić, J., Rock, G., Bil, C., Eds.; Springer: London, UK, 2013; pp. 741–752.
19. Huang, W.; Russell, R. *Adaptive Moving Mesh Methods*; Springer: Berlin/Heidelberg, Germany, 2011.
20. *ISO 10303-47*; Industrial Automation Systems and Integration—Product Data Representation and Exchange—Part 47: Integrated Generic Resource: Shape Variation Tolerances. International Organization for Standardization: Geneva, Switzerland, 2014.
21. *ISO 14306*; Industrial Automation Systems and Integration—JT File Format Specification for 3D Visualisation. International Organisation for Standardisation: Geneva, Switzerland, 2012.
22. Jauhar, T.A.; Han, S.; Kwon, S. Downstream Computer-Aided Design, Engineering, and Manufacturing Integration Using Exchangeable Persistent Identifiers in Neutral Re-imported Computer-Aided Design Models. *ASME J. Comput. Inf. Sci. Eng.* **2021**, *21*, 014501. [CrossRef]
23. Jian, C.; Miao, L.; Qiu, K.; Zhang, M. An Improved NBA-Based STEP Design Intention Feature Recognition. *Future Gener. Comput. Syst.* **2018**, *88*, 357–362. [CrossRef]
24. Li, X.; He, F.; Cai, X.; Zhang, D. CAD data exchange based on the recovery of feature modelling procedure. *Int. J. Comput. Integr. Manuf.* **2010**, *25*, 874–887. [CrossRef]
25. Kim, J.; Pratt, M.J.; Iyer, R.G.; Sriram, R.D. Standardized Data Exchange of CAD Models with Design Intent. *Comput.-Aided Des.* **2008**, *40*, 760–777. [CrossRef]
26. Kirkwood, R.; Sherwood, J.A. Sustained CAD/CAE Application Integration: Supporting Simplified Models. *ASME J. Comput. Inf. Sci. Eng.* **2021**, *21*, 011002. [CrossRef]
27. Kirkwood, R. CAD Integration through Virtual Persistent Identifiers and Design Change Requirements. U.S. Patent App. 15/456,475, 13 September 2018.

28. Varga, J.; Spišák, E.; Gajdoš, I.; Mulidrán, P. Comparison of Milling Strategies in the Production of Shaped Surfaces. *Adv. Sci. Technol. Res. J.* **2022**, *16*, 267–274. [CrossRef]
29. Kirkwood, R.; Sherwood, J.A. Sustained CAD/CAE integration: Integrating with successive versions of STEP or IGES files. *Eng. Comput.* **2018**, *34*, 1–13. [CrossRef]
30. Krause, F.L.; Stiel, C.; Lüddemann, J. Processing of CAD data-conversion, verification and repair. In Proceedings of the Fourth ACM Symposium on Solid Modeling and Applications, Atlanta, GA, USA, 14–16 May 1997; pp. 248–254.
31. Kreis, A.; Hirz, M.; Stadler, S.; Salchner, M.; Rossbacher, P. Convenient Connection Technology Data Model Supporting Optimized Information Exchange between CAx-Systems. *Comput. Aided Des. Appl.* **2018**, *15*, 771–778. [CrossRef]
32. Kumar, V.; Dutta, D. An assessment of data formats for layered manufacturing. *Adv. Eng. Softw.* **1997**, *28*, 151–164. [CrossRef]
33. Varga, J.; Tóth, T.; Kaščák, L'.; Spišák, E. The effect of the machining strategy on the surface accuracy when milling with a ball end cutting tool of the aluminum alloy AlCu4Mg. *Appl. Sci.* **2022**, *12*, 10638. [CrossRef]
34. Lee, K.S.; Kim, S.H. Non-uniform deformation of an STL model satisfying error criteria. *Comput. Aided Des.* **2010**, *42*, 238–247. [CrossRef]
35. Lee, K. *Principles of CAD/CAM/CAE Systems*; Addison-Wesley Longman Publishing Co., Inc.: Boston, MA, USA, 1999.
36. Lipman, R.; Lubell, J. Conformance Checking of PMI Representation in CAD Model STEP Data Exchange Files. *Comput. Aided Des.* **2015**, *66*, 14–23. [CrossRef]
37. Xiaoxia, L.; Fazhi, H.; Xiantao, C.; Zhang, D.; Chen, Y. A method for topological entity matching in the integration of heterogeneous CAD systems. *Integr. Comput. Aided Eng.* **2013**, *20*, 15–30. [CrossRef]
38. Lu, W.; Jiang, X.; Liu, X.; Qi, Q.; Scott, P.J. Modelling the integration between specifications and verification for cylindricity based on category theory. *Meas. Sci. Technol.* **2010**, *21*, 115107. [CrossRef]
39. Marsan, A.; Kumar, V.; Dutta, D.; Pratt, M.J. *An Assessment of Data Requirements and Data Transfer Formats for Layered Manufacturing*; National Institute of Standards and Technology: Gaithersburg, MD, USA, 1998.
40. Martin, W.; Cohen, E. Representation and extraction of volumetric attributes using trivariate splines: A mathematical framework. In *Proceedings of the 6th ACM Symposium on Solid Modelling and Applications, Ann Arbor, MI, USA, 4–8 June 2001*; Association for Computing Machinery: New York, NY, USA, 2001; pp. 234–240.
41. Varga, J.; Tóth, T.; Frankovský, P.; Dulebová, L'.; Spišák, E.; Zajačko, I.; Živčák, J. The influence of automated machining strategy on geometric deviations of machined surfaces. *Appl. Sci.* **2021**, *11*, 2353. [CrossRef]
42. Massarwi, F.; Elber, G. A B-spline based framework for volumetric object modelling. *Comput. Aided Des.* **2016**, *78*, 36–47. [CrossRef]
43. Matta, A.K.; Raju, D.R.; Suman, K.N. The integration of CAD/CAM and rapid prototyping in product development: A review. *Mater. Today* **2015**, *2*, 3438–3445. [CrossRef]
44. McMains, S. *Geometric Algorithms and Data Representation for Solid Freeform Fabrication*; University of California: Berkeley, CA, USA, 2000.
45. Khan, M.T.; Rezwana, S. A review of CAD to CAE integration with a hierarchical data format (HDF)-based solution. *J. King Saud Univ.-Eng. Sci.* **2020**, *33*, 248–258. [CrossRef]
46. Zhang, D.; He, F.; Han, S.; Zou, L.; Wu, Y.; Chen, Y. An efficient approach to directly compute the exact Hausdorff distance for 3D point sets. *Integr. Comput. Aided Eng.* **2017**, *24*, 261–277. [CrossRef]
47. Medichalam, M.S.; Shah, J.J.; D'Souza, R. N-Rep: A Neutral Feature Representation to Support Feature Mapping and Data Exchange across Applications. In Proceedings of the DETC2004-57712, International Design Engineering Technical Conferences, Salt Lake City, UT, USA, 28 September–2 October 2004.
48. Arkadiusz, M. SolidWorks'98 i SW'98 Plus. *Cadcam Forum* **1999**, *6*, 404.
49. Mun, D.; Han, S. Identification of topological entities and naming mapping for parametric CAD model exchanges. *Int. J. CAD/CAM* **2005**, *5*, 69–82.
50. Murshed, M.S.M.; Dixon, A.; Shah, J.J. Neutral Definition and Recognition of Assembly Features for Legacy Systems Reverse Engineering. In Proceedings of the DETC2009-86739, International Design Engineering Technical Conferences and Computers and Information in Engineering Conference, IDETC/CIE, San Diego, CA, USA, 30 August–2 September 2009; pp. 615–628.
51. Bernard, P. Problemy z Translacją Plików. *In TS Raport* **2002**, *12*, 26–27.
52. Pan, H.; Zhou, T. Generation and optimisation of slice profile data in rapid prototyping and manufacturing. *J. Mater. Process. Technol.* **2007**, *187*, 623–626.
53. Poya, R.; Sevilla, R.; Gil, A.J. A unified approach for a posteriori high-order curved mesh generation using solid mechanics. *Comput. Mech.* **2016**, *58*, 457–490. [CrossRef]
54. Pratt, M.J. Introduction to ISO 10303—The STEP standard for product data exchange. *J. Comput. Inf. Sci. Eng.* **2001**, *1*, 102–103. [CrossRef]
55. Adrian, Ł.; Piersa, P.; Szufa, S.; Romanowska-Duda, Z.; Grzesik, M.; Cebula, A.; Kowalczyk, S.; Ratajczyk-Szufa, J. Experimental research and thermographic analysis of heat transfer processes in a heat pipe heat exchanger utilizing as a working fluid R134A. In *Renewable Energy Sources: Engineering, Technology, Innovation*; Mudryk, K., Werle, S., Eds.; Springer Proceedings in Energy ICORES 2017; Springer: Berlin/Heidelberg, Germany, 2018; ISBN 978-3-319-72370-9. [CrossRef]
56. Rebaioli, L.; Fassi, I. A review on benchmark artifacts for evaluating the geometrical performance of additive manufacturing processes. *Int. J. Adv. Manuf. Technol.* **2017**, *93*, 2571–2598. [CrossRef]

57. Rodriguez, E.; Bonnard, R.; Alvares, A. Proposal of an advanced data model for step-nc compliant additive manufacturing. In Proceedings of the 24th ABCM International Congress of Mechanical Engineering, Curitiba, Brazil, 15 February 2017.
58. Ruiz-Gironés, E.; Roca, X.; Sarrate, J. High-order mesh curving by distortion minimization with boundary nodes free to slide on a 3D CAD representation. *Comput. Aided Des.* **2016**, *72*, 52–64. [CrossRef]
59. Ryou, M.S.; Jee, H.S.; Kwon, W.H.; Bang, Y.B. Development of a data interface for rapid prototyping in STEP-NC. *Int. J. Comput. Integr. Manuf.* **2006**, *19*, 614–626. [CrossRef]
60. Ramnath, S.; Haghighi, P.; Venkiteswaran, A.; Shah, J.J. Interoperability of CAD geometry and product production information for computer Integrated Manufacturing. *Int. J. Comput. Integr. Manuf.* **2020**, *33*, 116–132. [CrossRef]
61. Seo, T.; Lee, Y.; Cheon, S.; Patil, L. Sharing CAD Models Based on Feature Ontology of Commands History. *Int. J. CAD/CAM* **2005**, *5*, 39–47.
62. Shah, J.J.; Mäntylä, M. *Parametric and Feature-Based CAD/CAM: Concepts, Techniques, and Applications*; John Wiley & Sons: Hoboken, NJ, USA, 1995.
63. Šormaz, D.N.; Arumugam, J.; Harihara, R.S.; Patel, C.; Neerukonda, N. Integration of product design, process planning, scheduling, and FMS control using xml data representation. *Robot. Comput.-Integr. Manuf.* **2010**, *26*, 583–595. [CrossRef]
64. Starly, B.; Lau, A.; Sun, W.; Lau, W.; Bradbury, T. Direct slicing of STEP based NURBS models for layered manufacturing. *Comput. Aided Des.* **2005**, *37*, 387–397. [CrossRef]
65. Staves, D.R.; Salmon, J.L.; Red, W.E. Associative CAD references in the neutral parametric canonical form. *Comput. Aided Des. Appl.* **2017**, *14*, 408–421. [CrossRef]
66. Venkiteswaran, A. Interoperability of Geometric Dimension & Tolerance Data between CAD Systems. Master's Thesis, Arizona State University, Tempe, AZ, USA, 2016.
67. STEP Tools. *AdditiveNC: Make STEP-NC for Additive Manufacturing*; STEP Tools, Inc.: New York, NY, USA, 2015; Available online: https://github.com/steptools/AdditiveNC (accessed on 19 March 2021).
68. Turk, G. *The PLY Polygon File Format*; Leland Stanford Junior University: Palo Alto, CA, USA, 1994.
69. Joris, V. Product Data Exchange and Product Model Sharing. Delft University of Technology. Available online: www.pdteurope.com (accessed on 14 June 2001).
70. Vilmart, H.; Léon, J.C.; Ulliana, F. From CAD assemblies toward knowledge-based assemblies using an intrinsic knowledge-based assembly model. *Comput.-Aided Des. Appl.* **2018**, *15*, 300–317. [CrossRef]
71. Wu, Y.; Shah, J.J.; Davidson, J.K. Computer Modeling of Geometric Variations in Mechanical Parts and Assemblies. *J. Comput. Inf. Sci. Eng.* **2003**, *3*, 54. [CrossRef]
72. Adrian, Ł.; Szufa, S.; Piersa, P.; Kuryło, P.; Mikołajczyk, F.; Kurowski, K.; Pochwała, S.; Obraniak, A.; Stelmach, J.; Wielgosiński, G.; et al. Analysis and evaluation of heat pipe efficiency to reduce low emission with the use of working agents r134a, r404a and r407c, r410a. *Energies* **2021**, *14*, 1926. [CrossRef]
73. Yin, Z. Direct generation of extended STL file from unorganised point data. *Comput. Aided Des.* **2011**, *43*, 699–706. [CrossRef]
74. Zhang, Z.; Joshi, S. Slice data representation and format for multimaterial objects for additive manufacturing processes. *Rapid Prototyp. J.* **2017**, *23*, 149–161. [CrossRef]
75. Zhao, X.; Pasupathy, T.K.; Wilhelm, R.G. Modelling and representation of geometric tolerances information in integrated measurement processes. *Comput. Ind.* **2006**, *57*, 319–330. [CrossRef]
76. Zhou, M.Y. Modelling and representation of heterogeneous objects based on STEP for layered manufacturing. *Int. J. Prod. Res.* **2006**, *44*, 1297–1311. [CrossRef]
77. Zhou, M.Y. STEP-based approach for direct slicing of CAD models for layered manufacturing. *Int. J. Prod. Res.* **2005**, *43*, 3273–3285. [CrossRef]
78. Zhu, L.; Jayaram, U.; Jayaram, S.; Kim, O. Ontology-Driven Integration of CAD/CAE Applications: Strategies and Comparisons. In Proceedings of the International Design Engineering Technical Conferences and Computers and Information in Engineering Conference, San Diego, CA, USA, 30 August–2 September 2009.
79. Stelmach, J.; Kuncewicz, C.; Szufa, S.; Jirout, T.; Rieger, F. The Influence of Hydrodynamic Changes in a System with a Pitched Blade Turbine on Mixing Power. *Processes* **2021**, *9*, 68. [CrossRef]
80. MBE PMI Validation and Conformance Testing Project. Available online: www.nist.gov/ctl/smart-connected-systems-division/smart-connected-manufacturing-systems-group/mbe-pmi-validation (accessed on 23 December 2022).
81. Bozek, P.; Krenicky, T.; Prajova, V. Digital Induction Motor Model Based on the Finite Element Method. *Appl. Sci.* **2023**, *13*, 5124. [CrossRef]
82. Komák, M.; Pivarčiová, E. Creating a Simulation Environment for the Micromouse. *TEM J.* **2022**, *11*, 479–483. [CrossRef]
83. Ivanova, T.N.; Korshunov, A.I.; Peterka, J.; Ratnikov, I.A.; Pokorny, P. The Possibility of Modeling Parts in Production Technology. *Multidiscip. Asp. Prod. Eng.* **2020**, *3*, 359–373. [CrossRef]
84. Runde, S.; Fay, A. A data exchange format for the engineering of building automation systems. In Proceedings of the 2008 IEEE International Conference on Emerging Technologies and Factory Automation, Hamburg, Germany, 15–18 September 2008. [CrossRef]
85. Kuryło, P.; Frankovský, P.; Papacz, W.; Delyová, I.; Schurger, B. Design of an automated system for measuring car bodies. *Acta Mechatronica* **2021**, *6*, 11–20. [CrossRef]

86. Rappoport, A. An architecture for universal CAD data exchange. In *Proceedings of the SM03: 8th ACM Symposium on Solid Modeling and Applications, Seattle, WA, USA, 16–20 June 2003*; Association for Computing Machinery: New York, NY, USA, 2003; ISBN 978-1-58113-706-4.
87. Wu, Y.; He, F.; Zhang, D.; Li, X. Service-oriented feature-based data exchange for cloud-based design and manufacturing. *IEEE Trans. Serv. Comput.* **2018**, *11*, 341–353. [CrossRef]
88. Zhang, D.; He, F.; Han, S.; Li, X. Quantitative optimization of interoperability during feature-based data exchange. *Integr. Comput. Aided Eng.* **2015**, *23*, 31–50. [CrossRef]
89. Varga, J.; Stahovec, J.; Beňo, J.; Vrabeľ, M. Assessment of surface quality for chosen milling strategies when producing relief surfaces. *Adv. Sci. Technol. Res. J.* **2014**, *8*, 37–41.

applied sciences

MDPI

Article

Automation of the Edge Deburring Process and Analysis of the Impact of Selected Parameters on Forces and Moments Induced during the Process

Karol Falandys [1], Krzysztof Kurc [2,*], Andrzej Burghardt [2] and Dariusz Szybicki [2]

[1] Safran Aircraft Engines Poland, Południowa 23, 39-120 Sędziszów Małopolski, Poland; karol.falandys@safrangroup.com

[2] Faculty of Mechanical Engineering and Aeronautics, Rzeszow University of Technology, Aleja Powstancow Warszawy 8, 35-959 Rzeszow, Poland; andrzejb@prz.edu.pl (A.B.); dszybicki@prz.edu.pl (D.S.)

* Correspondence: kkurc@prz.edu.pl; Tel.: +48-(17)-8651814

Abstract: The article concerns the possibility of the automation and robotization of the process of deburring jet engine components. The paper presents the construction of a laboratory stand enabling the automation of selected production operations of typical low-pressure turbine blades. The work identifies important parameters and results of the technological process related to the removal of burrs that affect the exactness of the process. The results of the analysis of the impact of individual process parameters on the magnitude of forces and moments occurring during deburring were carried out and presented. The results of initial and detailed tests were presented. Based on the results obtained, it was noticed that doubling the rotational speed of the brush results in a linear increase in torque and an increase in the engagement of the detail in the disc brush, leading to a non-linear increase in torque. It has also been shown that with tool wear, the value of the torque generated by the rotating tool decreases. Based on the results of a comparison of manual and automated process and histogram analysis, results from an automated stand are centered more correctly inside of the required radius range. This means that the repeatability of the process is higher for an automated test stand, which is one of the key aspects of large-scale aviation component manufacturing. Additionally, it was confirmed by visual inspection that all burs had been removed correctly—the deburring operation for all tested work pieces was successful. Based on the results obtained, it was proven that introduction of an automated stand can improve working conditions (by the elimination of the progressive fatigue of employees and the possibility for injury) and allows for the elimination of the negative impact of the machining process on workers. Further areas in which the optimization of the process parameters of the edge deburring can be developed in order to reduce unit costs have also been indicated.

Keywords: automation; robot; edge deburring; production process improvements; analysis of production

Citation: Falandys, K.; Kurc, K.; Burghardt, A.; Szybicki, D. Automation of the Edge Deburring Process and Analysis of the Impact of Selected Parameters on Forces and Moments Induced during the Process. *Appl. Sci.* **2023**, *13*, 9646. https://doi.org/10.3390/app13179646

Academic Editors: Tibor Krenicky, Maros Korenko and Juraj Ruzbarsky

Received: 17 July 2023
Revised: 22 August 2023
Accepted: 23 August 2023
Published: 25 August 2023

1. Introduction

The production process of components used in aircraft engines is a complex and multi-stage operation. Starting with the production of a casting resembling a finished product, a ready component, e.g., a blade for a low pressure turbine, is prepared through a series of production operations such as grinding, thermal treatment, shot peening or locksmith operations. The multitude of production operations that are aimed at ensuring the appropriate performance properties translates into the time needed to produce the finished component, and thus the cost. Taking into consideration the economic conditions and the economic situation, companies operating in the aviation industry set the reduction in unit costs as one of their main goals, which enables them to compete in the aviation sector. This goal is achieved by reducing electricity consumption, reducing the number of tools used in the production process, and reducing the number of man-hours spent on the production of single components. One way to improve economic results by reducing costs directly related to production, and ensuring appropriate technological capabilities and repeatability, is through the automation of selected stages of production processes.

From the point of view of a company producing aviation components, the primary goal is to minimize operating costs, which translates into the economic parameters of the production process. Automation of the production process will enable wider control of the process by eliminating the impact of the human factor on the final product, and will also enable automatic control, which then allows for the optimization of the parameters of the brushing process, which will reduce production costs by reducing the time spent on the edge deburring process. In line with the above trend, research work was undertaken to robotize the brushing process and verify the impact of selected process parameters on the values of forces and moments. The main goal of the production process is to prepare components that meet specific design assumptions, such as defined dimensions, mechanical properties, or properly machined external edges. The condition of the outer edges has a direct impact on the durability and safety of component assembly. Therefore, it is important to remove all burrs and ensure the appropriate value of the radius of the outer edges of the work piece.

Burr is essentially a part of work material that has experienced plastic deformation owing to the compression exerted by the cutting tool but which ultimately failed to completely separate from the ductile workpiece [1]. Burrs limit the dimensional accuracy of the features, induce difficulty during assembly, prohibit sliding motion over the surface, and degrade its aesthetic value [2–4]. Burrs can also act as debris that accelerates wear and tear, resulting in the reduction in useful service life rendered by the components [5]. The research group of Dornfeld [6–8] also made extensive investigations to analytically model burr size, introducing the concept of the "negative shear plane". Effects of various machining parameters such as cutting speed, feed rate [9–13], and cutting environment on burr formation have been studied in detail. Deburring, breaking sharp edges, or chamfering are the most common finishing operations performed in the aero-engine manufacturing sector [14,15]. Burrs and sharp edges are not allowed [16]. Other removal methods cannot be used, such as the physical and technical methods involving ultrasound, electrohydropulse processing, thermal energy, thermal pulse, electrochemical, and electrocontact methods [17,18]. Manual mechanical removal (or robotic mechanical removal if possible) is the only approved process.

An important point of view on the reliability and durability of parts included in a jet engine is the need to ensure the appropriate value of the edge deburring radius. This radius has a direct impact on the local stress level, and thus on the durability of the parts and the entire engine. The need to predict fatigue strength is the most important challenge for airlines and passengers using the services of the abovementioned companies [19], along with monitoring trends [20,21]. From the point of view of a manufacturer of aviation components, the main goal is to minimize operating costs, which refers to the economic parameters of the production process. The automation of the production process will enable much greater control by eliminating the impact of the human factor on the final condition of the product, and enabling the control and optimization of the parameters of the brushing process, which will reduce production costs by reducing the time spent on the edge deburring process.

Many studies have been written regarding work related to the implementation of industrial robots for machining operations. These studies clearly indicate a number of advantages thanks to the introduction of robotization, in the form of increased repeatability or cost reduction. Industrial robots have been successfully used for milling, deburring, polishing, or machining thin-walled elements [22].

In the scientific literature, a number of articles dealing with various methods of optimizing the deburring process can be found. Over the years, various approaches to achieving the desired machining accuracy have been described. In order to increase the accuracy of machining with the application of a robot, work was carried out to optimize the trajectory of the robot's movement based on a dynamic model of the robot [23], as well as by real-time force control [24,25]. Laser trackers were used as an alternative to force control in order to track the required path of the robot, which led to an increase in machining accuracy [26]. Work was also carried out to optimize the deburring process in relation to the structural stiffness of the robot used [27]. The authors of [28] describe the process of optimizing the parameters in the process of belt grinding and deburring steel according to the surface roughness and material removal rate. Both indicators are considered, and experimental results have shown that the comprehensive target can be effectively used as a metric to control grinding performance and process improvement. In [29–32], the Taguchi design method was widely used to optimize the milling parameters for surface roughness. In [33], the authors presented and discussed the concept of the project of the sensor-less force control of automatic grinding and deburring through an adjustable mechanism. In [34,35], the authors show that the key to high-quality polishing and deburring is maintaining a constant contact stress between the grinding tool and the work piece. Excessive or insufficient clamping force significantly worsens the surface roughness. The study in [36] aims to provide insight into the development and evaluation of force-guided machining processes. The focus was on the deburring and grinding scenario. Deburring implements a force-dependent feed-rate control, while the grinding use case implements force (pressure) control. The authors of [37] indicate that controlling the clamping force on the work pieces is a difficult task in industrial deburring operations. To solve this problem, a constant force mechanism based on a combination of positive and negative stiffness mechanisms has been proposed.

The analyzed works clearly indicate the possible advantages related to the implementation of robots for individual production operations. Considering the multitude of studies, it was decided to prepare a laboratory stand in order to verify the feasibility of brushing the blades produced for the needs of modern jet engines. The analyzed works related to the optimization of the deburring process [23–27] indicate that in order to carry out the process of optimizing the brushing operation, it is necessary to determine the key parameters of the process and then to understand their mutual interaction. In addition, due to the unique pair of materials associated with each other (work piece material and brush material), it was necessary to carry out a number of tests to understand the interaction of process parameters, such as time, rotational speed, and brush infeed, with the work piece.

This article describes the preparation of a conceptual robotic station aimed at carrying out the edge deburring process with simultaneous control of the loads generated on the detail (blade) from the tool (brush) and the automatic verification of the edge condition after the process is completed. In the next section, a number of tests were carried out to examine the dependencies between the key parameters of the technological process (such as rotational speed, time of the process, or the depth to which the workpiece is moved to the rotating brush) and the forces generated during machining processing. In addition, blades from a jet engine low-pressure turbine are characterized by a complex shape and a multitude of edges, where for design reasons it is necessary to obtain the required value of edge rounding. Therefore, it was necessary to verify the results of the process automatically. The use of an automated device for the verification of the radius of edge rounding will increase the reliability of the process by eliminating the subjective assessment of the worker. The aforementioned visual inspection performed by a locksmith is subject to the possibility of human error. In conditions of high-volume production, taking measurements using a microscope with appropriate equipment significantly increases the cost of manufacturing work pieces. Keeping in mind the desire to reduce the unit costs of the manufactured components, it was decided to implement an in-process measuring device that would allow for the verification of the process results at any time.

The edge spinning process consists of bringing the manufactured work piece into contact with a rotating tool, such as a disc with bristles covered with abrasive material. As a result of the contact of the detail with the rotating disc, both the tool and the manufactured detail are worn at the point of contact with the mentioned objects. In the case that abrasion process takes place and the objects remain in contact with each other, it is possible to observe the interaction of the objects with each other through the creation of forces. In this paper, it was decided to examine how the parameters of the technological process affect the level of interaction between the detail and the tool. A small number of studies [38–41] have described in detail the relationship between the parameters of the technological process and the life of the tool. This is a key need, which is complemented by the subject matter taken up in the article. Requirements, such as improvement of working conditions, improvement of the company's financial result, or control of the correctness of the locksmith operation, were the inspiration to take up the topic of developing the concept of an automated deburring station, and later to enable considerations related to the impact of parameters on the impact forces.

2. Problem Definition

In order to achieve the proper quality of the outer edges of the detail, i.e., to remove all burrs and obtain the appropriate radius of edge deburring, the work piece should be brought into proper contact with the rotating brush. The current way of carrying out the operation is shown in Figure 1. The operator uses his hands to bring the detail into contact with the rotating tool and thus performs the brushing process. As a result of contact and friction occurring between the tool and the work piece, burrs are removed and the edges are rounded.

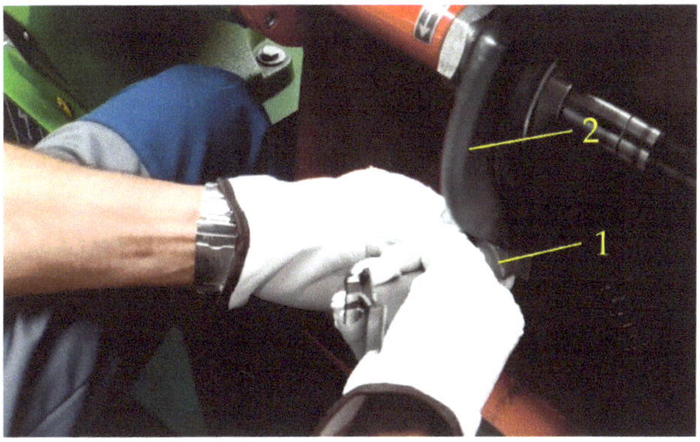

Figure 1. An operator during the deburring process (1—detail, 2—rotating tool).

The implementation of the brushing process with the direct participation of human operators in the process raises the following challenges, such as:

- Lack of repeatability of the brushing process;
- Lack of control of forces and moments occurring during the process (lack of objectivity), which prevents any work related to the optimization of tool wear;
- Progressive fatigue of operators during the processing of subsequent pieces;
- Costs of using appropriate individual and collective protective equipment, such as earmuffs and protective sleeves.

The lack of objectivity is directly related to the different pace of fatigue of individual operators or their different levels of rest at the time of starting work. This is a direct result of differences in the bodily makeup of individual operators, their psycho-physical condition, or their lifestyle. In connection with the conditions mentioned, before starting work related to the optimization of the edge deburring process, the concept of the workstation should first be developed. The stand should enable the measurement of the force between the detail and the tool and, furthermore, the verification of the impact of selected parameters on the mutual interaction of the detail and the tool.

The above actions are in line with the trend visible in the aviation industry, where efforts are made to minimize the number of expensive manual operations by automating and robotizing production processes. This approach is also in line with the ideas of Industry 4.0. Thanks to the implementation of industrial robots, the possibility of obtaining significantly greater repeatability and stability of technological process parameters for operations such as locksmithing or welding increases [42–45].

3. Solution—Test Stand

Due to the company's desire to optimize production, it is necessary to carry out works aimed at the robotization and automation of locksmith operations occurring in the production process of blades. Therefore, actions were taken to verify the possibility of performing such an improvement and then to identify the key parameters of the technological process affecting the size of forces and moments occurring in the process. The research conducted was aimed at developing the assumptions necessary to launch an automatic station that performs the process of edge breaking and enabling the process of optimizing the parameters in terms of reducing the costs associated with the tools (brushes) used.

The work linked to the test workstation is intended to help gain the experience necessary to implement the desired solutions in industrial conditions, as well as verifying the relationship between parameters such as tool rotational speed and the depth of insertion of the work piece into the tool (brush), as well as the induced forces and moments during the analyzed operation.

3.1. Test Stand

In order to verify the possibility of conducting the process of breaking selected edges of low-pressure turbine blades in an automated manner, a dedicated test stand was prepared. The test stand was located in the laboratory of the Department of Applied Mechanics and Robotics of the Rzeszow University of Technology (Figure 2).

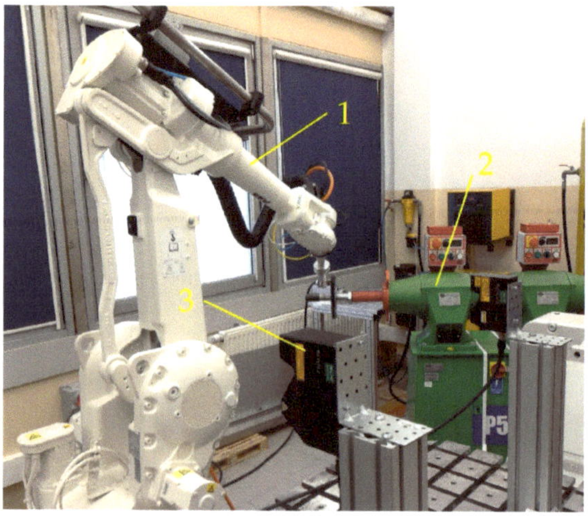

Figure 2. Test stand (1—industrial robot, 2—industrial grinder, 3—laser profilometer).

The following components were used to build the test stand:

- ABB IRB 2400 industrial robot with ABB IRC5 controller and Force Control system;
- An industrial grinder with a control system that allows one to change the rotational speed of the tool;
- Keyence LJ-V7060 laser profilometer.

The ABB IRB 2400 robot is characterized by a maximum payload of 16 kg, a path repeatability of 0.15 mm, and a position repeatability of 0.03 mm. The aforementioned repeatability is very sufficient in relation to the process being carried out.

The laser profilometer is a sensor that enables 2D or 3D measurement (with additional equipment) of selected dimensions of the inspected object.

In the work, related to the preparation of the test stand, a KEYENCE LJ-V7060 device with the technical parameters presented in Table 1 was used.

Table 1. Technical data of laser profilometer Keyence LJ-V7060.

Model	LJ-V7060
Reference distance	60 mm
Measurement range	+/−8 mm—z-axis
Repeatability	0.4 μm
Temperature characteristic	0.01% full scale/°C

An important feature, from the point of view of the aviation industry, is that the measuring device makes non-contact measurements. This is a key detail due to the fact that all kinds of defects may arise when the detail comes into contact with the measuring device, directly affecting the quality of the manufactured work pieces. An undoubted advantage of implementing such a solution is also the elimination of human error (the human factor) in activities related to making all kinds of measurements using various tools. In addition, the use of a pattern (reference detail) also does not eliminate the possibility of human error because it is the operator who compares the detail with the reference part. In the case that one uses this type of control, the final decision depends on the objective assessment of the operator and is also subject to error. Undoubtedly, the introduction of a laser profilometer to the workstation increases the perfection of the production process, and due to the shortening of the control time (the difference between the measurement or comparison made by a human and the use of a robotic station), it can have a positive impact on the unit cost of producing components. The laser profilometer has been installed in the working space of the industrial robot (Figure 2), and the measurement is carried out at a reference distance of 60 mm. In order to know the impact of individual key parameters of the technological process on the value of forces and moments occurring during the grinding process, it was necessary to use an appropriate sensor. The measurements were carried out using the Force Control Sensor 660, which is used in areas such as assembly, machining (e.g., grinding), and product testing. It is characterized by the following parameters (Figure 3).

This sensor makes it possible to adjust the TCP (tool center point) feed rate of the robot depending on the measured forces and moments during the process, which come from various machining parameters, e.g., the rotational speed of the tool. In practice, this means that the robot can modify the path or speed at which it moves to ensure that forces are maintained between the tool and the work piece. As the result of this functionality, the discussed sensor can be successfully used to control the deburring processes of elements with complex shapes, such as blades from a turbine engine. As a consequence, after using the Force Control sensor [46–49], it is possible to control the forces and moments generated during the grinding process, which has a direct impact on extending the life of the tools used in the process. The final result of these activities is an improvement in the company's financial results. The tools used in the brushing process are disc brushes, made of a metal disc and bristles made of Abralon612 SiC material. The current consumption of brushes is about 6000–7000 pieces per year.

Figure 4 shows the Force Control sensor mounted on the robot's arm with a handle and a detail.

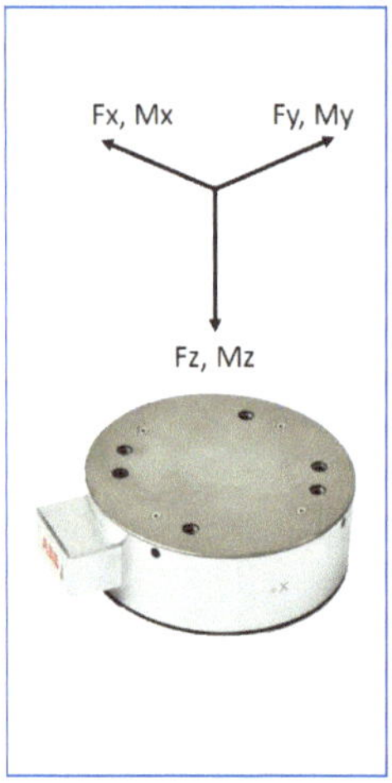

Specification	Sensor 660
Capacity	
Fx, Fy	660 N
Fz	1980 N
Mx, My, Mz	60 Nm
Overload capacity	
Fx, Fy	6600 N
Fz	19800 N
Mx, My, Mz	600 Nm
Operating temperature	−40 to +100°C
IP rating	IP65
Dimensions	
Height	40 mm
Diameter	Ø 104 mm
Weight	1.25 kg
Suitable robots	IRB 2400
	IRB 2600
	IRB 4400
	IRB 4600

Figure 3. Technical data of the applied force sensor.

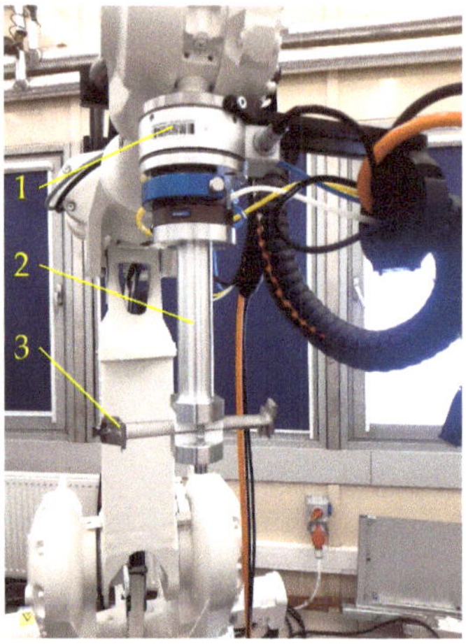

Figure 4. Force sensor mounted on the manipulator (1—Force Control sensor, 2—griper, 3—detail).

The other device included in the prepared workstation is a brush used in the deburring process, mounted on an industrial grinder that has a built-in control system that allows it to control the tool rotation speed.

3.2. Identification of Key Process Parameters

Every production process is characterized by a significant number of variables and parameters, and their definition requires thorough familiarization with the construction and usage of the devices. To correctly assess the results obtained from individual operations, it is necessary to understand the technical requirements for the machined work piece. Before starting work on the automation of the deburring operation, an analysis of the process and station equipment was carried out. This made it possible to determine the parameters affecting the execution of the process. Parameters directly determining the result of the process were also identified. In this way, two groups of data were created, referred to as input data and output data. These data are presented in Table 2.

Table 2. Input and output data for deburring process.

Input Data	Output Data
Tool rotational speed, rpm Time of the operation, s Depth of engage—tool and work piece, mm	Radius of edge break, mm All burrs removed, yes/no

4. Tests

The deburring process carried out for the production of engine blades is aimed at achieving two results: removing burrs that are harmful during operation and giving the appropriate radius to all machined edges. Such action will guarantee the safety of the worker during turbine assembly and the safe operation of the entire engine during its life. In order to carry out the deburring process, the work piece being processed should be brought into contact with the rotating brush. During the tests (Figure 5), the work piece placed in the holder is moved by the robot along the programmed trajectory against the rotating brush mounted on the industrial grinder. Measurement values such as torque, force, and blunting radius are acquired by the force sensor and laser profileometer and then recorded in the database.

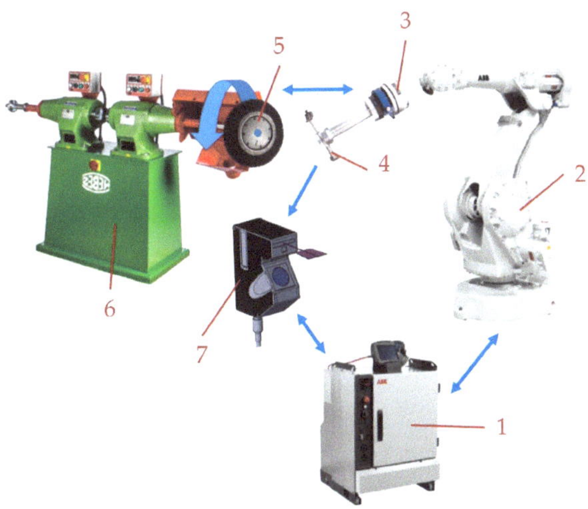

Figure 5. Test stand (1—controller ABB IRC5, 2—industrial robot ABB IRB 2400, 3—Force Control, 4—work piece, 5—tool, 6—industrial grinder, 7—laser profilometer).

In order to carry out the correct brushing process, it is necessary to remove the burr and give the appropriate radius to all machined edges. For this purpose, it is necessary to prepare an appropriate sequence of movements of the work piece in relation to the brush. Based on the experience related to the implementation of the manual process, it was decided to copy the movements of the locksmith and program the manipulator in such a way that it would perform the movements of the locksmith as accurately as possible. The movements performed by the locksmith are presented in Figure 6.

In the manual process, the operator, after completing the first sequence shown schematically in the figure, rotates the work piece by 180 degrees and repeats the entire sequence. It was noticed during the tests that the mentioned sequence can be implemented more simply by changing the direction of rotation of the tool.

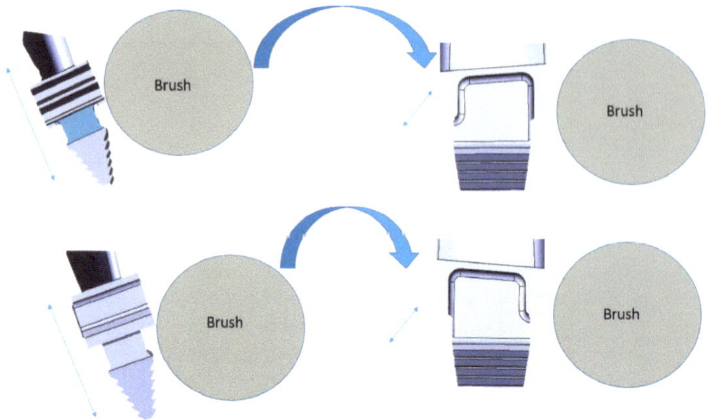

Figure 6. Schematic of piece movement during brushing operation.

In the initial phase of the tests, the aim was to examine the values of forces and moments occurring during the process of edge deburring. For this purpose, a number of tests were carried out at different values of the rotational speed of the brush. In the end, two speeds were selected. For the initial test, it was decided to cover the potential maximum scope of the rotational speed. Based on experience from manual process, it was chosen to start from maximum rotational speed of brush used during manual deburring operation—1600 rpm. At higher speeds, the worker is unable to hold the work piece for the entire shift (great fatigue, pain in the wrists). As a maximum limit, for automated test stand tests it was decided to use 3000 rpm—this is maximum speed allowed by available industrial grinders. Data were recorded using the Force Control package [43,46,49,50]. Table 3 presents the observed differences between the initial state and the change in force and torque during the operation.

Table 3. Results of initial test.

Brush Rotational Speed, rpm	Time of Contact Work Piece—Tool, s	Force Difference, N	Moment Difference, Nm
1600	5	30	5
3000	5	40	11.5

The results of the first trials, i.e., the values of the observed forces and moments, confirm the correctness of the sensor selection as they are within the measurement range according to the specification. Due to the negligibly small difference between the initial state and the forces obtained during the tests (a change of about 25% between the two rotational speeds used) in relation to the changes of moments (a change of about 130% between two used rotational speeds), it was decided to focus on the values of moments in further tests considered.

In the second stage of the tests, a number of tests were carried out (each test was repeated three times) aimed at verifying the moments occurring during the deburring process depending on the rotational speed of the tool and the depth of insertion of the work piece into the brush (Figure 7).

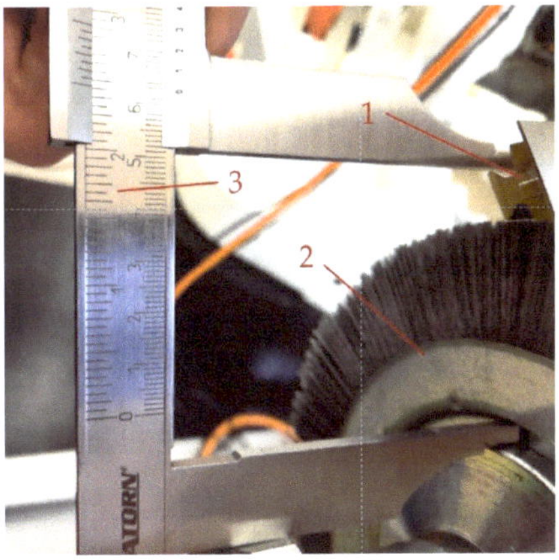

Figure 7. Measurement of inserting the detail into the tool (1—work piece, 2—tool, 3—measurement).

The results of the recorded values are presented in Table 4. The tests were carried out in two series. The first series included tests using a new (unused) brush. The second series of tests (Table 5) used a tool with a certain state of wear.

Table 4. Test results—new brush.

Rotational Speed, rpm	Depth of Engagement, mm	Average Value of Moment during Deburring Process, Nm
1600	0 ("touch")	2.4
3000	0 ("touch")	2.9
1600	2	5.3
3000	2	9.0
1600	4	9.0
3000	4	10.0

Table 5. Test results—used brush.

Rotational Speed, rpm	Depth of Engagement, mm	Average Value of Moment during Deburring Process, Nm
1600	0 ("touch")	2.2
3000	0 ("touch")	2.9
1600	2	4.2
3000	2	8.0
1600	4	6.1
3000	4	9.0

Figures 8 and 9 show the results obtained for the two tested rotational speeds, with the torque value depending on the depth of insertion of the work piece into the tool and brush condition (new and used brushes).

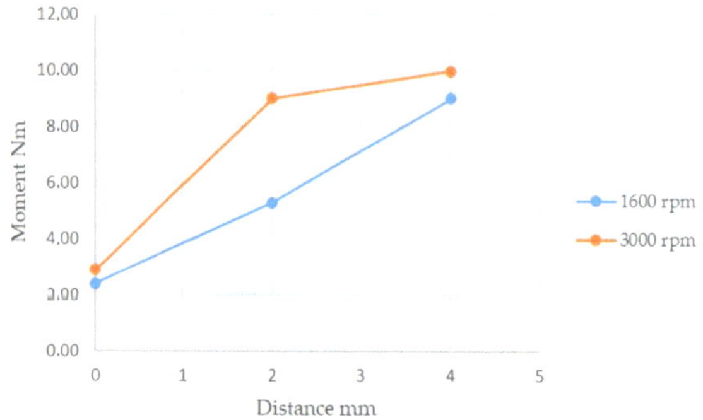

Figure 8. Dependence of the moment on the depth of insertion of the detail into the tool for two rotational speeds, new brushes.

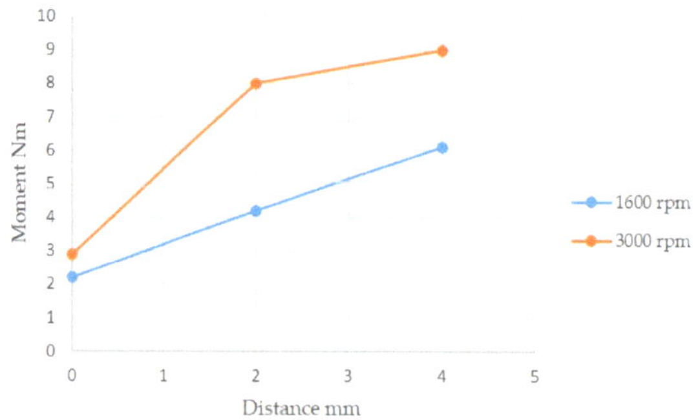

Figure 9. Dependence of the moment on the depth of insertion of the detail into the tool for two rotational speeds using brushes.

The measurement of the edge radius using a laser profilometer (Figure 10) allowed for the verification of the correctness of the brushing process.

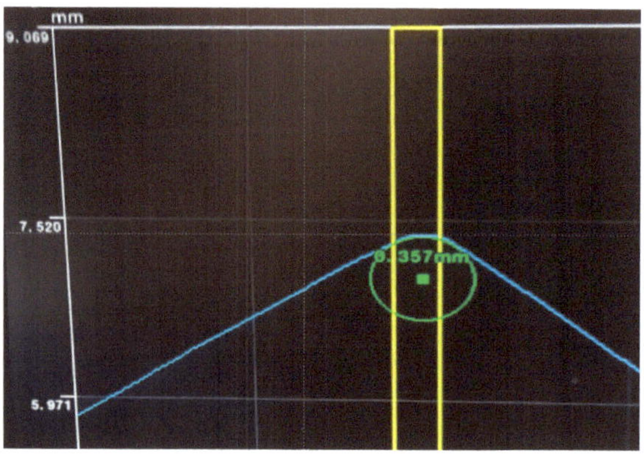

Figure 10. Results of deburring operations measured by a laser profilometer.

Edge break radius measurements for 30 different work pieces are shown in Table 6.

Table 6. Edge break measurements for selected work pieces.

Measurement Number	Edge Break Value, mm
1	0.23
2	0.23
3	0.22
4	0.3
5	0.32
6	0.26
7	0.24
8	0.25
9	0.29
10	0.3
11	0.28
12	0.25
13	0.22
14	0.24
15	0.26
16	0.23
17	0.26
18	0.27
19	0.25
20	0.28
21	0.29
22	0.25
23	0.19
24	0.26
25	0.25
26	0.27
27	0.27
28	0.25
29	0.25
30	0.25

Based on the measurements carried out, the arithmetic mean was calculated according to the following dependence:

$$\overline{\mathrm{X}} = \frac{x_1 + x_2 + x_3 + \cdots x_n}{n} = 0.26\ \mathrm{mm} \tag{1}$$

and consequently the standard deviation according to the relationship:

$$a = \sqrt{\frac{\left(x_1 - \overline{\mathrm{X}}\right)^2 + \left(x_2 - \overline{\mathrm{X}}\right)^2 + \cdots + \left(x_n - \overline{\mathrm{X}}\right)^2}{n}} = 0.027\ \mathrm{mm} \tag{2}$$

and the $+/-3\sigma$ limits were calculated:

$$+3\sigma = \overline{\mathrm{X}} + 3 \times a = 0.34\ \mathrm{mm} \tag{3}$$

$$-3\sigma = \overline{\mathrm{X}} - 3 \times a = 0.18\ \mathrm{mm} \tag{4}$$

Obtaining the edges, characterized by an average radius of about 0.26 mm, proves the correctness of the deburring operation (design requirements have been met). In order to illustrate the obtained results in the form of the edge radius, a histogram (normal distribution) was prepared (Figure 11). From the design-requirement point of view, it is desire to obtain edge radius values in the range from 0.15 to 0.35 mm. The calculated limit of $\pm 3\sigma$ stays within the limit defined by the design requirements. Based on the above automated process of deburring, this operation is adequate to meet the quality standards of the aviation industry.

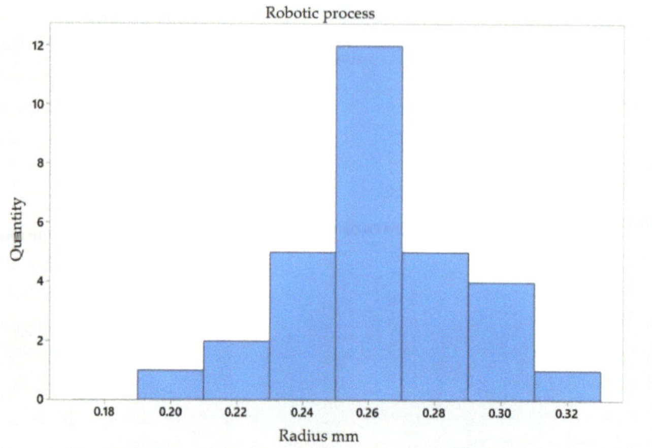

Figure 11. Histogram—values of edge break radius resulting from the robotic process.

To prepare a histogram, a population of 30 measurements was used. Parts were prepared with different process parameters (different values of the rotational speed of the tool, different values of the detail depth in the tool).

To compare the results of the automatic process, radius measurements for 30 random details with deburring performed manually are presented in Table 7.

Table 7. Edge break measurements for selected work pieces—manual process.

Measurement Number	Edge Break Value, mm
1	0.11
2	0.34
3	0.31
4	0.11
5	0.18
6	0.14
7	0.13
8	0.19
9	0.20
10	0.13
11	0.16
12	0.12
13	0.12
14	0.25
15	0.19
16	0.11
17	0.15
18	0.10
19	0.15
20	0.17
21	0.13
22	0.13
23	0.24
24	0.13
25	0.16
26	0.15
27	0.11
28	0.13
29	0.21
30	0.10

Based on the measurements carried out, the arithmetic mean was calculated according to the following dependence:

$$\overline{X}_1 = \frac{x_1 + x_2 + x_3 + \cdots x_n}{n} = 0.16 \text{ mm} \tag{5}$$

and consequently the standard deviation according to the relationship:

$$a_1 = \sqrt{\frac{(x_1 - \overline{X})^2 + (x_2 - \overline{X})^2 + \cdots + (x_n - \overline{X})^2}{n}} = 0.058 \text{ mm} \tag{6}$$

and the $+/-3\sigma$ limits were calculated:

$$+3\sigma = \overline{X}_1 + 3 \times a_1 = 0.34 \text{ mm} \tag{7}$$

$$-3\sigma = \overline{X}_1 - 3 \times a_1 = -0.02 \text{ mm} \tag{8}$$

In order to illustrate the obtained results in the form of an edge radius, a histogram (normal distribution) was prepared as a result of the manual process (Figure 12).

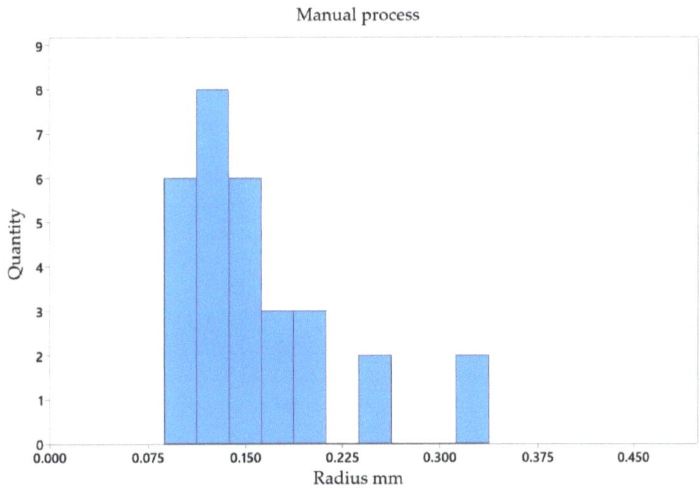

Figure 12. Histogram—values of edge break radius resulted from the manual process.

The typical work-piece condition is presented on Figure 13. The left side represents a work piece with sharp edges and burrs present—i.e., the work piece immediately after the grinding operation. The right side of Figure 13 represents the typical condition of a work piece after the deburring process is performed on a built automated test stand.

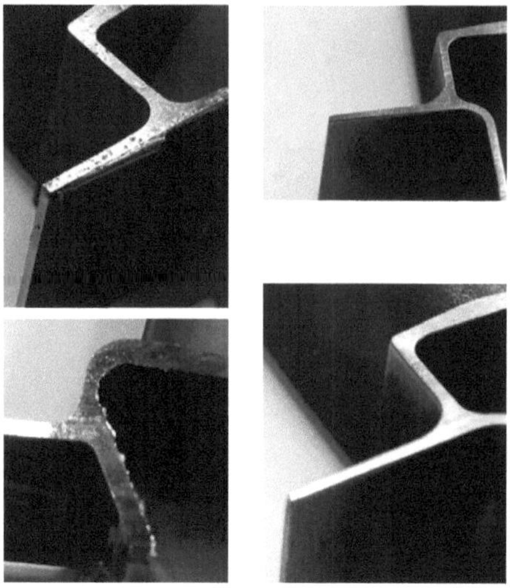

Figure 13. Typical work piece condition—before and after deburring operation.

5. Discussion

Based on the calculations of the standard deviation and the average, it can be predicted that the radius value will be, in the vast majority of cases, in the range from 0.18 to 0.34 mm (range $+/-3\sigma$). From the point of view of design requirements, these values are considered satisfactory. Based on experience related to the results of the manual process and histogram analysis, results from the automated stand are centered more correctly—the average value for the automated process is 0.26 mm, while for the manual process it is 0.16 mm.

We carried out a feasibility study of robotizing the process; the possibility of human error during the evaluation of the process results was eliminated, and the entire brushing process was carried out with minimal use of human force.

Based on the tests carried out and the results obtained, it can be stated:

- The deburring operation performed on built test stand was performed correctly—based on the observed work-piece condition, burrs were removed successfully—additionally, we created a histogram for the edge radius that confirms that the process is well centered;
- The preparation of an automated stand eliminates a lot of challenges related to human factor, process stability and repeatability, and allows us to eliminate the process of operator fatigue;
- Doubling the rotational speed of the brush results in a linear increase in torque. The obtained results confirm that the above relation is correct for different rotational speeds;
- Compared to manual processing, the operator (locksmith) was not able to work within the range of rotational speed tested. This is directly related to the physical endurance of a human, and directly translates into much faster wear of the brush in a manual process. In connection with these observations, it is important to introduce an industrial manipulator to the process in order to reduce tool wear;
- The increase in the engagement of the detail in the disc brush leads to a non-linear increase in torque. The observed effect is much smaller at higher rotational speeds;
- During the tests, tools that were not brand new were also used. Based on the tests carried out, it was observed that for the new disc, the measured moment during brushing was about 20% higher;
- During all the tests carried out, burr removal was recorded in a way that meets the design requirements.

6. Conclusions

With reference to the obtained results, it can be concluded that the introduction of robotization to the deburring process will contribute to improving the working conditions of employees by eliminating harmful factors such as progressive fatigue or the possibility of injury. Based on the results obtained for 30 work pieces, it was shown that the large-scale introduction of robotization to the edge deburring process leads to an increase in the repeatability of results and improves the quality of the work pieces produced, significantly increasing safety during the use of devices containing brushed work pieces in the automatic process.

Based on the measurements, it can be concluded that the individual key parameters of the brushing process have a different impact on the load values. In the case of rotational speed, the influence is linear, i.e., the increase in rotational speed translates directly into an increase in loads. On the other hand, a change in the depth of insertion of the workpiece into the tool causes a non-linear increase in loads. It has also been shown that with tool wear, the value of the torque generated by the rotating tool decreases. In connection with the above, it is reasonable to carry out further work aimed at a selection of technological parameters that will lead to a reduction in tool wear in the technological process while meeting the design requirements for the blades.

Author Contributions: Conceptualization, A.B., D.S., K.F. and K.K.; methodology, K.F., K.K. and D.S.; software, A.B., D.S. and K.K.; validation, K.F., D.S. and K.K.; formal analysis, K.F. and A.B.; investigation, K.F., D.S. and K.K.; resources, A.B., K.K. and K.F.; data curation, D.S., K.K. and K.F.; writing—original draft preparation, A.B., D.S., K.K. and K.F.; writing—review and editing, K.K. and K.F.; visualization, K.K., K.F. and D.S.; supervision, A.B., K.K. and K.F.; project administration, A.B.; funding acquisition, A.B., K.K. and D.S. All authors have read and agreed to the published version of the manuscript.

Funding: This research received no external funding.

Institutional Review Board Statement: Not applicable.

Informed Consent Statement: Not applicable.

Data Availability Statement: Data are contained within the article.

Conflicts of Interest: The authors declare no conflict of interest.

References

1. Hashimura, M.; Hassamontr, J.; Dornfeld, D.A. Effect of in-plane exit angle and rake angles on burr height and thickness in face milling operation. *J. Manuf. Sci. Eng.* **1999**, *121*, 13–19. [CrossRef]
2. Mathai, G.; Melkote, S. Effect of process parameters on the rate of abrasive assisted brush deburring of microgrooves. *Int. J. Mach. Tools Manuf.* **2012**, *57*, 46–54. [CrossRef]
3. Alagan, N.T.; Hoier, P.; Zeman, P.; Klement, U.; Beno, T.; Wretland, A. Effects of high-pressure cooling in the flank and rake faces of WC tool on the tool wear mechanism and process conditions in turning of alloy 718. *Wear* **2019**, *434*, 102922. [CrossRef]
4. Saha, S.; Ansary, S.I.; Deb, S.; Bandyopadhyay, P.P. Influence of tool wear on chip-like burr formation during micro-milling, and image processing based measurement of inwardly-deflected burrs. *Wear* **2023**, *530–531*, 205024. [CrossRef]
5. Niknam, S.A.; Songmene, V. Milling burr formation, modeling and control: A review. *Proc. Inst. Mech. Eng. Part B J. Eng. Manuf.* **2015**, *229*, 893–909. [CrossRef]
6. Ko, S.L.; Dornfeld, D.A. A study on burr formation mechanism. *J. Eng. Mater. Technol.* **1991**, *113*, 75–87. [CrossRef]
7. Chern, G.L.; Dornfeld, D.A. Burr/breakout model development and experimental verification. *J. Eng. Mater. Technol.* **1996**, *118*, 201–206. [CrossRef]
8. Hashimura, M.; Chang, Y.P.; Dornfeld, D. Analysis of burr formation mechanism in orthogonal cutting. *J. Manuf. Sci. Eng.* **1999**, *121*, 1–7. [CrossRef]
9. Gaitonde, V.N.; Karnik, S.R.; Achyutha, B.T.; Siddeswarappa, B. Methodology of Taguchi optimization for multi-objective drilling problem to minimize burr size. *Int. J. Adv. Manuf. Technol.* **2007**, *34*, 1–8. [CrossRef]
10. Dhanke, V.D.; Phafat, N.G.; Deshmukh, S.D. Multi response optimization of process parameters in drilling of AISI 1015 steel for exit burr using combined PCA and grey relational analysis. *Int. J. Mech. Prod. Eng. Res. Dev* **2013**, *3*, 91–98.
11. Kilickap, E.; Huseyinoglu, M.; Yardimeden, A. Optimization of drilling parameters on surface roughness in drilling of AISI 1045 using response surface methodology and genetic algorithm. *Int. J. Adv. Manuf. Technol.* **2011**, *52*, 79–88. [CrossRef]

12. Singh, H.; Kamboj, A.; Kumar, S. Multi response optimization in drilling Al6063/SiC/15% metal matrix composite. *Int. J. Chem. Nucl. Mater. Met. Eng.* **2014**, *8*, 281–286.
13. Raj, A.M.; Das, S.L.; Palanikumar, K. Influence of drill geometry on surface roughness in drilling of Al/SiC/Gr hybrid metal matrix composite. *Indian J. Sci. Technol.* **2013**, *6*, 5002–5007. [CrossRef]
14. Ramachandran, N.; Pande, S.S.; Ramakrishnan, N. The role of deburring in manufacturing: A state-of-the-art survey. *J. Mater. Process. Technol.* **1994**, *44*, 1–13. [CrossRef]
15. Thakre, A.A.; Soni, S. Modeling of burr size in drilling of aluminum silicon carbide composites using response surface methodology. *Eng. Sci. Technol. Int. J.* **2016**, *19*, 1199–1205. [CrossRef]
16. STD-01. WBTC. Draft. Kansas City, MO: Deburring Technology International. Gillespie, LaRoux, K., Ed, 1996. "Burr and Edge Terminology: Definitions." WBTC STD 01. Draft. Available online: https://www.larouxgillespie.com/burrhist.pdf (accessed on 16 July 2023).
17. Karlina, Y.I. Improvement of the technological process of processing parts of coaxial radio components using thermal impulse deburring. *J. Phys. Conf. Ser.* **2021**, *2032*, 012066. [CrossRef]
18. Martyushev, N.V.; Bublik, D.A.; Kukartsev, V.V.; Tynchenko, V.S.; Klyuev, R.V.; Tynchenko, Y.A.; Karlina, Y.I. Provision of Rational Parameters for the Turning Mode of Small-Sized Parts Made of the 29 NK Alloy and Beryllium Bronze for Subsequent Thermal Pulse Deburring. *Materials* **2023**, *16*, 3490. [CrossRef]
19. *Aircraft Accident Report—United Airlines Flight 232 McDonnell Douglas DC-10-10 Sioux Gateway Airport*; National Transportation Safety Board: Sioux City, IA, USA, 1989.
20. Śliwa, R.E.; Dymora, P.; Mazurek, M.; Kowal, B.; Jurek, M.; Kordos, D.; Rogalski, T.; Flaszynski, P.; Doerffer, P.; Doerffer, K.; et al. The Latest Advances in Wireless Communication in Aviation, Wind Turbines and Bridges. *Inventions* **2022**, *7*, 18. [CrossRef]
21. Rzucidło, P.; Jaromi, G.; Kapuściński, T.; Kordos, D.; Rogalski, T.; Szczerba, P. In-Flight Tests of Intruder Detection Vision System. *Sensors* **2021**, *21*, 7360. [CrossRef]
22. Verl, A.; Valente, A.; Melkote, S.; Brecher, C.; Ozturk, E.; Tunc, L.T. Robots in machining. *CIRP Ann.* **2019**, *68*, 799–822. [CrossRef]
23. Abele, E.; Bauer, J.; Rothenbücher, S.; Stelzer, M.; Von Stryk, O. Prediction of the Tool Displacement by Coupled Models of The Compliant Industrial Robot and The Milling Process. In Proceedings of the International Conference on Process Machine Interactions, Technische Universität Darmstadt, Berlin, Germany, 3–4 September 2008; pp. 223–230.
24. Sörnmo, O.; Olofsson, B.; Robertsson, A.; Johansson, R. Increasing time-efficiency and accuracy of robotic machining processes using model-based adaptive force control. *IFAC Proc. Vol.* **2012**, *45*, 543–548. [CrossRef]
25. Lehmann, C.; Halbauer, M.; Euhus, D.; Overbeck, D. Milling with Industrial Robots: Strategies to Reduce and Compensate Process Force Induced Accuracy Influences. In Proceedings of the 2012 IEEE 17th International Conference on Emerging Technologies & Factory Automation (ETFA 2012), Krakow, Poland, 17–21 September 2012; pp. 1–4. [CrossRef]
26. Moeller, C.; Schmidt, H.C.; Koch, P.; Boehlmann, C.; Kothe, S.; Wollnack, J.; Hintze, W. Real Time Pose Control of an Industrial Robotic System for Machining of Large Scale Components in Aerospace Industry Using Laser Tracker System. *SAE Int. J. Aerosp.* **2017**, *10*, 100–108. [CrossRef]
27. Gotlih, J.; Brezocnik, M.; Karner, T. Stiffness-Based Cell Setup Optimization for Robotic Deburring with a Rotary Table. *Appl. Sci.* **2021**, *11*, 8213. [CrossRef]
28. Li, F.; Xue, Y.; Zhang, Z.; Song, W.; Xiang, J. Optimization of Grinding Parameters for the Workpiece Surface and Material Removal Rate in the Belt Grinding Process for Polishing and Deburring of 45 Steel. *Appl. Sci.* **2020**, *10*, 6314. [CrossRef]
29. Abdur-RASHEED, A.; Konneh, M. Optimization of Precision Grinding Parameters of Silicon for Surface Roughness Based on Taguchi Method. *Adv. Mater. Res.* **2011**, *264*, 997–1002. [CrossRef]
30. Zheng, X.F.; Yuan, J.L.; Wen, D.H. Parameters optimization on the lapping process of 9Cr18 with Taguchi method. *Key Eng. Mater.* **2008**, *359*, 158–161. [CrossRef]
31. Yuan, J.L.; Lv, B.H.; Zhou, Z.Z. Parameters optimization on the lapping process for advanced ceramics by applying Taguchi method. *Mater. Sci. Forum* **2006**, *532*, 488–491. [CrossRef]
32. Guven, O. Application of the Taguchi method for parameter optimization of the surface grinding process. *Mater. Test.* **2015**, *57*, 43–48. [CrossRef]
33. Kuo, Y.L.; Huang, S.Y.; Lan, C.C. Sensorless Force Control of Automated Grinding/Deburring Using an Adjustable Force Regulation Mechanism. In Proceedings of the 2019 International Conference on Robotics and Automation (ICRA), Montreal, QC, Canada, 20–24 May 2019; pp. 9489–9495. [CrossRef]
34. Liao, L.; Xi, F.J.; Liu, K. Modeling and control of automated polishing/deburring process using a dual-purpose compliant toolhead. *Int. J. Mach. Tools Manuf.* **2008**, *48*, 1454–1463. [CrossRef]
35. Tsai, M.J.; Huang, J.F.; Kao, W.L. Robotic polishing of precision molds with uniform material removal control. *Int. J. Mach. Tools Manuf.* **2009**, *49*, 885–895. [CrossRef]
36. Domroes, F.; Krewet, C.; Kuhlenkoetter, B. Application and analysis of force control strategies to deburring and grinding. *Mod. Mech. Eng.* **2013**, *3*, 11–18. [CrossRef]
37. Ding, B.; Zhao, J.; Li, Y. Design of a spatial constant-force end-effector for polishing/deburring operations. *Int. J. Adv. Manuf. Technol.* **2021**, *116*, 3507–3515. [CrossRef]
38. Muszyńska, M.; Burghardt, A.; Kurc, K.; Szybicki, D.; Szczęch, T. Monitoring the Parameters of the Robot-Operated Quality Control Process. *Adv. Sci. Technol. Res. J.* **2017**, *11*, 232–236. [CrossRef]

39. Ramsauer, C.; Oswald, R.; Schörghofer, P.; Leder, N.; Schmitz, T.; Bleicher, F. Primary testing of an instrumented tool holder for brush deburring of milled workpieces. *J. Mach. Eng.* **2022**, *22*, 99–107. [CrossRef]
40. Rodríguez, A.; Fernández, A.; López de Lacalle, L.N.; Sastoque Pinilla, L. Flexible Abrasive Tools for the Deburring and Finishing of Holes in Superalloys. *J. Manuf. Mater. Process.* **2018**, *2*, 82. [CrossRef]
41. Uhlmann, E.; Hoyer, A. Modeling of Contact Forces for Brushing Tools. *Ceramics* **2021**, *4*, 397–407. [CrossRef]
42. Caggiano, A.; Teti, R. Digital factory technologies for robotic automation and enhanced manufacturing cell design. *Cogent Eng.* **2018**, *5*, 1426676. [CrossRef]
43. Burghardt, A.; Szybicki, D.; Gierlak, P.; Kurc, K.; Muszyńska, M. Robotic Grinding Process of Turboprop Engine Compressor Blades with Active Selection of Contact Force. *Teh. Vjesn.* **2022**, *29*, 15–22. [CrossRef]
44. Kurc, K.; Burghardt, A.; Gierlak, P.; Muszyńska, M.; Szybicki, D.; Ornat, A.; Uliasz, M. Application of a 3D Scanner in Robotic Measurement of Aviation Components. *Electronics* **2022**, *11*, 3216. [CrossRef]
45. Zhao, X.; Lu, H.; Yu, W.; Tao, B.; Ding, H. Robotic grinding process monitoring by vibration signal based on LSTM method. *IEEE Trans. Instrum. Meas.* **2022**, *71*, 1–10. [CrossRef]
46. Burghardt, A.; Gierlak, P.; Kurc, K.; Szybicki, D. Automatic Detection of Industrial Robot Tool Damage Based on Force Measurement. *Teh. Vjesn.* **2020**, *27*, 1385–1393. [CrossRef]
47. Villani, L.; De Schutter, J. Force Control. In *Springer Handbook of Robotics*; Siciliano, B., Khatib, O., Eds.; Springer: Cham, Switzerland, 2016; pp. 195–220. [CrossRef]
48. Pires, J.N.; Afonso, G.; Estrela, N. Force control experiments for industrial applications: A test case using an industrial de-burring example. *Assem. Autom.* **2007**, *27*, 148–156. [CrossRef]
49. Szybicki, D.; Burghardt, A.; Kurc, K.; Pietruś, P. Calibration and verification of an original module measuring turbojet en-gine blades geometric parameters. *Arch. Mech. Eng.* **2019**, *66*, 97–109. [CrossRef]
50. Bilancia, P.; Schmidt, J.; Raffaeli, R.; Peruzzini, M.; Pellicciari, M. An Overview of Industrial Robots Control and Programming Approaches. *Appl. Sci.* **2023**, *13*, 2582. [CrossRef]

applied sciences

MDPI

Article

Development of a Magnetic Levitation Wafer Handling Robot Transfer System with High-Accuracy and High-Cleanliness: Experimental Evaluation

Chang-Wan Ha, Sungho Jung, Jinseong Park and Jaewon Lim *

Department of AI Machinery, Korea Institute of Machinery & Materials (KIMM), Daejeon 34103, Republic of Korea; hawan@kimm.re.kr (C.-W.H.); sungho@kimm.re.kr (S.J.); jspark2090@kimm.re.kr (J.P.)
* Correspondence: einses@kimm.re.kr

Abstract: Magnetic levitation can reduce particulate contamination that occurs during wafer transportation in the semiconductor manufacturing process. This technology radically eliminates contact between the wafer and the transport system, reducing friction, wear, and particle generation. Therefore, it is suitable for achieving high cleanliness in the ultra-fine line-width semiconductor production process and solving the need for particle removal in a vacuum environment. In this study, the roller and linear motion guide components of the wafer transfer system were replaced with a magnetic levitation module, and a robot arm was installed on top to transport a single wafer. A posture controller and a current controller were designed, and test equipment simulating the wafer transfer system was also manufactured and tested. Regarding mover and system identification, a sine sweep test was performed on the motion axis of the five degrees of freedom. Through the obtained system identification, it was possible to design the posture controller more precisely. Moreover, through levitation in standstill experiments and high-speed operation experiments, the wafer transport system can be used to verify dust-free high-speed transport and accurate positioning performance.

Keywords: magnetic levitation; wafer transfer system; high accuracy; high cleanness; system modeling; control

Citation: Ha, C.-W.; Jung, S.; Park, J.; Lim, J. Development of a Magnetic Levitation Wafer Handling Robot Transfer System with High-Accuracy and High-Cleanliness: Experimental Evaluation. *Appl. Sci.* **2023**, *13*, 9482. https://doi.org/10.3390/app13169482

Academic Editors: Tibor Krenicky, Juraj Ruzbarsky and Maros Korenko

Received: 31 July 2023
Revised: 18 August 2023
Accepted: 19 August 2023
Published: 21 August 2023

1. Introduction

As semiconductor processes continue to be refined, the requirements for the equipment used in semiconductor manufacturing are becoming increasingly demanding. Since even fine particles can directly cause defects in semiconductors, various efforts have been made to reduce particle contamination.

The main factors contributing to particle generation in wafer transport systems can be summarized as linear motion (LM) guides, belts, cables, cable bays, and lubrication [1,2]. In other words, particle generation in the wafer transport system is caused by friction, abrasion, and lubrication. Particle generation cannot be completely eliminated with contact-based transport systems. Therefore, interest in high-clean transfer systems is increasing in the field of ultra-fine semiconductor manufacturing.

Recently, in semiconductor manufacturing processes, the development of contactless delivery systems has been actively developed due to the importance of particle-free and highly accurate delivery operations [3–6]. Magnetic levitation technology is considered a highly promising alternative to meet these requirements [7]. Magnetic levitation eliminates contact in the conveying system, minimizing friction, wear, and particle generation. It can also operate stably in a vacuum environment, making it suitable for various semiconductor manufacturing processes [8]. In a study by Yamakawa et al. [9], experiments were conducted to measure the contamination of dust particles during wafer delivery in a vacuum environment. The results showed that the magnetic levitation transport system is highly beneficial for particle-free transportation and exhibits very low outgassing.

Conventional research on magnetic levitation equipment has focused on implementing stages with levitation masses within several kilograms to achieve exceptional motion control accuracy [10–13]. These advanced magnetic levitation stages hold significant promise for critical semiconductor manufacturing processes, such as photolithography, which demand nanometer-level precision [14,15].

Recently, there have been notable developments in the field, including a high-precision magnetic levitation transfer system capable of enabling long-stroke movements and transporting movers weighing more than hundreds of kilograms [16,17]. Moreover, researchers are exploring various configurations for magnetic levitation transport systems, such as attaching electromagnets and linear propulsion motors to the moving part or fixed rails [18–20]. Additionally, real-scale prototypes are being developed to experimentally validate the effectiveness of proposed methods. These efforts aim to further enhance the capabilities and applications of magnetic levitation technology in the semiconductor manufacturing industry.

In this study, magnetic levitation was applied to a wafer transport system by replacing the roller and LM guide. A multi-joint wafer handling robot was installed at the top of the wafer transfer system, and this robot performed precise work, enabling rapid and accurate transfer even under harsh working conditions with extremely high acceleration and deceleration. To ensure control performance even under these challenging operational conditions, a robust controller has been designed based on loop shaping method. Therefore, a controller that precisely controls the five degrees of freedom (5-DOF) of the wafer transfer system was designed to enable high-speed transfer and precise position control by replacing wheels with magnetic levitation modules installed at the four corners. In addition, the stability and reliability of the magnetic levitation was experimentally verified to confirm its applicability in the actual manufacturing environment with a real-scale prototype of a wafer transfer system.

This paper is organized as follows. In Section 2, the system configuration of a magnetic levitation wafer transfer system is presented. The system modeling and control design of magnetic levitation are proposed, and the experimental evaluation of a real-scale prototype of a magnetic levitation wafer transfer system is presented in Section 3. In Section 4, a discussion is presented.

2. System Configuration

The magnetic levitation wafer transfer system controls the 5-DOF motion, excluding propulsion, and the linear synchronous motor (LSM) handles the linear drive. Figure 1 illustrates the concept of the wafer transfer platform, multi-joint robot arm, and magnetic levitation transfer system. The robot arm is installed on the top, and the magnetic levitation units and propulsion system are positioned at the bottom relative to the aluminum base. On the left side of the circled diagram, the current controller, the posture controller, the isolation board, and the oscilloscope for real-time magnetic levitation control and measurement are located. To mitigate the influence of wire tension from electromagnets (EMs) and gap sensors, the wires were configured with extended lengths to ensure minimal force transmission. Consequently, the impact of wire tension can be considered negligible. In the forthcoming system operation, electric power will be provided through the wireless power supply method, allowing the effects resulting from wiring to be disregarded.

To control magnetic levitation, magnetic levitation modules are placed at the four corners of the aluminum base. As shown in Figure 2a, four magnetic levitation modules generate four attractive forces ($f_1 \sim f_4$) for vertical control and four attractive forces ($f_5 \sim f_8$) for horizontal control. Figure 2b illustrates four gap sensors ($c_1 \sim c_4$) for measuring the z-axis airgaps and four gap sensors ($c_5 \sim c_8$) for the x-axis airgaps measurement. Based on the structural analysis of the base, the center of gravity (CG) of the mover can be calculated, and the 5-DOF posture can be calculated. The posture controller generates a 5-DOF force reference (F_x, F_z, M_x, M_y, M_z) based on the 5-DOF posture information. The generated 5-DOF force is converted into eight force ($f_1 \sim f_8$) reference inputs and applied to each electromagnet module.

Figure 1. Prototype of a magnetic levitation wafer transfer system.

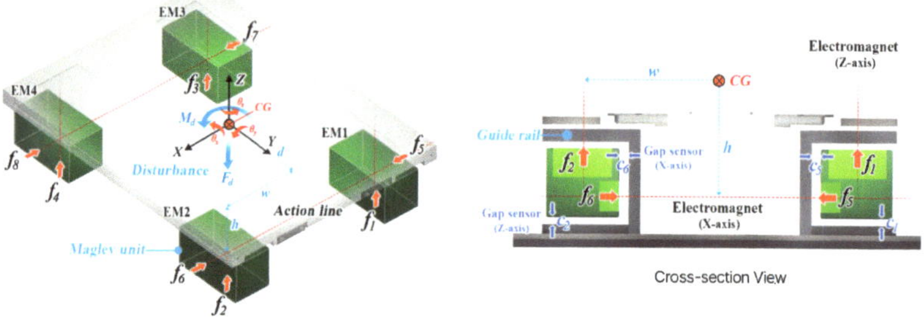

(a) (b)

Figure 2. Magnetic levitation forces and airgap measurements: (**a**) Magnetic levitation forces; (**b**) Airgap measurements.

The proposed magnetic levitation module, which includes x- and z-axis EMs and gap sensors, and a guide rail, is shown in Figure 3a. To implement the x- and z-axis EMs in a single magnetic levitation module, each flux path is orthogonalized to eliminate the effects of magnetic interference. The x-axis EM generates a closed flux loop in the XY plane, and the z-axis EM produces a closed flux loop in the XZ plane. The z-axis EM was designed in a dual-coil configuration, and the x-axis EM was divided into three coils to enable a simultaneous reduction in backcore thickness, core volume, and weight. The maximum force required by the EM was determined through dynamic analysis by simulation. The experimental results demonstrated that the power consumption of the EM remained within 150 W during stationary levitation. Additionally, even after extended periods of operation, the temperature was maintained below 40 °C. The magnetic flux density distribution in the landing state, which is the condition for applying the maximum current of the z-axis EM, is shown in Figure 3b. In this case, the magnetic flux generated by the z-axis current constitutes a closed loop in the ZX plane, and it can be stably driven without magnetic saturation within the maximum magnetic flux density of 1.5 T. The total mass, moment of inertia, and distances from the CG to the EM modules are presented in Table 1.

To control the magnetic levitation, both a higher-level controller for posture control and a lower-level controller for current control of the EMs are necessary. Figure 4 illustrates the arrangement of power and signal lines for connecting the controllers and the test plant. The main power line of $180V_{dc}$ is supplied to the current controller, and the control power of

$24V_{dc}$ is supplied to the isolation board and gap sensor converter. An additional $5V_{dc}$ power is supplied to measure the acceleration of the mover, and the acceleration sensor is placed at the center of the mover. The host PC controls the posture controller (dSPACE) using the MATLAB/Simulink-based plant model. Thanks to the utilization of a validated and user-friendly MATLAB/Simulink-based experimental environment, the development time of control systems can be significantly expedited. The force references ($f_1 \sim f_8$) calculated by the posture controller are converted into current commands and transmitted to the current controller, and the force of the EM is produced by the applied current of the current controller. In terms of airgap sensing, the inductive gap measuring sensor (AEC's PU-05) has been selected. An interface board was also developed to facilitate the transmission of current references to the current controller (Copley's JSP-180-30).

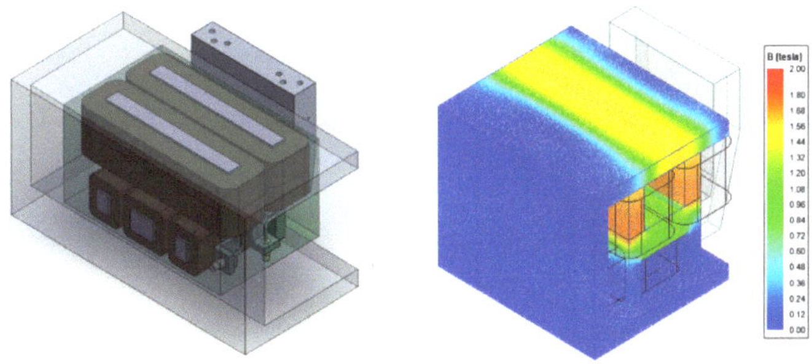

(a) (b)

Figure 3. Electromagnetic field analysis of the magnetic levitation unit: (**a**) EM module; (**b**) Magnetic field analysis.

Table 1. Major mechanical parameters of the magnetic levitation wafer transfer system.

Parameters	Description	Values
m	Total mass	213.9 kg
I_x I_y I_z	Moment of inertia	14.8 kg·m^2 12.7 kg·m^2 7.7 kg·m^2
w d h	Relative distance between the mass center and EM	164.0 mm 233.5 mm 240.4 mm

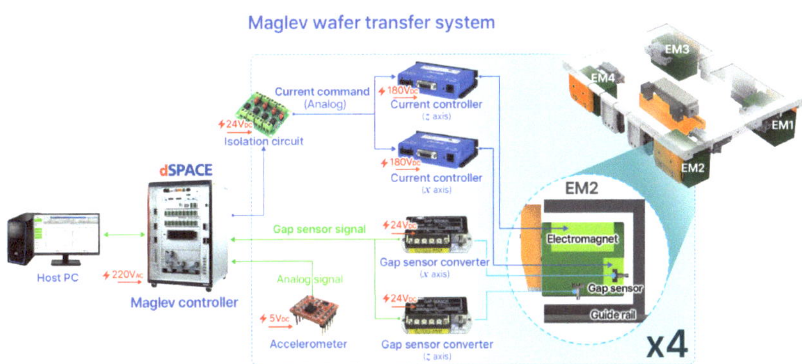

Figure 4. Diagram of the power and signal line connections.

3. Magnetic Levitation Control System

3.1. System Modeling

- **EM Model**
According to previous studies [21,22], the attractive force generated by the EM can be written as

$$f_n = \frac{\mu_0 N^2 A}{4}\left(\frac{i_n}{c_n}\right)^2 \tag{1}$$

where $\mu_0 = 4\pi \times 10^{-7}$ H/m is the vacuum permeability, i is the applied current, and c is the airgap. The values of the number of turns N and pole area of EM A that are used are summarized in Table 2.

Table 2. Major design parameters of the magnetic levitation units.

Parameters	Description	x-axis	z-axis
c_0	Nominal airgap	1 mm	1 mm
f_0	Nominal force	0 N	524 N
f_{Max}	Maximum force	110 N	1100 N
N	Number of turns	266 turn	396 turn
A	Pole area	1440 mm^2	352 mm^2

- **Dynamic Equation**
The dynamic equation of the magnetic levitation wafer transfer system can be described by Newton's equation for translation and Euler's equation for rotation. Newton's equation represents the relationship between the force acting on the center of mass of a rigid body and its acceleration, while Euler's equation represents the relationship between the moments acting on the center of mass of a rigid body and the time derivative of the angular momentum.

$$\begin{aligned} &\sum \mathbf{F} = m\mathbf{a} \\ &\sum \mathbf{M_G} = \dot{\mathbf{H}}_G \end{aligned} \tag{2}$$

The developed magnetic levitation wafer transfer system is mechanically symmetric about the x- and y-axes, so the coordinate system used is the principal axes of inertia, and the products of inertia can be neglected: $I_{xy} = I_{xz} = I_{yz} = 0$. Therefore, the angular momentum is expressed as follows: $\mathbf{H_G} = I_x\omega_x\hat{i} + I_y\omega_y\hat{j} + I_z\omega_z\hat{k}$. By substituting the time derivative of the angular momentum and the attractive force generated by EMs into Equation (2), the dynamic equation of the magnetic levitation wafer transfer system can be summarized.

$$\begin{aligned} &m\ddot{x} = f_5 - f_6 + f_7 - f_8 + F_{dx} \\ &m\ddot{z} = f_1 + f_2 + f_3 + f_4 + F_{dz} \\ &I_x\ddot{\theta}_x - (I_y - I_z)\dot{\theta}_y\dot{\theta}_z = d(f_1 + f_2 - f_3 - f_4) + M_{dx} \\ &I_y\ddot{\theta}_y - (I_z - I_x)\dot{\theta}_z\dot{\theta}_x = w(f_1 - f_2 + f_3 - f_4) + h(-f_5 + f_6 - f_7 + f_8) + M_{dy} \\ &I_z\ddot{\theta}_z - (I_x - I_y)\dot{\theta}_x\dot{\theta}_y = d(-f_5 + f_6 + f_7 - f_8) + M_{dz} \end{aligned} \tag{3}$$

The y-axis motion of the magnetic levitation wafer transfer system is controlled by an LSM, and the 5-DOF motion excluding the y-axis motion is precisely controlled by the magnetic levitation units attached to the four corners of the magnetic levitation wafer transfer system.
To design a linear controller, the dynamic equation Equation (3) is linearized. The simulation results show that the nonlinear terms are relatively small compared to the linear terms ($I_x\ddot{\theta}_x \gg (I_y - I_z)\dot{\theta}_y\dot{\theta}_z$). Thus, the linearized dynamic equation is sufficient

for obtaining meaningful interpretation results. The linearized motion equation can be expressed in the following matrix form:

$$\begin{aligned}\ddot{\mathbf{q}} &= \mathbf{M}^{-1}\mathbf{B}\mathbf{f} + \mathbf{M}^{-1}\mathbf{d} \\ &= \mathbf{M}^{-1}\mathbf{u} + \mathbf{M}^{-1}\mathbf{d}\end{aligned} \tag{4}$$

where

$$\begin{aligned}
&\mathbf{q} = \begin{bmatrix} x & z & \theta_x & \theta_y & \theta_z \end{bmatrix}^T \\
&\mathbf{f} = \begin{bmatrix} f_1 & f_2 & f_3 & f_4 & f_5 & f_6 & f_7 & f_8 \end{bmatrix}^T \\
&\mathbf{d} = \begin{bmatrix} F_{dx} & F_{dz} & M_{dx} & M_{dy} & M_{dz} \end{bmatrix}^T \\
&\mathbf{M} = \mathrm{diag}\begin{pmatrix} m & m & I_x & I_y & I_z \end{pmatrix} \\
&\mathbf{B} = \begin{bmatrix} & & & & 1 & -1 & 1 & -1 \\ 1 & 1 & 1 & 1 & & & & \\ d & d & -d & -d & & & & \\ w & -w & w & -w & -h & h & -h & h \\ & & & & -d & d & d & -d \end{bmatrix} \\
&\mathbf{u} = \mathbf{B}\,\mathbf{f} = \begin{bmatrix} u_x & u_z & u_{\theta x} & u_{\theta y} & u_{\theta z} \end{bmatrix}^T.
\end{aligned}$$

The **q** matrix represents the movements of the 5-DOF of the magnetic levitation wafer transfer system, the **f** matrix represents the attractive forces generated by the EMs, the **d** matrix represents the disturbances, the **M** matrix is the inertia matrix, the **B** matrix represents the geometric installation information of the EMs, and the **u** matrix is the decoupled control input that is used to achieve a desired output.

- **Kinematics**
  The 5-DOF motion of the magnetic levitation wafer transfer system is measured by the gap sensors installed at the four corners. To control the system, a process of reconstructing the 5-DOF motion based on the measured sensor signals is carried out. The motion of the electromagnet module 1 (EM1) can be expressed as $\vec{\mathbf{r}}_{EM1}$ based on the inertial reference coordinate. And, it can also be expressed as the sum of the center of gravity motion $\vec{\mathbf{r}}_{CG}$ and the relative motion with respect to center of gravity $\vec{\boldsymbol{\theta}} \times \vec{\mathbf{r}}_{EM1/CG}$ [23]. The relationship between the 5-DOF motion of the magnetic levitation wafer transfer system and the gap sensor measurements installed at the four corners is determined by the following kinematics.

$$\vec{\mathbf{r}}_{EM1} = \vec{\mathbf{r}}_{CG} + \vec{\boldsymbol{\theta}} \times \vec{\mathbf{r}}_{EM1/CG} \tag{5}$$

For example, the relationship between the 5-DOF motion of the magnetic levitation wafer transfer system and the gap sensor measurements in EM1 is

$$\begin{aligned}\vec{\mathbf{r}}_{EM1} &= (x\hat{i} + y\hat{j} + z\hat{k}) + (\theta_x\hat{i} + \theta_y\hat{j} + \theta_z\hat{k}) \times (-w\hat{i} + d\hat{j} - h\hat{k}) \\ &= (x - h\theta_y - d\theta_z)\hat{i} + (y + h\theta_x - w\theta_z)\hat{j} + (z + d\theta_x + w\theta_y)\hat{k}\end{aligned} \tag{6}$$

As shown in Figure 3, the z-axis motion measured by Gap Sensor 1 in EM1 can be expressed as follows: $c_1 = c_1 - c_{0z} = z + d\theta_x + w\theta_y$. Similarly, the x-axis motion measured by Gap Sensor 5 in EM1 can be expressed as follows: $c_5 = c_5 - c_{0x} = -x + h\theta_y + d\theta_z$. By repeating the same process for EM2, EM3, and EM4, the relationship between the 5-DOF motion of the magnetic levitation wafer transfer system and the gap sensor measurements installed at the four corners can be obtained as follows:

$$\Delta\mathbf{c} = \mathbf{T}\mathbf{q} \tag{7}$$

where

$$\mathbf{\Delta c} = \begin{bmatrix} \Delta c_1 & \Delta c_2 & \Delta c_3 & \Delta c_4 & \Delta c_5 & \Delta c_6 & \Delta c_7 & \Delta c_8 \end{bmatrix}^T$$

$$\mathbf{T} = \begin{bmatrix} & 1 & d & w & \\ & 1 & d & -w & \\ & 1 & -d & w & \\ & 1 & -d & -w & \\ -1 & & & h & d \\ 1 & & & -h & -d \\ -1 & & & h & -d \\ 1 & & & -h & d \end{bmatrix}$$

The $\mathbf{T}$ matrix represents the geometric installation information of the gap sensors. Conversely, the 5-DOF motion of the magnetic levitation wafer transfer system can be reconstructed from the measured gap sensor signals using the pseudo inverse.

$$\mathbf{q} = \mathbf{T}^{+}\mathbf{\Delta c} \tag{8}$$

- **Block diagram**
Figure 5 shows the block diagram of the feedback control system of the magnetic levitation wafer transfer system. The input of the dynamic equation is the decoupled control input $\mathbf{u}$ and disturbance $\mathbf{d}$, and the output is the 5-DOF motion $\mathbf{q}$. The 5-DOF motion of the magnetic levitation wafer transfer system is measured by the gap sensors $\mathbf{\Delta c}$ installed at the four corners, and the sensor noise is included in the measurement process. The measured sensor signal is reconstructed into the 5-DOF motion, and the difference $\mathbf{e}$ between the reference and the measured 5-DOF motion is calculated and utilized as the input of the feedback controller. The output of the feedback controller is the required force and momentum $\mathbf{u_{ref}}$ that are used to achieve the desired 5-DOF motion. By using the pseudo inverse of the $\mathbf{B}$ matrix and the EM model, the required current for each EM $\mathbf{i_{ref}}$ is calculated. This calculated current is utilized as the input of the current controller and affects the 5-DOF motion of the magnetic levitation transfer system through the EM model and $\mathbf{B}$ matrix related to the installation information of the EMs. The left green box is the discrete control part, and the right pink box is the continuous physical system.

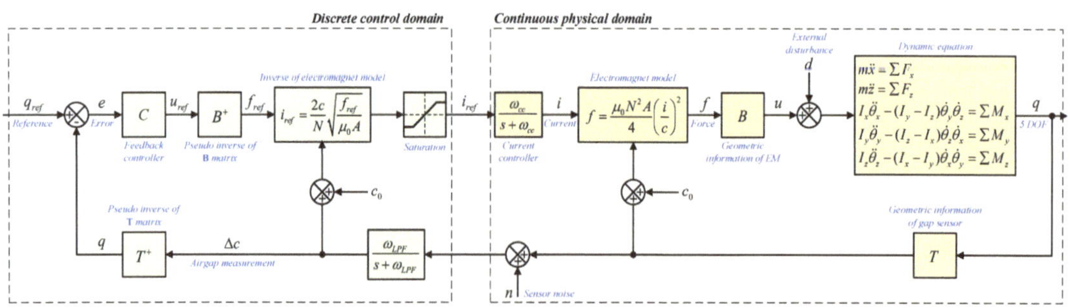

Figure 5. Block diagram of the feedback control system of the magnetic levitation wafer transfer system.

3.2. Feedback Controller Design

Assuming that the control bandwidth of the current controller is sufficiently fast compared to the magnetic levitation control bandwidth, the sensor noise is small, and the fluctuation of the 5-DOF motion is not large so that the nonlinear term of the dynamic equation can be ignored, and the block diagram can be simplified as shown in Figure 6. Furthermore, the magnetic levitation feedback controller is designed as follows:

$$\mathbf{u} = \mathbf{C}(\mathbf{q_{ref}} - \mathbf{q}) \tag{9}$$

where

$$\mathbf{C} = diag(C_x \quad C_z \quad C_{\theta_x} \quad C_{\theta_y} \quad C_{\theta_z})$$

A key feature of this controller structure is that there is no coupling term between the DOFs, so it can easily implement independent control characteristics for each DOF. In this study, a proportional–integral–derivative (PID) with a low-pass filter control structure is used. By combining the dynamic equation Equation (4) and the feedback controller Equation (9), this model can be expressed as follows by utilizing the Laplace transform.

$$\ddot{\mathbf{q}} = \mathbf{M}^{-1}\mathbf{C}(\mathbf{q_{ref}} - \mathbf{q}) + \mathbf{M}^{-1}\mathbf{d} \longrightarrow s^2\mathbf{Q} = \mathbf{M}^{-1}\mathbf{C}(\mathbf{Q_{ref}} - \mathbf{Q}) + \mathbf{M}^{-1}\mathbf{D} \tag{10}$$

The above equation can be expressed as

$$\mathbf{Q} = [s^2\mathbf{I} + \mathbf{M}^{-1}\mathbf{C}]^{-1}\mathbf{M}^{-1}\mathbf{C}\mathbf{Q_{ref}} + [s^2\mathbf{I} + \mathbf{M}^{-1}\mathbf{C}]^{-1}\mathbf{M}^{-1}\mathbf{D} \tag{11}$$

where

$$C(s) = k_p + k_i\frac{1}{s} + k_d\frac{Ns}{s+N}$$

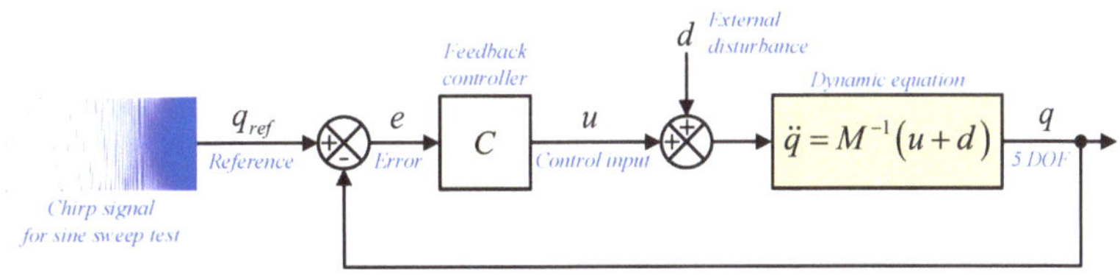

Figure 6. Simplified block diagram of feedback control for system identification.

Equation (11) can be used to obtain the transfer functions for each DOF. For example, the transfer function for the x-axis is

$$x = \frac{\#2}{\#1}x_{ref} + \frac{\#3}{\#1}F_{dx} \tag{12}$$

where

$$\begin{aligned}
\#1 &= ms^4 + mNs^3 + (k_p + k_dN)s^2 + (k_pN + k_i)s + k_iN \\
\#2 &= (k_p + k_dN)s^2 + (k_pN + k_i)s + k_iN \\
\#3 &= s^3 + Ns
\end{aligned}$$

The equation labeled #1 is called the characteristic equation, which reflects various feedback control characteristics such as the stability and response. In this study, the controller is designed to achieve a control bandwidth of approximately 10 Hz, a phase margin of more than 30 degrees, and a gain margin as large as possible. The feedback controller is designed to meet the desired system requirements by adjusting the control gain using the loop shaping technique. Loop shaping is a well-known systematic control system design method that can be used to achieve the desired control performance, stability, and robustness by shaping the loop transfer function of the control system [24]. In the low-frequency range, the magnitude of loop gain is closely related to control performance in terms of external disturbance attenuation and tracking performance. In the high-frequency range, the magnitude of loop gain is closely related to control performance in terms of sensor noise attenuation.

After completing the controller design on a MATLAB simulation, a sine sweep test was performed to verify that the developed feedback controller works properly in the actual physical system. The sine sweep test is an experiment that analyzes the frequency response of the developed feedback control system by examining the relationship between the input and output. In the sine sweep test, a chirp signal is added to the reference. The amplitude of the chirp signal applied was ±0.01 mm, and the frequency was set to gradually change from 0.1 Hz to 100 Hz over 300 s. During the test, the reference, error, and output were measured.

Figure 7 shows the reference and output measured by conducting a sine sweep test on the x-axis. Figure 8 shows the frequency response characteristics based on the measured reference, error, and output estimated by the MATLAB System Identification Toolbox. Figure 8a represents the frequency response curve between the reference and output, which corresponds to the complementary sensitivity function. This graph shows that the control bandwidth of the designed feedback control system is approximately 10 Hz. Figure 8b represents the frequency response curve between the error and output, which corresponds to the loop gain. This graph shows that the gain margin is infinite and that the phase margin is approximately 30 degrees. The reason for the negative gain margin is the presence of an unstable zero within the transfer function. Through various trial and error processes, the desired frequency response characteristics and satisfactory control performance were achieved. Similar processes were performed for other DOFs (z, θ_x, θ_y, θ_z), and feedback controllers were designed to achieve control bandwidths in the range of 10-20 Hz, similar to the x-axis case.

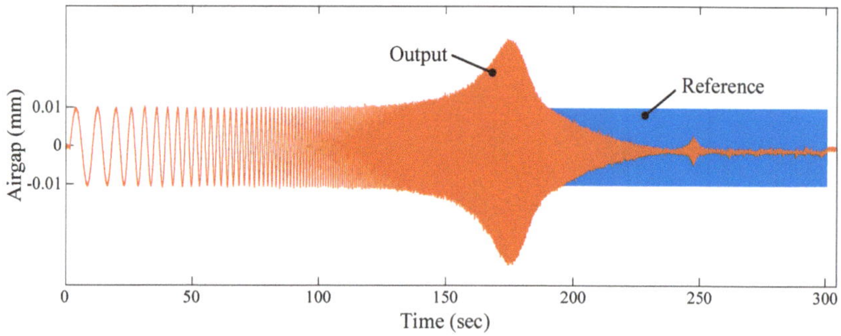

Figure 7. Experimental results of the sine sweep test in the x-axis: reference and output measurements.

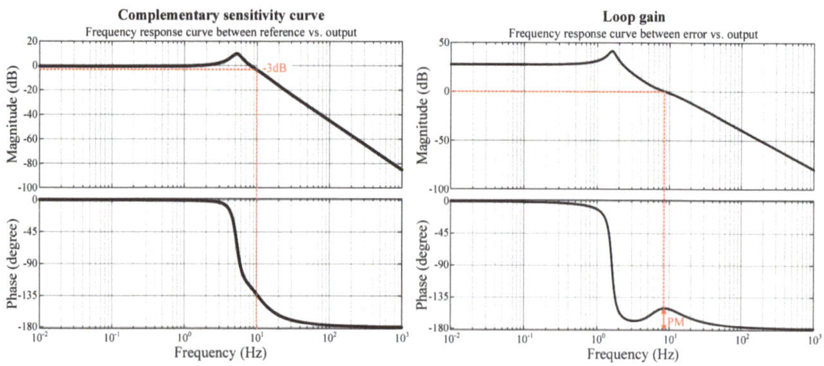

(**a**) (**b**)

Figure 8. Estimated frequency response curves based on measurement data from the sine sweep test in the x-axis: (**a**) Reference vs. Output; (**b**) Error vs. Output.

3.3. Experimental Evaluation

3.3.1. Magnetic Levitation at Standstill

The control performance of the developed magnetic levitation wafer transfer system was verified through experiments under various conditions. Figure 9a is the transient response characteristic result for the magnetic levitation under the conditions of the initial-take-off-landing state. In the case of a linear controller, the levitation control performance in a transient region may deteriorate since the operating point is further away from the equilibrium point. To minimize this effect, ensuring robustness against the modeling errors and ensuring sufficient relative stability, such as gain margins, are important. The control inputs for the 5-DOF motion of the system are shown in Figure 9b. To achieve a smooth transient response, the reference signal from the landing state to the levitation state was set to gradually change in the form of a ramp rather than a step. As a result, as shown in the figure, the system smoothly converges to the steady state without significant overshoot during the state transition. After the state transition is complete, the 5-DOF variation is ± 0.0048 mm, ± 0.0072 mm, $\pm 1.29 \times 10^{-5}$ rad, $\pm 1.40 \times 10^{-5}$ rad, and $\pm 8.54 \times 10^{-6}$ rad for the x- , z-, θ_x-, θ_y-, and θ_z-axes, respectively.

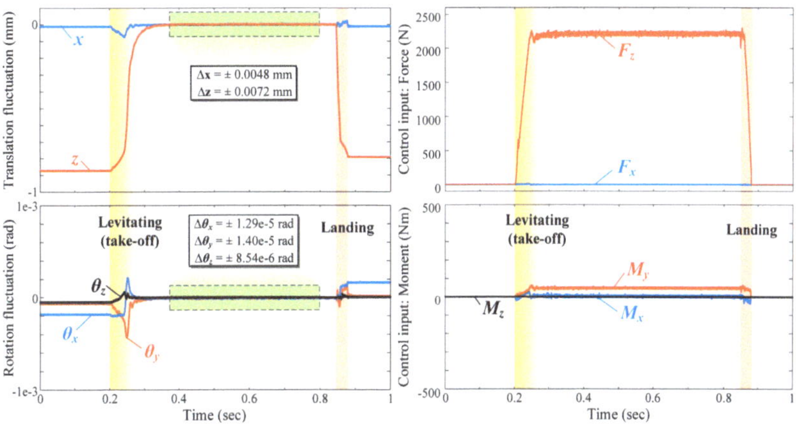

(a) (b)

Figure 9. Experimental results related to the transient response from the landing state to the magnetic levitation state: (**a**) 5-DOF fluctuations; (**b**) Control inputs.

System characteristics can be analyzed based on the force and moment applied to each DOF in the steady state. For example, the weight of the system can be estimated from F_z, and the CG can be estimated from the applied moment acting in the steady state. An analysis of F_z confirmed that the actual weight was approximately 225 kg, similar to the design values specified in Table 2. Additionally, the analysis of M_y confirms that the CG is slightly shifted in the x-direction. In fact, this system has a center of gravity deflected by approximately 20 mm in the $-x$-direction according to the robot's posture and mass distribution inside the control box, which were not considered during the initial design.

3.3.2. Magnetic Levitation under Driving Conditions

An experiment was conducted to observe the control performance of the magnetic levitation while the system maintained magnetic levitation and moved in the y-axis driven by the LSM. Figure 10 shows the 5-DOF fluctuation of the magnetic levitation wafer transfer system when it travels a distance of D = 1100 mm with an S-curve motion profile [25,26] under a maximum speed of $V_{\max}$ = 1500 mm/s and a maximum acceleration of $A_{\max}$ = 3060 mm/s. During the driving process, the maximum fluctuations in the x, z, θ_x, θ_y, and θ_z directions were ± 0.097 mm, ± 0.101 mm, $\pm 6.37 \times 10^{-4}$ rad, $\pm 2.97 \times 10^{-4}$ rad, and $\pm 7.48 \times 10^{-4}$ rad, respectively. These results represent a very satisfactory control performance. In this system, during the acceleration and deceleration periods, inertial moments are generated due to the height difference (h) between the point in which the thrust force is generated and the center of gravity, as shown in Figure 11. This leads to significant oscillations in the θ_x-direction of the magnetic levitation wafer transfer system, and these oscillations are considered external disturbances. To overcome these external disturbances, we attached an accelerometer to the moving base as shown in Figure 4, to measure the actual acceleration in the y-axis. Furthermore, we took into account the height difference between the point at which the thrust force is generated and the center of gravity, which we know in advance. By multiplying the actually measured acceleration and height difference, external disturbances were roughly predicted to minimize the effects of external disturbances. Thanks to this, it was experimentally confirmed that control performance was significantly improved.

Table 3 summarizes the maximum fluctuations in the 5-DOF of the magnetic levitation wafer transfer system. Through these experimental results, it was confirmed that the developed magnetic levitation wafer transfer system with a robust feedback control design can achieve a high-accuracy magnetic levitation control performance during the state transition initial takeoff from the landing state and high-speed operation conditions.

Table 3. Maximum fluctuations in the 5 DOF of the magnetic levitation wafer transfer system.

Description	Standstill	Driving Condition
Max. fluctuation in the x-axis	$\pm$0.0048 mm	$\pm$0.097 mm
Max. fluctuation in the z-axis	$\pm$0.0072 mm	$\pm$0.101 mm
Max. fluctuation in the θ_x-axis	$\pm 1.29 \times 10^{-5}$ rad	$\pm 6.37 \times 10^{-4}$ rad
Max. fluctuation in the θ_y-axis	$\pm 1.40 \times 10^{-5}$ rad	$\pm 2.97 \times 10^{-4}$ rad
Max. fluctuation in the θ_z-axis	$\pm 8.54 \times 10^{-6}$ rad	$\pm 7.48 \times 10^{-4}$ rad

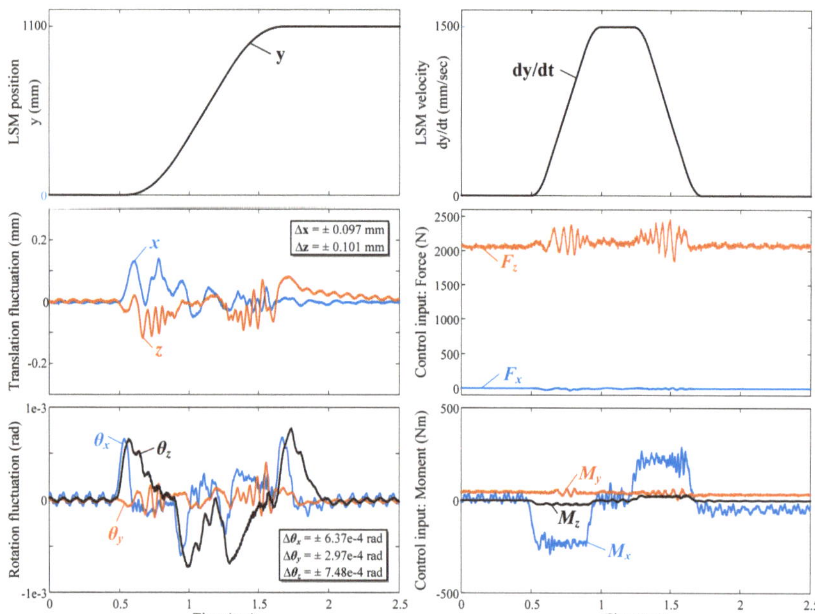

Figure 10. Experimental results of the magnetic levitation wafer transfer system under driving conditions.

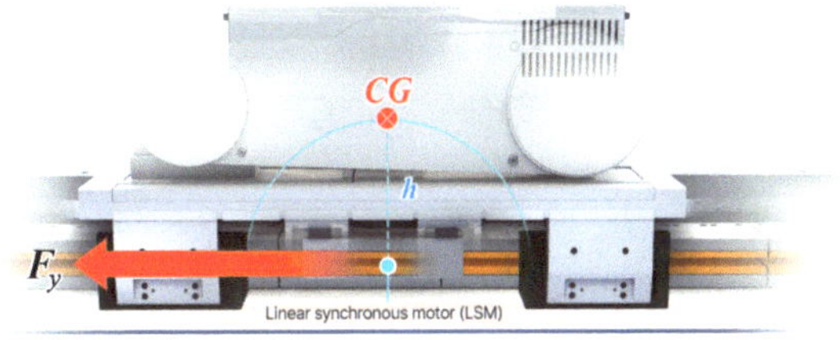

Figure 11. External disturbance caused by the height difference between the point in which the thrust force is generated and the center of gravity.

4. Discussion

Research on conventional magnetic levitation equipment has focused on implementing stages in which the levitation mass is within several kilograms. The model developed in this study is a high-precision transfer system that can transport a mover weighing more than 200 kg by implementing magnetic levitation without particle contamination. A magnetic levitation module, a posture controller, and a current controller were designed and manufactured, and a robot arm was installed on top to transfer single wafers. To verify the performance of the magnetic levitation system, system identification for each 5-DOF axis was performed through the sine sweep test of the mover. The results demonstrated that the system achieved a control bandwidth of 10–20 Hz for 5-DOF axes and precise operation for all 5-DOF in both fixed and high-speed tests. The maximum z-axis pore displacement during static levitation was measured at ± 0.0072 mm and slightly increased to ± 0.101 mm during high-speed operation. Gap displacement during static levitation will contribute to the precise positioning of the wafer, and gap displacement during high-speed levitation is sufficient to transport the wafer without slipping.

5. Conclusions

In conclusion, this study developed a high-precision magnetic levitation transport system that can safely and efficiently transport movers weighing more than 200 kg without particle contamination. The developed magnetic levitation module, posture controller, and current controller proved to be very effective in achieving stable levitation and precise control. In addition, the 5-DOF model of our system showed excellent performance in both static and high-speed operation, allowing accurate control of each axis. As a next step, it is necessary to study the stability of the levitation control according to the confirmation of the magnetic levitation characteristics that are based on the movement of the robot arm and the load change, which will contribute to improving the practical applicability and productivity of wafer transportation in the semiconductor manufacturing process.

Author Contributions: Conceptualization, J.L.; methodology, J.L. and S.J.; software, S.J. and C.-W.H.; validation, C.-W.H. and J.P.; formal analysis, J.L. and C.-W.H.; investigation, C.-W.H. and J.P.; resources, J.L. and C.-W.H.; data curation, C.-W.H. and J.L.; writing—original draft preparation, C.-W.H.; writing—review and editing, C.-W.H. and J.L.; visualization, C.-W.H. and J.L.; supervision, C.-W.H.; project administration, C.-W.H.; funding acquisition, C.-W.H. All authors have read and agreed to the published version of the manuscript.

Funding: This research was funded by the National Research Council of Science & Technology under Project "Development of core machinery technologies for autonomous operation and manufacturing".

Institutional Review Board Statement: Not applicable.

Informed Consent Statement: Not applicable.

Data Availability Statement: Not applicable.

Conflicts of Interest: The authors declare no conflict of interest.

Abbreviations

The following abbreviations are used in this manuscript:

DOF	Degree of Freedom
EM	Electromagnet
CG	Center of Gravity
LSM	Linear Synchronous Motor
LM	Linear Motion

References

1. Reinhardt, K.; Kern, W. *Handbook of Silicon Wafer Cleaning Technology*, 3rd ed.; William Andrew: Oxford, UK, 2018; pp. 87–149.
2. Srinivasan, R.S. Models for Estimating Wear Particles in Preliminary Design of Semiconductor Equipment. In Proceedings of the 1997 IEEE International Symposium on Semiconductor Manufacturing Conference, San Francisco, CA, USA, 6–8 October 1997; pp. 45–48.
3. Schmulling, B.; Appunn, R.; Hameyer, K. Electromagnetic Guiding of Vertical Transportation Vehicles: State Control of an Over-determined System. In Proceedings of the 2008 18th International Conference on Electrical Machines, Vilamoura, Portugal, 6–9 September 2008.
4. Appunn, R.; Schmulling, B.; Hameyer, K. Electromagnetic Guiding of Vertical Transportation Vehicles: Experimental Evaluation. *IEEE Trans. Ind. Electron.* **2009**, *57*, 335–343. [CrossRef]
5. Liu, G.; Lu, Y.; Xu, J.; Cui, Z.; Yang, H. Magnetic Levitation Actuation and Motion Control System with Active Levitation Mode based on Force Imbalance. *Appl. Sci.* **2023**, *13*, 740. [CrossRef]
6. Hu, K.; Jiang, H.; Zhu, Q.; Qian, W.; Yang, J. Magnetic Levitation Belt Conveyor Control System based on Multi-sensor Fusion. *Appl. Sci.* **2023**, *13*, 7513. [CrossRef]
7. Han, H.S.; Kim, D.S. *Magnetic Levitation*; Springer Tracts on Transportation and Traffic: Dordrecht, The Netherlands, 2016.
8. Kim, J.; King, G.B.; Kim, C.-H.; Ha, C.-W. Experimental Validation of Deadzone Compensation for a Magnetic Levitation Transporting OLED Displays System. *Int. J. Control. Autom. Syst.* **2022**, *20*, 2937–2947. [CrossRef]
9. Yamakawa, H.; Moriyama, I.; Minamigawa, Y.; Maeba, Y.; Takematsu, T.; Nishitsuji, M.; Fujiki, O.; Asaishi, T.; Koike, T. Contamination-free Transfer of Silicon Wafers with a Magnetic Levitation Transport System in Vacuum. *Vacuum* **1990**, *41*, 1843–1845. [CrossRef]
10. Peijnenburg, A.T.A.; Vermeulen, J.P.M.; Van Eijk, J. Magnetic Levitation Systems compared to Conventional Bearing Systems. *Microelectron. Eng.* **2006**, *83*, 1372–1375. [CrossRef]
11. Lim, J.S.; Lee, H.W. Movement Control Method of Magnetic Levitation System using Eccentricity of Non-Contact Position Sensor. *Appl. Sci.* **2021**, *11*, 2396. [CrossRef]
12. Kim, W.J.; Verma, S.; Shakir, H. Design and Precision Construction of Novel Magnetic-levitation-based Multi-axis Nanoscale Positioning Systems. *Precis. Eng.* **2007**, *31*, 337–350. [CrossRef]
13. Kim, O.-S.; Lee, S.-H.; Han, D.-C. Positioning Performance and Straightness Error Compensation of the Magnetic Levitation Stage supported by the Linear Magnetic Bearing. *IEEE Trans. Ind. Electron.* **2003**, *50*, 374–378.
14. Kim, W.J.; Trumper, D.L. High-precision Magnetic Levitation Stage for Photolithography. *Precis. Eng.* **1998**, *22*, 66–77. [CrossRef]
15. Williams, M.E.; Trumper, D.L.; Hocken, R. Magnetic Bearing Stage for Photolithography. *CIRP Ann.* **1993**, *42*, 607–610. [CrossRef]
16. Gabriel G.-G.; Diego, A.-A.; Enrique, V.C.; Francesc, G.; Emilia, M.; Paul, A.; Alexander, I. Fuzzy Logic Controller Parameter Optimization using Metaheuristic Cuckoo Search Algorithm for a Magnetic Levitation System. *Appl. Sci.* **2019**, *9*, 2458. [CrossRef]
17. Hong, D.-K.; Woo, B.-C.; Koo, D.-H.; Lee, K.-C. Electromagnet Weight Reduction in a Magnetic Levitation System for Contactless Delivery Applications. *Sensors* **2010**, *10*, 6718–6729. [CrossRef] [PubMed]
18. Lee. K.-C.; Moon, J.-W.; Lee, M.-C.; Kim, J.-M.; Kim, J.-W.; Koo, D.-H. Electric Monorail System with Magnetic Levitations and Linear Induction Motors for Contactless Delivery Applications. In Proceedings of the 8th International Conference on Power Electronics, Jeju, Republic of Korea, 30 May–3 June 2011.
19. Lee, K.-C.; Moon, S.; Ha, H.; Park, B.-G.; Kim, J.-W.; Baek, J.-Y.; Lee, M.-C. A Novel High Precision Electromagnetic Suspension for Long-stroke Movement and Its Performance Evaluation. *J. Electr. Eng. Technol.* **2014**, *9*, 514–522. [CrossRef]
20. Lee, K.-C.; Moon, J.-W.; Koo, D.-H.; Lee, M.-C. Magnetic Levitated Electric Monorail System for Flat Panel Display Glass Delivery Applications. *J. Inst. Control. Robot. Syst.* **2011**, *17*, 566–572. [CrossRef]
21. Kim, J.; Ha, C.-W.; King, G.B.; Kim, C.-H. Experimental Development of Levitation Control for a High-accuracy Magnetic Levitation Transport System. *ISA Trans.* **2020**, *101*, 358–365. [CrossRef] [PubMed]
22. Ha, C.-W.; Lim, J.; Kim C.-H. Development of a High-accuracy Magnetic Levitation Transport System for OLED Evaporation Process. *Trans. Korean Soc. Mech. Eng. A* **2018**, *42*, 1111–1118. [CrossRef]
23. Beer, F.P.; Johnston, E.R.; Eisenberg, E.R.; Mazurek, D.F.; Clausen, W.E.; Cornwell, P. J. *Vector Mechanics for Engineers*, 4th ed.; McGraw-Hill: New York, NY, USA, 1977; Chapter 15.
24. Morris, K.A. *Introduction to Feedback Control*; Academic Press: Boca Raton, FL, USA, 2000.

25. Ha, C.W.; Lee, D. Analysis of Embedded Pre-filters in Motion Profiles. *IEEE Trans. Ind. Electron.* **2017**, *65*, 1481–1489. [CrossRef]
26. Lee, D.; Ha, C.-W. Optimization Process for Polynomial Motion Profiles to Achieve Fast Movement with Low Vibration. *IEEE Trans. Control. Syst. Technol.* **2020**, *28*, 1892–1901. [CrossRef]

applied sciences

MDPI

Article

New Possibilities of Using DEMATEL and ERPN in the New PFMEA Hybrid Model

Marián Bujna [1], Chia Kuang Lee [2], Milan Kadnár [1,*], Maroš Korenko [1] and Juraj Baláži [1]

[1] Institute of Design and Engineering Technologies, Faculty of Engineering, Slovak University of Agriculture in Nitra, Tr. A. Hlinku 2, 949 76 Nitra, Slovakia
[2] Faculty of Industrial Management, Universiti Malaysia Pahang, Lebuhraya Tun Razak, Gambang 26300, Pahang, Malaysia
* Correspondence: milan.kadnar@uniag.sk; Tel.: +421-37-641-4107

Abstract: The aim of the paper is to examine the requirements of producers in post-communist countries with lower economic level. The first requirement was how to overcome the limitations of conventional PFMEA to propose measures effectively. The second requirement solved the economic effect of failure modes. The aim of the paper was to create a new hybrid PFMEA–DEMATEL–ERPN model to manage failure modes to resolve the requirements. The DEMATEL model overcame the limitations of PFMEA. DEMATEL data were used to estimate the functionality of the proposed models. Criteria such as the occurrence of defective products and the probability of their occurrence (O and RPN) were monitored. ERPN also overcame the limitations of PFMEA. Internal and external costs arise as effects of failure modes. The costs were included in the economic evaluation of the models. We validated the models in a transfer pressing process. The estimation of models' functionality proved to be correct. The economic evaluation refined the research results and resolved the second requirement of the manufacturers. The DEMATEL and ERPN models (compared to PFMEA) proved their validity when the use of PFMEA was limited. By using DEMATEL, we registered the lowest number of defective products and the lowest costs.

Keywords: PFMEA; DEMATEL; ERPN; hybrid model; prediction model; production quality

Citation: Bujna, M.; Lee, C.K.; Kadnár, M.; Korenko, M.; Baláži, J. New Possibilities of Using DEMATEL and ERPN in the New PFMEA Hybrid Model. *Appl. Sci.* **2023**, *13*, 3627. https://doi.org/10.3390/app13063627

Academic Editor: Nikos D. Lagaros

Received: 10 February 2023
Revised: 6 March 2023
Accepted: 10 March 2023
Published: 12 March 2023

1. Introduction

Continuous improvement is an important global aspect of the growth and competitiveness of manufacturing organizations. Improving the quality of production implies the elimination of failure modes in the production process. For each single failure mode, it is important to know and research the causes of that failure [1].

FMEA (Failure Mode and Effect Analysis) is a tool used to identify failure modes of either a manufacturing process (PFMEA), design (DFMEA), system (SFMEA), or service [2]. It also has its applications in production quality and in solving the reliability of technical systems [3,4]. The goal of a process FMEA is to improve the manufacturing process, ensure the product is of the required quality, minimize downtime, and minimize defective products and scrap. The focus of the PFMEA is on the manufacturing and assembly process [5].

The risk priority number (RPN) is used to determine the risk or criticality of failure modes. The RPN is obtained by simply multiplying the risk factors such as occurrence (O), severity (S), and detection (D). The problem is that this determination of the RPN can be highly subjective and introduces many shortcomings. This invalidates its full use in practice [6].

Many authors have been devoted to solving the problems of conventional PFMEA. Authors have proposed various alternative approaches to solve this problem. Liu [2] proposed a new framework to improve the execution of the PFMEA while improving risk assessment.

The new FMEA trends offer an opportunity to refine the results of the method. However, there is still a question of how important these parameters are to the size of the RPN and the resulting threats. Several authors consider the S and O factors as the two most important risk factors [7]. The limits of PFMEA can be explained as follows: we identified three failure modes, A, B and C, which we ranked as follows: (SA = 8, OA = 5, DA = 4), (SB = 10, OB = 4, DB = 4), and (SC = 5, OC = 4, DC = 8). Although each factor is rated differently, the resulting RPN is 160 for all failure modes. This poses a problem in determining the prioritization for the measures proposed. This is a metric where equal weight is given to each of the risk factors (S, O, and D) [8].

As summarized by Liu [9] in his paper, many authors have proposed many ways to enhance the effectiveness of FMEA. The most widely used techniques are cognitive mapping [10]; mathematical programming methods such as data envelopment analysis (DEA) [11]; adaptive resonance theory [12]; linear programming [13]; fuzzy rule-base [14,15]. Often used are multi-criteria decision-making (MCDM) methods such as a technique for order preference by similarity to ideal solution (TOPSIS) [16–18]; grey relational analysis (GRA) [19,20]; and decision-making trial and evaluation laboratory (DEMATEL) [9,21,22]. Hybrid models have also found their use as a combination of the baseline method (PMEA) with fuzzy analytic hierarchy process (AHP) and fuzzy TOPSIS [23]; integrated approaches such as similarity measure and adaptive resonance theory [24], D numbers, and grey relational projection method [25]; extended FMEA [8]; PFDA-FMEA [26], and other methods and techniques [27,28].

The Battelle Memorial Institute developed the DEMATEL method to solve complex and interrelated problems [29,30]. The DEMATEL methodology is based on pairwise comparisons of individual factors, the definitions and descriptions of which will be available to experts to assess the interrelations between the factors (quantitatively using scales). Its strength is its ability to map all the influences and factors of a decision problem, including their representation in a clear influence map. In practice, it involves one or two rounds of questioning via a paper-based or online questionnaire. It is possible to include a larger number of experts, resulting in a map that depicts the relations among them as well as the significance of each step [31].

DEMATEL (Decision-Making Trial and Evaluation Laboratory) clarifies the complex relations between failure modes in the manufacturing process and provides solutions by comparing these failure modes in the manufacturing process. The solution uses matrices to calculate direct and indirect causal relations, influence rates, and quantify the degree of mutual influence between factors. By doing so, the FMEA process' limitations about the same RPN should be overcome [21].

In Table 1, we can see the studies that have considered the issue of combining FMEA and DEMATEL. Most of the authors used the DEMATEL model to prioritize the failure modes.

Table 1. Review of PFMEA and DEMATEL usage.

Authors and Studies	Models Applied	DEMATEL Usage
Tsai et al. (2017) [21]	PFMEA + DEMATEL	Failure mode priority
Liu et al. (2022) [31]	PFMEA + DEMATEL	Failure mode priority
Cheshmberah et al. (2020) [32]	FMEA + DEMATEL	Selection of the most important roots
Díkmen et al. (2018) [33]	SERVQUAL + DEMATEL	Factor priority
Mzougui et al. (2019) [34]	FMEA + FUZZY DEMATEL	Prioritisation of failure causes
Fei et al. (2022) [35]	FMEA + DEMATEL + GRA	Identifying root causes of failure
Hua et al. (2023) [36]	FMEA + ELICIT + GRA + DEMATEL	Failure mode priority
Lo et al. (2022) [4]	FMEA + DEMATEL + TOPSIS	Identifying the critical factors
Jin (2023) [37]	FMEA + DEMATEL + TOPSIS	Analyze the direct and indirect effects
Song et al. (2022) [38]	FMEA + DEMATEL + WINGS	Failure mode ranking

Nowadays, in the context of globalization and international integration, competition in the marketplace becomes more and more severe. Hence, focusing on improving the quality of a product/service with high speed and low waste production has been highly appreciated over the last few years, especially with the current prices of materials and energy. With a good quality scheme, a company can sustain its development by gaining critical competitive advantages over its competitors: high yield and low wastage [8].

ERPN is an extended risk priority number created from conventional RPN. With the identification of failure modes, costs are also incurred because of these modes. The controllable costs are related to the preventive approach and the evaluation costs. The ERPN mainly examines the costs of internal and external errors that are directly related to the failure modes in question [8].

The aim of the paper is to create a new hybrid model of PFMEA, DEMATEL, and ERPN. This model reflects the requirements of producers in the post-communist countries of Europe, where the standard of living is lower compared to the developed countries of Europe, America, and Asia. This is accomplished by the extended risk number ERPN. This is extended by internal and external costs, which, given the standard of living in these countries, are a very important indicator. The DEMATEL model, in addition to its classical use as a multi-criteria analysis model (MCDM), will be used as a design indicator that will help predict the functionality of the models and methods used in the hybrid model. For the comparison parameter, we will select the number of defective products, probability of occurrence (O), and risk priority number (RPN). The hybrid model will be applied to the production process of transfer pressing in the production of a brake element for the automotive industry. An economic model will also be created to evaluate the contribution of individual models and fulfil the requirements of manufacturers.

The DEMATEL model has been used by many authors [7,21,32–36]. It is part of many scientific studies. DEMATEL methodology has been used in our research. We directly followed the research of Tsai et al. [21]. In contrast to that study, we applied these results in practice and proposed measures according to the resulting DEMATEL digraph. The results of the digraph, which shows the relations between failure modes, were used to outline which model is predicted to have the lowest number of defective products.

The ERPN model was created by Nguyen [8], and is a little-known extension of PFMEA. In our study, we validated the effectiveness of this model in practice by prioritizing the results to propose measures. Moreover, we used the data obtained in the ERPN model for the economic evaluation of all the proposed models.

The intended outcome of the study was the proposal of a hybrid model where the advantages of each of the PFMEA, DEMATEL, and ERPN models were exploited.

2. Materials and Methods

2.1. Solution Procedure of the Examined Task

1. Selection of the research object;
2. Creation of an evaluation team;
3. Failure mode identification and development of PFMEA;
4. Creation of the DEMATEL model;
5. Creation of the ERPN model;
6. Applying the models of the hybrid model PFMEA, DEMATEL, and ERPN to the production process;
7. Prediction of the functionality of the models based on the DEMATEL model;
8. Evaluation.

2.2. Object of Research

The monitored product was a component for the brake system of a car (Figure 1), produced by transfer pressing technology in a production organization focused on the production of components for the automotive industry. The material of the component is steel 5355 MC (EN10149). The production organization aims to produce a product of light construction for maximum weight savings. This saves raw materials, costs, and energy when using the products. The brake component is produced on the ARISA pressing line. We will focus on transfer production within the transfer pressing technology. Transfer production is a type of pressing process. In transfer pressing, a blank is cut at the beginning of the process, which is then moved to the next steps of the tool via the transfer system, where it is subsequently shaped. In transfer pressing, a semi-finished product is cut at the beginning of the process and then moved to the next steps of the tool via the transfer system, where it is subsequently shaped. The component is gradually created with seven strikes of the ram.

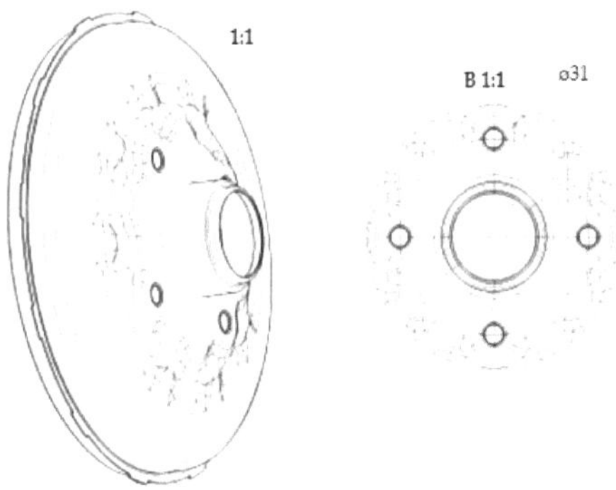

Figure 1. The brake component.

2.3. Creation of an Evaluation Team

We created an evaluation team that participated in the presented analyses and research. The evaluation team consists of three authors of the research paper, three project managers, two quality managers, and one production manager. Although the proposed number (nine members) does not meet the requirements of the conventional FMEA for team size (four to six members), it does meet the requirements of the DEMATEL model. With the DEMATEL model, it is necessary to involve as many experts as possible in the evaluation. We selected experts from the domestic sources of the production organization and scientific and research workers at universities.

2.4. Identifying Failure Modes and Creating an Initial PFMEA

Initial PFMEA progression:

- Identification of failure modes in the selected process;
- Identification of the effects of failure modes;
- Classification of the severity (S) of the failure mode concerning the failure mode effect;
- Severity scale (S) (1–10), where S = 1 means the lowest severity, and S = 10 means the highest severity;
- Identification of the cause of the failure mode;
- Probability of occurrence (O) scale (1–10), with 1 indicating a very unlikely occurrence and 10 an almost certain occurrence;
- Description of the current process controls to prevent the failure mode method—controls that will either prevent the failure mode from occurring or detect the failure mode should it occur;
- Evaluation of the probability of detection (D) that the failure mode cause will be detected;
- Detection scale (1–10), with 1 being "almost certain" detection and 10 being "almost impossible";
- Multiplying S, O, and D to determine the Risk Priority Number (RPN) for each potential failure mode;

- Determination of the average number of defective products on average per week of production in continuous operation. Observation period—3 weeks.

2.5. Creation of DEMATEL Model

We used the well-known DEMATEL methodology, primarily following Tsai et al. [21]. This methodology [21] was used to develop the model on which we built our research (Section 2.7).

Establishing a direct-relation matrix (Equation (1)):

Each team member will determine the intensity of the relations between the selected failure modes as an expert, and thus the basis of the initiation matrix for the DEMATEL model will be created. The coupling between failure modes (FM's) will be evaluated on a scale of 0–6, where 0 means no relations and 6 is the maximum number of relations (Figure 2).

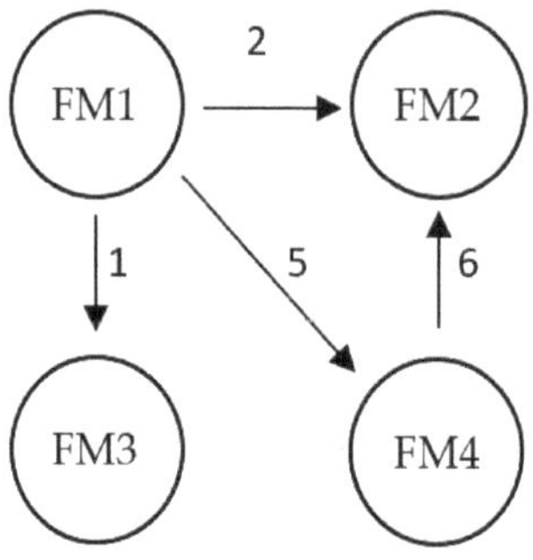

Figure 2. Relations by DEMATEL.

Subsequently, a direct-relation matrix is created (Equation (1))

$$x = \begin{bmatrix} 0 & x_{12} & \cdots & x_{1n} \\ \vdots & 0 & \cdots & \vdots \\ \vdots & \vdots & \ddots & \vdots \\ x_{n1} & x_{n2} & \cdots & 0 \end{bmatrix} \quad (1)$$

Determination of Cronbach's alpha and CITC:

Cronbach's alpha (Equation (2)) will be used to assess the reliability of the survey data, and CITC will be calculated. The test results will show how reliable the questionnaires used in the DEMATEL analyses are. The acceptable limit value of Cronbach's alpha is 0.7 [33,39].

$$\alpha = \frac{k}{k-1}\left(1 - \frac{V_i}{V_t}\right) \quad (2)$$

where k is the number of experts, V_i is the sum of item variances, and V_t response variance of total scores.

A corrected total item correlation (CITC) score is determined for each item and each expert. The CITC expresses how well each item contributes to the internal consistency of a particular construct, as measured by Cronbach's alpha coefficient [40]. CITC was evaluated by Pearson's product-moment correlation coefficient.

Determination of the normalized direct-relation matrix N:

Using column vectors and maximum values (Equation (3)) as the basis for matrix normalization (Equation (4)) [21].

$$\lambda = 1/max_{1 \leq i \leq n} \sum\nolimits_{j=1}^{n} x_{ij} \quad (3)$$

$$N = \lambda X \tag{4}$$

Determination of the direct/indirect relation matrix (*T*) or the total-relation matrix (Equation (5)):

$$T = \lim_{n \to \infty} \left(N + N^2 + \ldots + N^k \right) = N(I - N) \tag{5}$$

where *I* represents the identity matrix.

Determination of the values in each row and column:

This is followed by performing a mathematical summation of the values in each row and column of the total relation matrix (*T*). We denote D_i (Equation (6)) as the sum of the *i*-th column and R_i (Equation (7)) as the sum of the *j*-th row. These values of D_i and R_i consider both indirect and direct influences [21].

$$D_i = \sum_{j=1}^{n} t_{ij} (i = 1, 2, \ldots, n) \tag{6}$$

$$R_i = \sum_{j=1}^{n} t_{ij} (i = 1, 2, \ldots, n) \tag{7}$$

Creating the DEMATEL cause and effect diagram:

The horizontal axis (D + R) and vertical axis (D − R) are used to represent the cause and effect diagram, respectively. The diagram simplifies complex causal relations into an easy-to-understand visual structure. Decision-makers can determine factor types based on the characteristics of the factors and formulate appropriate solutions based on the extent of influence of each factor [21].

Attribute k is either a cause or effect attribute when ($D_k - R_k$) is a positive or negative value, respectively. The size of ($D_k + R_k$) represents the extent of the attribute's cause or effect [21].

Based on the coordinates in ($D_k + R_k$) and ($D_k - R_k$), *k* can be categorized into four categories:

- Positive ($D_k - R_k$) and large ($D_k + R_k$) values: *k* is a cause factor and an actuating factor for solving the problem;
- Positive ($D_k - R_k$) and small ($D_k + R_k$) values: *k* is an independent factor and influences only a small number of other factors;
- Negative ($D_k - R_k$) and small ($D_k + R_k$) values: *k* is an independent factor and is influenced by only a small number of factors;
- Negative ($D_k - R_k$) and large ($D_k + R_k$) values: *k* is a core problem that requires resolution. However, it is an effect attribute; thus, it cannot be directly improved [21,41,42].

2.6. Determination of ERPN

It is an extended risk number that is created by including internal and external costs as a result of failure modes in the model. We used the methodology proposed by Nguyen [8]. This methodology [8] was used to develop the model on which we built our research (Section 2.7).

Internal Failure Cost (*IFC*) is the cost incurred as a result of product quality nonconformance, assuming that defective products are detected at the production stage. If no defective product is detected, then *IFC* = 0 [8].

External Failure Costs (*EFC*) are the costs that are incurred by a non-conformity detected by the customer. If they do not occur, *EFC* = 0. *EFC* is divided into two parts:

1. Without-casualty costs (*WoC*): field servicing and handling complaints; recalls, returns, and replacements; warranty; other indirect costs, etc.;
2. Casualty costs (*CC*): the cost of compensating customers for damage caused by a defective product. Then, *IFC* = *WoC* + *CC*.

Additionally, *IFC* and *EFC* can be converted to SI and SE as follows [8]:

1. We follow up on the conventional PFMEA and the determined RPN.
2. Internal costs are determined—*IFC* (for 1 defective product).

3. External costs—*EFC* are determined, considering *WoC* costs and *CC* costs (per one defective product).
4. The probability of occurrence of the mode—*PO* is determined. *PO* (Equation (8) is based on the probability of occurrence of *O*:

$$PO = \frac{O}{10} \tag{8}$$

5. We determine the detection probability of the mode—*PD*. The *PD* (Equation (9)) is based on the detection parameter *D*.

$$PD = \frac{(10 - D)}{9} \tag{9}$$

6. The internal costs of the *IFC* will be taken into account when determining the severity level (*SI*) (Equation (10)).

$$SI = \frac{IFC}{FCmin} \tag{10}$$

where *FCmin* response minimum costs by all failures and causes.

7. The *IFC*'s internal expenditures will be taken into consideration when establishing a severity level (*SI*) (Equation (10)).

$$SE = \frac{WoC}{FCmin} \tag{11}$$

8. The external costs of the *CC* will be taken into account when establishing a severity level (*SC*) (Equation (12)).

$$SC = \frac{CC}{FCmin} \tag{12}$$

9. We determine the poc—the probability of the failure mode. For *WoC* costs, we consider poc = 0. For *CC* costs, we consider *poc* = 1.
10. The *ERPN* calculation for the extended FMEA is calculated according to Equation (13):

$$ERPN = PO.ST.[PD.SI + (1 - PD).(poc.SC + (1 - poc)SE)] \tag{13}$$

11. The comparison of the RPN calculated for conventional PFMEA with the calculated *ERPN*.

2.7. Creation of a Hybrid Model

After the development of the initial state of the PFMEA, a hybrid PFMEA–DEMATEL–ERPN model is created.

We have proposed three models to solve the research problem:

1. Conventional PFMEA;
2. Combined PFMEA with the DEMATEL model;
3. Combined PFMEA with the ERPN model.

For all models, the procedure will be as follows:

- Monitored criteria—number of defective products, probability of occurrence—O, risk priority number—RPN;
- Proposal of measures for the four most important failure modes;
  - With conventional PFMEA for failure modes with the highest RPN;
  - With the PFMEA and DEMATEL model for the most significant failure modes according to the basic DEMATEL model;
  - In the hybrid model for failure modes with the highest ERPN;

- Measures will be designed to reduce the number of defective products; regarding reducing the occurrence of O.
- Prediction and estimation of models' functionality based on defined relations in the DEMATEL model;
  - Estimation of monitored criteria based on four proposed measures for each model;
  - Based on the results of the digraph of the DEMATEL model, we have proposed a procedure where we show with a bold line the relations between the failure modes of the significant intensity of sizes 5 and 6, which are the values from experts for creating the initial matrix of the direct relation, and we show with a thin line the relations of average intensity (size 3 and 4). Relations of low intensity (size 1–2) will not be displayed. The location of the failure modes in the quadrants of the DEMATEL model digraph will also be crucial.
- The data collection process for each model will take three weeks;
- The average value of the occurrence of defective products for one week is considered reliable for obtaining results and defining random phenomena;
- A comparison of proposed models based on the modification of monitored criteria;
- The economic evaluation of the models is based on the determined costs of *IFC* and *EFC* for the elimination of the effect of each mode of failure according to Formulas (14) and (15).

$$N_{FMi} = n_n.FC = n_n(IFC + EFC)[€] \tag{14}$$

$$N_{Mi} = \sum_{i=1}^{n} N_{FMi}[€] \tag{15}$$

where N_{Mi} is the total cost of the model; N_{FMi} is the cost caused by the effect of the i-th failure mode; n_n is the number of defective products; and FC is the failure cost.

3. Results and Discussion

3.1. Initial PFMEA

The evaluation team identified 54 failure modes that arise during a transfer manufacturing operation. These have been included in Table 2.

Table 2. Initial PFMEA.

Failure	S	O	D	RPN
FM1 Diameter out of tolerance after maintenance	8	10	1	80
FM2 Shape improper	2	7	6	84
FM3 Deformation	6	2	7	84
FM4 Roundness at a diameter	8	2	5	80
FM5 Hole rim deformation	7	2	5	80
FM6 Burr	6	2	7	84
FM7 Loosen chips on the rim of the neck	6	2	7	84
FM8 Cracks caused by production	8	2	7	112
FM9 V crack	8	2	6	96
FM10 Crack on neck	8	2	4	64
FM11 Insufficient remaining material thickness	8	2	7	112

Legend: S—severity, O—occurrence, D—detection, RPN—risk priority number.

The average number of products produced per week in the observed period (at high OEE) was 41,200. We recorded an average of 2% of defective products (824 pieces) per week (stable value at the end of the monitored period). This was a high value that needed to be reduced.

3.2. Model DEMATEL

Table 3 shows the direct relation matrix, which is the same as the expert pairwise comparison matrix. Before that, we calculated Cronbach's alpha and CITC based on the questionnaires from the experts.

Table 3. Initial direct-relation matrix X.

	FM1	FM2	FM3	FM4	FM5	FM6	FM7	FM8	FM9	FM10	FM11	FM12
FM1	0	2	1	3	1	0	0	0	1	0	0	1
FM2	5	0	4	5	3	1	0	3	2	3	3	1
FM3	5	6	0	5	3	3	1	3	2	3	3	2
FM4	4	4	2	0	1	0	0	1	1	1	1	0
FM5	3	2	6	1	0	4	3	1	3	1	1	4
FM6	3	0	0	2	2	0	2	0	0	0	0	0
FM7	0	0	4	0	2	1	0	1	0	1	0	3
FM8	0	3	4	0	0	0	0	0	1	6	6	2
FM9	2	3	4	1	6	0	0	6	0	0	6	0
FM10	0	3	4	0	0	0	0	6	1	0	6	0
FM11	0	0	0	0	0	0	0	5	5	5	0	0
FM12	3	4	5	3	2	0	0	3	0	3	1	0

Legend: FM—failure mode.

The reliability of the questionnaire data was of a high level and reached 0.9971.

The survey results were very consistent. All the adjusted values were greater than 0.3. The values of the corrected item-total correlation for the nine experts were listed in Table 4.

Table 4. Corrected item-total correlation value of validated respondents.

Respondents	Pearson
Expert 1	0.9965
Expert 2	0.9879
Expert 3	0.9809
Expert 4	0.9863
Expert 5	0.9793
Expert 6	0.9804
Expert 7	0.9902
Expert 8	0.9839
Expert 9	0.9870

The normalized direct relation matrix was constructed based on Equation (3) and can be found in Table 5.

Table 5. Use the column vectors and maximum values as the baseline for normalization.

Σx_i	$\max \Sigma x_{ij}$	λ
9		
30		
36		
15		
29		
9		
12	36	0.027778
22		
28		
20		
15		
24		

The normalized direct-relation matrix N was determined based on Equation (4) and can be found in Table 6.

Table 6. Normalized direct-relation matrix N.

	FM1	FM2	FM3	FM4	FM5	FM6	FM7	FM8	FM9	FM10	FM11	FM12
FM1	0	0.056	0.028	0.083	0.028	0	0	0	0.028	0	0	0.028
FM2	0.139	0	0.111	0.139	0.083	0.028	0	0.083	0.056	0.083	0.083	0.028
FM3	0.139	0.167	0	0.139	0.083	0.083	0.028	0.083	0.056	0.083	0.083	0.056
FM4	0.111	0.111	0.056	0	0.028	0	0	0.028	0.028	0.028	0.028	0
FM5	0.083	0.056	0.167	0.028	0	0.111	0.083	0.028	0.083	0.028	0.028	0.111
FM6	0.083	0	0	0.056	0.056	0	0.056	0	0	0	0	0
FM7	0	0	0.111	0	0.056	0.028	0	0.028	0	0.028	0	0.083
FM8	0	0.083	0.111	0	0	0	0	0	0.028	0.167	0.167	0.056
FM9	0.056	0.083	0.111	0.028	0.167	0	0	0.167	0	0	0.167	0
FM10	0	0.083	0.111	0	0	0	0	0.167	0.028	0	0.167	0
FM11	0	0	0	0	0	0	0	0.139	0.139	0.139	0	0
FM12	0.083	0.111	0.139	0.083	0.056	0	0	0.083	0	0.083	0.028	0

The overall relation matrix T was determined from Equation (5) and is given in Table 7. It is an n × n matrix.

Table 7. Total-relation matrix T.

	FM1	FM2	FM3	FM4	FM5	FM6	FM7	FM8	FM9	FM10	FM11	FM12
FM1	0	0	0	0	0	0	0	0	0	0	0	0
FM2	0.257	0.150	0.255	0.237	0.166	0	0	0.227	0.149	0.206	0.228	0
FM3	0.282	0.318	0.181	0.259	0.184	0.129	0	0.247	0.161	0.224	0.246	0
FM4	0.178	0.182	0.134	0	0	0	0	0	0	0	0	0
FM5	0.215	0.199	0.308	0.146	0	0.157	0	0.167	0.161	0.144	0.161	0.170
FM6	0	0	0	0	0	0	0	0	0	0	0	0
FM7	0	0	0.184	0	0	0	0	0	0	0	0	0
FM8	0	0.195	0.227	0	0	0	0	0.155	0	0.282	0.301	0
FM9	0.175	0.222	0.266	0.129	0.243	0	0	0.313	0	0.154	0.316	0
FM10	0	0.184	0.214	0	0	0	0	0.288	0	0.130	0.293	0
FM11	0	0	0	0	0	0	0	0.244	0.186	0.217	0.126	0
FM12	0.198	0.241	0.267	0.183	0.129	0	0	0.209	0	0.197	0.162	0

We then created Table 8 according to Equations (6) and (7).

Table 8. Summary of the prominence and relation of the 12 FM.

	R	D	D + R	D − R
FM1	1.758	0.686	2.444	−1.071
FM2	1.993	2.058	4.051	0.066
FM3	2.258	2.407	4.665	0.148
FM4	1.475	1.09	2.565	−0.384
FM5	1.319	2.041	3.36	0.723
FM6	0.65	0.535	1.185	−0.116
FM7	0.375	0.966	1.341	0.591
FM8	2.122	1.652	3.774	−0.47
FM9	1.277	2.092	3.369	0.815
FM10	1.796	1.519	3.316	−0.277
FM11	2.072	1.148	3.22	−0.925
FM12	0.886	1.787	2.674	0.901

According to Figure 3, each factor can be assessed based on the following aspects:

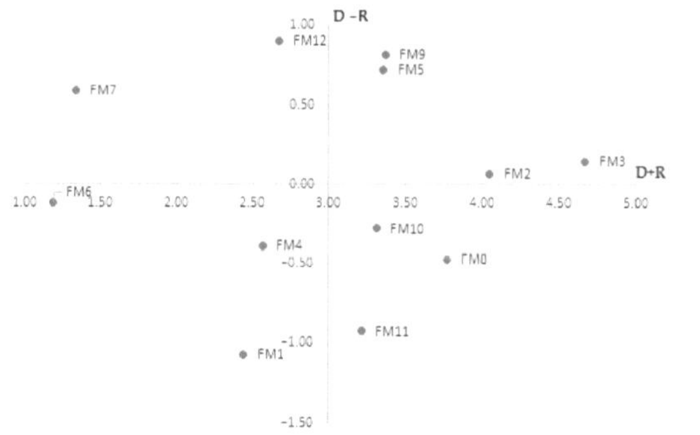

Figure 3. Relational diagram of the 12 failure modes.

In terms of degree of importance, FM3 is ranked in first place, followed by FM2, FM8, FM9, FM5, FM10, FM11, FM12, FM4, FM1, FM7, and FM6.

In this study, FM2, FM3, FM5, FM7, FM9, FM12 are regarded as causal variables, and FM1, FM4, FM6, FM8, FM10, FM11 are regarded as an effect.

The division of failure modes according to their significance (Figure 3):

1. High significance and high correlation: This quadrant include failure modes FM3, FM2, FM9 and FM5. These failure modes are the main causal factors influencing the other items. They are therefore triggers for problem-solving.
2. Failure modes FM12 and FM7 were in the second quadrant. These are characterized by low significance and high relation. They are referred to as moderate causal. They moderately affect the other failure modes. They are thus relatively independent.
3. In the third quadrant, failure modes FM4 and FM1 were found. These are characterized by low significance and low relation. They are moderately influenced by the other failure modes. Thus, they are relatively independent.
4. In the fourth quadrant, failure modes FM8, FM10, and FM11 were located. These are characterized by low significance and low relation. They are the effects of other failure modes and significantly influenced by other failure modes. By being effect factors, they cannot be directly improved.

3.3. Model ERPN

Based on Section 2.6, costs have been assigned to each failure mode. These costs were incurred as a consequence of these failure modes and are shown in Table 9.

Table 9. Determination of internal and external costs for each failure mode per one piece of a defective product.

Failure Mode	*IFC* [€]	*WoC* [€]
FM1 Diameter out of tolerance after maintenance	0.1	0.1
FM2 Shape improper	3.55	8.61
FM3 Deformation	0.31	0.31
FM4 Roundness at a diameter	0.2	0.2
FM5 Hole rim deformation	6.79	8.86
FM6 Burr	0.15	3.83
FM7 Loosen chips on the rim of the neck	0.1	0.11
FM8 Cracks caused by production	0.53	2.93
FM9 V crack	0.2	3.10
FM10 Crack on neck	2.31	4.34
FM11 Insufficient remaining material thickness	0.68	6.49
FM12 Diameter out of tolerance after maintenance	2.14	7.9

Legend: FM—failure mode; *IFC*—internal costs; *WoC*—without-casualty costs.

Subsequently, based on the instructions in Section 2.6 and Formulas (8)–(13), we compiled Table 10.

Table 10. Representation of PFMEA with the ERPN model.

Failure Mode	Initial PFMEA					PFMEA with ERPN						
	S(ST)	O	D	RPN	Cl	PO	PD	SI	SE	poc	ERPN	Cl
FM1	8	10	1	80	9–10	1	1.000	1	1	0	8.00	9
FM2	2	7	6	84	5–8	0.7	0.444	35.5	86.1	0	89.06	2
FM3	6	2	7	84	5–8	0.2	0.333	3.1	3.1	0	3.72	10
FM4	8	2	5	80	9–10	0.2	0.556	2	2	0	3.20	11
FM5	7	2	5	70	11	0.2	0.556	67.9	88.6	0	107.94	1
FM6	6	2	7	84	5–8	0.2	0.333	1.5	38.3	0	31.24	7
FM7	6	2	7	84	5–8	0.2	0.333	1	1.1	0	1.28	12
FM8	8	2	7	112	1/2	0.2	0.333	5.3	29.3	0	34.08	6
FM9	8	2	6	96	4	0.2	0.444	2	31	0	28.98	8
FM10	8	2	4	64	12	0.2	0.667	23.1	43.4	0	47.79	5
FM11	8	2	7	112	1/2	0.2	0.333	6.5	64.9	0	72,85	4
FM12	7	2	7	98	3	0.2	0.333	21.4	79	0	83.72	3

Legend: FM—Failure Mode; S(ST)—severity; O—occurrence; D—detection; RPN—risk priority number; Cl—classification; PO—the probability of occurrence, PD—the probability of detection; SI—severity level from an economic perspective in internally dealing; SE—severity level from an economic perspective in externally dealing; poc—the probability of a casualty caused; ERPN—extended risk priority number.

3.4. Monitoring of Indicators at the Initial Status

Table 11 summarizes the results of the proposed models. We used these data as a basis for the innovative designs for the hybrid model.

Table 11. Analysis of important parameters after creating conventional FMEA, DMEATEL model, and ERPN.

	INITIAL PFMEA					DEMATEL			ERPN	
	n	O	RPN	Cl	n_n	Kl	Significance	Influence	ERPN	Cl
FM1	18,100	10	80	9–11	362	12	12	effect	8.00	9
FM2	2000	2	84	5–8	40	2	2	casual	89.06	2
FM3	2100	2	84	5–8	42	1	1	casual	3.72	10
FM4	2000	2	80	9–11	40	11	9	effect	3.20	11
FM5	2200	2	80	9–11	44	4	5	casual	107.94	1
FM6	2200	2	84	5–8	44	10	11	effect	31.24	7
FM7	2100	2	84	5–8	42	6	10	casual	1.28	12
FM8	2200	2	112	1–2	44	7	3	effect	34.08	6
FM9	2100	2	96	4	42	3	4	casual	28.98	8
FM10	2100	2	64	12	42	8	6	effect	47.79	5
FM11	2000	2	112	1–2	40	9	7	effect	72.85	4
FM12	2100	2	98	3	42	5	8	casual	83.72	3
Σ	41,200				824					

3.5. Hybrid Model PFMEA–DEMATEL–ERPN

3.5.1. Proposal of Measures

By proposing four measures, we want to improve the original state achieved by the conventional PFMEA.

An important parameter for meeting the objectives of the work is the probability of occurrence of failure modes O. The failure modes have 11 equal values (for FM2–12 is O = 2) and 1 abnormally high value (FM1 = 8). Therefore, the FM1 mode has been a matter of priority, as there was a human error in updating the data. Failure mode 1 was not considered in the final statistical evaluation in the other models. By solving the FM1 failure mode, the incidence of failure modes was reduced to 462, which represents 1.12%.

Measures for the model of the conventional PFMEA (according to the highest RPN):

- FM8—Increasing the amount of oil in the event of the formation of a vacuum bubble (the part remains stuck in the tool), the holes in the affected area responsible for the occurrence must be blinded, the release of tension elements if necessary;
- FM9—Tool setting (position of guiding elements);
- FM11—Tool maintenance;
- FM12—Production of a longer punch/underlaying of the existing punch; optimization of the internal diameter of the shear to a larger value.

Measures for the DEMATEL model (according to the first quadrant of the Figure 3):

- FM3—Resharpening the cutting knives in the trimming step, installing the waste brake during the process of cutting the central hole;
- FM2—Correct use of tools—retraining;
- FM9—Tool setting (position of guiding elements);
- FM5—The arresting area of the elements must be made with a wider diameter so that the clearance is as small as the process allows.

Measures for the ERPN model (according to the highest ERPN):

- FM5—The arresting area of the elements must be made with a wider diameter so that the clearance is as small as the process allows;
- FM2—Correct use of tools—retraining;
- FM12—Production of a longer punch/underlaying of the existing punch, optimization of the internal diameter of the shear to a larger value;
- FM11—Tool maintenance.

3.5.2. Prediction and Estimation of the Functionality of the Proposed Models

Based on the location of the failure modes in the individual quadrants of the DEMATEL model digraph (Figure 2), we estimated the success of the used models that make up the resulting hybrid model.

The location of the failure modes for the conventional PFMEA for which we proposed measures is shown in Figure 4. For each model, we divided the failure modes (for which we proposed measures) into cause and effect. At the same time, For each failure mode, we displayed the intensity of the relation to other failure modes. We have shown their quantity and intensity in Figure 4 (highlighted by blue arrows) and Table 12.

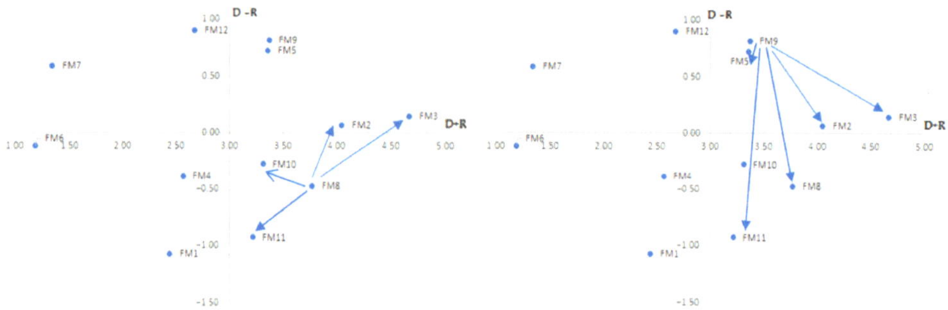

Figure 4. *Cont.*

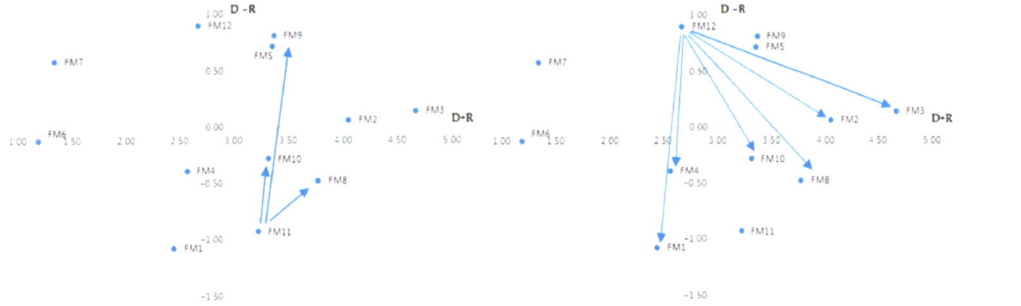

Figure 4. Main failure modes of the conventional PFMEA and their relations on other failure model.

Table 12. The number of bonds and their intensity for each model.

Failure Mode Model	Casual	Effect	Number of Bonds		
			Significant Intensity	Average Intensity	Low Intensity
Conventional PFMEA	2	2	9	9	6
DEMATEL	4	0	9	17	13
ERPN	3	1	7	15	10

The location of the failure modes for the DEMATEL model for which we proposed measures is shown in Figure 5.

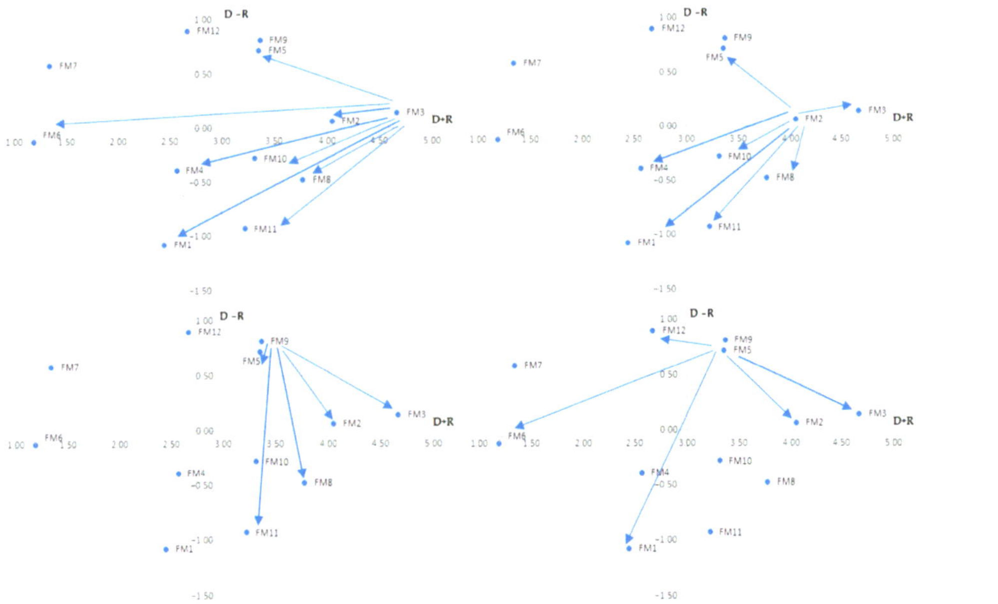

Figure 5. Main failure modes in PFMEA combined with DEMATEL and their relations with the other failure modes.

The location of the failure modes for the model with ERPN for which we proposed measures is shown in Figure 6.

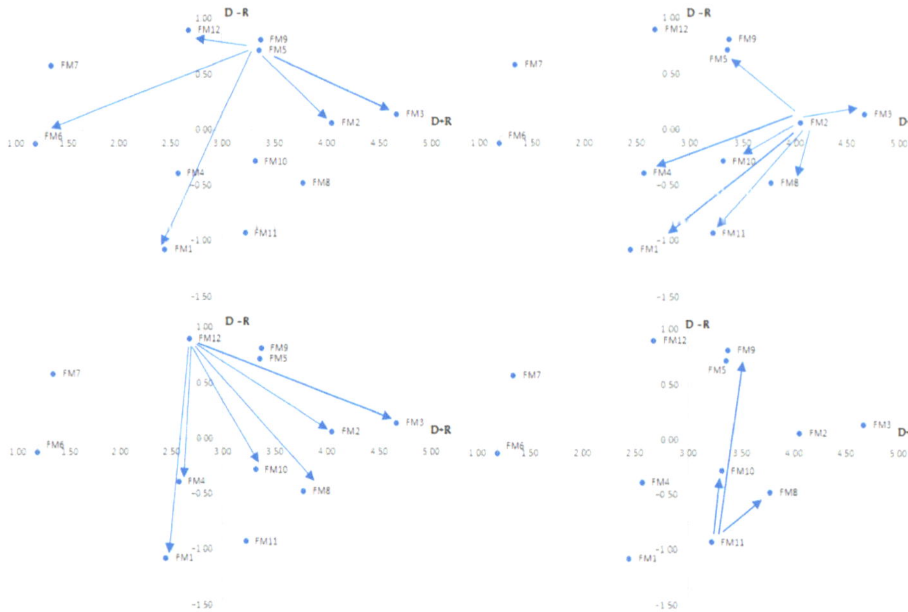

Figure 6. Main failure modes in PFMEA combined with ERPN and their relations with the other failure modes.

Based on Figures 4–6, we created Table 12.

Discussion of Table 12:

The DEMATEL model has all four failure modes (FM2, FM3, FM5, FM9) for which measures have been proposed, located in the first quadrant (Figure 4). All four failure modes are significant and causal, which means they also influence the emergence of other failure modes. If we, as an FMEA team, made the right decisions when creating the initial matrix of a given model, it should manifest itself as the model that manifests itself with the highest reduction in the monitored criteria. This is also indicated by the fact that these four failure modes influence the other 9 failure modes (with repetition) with a significant relation and influence 17 failure modes by the average intensity relation.

The ERPN model has two failure modes (FM2, FM5) in the first quadrant (Figure 5). The FM12 failure mode is in the second quadrant, so it is relatively independent, but it is very close to the border of the first quadrant. These three failure modes, FM2, FM5, and FM12, are causative and significant. Although FM11 failure mode is more influenced by other failure modes, its relations with other failure modes are significant. We estimate that the ERPN model will be successful and significantly reduce the monitored criteria. This is also indicated by the auxiliary criteria, as the given 4 failure modes influence the other 7 failure modes (with repetition) with a significant relation and influence 15 by the average intensity relation.

The conventional PFMEA model has only one failure mode (FM9) in the first quadrant, FM12 in the second quadrant (close to the first), and two failure modes (FM8, FM11) only in the third, although both with higher significance (Figure 6). We refer to only two failure modes as causative, namely FM9 and FM12. We assume that this model will also reduce the monitored criteria, though as compared to the DEMATEL and ERPN models with a lower success rate. This is also indicated by auxiliary criteria regarding the number of affected failure modes with average and lower intensity.

In our case, the success of the models is determined by the relations of average and low intensity, but also by the distribution of failure modes in the quadrants of the digraph (Figure 2) of the DEMATEL model.

3.5.3. Evaluation of Models

In Table 13, we can see the results obtained after application in the transfer molding process.

Table 13. Analysis of the monitored criteria after application of all proposed models.

	PFMEA Initial State					Conventional PFMEA			DEMATEL			ERPN		
	n	n_n	O	RPN	KI	n_n	O	RPN	n_n	O	RPN	n_n	O	RPN
FM1	18,100	362 *	10	80	9–11	-	-	-	-	-	-	-	-	-
FM2	2000	40	2	84	5–8	29	2	84	14	1	42	17	1	42
FM3	2100	42	2	84	5–8	29	2	84	12	1	42	29	2	84
FM4	2000	40	2	80	9–11	35	2	80	29	2	80	32	2	80
FM5	2200	44	2	80	9–11	38	2	80	14	1	40	31	2	80
FM6	2200	44	2	84	5–8	43	2	84	35	2	84	37	2	84
FM7	2100	42	2	84	5–8	44	2	84	38	2	84	39	2	84
FM8	2200	44	2	112	1–2	16	1	56	31	2	112	30	2	112
FM9	2100	42	2	96	4	23	1	48	17	1	48	31	2	96
FM10	2100	42	2	64	12	28	2	64	34	2	64	29	2	58
FM11	2000	40	2	112	1–2	18	1	56	29	2	112	15	1	56
FM12	2100	42	2	98	3	24	1	49	30	2	98	18	1	49
Σ	41,200	462				327			283			308		

* Failure mode 1 was not considered in the final statistical evaluation in the other models. See Section 3.5.1.

We converted the monitored criteria into a graphical representation for better visualization (Figure 7).

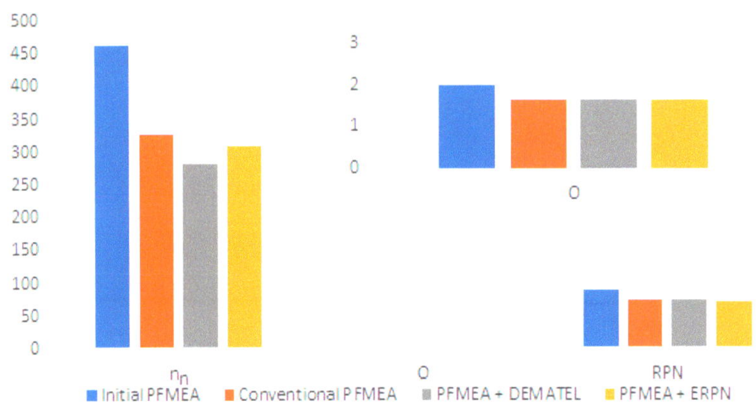

Figure 7. Comparison of the different models and methods used based on the monitored criteria. Legend: n_n—number of defective products; O—occurrence; RPN—risk priority number.

Based on Figure 7, we can conclude that our prediction of the estimation of the model's functionality was correct. We recorded the lowest number of defective products with the DEMATEL model. The categorization of occurrence O was at the same level for all proposed models. Improvements occurred only in failure modes to which measures were specifically applied. We recorded the lowest average RPN value for the ERPN model. However, it should be emphasized that criteria O and RPN do not have such a telling value and are not as specific as the number of defective products.

The conventional PFMEA reduced the number of defective products the least.

It is important to emphasize that with the DEMATEL and ERPN models, we have a clearly defined prioritization for the possible further proposal of measures. PFMEA groups failure modes with the same RPN and O, which is one of the most common limitations of PFMEA.

3.5.4. Economic Evaluation of Models

Based on the costs of a specific failure mode for one component listed in Table 9, we compiled an economic evaluation of the number of defective products. This was reflected in Table 14.

Table 14. The economic evaluation of models.

Failure Mode	*IFC* [€]	*EFC* [€]	PFMEA Initial State		PFMEA Conventional		DEMATEL		ERPN	
			n_n	*FC* [€]	n_n	*FC* [€]	n_n	*FC* [€]	n_n	*FC* [€]
FM2	3.55	8.61	40	486.4	29	352.64	14	170.24	17	206.72
FM3	0.31	0.31	42	26.04	29	17.98	12	7.44	29	17.98
FM4	0.2	0.2	40	16	35	14	29	11.6	32	12.8
FM5	6.79	8.86	44	688.6	38	594.7	14	219.1	31	485.15
FM6	0.15	3.83	44	175.12	43	171.14	35	139.3	37	147.26
FM7	0.1	0.11	42	8.82	44	9.24	38	7.98	39	8.19
FM8	0.53	2.93	44	152.24	16	55.36	31	107.26	30	103.8
FM9	0.2	3.1	42	138.6	23	75.9	17	56.1	31	102.3
FM10	2.31	4.34	42	279.3	28	186.2	34	226.1	29	192.85
FM11	0.68	6.49	40	286.8	18	129.06	29	207.93	15	107.55
FM12	2.14	7.9	42	421.68	24	240.96	30	301.2	18	180.72
Σ			462	2679.6	327	1847.18	283	1454.25	308	1565.32

We evaluated the individual models economically. As can be seen (Table 14), the lowest cost of dealing with the consequences of failure modes was achieved by using DEMATEL. We observed higher costs when using the ERPN model. We found the highest costs when we used conventional PFMEA. It might seem that this is only due to the number of defective products, but this is not a correct interpretation. Each failure mode is uniquely costly to correct and this as a significant factor determines the final economic evaluation of the model.

3.6. Theoretical Analysis of the Proposed Models

In our research, we started with a real-world problem that often occurs when PFMEA is used in practice. PFMEA reaches its limits if we identify many failure modes with the same RPN. This was evident in one of the proposals for solving this problem. This was the classical continuation analysis of PFMEA (conventional PFMEA). For time, organizational reasons, and economic reasons, it is not possible to apply the measures and solve all the problems at once. It is therefore very important to establish a clear prioritization for the proposal of measures. We used the conventional PFMEA as a benchmark for comparison with other proposals, even though it was clear that the PFMEA would be ineffective and unable to reduce the monitored criteria. This was also confirmed.

Many authors have used the DEMATEL model to overcome these limita tions [21,31–33,35–39,43,44]. This was also demonstrated in our research. Moreover, we confirmed these results by monitoring the proposed criteria, which we were able to reduce. Among the requirements of manufacturing organizations, the main aim is to reduce the number of defective products. The DEMATEL model results in a digraph that gives us information about the interaction between failure modes and categorizes the modes into influencing and influenced modes. We have used these results to estimate whether and how each of the proposed models can reduce the monitored criteria. In doing so, we also considered the strength of the interaction between the failure modes that were obtained from the experts' evaluations and is presented in Table 3. The results showed that our estimation was correct. It should be emphasized that DEMATEL is highly dependent on the input values. This requires quality experts knowledgeable about the system being evaluated. If the input data are not exact, the DEMATEL results will be misleading and the estimation of the models functionality will be incorrect.

Based on the research by Nguyen et al. [8], we constructed the ERPN model. We confirmed the results of that research because we were able to determine the prioritization of failure modes. In our study, we again confirmed these results by validating them in practice. Subsequently, we proposed a model for the economic evaluation of the benefit of the models, where we used internal and external costs (they are created as an economic effect of the failure modes and are identified in the ERPN model). These data were the basis for the determination of the ERPN. At the same time, these data should be recorded by each production organization. We pointed out that determining the most successful model by the lowest number of defective products may not be correct since each failure mode produces different costs. This does not change the fact that in our research, the model with the lowest number of defective products also produced the lowest costs.

We hypothesize that based on the methodology set out in the paper, manufacturing organizations may be able to address the situation if PFMEA achieves its limits.

4. Conclusions

The aim of the paper was to create a new hybrid model PFMEA–DEMATEL–ERPN for practical requirements. One of the requirements was how to overcome the limitations of conventional PFMEA for the effective design of measures. The second request contained a proposal on whether it is possible to incorporate the costs arising as an effect of the failure mode into the PFMEA. DEMATEL was used as a tool to estimate the functionality of the proposed models. This required high-quality experts in the field of process knowledge. When used as an estimation tool, it requires high-quality experts in the area of the process under study. The extended risk priority number ERPN ensured the inclusion of costs as consequences of failure in the model, and the inputs to the model were used during the construction of the economic model to evaluate the models from an economic point of view as well. In addition, the ERPN model also ensured that the limitations of the conventional PFMEA were overcome. Both models reduced the monitored criteria proposed by the aim of the paper.

The conventional PFMEA reduced the monitored criteria the least; moreover, in practice, it brought a few critical situations when it was difficult to prioritize failure modes for the design of measures. Our research shows that for the 12 failure modes, we identified several failure modes with the same RPN and the same occurrence of O. We managed to solve this problem with both the DEMATEL model and the ERPN model.

Our research paper deals with real problems in practice (operations, technology, and product) and guides on how to effectively use DEMATEL in managing these problems. In our paper, we have proved both the positives of the DEMATEL model in combination with PFMEA, which many other authors have completed as well [21,31–33,35–39,45,46], and the ability to specific situations where PFMEA is unusable, where it is necessary to use the DEMATEL model.

We achieved the following benefits in our work:

- Use of an innovative combination of models within the hybrid model;
- Defining border situations where conventional PFMEA is of limited applicability;
- Defining border situations when it is important to use the DEMATEL and ERPN model for PFMEA;

- Using the prediction of model functionality estimation based on the DEMATEL model and confirming its success (extended innovative use of the well-known DEMATEL model);
- Reduction in monitored criteria (occurrence of defective products, probability of occurrence of failure modes, and resulting risk number) for designed models;
- Designing and evaluating the economic benefits of the models (extended innovative use of ERPN);
- Proving the economic benefits of the models.

The proposed models solved real problems in practice and provided instructions on how to effectively use them to solve problems in the production process. The proposed ERPN and DEMATEL models combined with conventional PFMEA have shown their validity in solving various requirements of manufacturers.

It is important to emphasize that the results are applicable if we have identified failure modes with the same RPN in the PFMEA analysis. This means that the PFMEA has reached a limit regarding the prioritization of the proposed measures.

Author Contributions: Conceptualization, M.B.; methodology, M.B., M.K. (Milan Kadnár) and C.K.L.; validation, M.B., M.K. (Milan Kadnár) and C.K.L.; formal analysis, M.B. and C.K.L.; investigation, M.B.; resources, M.B.; data curation, M.B., J.B. and M.K. (Maroš Korenko); writing—original draft preparation, M.B.; writing—review and editing, M.K. (Milan Kadnár) and M.B.; visualization, M.B.; supervision, M.B.; funding acquisition, M.B. All authors have read and agreed to the published version of the manuscript.

Funding: This research work has been supported by Ministry of Education, Science, Research and Sport of the Slovak Republic by the grant VEGA: 1/0236/21 and by Drive4SIFood 313011V336, co-financed by the European Regional Development Fund.

Institutional Review Board Statement: Not applicable.

Informed Consent Statement: Not applicable.

Data Availability Statement: The data presented in this study are available on request from the corresponding author after obtaining permission of authorized person.

Conflicts of Interest: The authors declare no conflict of interest.

References

1. Markulik, S.; Šolc, M.; Petrík, J.; Balážiková, M.; Blaško, P.; Kliment, J.; Bezák, M. Application of Fta Analysis for Calculation of the Probability of the Failure of the Pressure Leaching Process. *Appl. Sci.* **2021**, *11*, 6731. [CrossRef]
2. Liu, H.-C. FMEA Using Uncertainty Theories and MCDM Methods. In *FMEA Using Uncertainty Theories and MCDM Methods*; Springer: Berlin/Heidelberg, Germany, 2016; pp. 13–27. [CrossRef]
3. Kamp, B.; Ochoa, A.; Diaz, J. Smart Servitization within the Context of Industrial User–Supplier Relationships: Contingencies According to a Machine Tool Manufacturer. *Int. J. Interact. Des. Manuf.* **2017**, *11*, 651–663. [CrossRef]
4. Lo, H.W.; Shiue, W.; Liou, J.J.H.; Tzeng, G.H. A Hybrid MCDM-Based FMEA Model for Identification of Critical Failure Modes in Manufacturing. *Soft Comput.* **2020**, *24*, 15733–15745. [CrossRef]
5. Carlson, C. *Effective FMEAs: Achieving Safe, Reliable, and Economical Products and Processes Using Failure Mode and Effects Analysis*; John Wiley and Sons: Hoboken, NJ, USA, 2012.
6. Rimantho, D.; Noor, E.; Eriyatno; Effendi, H. Risk Assessment on Failure Factors of E-Waste Management Process Using FMEA Method. *Int. J. Adv. Sci. Eng. Inf. Technol.* **2020**, *10*, 2504–2510. [CrossRef]
7. Bujna, M.; Matušeková, E. *New Trends by FMEA*; Richter-Altman, J., Ed.; RAM-Verlag: Lüdenscheid, Germany, 2018; ISBN 9783942303781.
8. Nguyen, T.L.; Shu, M.H.; Hsu, B.M. Extended FMEA for Sustainable Manufacturing: An Empirical Study in the Non-Woven Fabrics Industry. *Sustainability* **2016**, *8*, 939. [CrossRef]
9. Liu, H.C.; You, J.X.; Lin, Q.L.; Li, H. Risk Assessment in System FMEA Combining Fuzzy Weighted Average with Fuzzy Decision-Making Trial and Evaluation Laboratory. *Int. J. Comput. Integr. Manuf.* **2015**, *28*, 701–714. [CrossRef]
10. Salmeron, J.L.; Gutierrez, E. Fuzzy Grey Cognitive Maps in Reliability Engineering. *Appl. Soft Comput. J.* **2012**, *12*, 3818–3824. [CrossRef]
11. Netto, T.A.; Honorato, H.J.; Qassim, R.Y. Prioritization of Failure Risk in Subsea Flexible Pipes via Data Envelopment Analysis. *Mar. Struct.* **2013**, *34*, 105–116. [CrossRef]

12. Keskin, G.A.; Özkan, C. An Alternative Evaluation of FMEA: Fuzzy ART Algorithm. *Qual. Reliab. Eng. Int.* **2009**, *25*, 647–661. [CrossRef]
13. Wang, Y.M.; Chin, K.S.; Poon, G.K.K.; Yang, J.B. Risk Evaluation in Failure Mode and Effects Analysis Using Fuzzy Weighted Geometric Mean. *Expert Syst. Appl.* **2009**, *36*, 1195–1207. [CrossRef]
14. Kumar, R. Systematic Failure Mode Effect Analysis (FMEA) Using Fuzzy Linguistic Modelling. *Int. J. Qual. Reliab. Manag.* **2005**, *22*, 986–1004.
15. Tay, K.M.; Lim, C.P. Fuzzy FMEA with a Guided Rules Reduction System for Prioritization of Failures. *Int. J. Qual. Reliab. Manag.* **2006**, *23*, 1047–1066. [CrossRef]
16. Ahmed, U.; Carpitella, S.; Certa, A. An Integrated Methodological Approach for Optimising Complex Systems Subjected to Predictive Maintenance. *Reliab. Eng. Syst. Saf.* **2021**, *216*, 108022. [CrossRef]
17. Song, W.; Ming, X.; Wu, Z.; Zhu, B. A Rough TOPSIS Approach for Failure Mode and Effects Analysis in Uncertain Environments. *Qual. Reliab. Eng. Int.* **2014**, *30*, 473–486. [CrossRef]
18. Liu, H.C.; You, J.X.; Shan, M.M.; Shao, L.N. Failure Mode and Effects Analysis Using Intuitionistic Fuzzy Hybrid TOPSIS Approach. *Soft Comput.* **2015**, *19*, 1085–1098. [CrossRef]
19. Chang, C.L.; Liu, P.H.; Wei, C.C. Failure mode and effects analysis using grey theory. *Integr. Manuf. Syst.* **2001**, *12*, 211–216. [CrossRef]
20. Liu, H.C.; Li, P.; You, J.X.; Chen, Y.Z. A Novel Approach for FMEA: Combination of Interval 2-Tuple Linguistic Variables and Gray Relational Analysis. *Qual. Reliab. Eng. Int.* **2015**, *31*, 761–772. [CrossRef]
21. Tsai, S.B.; Zhou, J.; Gao, Y.; Wang, J.; Li, G.; Zheng, Y.; Ren, P.; Xu, W. Combining FMEA with DEMATEL Models to Solve Production Process Problems. *PLoS ONE* **2017**, *12*, e0183634. [CrossRef]
22. Seyed-Hosseini, S.M.; Safaei, N.; Asgharpour, M.J. Reprioritization of Failures in a System Failure Mode and Effects Analysis by Decision Making Trial and Evaluation Laboratory Technique. *Reliab. Eng. Syst. Saf.* **2006**, *91*, 872–881. [CrossRef]
23. Kutlu, A.C.; Ekmekçioğlu, M. Fuzzy Failure Modes and Effects Analysis by Using Fuzzy TOPSIS-Based Fuzzy AHP. *Expert Syst. Appl.* **2012**, *39*, 61–67. [CrossRef]
24. Tay, K.M.; Jong, C.H.; Lim, C.P. A Clustering-Based Failure Mode and Effect Analysis Model and Its Application to the Edible Bird Nest Industry. *Neural Comput. Appl.* **2015**, *26*, 551–560. [CrossRef]
25. Liu, H.C.; You, J.X.; Fan, X.J.; Lin, Q.L. Failure Mode and Effects Analysis Using D Numbers and Grey Relational Projection Method. *Expert Syst. Appl.* **2014**, *41*, 4670–4679. [CrossRef]
26. Aguirre, P.A.G.; Pérez-Domínguez, L.; Luviano-Cruz, D.; Noriega, J.J.S.; Gómez, E.M.; Callejas-Cuervo, M. PFDA-FMEA, an Integrated Method Improving FMEA Assessment in Product Design. *Appl. Sci.* **2021**, *11*, 1406. [CrossRef]
27. Safari, H.; Faraji, Z.; Majidian, S. Identifying and Evaluating Enterprise Architecture Risks Using FMEA and Fuzzy VIKOR. *J. Intell. Manuf.* **2016**, *27*, 475–486. [CrossRef]
28. Sharma, R.K.; Sharma, P. Integrated Framework to Optimize RAM and Cost Decisions in a Process Plant. *J. Loss Prev. Process. Ind.* **2012**, *25*, 883–904. [CrossRef]
29. Alrifaey, M.; Hong, T.S.; Supeni, E.E.; As'arry, A.; Ang, C.K. Identification and Prioritization of Risk Factors in an Electrical Generator Based on the Hybrid FMEA Framework. *Energies* **2019**, *12*, 649. [CrossRef]
30. Shieh, J.I.; Wu, H.H.; Huang, K.K. A DEMATEL Method in Identifying Key Success Factors of Hospital Service Quality. *Knowl. Based Syst.* **2010**, *23*, 277–282. [CrossRef]
31. Liu, Z.; Bi, Y.; Liu, P. An Evidence Theory-Based Large Group FMEA Framework Incorporating Bounded Confidence and Its Application in Supercritical Water Gasification System. *Appl. Soft Comput.* **2022**, *129*, 109580. [CrossRef]
32. Cheshmberah, M.; Naderizadeh, A.; Shafaghat, A.; Nokabadi, M.K. An Integrated Process Model for Root Cause Failure Analysis Based on Reality Charting, FMEA and Dematel. *Int. J. Data Netw. Sci.* **2020**, *4*, 225–236. [CrossRef]
33. Díkmen, C.F.; Taş, Y. Applying DEMATEL Approach to Determine Factors Affecting Hospital Service Quality in a University Hospital: A Case Study. *J. Adm. Sci. Cilt.* **2018**, *16*, 11–28.
34. Mzougui, I.; el Felsoufi, Z. Proposition of a Modified FMEA to Improve Reliability of Product. *Procedia CIRP* **2019**, *84*, 1003–1009. [CrossRef]
35. Fei, H.; Zhu, Y.; Kang, Y.; Shi, S.; Xu, X. Identifying Root Causes of Important Service Failures across Medical Examination Processes with Integration of 4M1E, ITLV, GRA, DEMATEL and FMEA. *Healthcare* **2022**, *10*, 2283. [CrossRef] [PubMed]
36. Hua, Z.; Jing, X.; Martínez, L. An ELICIT Information-Based ORESTE Method for Failure Mode and Effect Analysis Considering Risk Correlation with GRA-DEMATEL. *Inf. Fusion* **2023**, *93*, 396–411. [CrossRef]
37. Jin, G. Selection of Virtual Team Members for Smart Port Development Projects through the Application of the Direct and Indirect Uncertain TOPSIS Method. *Expert Syst. Appl.* **2023**, *217*, 119555. [CrossRef]
38. Song, W.; Zheng, J.; Niu, Z.; Wang, Q.; Tang, Y.; Zheng, P. Risk Evaluation for Industrial Smart Product-Service Systems: An Integrated Method Considering Failure Mode Correlations. *Adv. Eng. Inform.* **2022**, *54*, 101734. [CrossRef]
39. Wang, W.C.; Lin, Y.H.; Lin, C.L.; Chung, C.H.; Lee, M.T. DEMATEL-Based Model to Improve the Performance in a Matrix Organization. *Expert Syst. Appl.* **2012**, *39*, 4978–4986. [CrossRef]
40. Al-Shboul, M.A.R.; Barber, K.D.; Garza-Reyes, J.A.; Kumar, V.; Abdi, M.R. The Effect of Supply Chain Management Practices on Supply Chain and Manufacturing Firms' Performance. *J. Manuf. Technol. Manag.* **2017**, *28*, 577–609. [CrossRef]

41. Chen, H.M.; Wu, C.H.; Tsai, S.B.; Yu, J.; Wang, J.; Zheng, Y. Exploring Key Factors in Online Shopping with a Hybrid Model. *Springerplus* **2016**, *5*, 2046. [CrossRef]
42. Deng, X.; Hu, Y.; Deng, Y.; Mahadevan, S. Environmental Impact Assessment Based on D Numbers. *Expert Syst. Appl.* **2014**, *41*, 635–643. [CrossRef]
43. Rouhani, S.; Ashrafi, A.; Afshari, S. Fuzzy DEMATEL Model for Evaluation Criteria of Business Intelligence. In Proceedings of the ICEIS 2014 16th International Conference on Enterprise Information Systems, Lisbon, Portugal, 27–30 April 2014; SciTePress: Setubal, Portugal, 2014; Volume 1, pp. 456–463.
44. Wang, X.; Zhang, Y.; Shen, G. An Improved FMECA for Feed System of CNC Machining Center Based on ICR and DEMATEL Method. *Int. J. Adv. Manuf. Technol.* **2016**, *83*, 43–54. [CrossRef]
45. Tseng, M.L. A Causal and Effect Decision Making Model of Service Quality Expectation Using Grey-Fuzzy DEMATEL Approach. *Expert Syst. Appl.* **2009**, *36*, 7738–7748. [CrossRef]
46. Wu, W.W. Choosing Knowledge Management Strategies by Using a Combined ANP and DEMATEL Approach. *Expert Syst. Appl.* **2008**, *35*, 828–835. [CrossRef]

applied sciences

MDPI

Article

The Use of Mobile Applications for Sustainable Development of SMEs in the Context of Industry 4.0

Angelina Iakovets [1,*], Michal Balog [2] and Kamil Židek [1]

[1] Department of Industrial Engineering and Informatics, Faculty of Manufacturing Technologies with a Seat in Presov, Technical University of Kosice, Bayerova 1, 08001 Presov, Slovakia
[2] Department of Logistics and Transport, Faculty of Mining, Ecology, Process Control and Geotechnologies, Technical University of Kosice, Komenskeho 19, 04001 Kosice, Slovakia
* Correspondence: angelina.iakovets@tuke.sk

Abstract: Sustainable development of technology in manufacturing motivates entrepreneurs to increasingly introduce technical innovations into their production process. In times of technical progress, the selection and support of qualified personnel are especially important. The process of introducing new technologies or innovations in an enterprise is closely related to the personnel involved in this process. Sustainable development of the enterprise is possible, subject to the sequence of processes, namely, ensuring the adaptation of personnel and collecting feedback. A mobile application can be innovative for a particular enterprise and ensure sustainable development and adaptation to Industry 4.0 technologies. Given the pressure of technology and the environment on small- and medium-sized manufacturing enterprises, it was proposed to conduct a practical study in the conditions of a typical enterprise of this category. To explore the attractiveness of mobile applications as a tool for enterprise sustainable development, an application with basic features that should help provide a framework for integrating Industry 4.0 technologies into the manufacturing process was proposed. In the course of the study, a cycle of innovations and a set of evaluation methods for some of them were proposed. With the help of the proposed assessment methods, as well as the practical use of the mobile application, it became clear that the proposed solution can create a positive effect. The success of this kind of innovation and the further sustainable development of the enterprise is possible if the gradual adaptation of employees will be ensured; it will enable further innovation.

Keywords: mobile application; Industry 4.0; sustainability; management; manufacturing

Citation: Iakovets, A.; Balog, M.; Židek, K. The Use of Mobile Applications for Sustainable Development of SMEs in the Context of Industry 4.0. *Appl. Sci.* **2023**, *13*, 429. https://doi.org/10.3390/app13010429

Academic Editor: Antonella Petrillo

Received: 20 October 2022
Revised: 20 December 2022
Accepted: 23 December 2022
Published: 29 December 2022

1. Introduction

The economic growth of a country largely depends on the ability of this particular country to support the development of entrepreneurship. Small- and medium-sized enterprises (SMEs) play a key role in the development of entrepreneurship. There is a special demand for individual entrepreneurial activity in the European Union, where 37% of the population wants to engage in entrepreneurial activity [1].

SMEs contribute 40% to the GDP in countries with emerging economies [2]. The income received by SMEs and research has shown the importance of SMEs in the national economy of each country. SMEs occupy the biggest part of firms in most countries, especially in countries with developing economies, (about 80–95%) [1,2].

According to the World Trade Organization, SMEs represent over 90% of the business population, 60–70% of employment, and 55% of GDP in developed economies. As the world economy faces prevailing challenges, governments increasingly start to turn to SMEs as a significant element of sustainable and inclusive economic growth, namely for economic growth, poverty reduction, innovation, and job creation [3].

Modern society is undergoing a profound transformation under the influence of new technologies, and the manufacturing sector has undergone special changes. It is especially difficult for small and medium-sized enterprises to maintain the required level of competitiveness as they need to pay special attention to modernization, the development of new technologies, products, and services, reduce the amount of waste, reduce resource consumption, and also invest in personnel with the right skills [4]. The introduction of the latest technologies can be a tool for solving the problem of SME competitiveness. Replacing paper carriers with digital ones opens up opportunities to ensure the continued operation of enterprises even during a pandemic, as well as accelerate the exchange of information within the enterprise and outside it. The most popular solution for manufacturing enterprises is the implementation of technologies of Industry 4.0 [5], which implies the use of wireless mobile technologies and the digitalization of processes. Mobile technologies can be represented by mobile devices that make possible the mobility of employees and automate the processes of obtaining and processing data by mobile applications. New technology implementation requires vertical and horizontal integration, which includes a lot of activities and most importantly time.

The widespread adoption of Industry 4.0 devices and technologies has increased the demand for technological innovations not only by manufacturers but to a greater extent by consumers. In the modern economy, the consumer forms the demand and characteristics of new products, and therefore enterprises try to organize their production to be able to flexibly respond to the changing consumers' needs. Features of the modern market can be summarized as:

- a consumer-oriented market,
- a highly competitive market,
- a market for a wide range of goods with a life cycle but a high level of technology,
- a market where production is characterized by full utilization of production capacities and use of low consumption technologies,
- a market characterized by a high level of service [6].

Thus, the main obstacles that need to be overcome by small- and medium-sized enterprises to take a competitive position in the market are the following:

- preparation and provision benefits of new technologies to employees,
- preparing the basis for the successful implementation of innovations,
- selection and preparation of the courses for personnel training programs,
- design a high-quality organizational structure of the enterprise to ensure the quality of services and products,
- effective planning.

Such criteria Chang et al. [7] define and evaluate by technology TOE (technology–organization–environment), multiple-criteria decision-making approach (MCDM), and influence network relations digraph (INRD) with accordance to DEMATEL analytic network process [7]. The research of Shiaw Tong Ha et al. [8] showed that three of four knowledge management (KM) dimensions, knowledge acquisition, knowledge conversion, and knowledge protection are positively related to enterprise performance. Summarizing the research of specialists in this field [5–7,9], it can be concluded that modernization according to Industry 4.0 depends on TOE and requires MCDM analysis to ensure a positive result from the implementation of the components of Industry 4.0.

E-business and information technology (IT) are critical components of improving logistics functions, Kamariotou et al. [9] confirm in their study, where they also note the importance of creating a high-quality enterprise information system. The importance of high-quality organization of enterprise information systems is due to its main task—continuous data exchange [9].

Investing large amounts of energy and money in major software updates and technological changes does not always bring the expected results. The automation process by software can be carried out in stages by small programs in the form of mobile applications. Although modern enterprises are increasingly replacing manual labor with automated, sustainable development of HR management and support of human resources has become especially important.

Earlier, Jerry S.F. Lee [10] wrote about the importance of technology adoption. The importance of adaptation is very acute, even in fully automated production processes, because it still involves people, in one or another way. Moreover, it should be highlighted that the more high-tech process, the higher the requirements for the employee of the enterprise. The main factors that affect the process of adaptation to new technologies mainly associated with the human factor can be seen in Figure 1 [10].

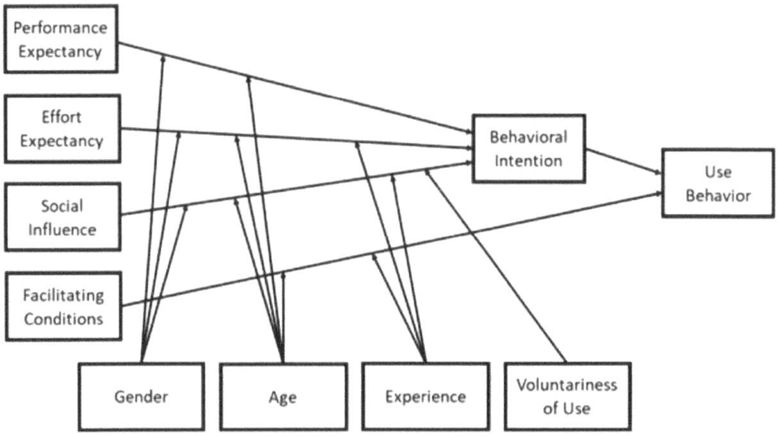

Figure 1. Theory of acceptance and use of technology [10].

Increasing requirements for employees are always associated with stress, which directly affects the quality of the enterprise [11]. The level and magnitude of stress depend not only on the technologies themselves but also on many additional factors, such as age, gender, experience, and voluntariness of use (Figure 1). Human resource management should be organized so that employees will be interested in developing not only their skills but also in the stable development of the entire enterprise [12]. Enterprises can partially relieve pressure on staff using covert data collection in the mobile application, which in turn will provide indirect feedback from employees. Data about quality of work, operating time and downtime, and equipment breakdowns can be quite easily collected by special sensors for automated reporting realized in the mobile applications [5].

The collection and timely processing of personnel data is especially important in the process of introducing new technologies to the enterprise. The work of employees is a powerful and significant potential of the company in terms of economic indicators, especially in SMEs [13].

Thanks to automation and the presence of a unified information base, all actions are carried out quickly, and most importantly, calculations, statistics, and analytics are always at hand. These conditions can be met by using mobile applications since their feature is the ability to integrate into enterprise information systems. Mobile devices can enable mobility of employees during the work day and help to save time on studying materials, reporting, analyzing and tracking processes, and also communication.

Automation of personnel management processes and training by mobile applications is also available and getting feedback by application is easier and faster than by traditional methods. The monetary drain of employee turnover costs manufacturers hundreds of thousands and sometimes millions of dollars annually [11,14].

A lot of enterprises face the challenge of finding new skilled workers. This is a multifaceted problem, as it relates to the rapid retirement rate of opulent employees, and a shortage of qualified young workers due to the lack of graduate programs in intelligent manufacturing (robotics, automation, big data, analytics, and additive manufacturing) [5,14,15].

Manufacturing enterprises every day are faced with the problem of ensuring all these three processes because they are difficult to provide during work processes.

Implementation of technologies of Industrial Revolution (IR) 4.0 and HR development depends on:

- IR 4.0 training
- IR 4.0 company strategy
- IoT and cyber-physical systems awareness
- New skills aligned to IR 4.0
- Implementation of the latest emerging technology
- Training content satisfaction
- Skill and knowledge utilization
- Increased employee efficiency [16]

The introduction of IR technologies and staff training are closely interconnected, and therefore it is worth exploring the possibility of using the mobile application as an integrating tool of Industry 4.0 technologies.

2. Materials and Methods

2.1. Mobile Application Integration Stages

To ensure the equal and efficient implementation of new technologies, the so-called "vertical integration" should be used, which means ensuring uninterrupted communication and the flow of technological infrastructure used in all processes. A tool for "vertical integration" of technologies can be a mobile application that contains a knowledge base for employee training, and accountability, as well as collecting feedback from employees. The need for reporting and feedback is grounded on the need of enterprise management to have an overview of the current situation and to understand how well employees are trained, the level of study materials, and to get the overall picture of the correctness of the redistribution of duties among employees to achieve the expected labor productivity [17,18]. Feedback from users of the application as well as a group of experts can be collected using tests from IT as well as a general survey using the coefficient of importance of factors and points.

The introduction of something new always affects the mental mood of the team and that is why a priority task of managers is to design a learning process to encourage employees to develop themselves and learn something new. To ensure the comfortable adaptation of employees and sustainable management of human resources of the enterprise, it is worth organizing an adaptation plan. Enterprises under Industry 4.0 implementation have to go through a cycle of innovation every time (Figure 2) [19,20].

The proposed cycle can provide sustainable development of the manufacturing enterprise, due to implementing innovations oriented not only on automating but also on people involved in the manufacturing process.

The "Innovation" stage of the proposed study lies in using the mobile application as Industry 4.0 tool of vertical integration and a tool for adapting to new technologies.

Under implementing new solutions, it is important to explain (Figure 2, "Motivation") employees' essence of the changes, their importance for employees and the enterprise, and other changes associated with it, to ensure sustainable personnel management [19–21].

Using mobile applications is very similar to the M-learning process (mobile learning), where new technologies become a helpful tool for the adaptation process to new technologies [12]. The M-learning process is very simple due to its way of providing information and graphical illustration of the information [22].

The introduction of mobile IT applications in SMEs is associated with a combination of the technology–organization–environment (TOE) structure and is aimed at improving IT innovation and the implementation of Industry 4.0 technologies. That is why it is necessary to evaluate the effect of aspects of technology (technology readiness, technology security, and technology integration), or generation (financial commitment, organization readiness, and top management support), and environmental criteria (regulatory support, competitive pressure, and environmental uncertainty) on Industry 4.0 technology application [7].

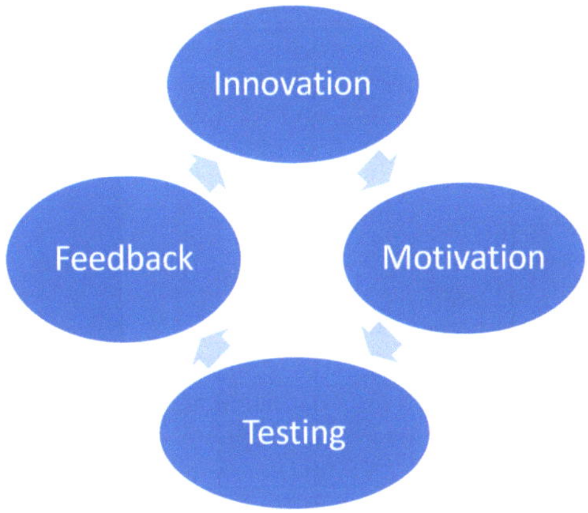

Figure 2. The enterprise's cycle of innovation.

The mobile application is a tool for implementing technologies of Industry 4.0 and is also a tool for M-learning [22–29]. According to this, there is also another aspect of evaluating such innovation—this is an assessment by the user or relevance for a particular group of users.

The enterprise under study has all features of manufacturing SMEs in the Slovak Republic:

- wide and non-constant range of products,
- training of personnel on the mentoring method using technical documentation and paper media,
- under the introduction of an increasing number of automated systems and mechanisms into production,
- quantity of employees is up to 250,
- two work shifts,
- main staff age group range is 25–70 years old,
- seasonal employees are hired with an increase in the number and volume of orders.

As part of the study, the main test group was the work shift of welders, since welding work had the greatest impact on the quality of the final product.

The stage of "Motivation" (Figure 2) should pick up arguments for employees. At this stage, it is important to explain that mobile applications, for users, have become commonplace for a long time and mobile work applications will not be something difficult. To ensure this stage, it is worth studying the characteristics of the personnel of manufacturing enterprises.

2.2. "Motivation" Stage

Since the proposed research was aimed at practical testing of the mobile application, as employee training and supporting tool in manufacturing enterprises, "Motivation" had to be supported not only by theory but also by the functionality of the application, which determined the main requirements for the mobile application sample:

- access to digital work instructions [29,30] and tests,
- two-level access to learning materials: for managers and operators,
- integrated data collection process about employees,
- automatic data exchange with the enterprise's database,
- feedback [31].

Such requirements were proposed to reduce requirements on users and reduce the engagement of humans in the data collection and processing process [29,30]. The emphasis on placing work instructions in the mobile application was justified by the fact that the company has mass production and the product range changes frequently. Changes in the range of products at the enterprise lead to constant training of employees and the creation of new work instructions.

All employees' activities and test results automatically will be put into the general table of personnel assessments on the server. The managers' interface includes some more options such as:

- quality reporting,
- adding tests,
- adding instructions,
- access to collected data from welders.

The structure of the mobile application was designed in accordance with the ERP program to reduce the time of data collection and proceeding.

ERP program is used by all employees; therefore, it was important to connect the data collected by the mobile application and the ERP program within the database. The connection was realized on the level of the server.

In theory and practice, a quantitative indicator of labor productivity is calculated with the use of strict values [32,33]. To ensure soft adaptation to new technologies, a combination of classical methods and data collected from the application, namely the start and end times of work shifts, was proposed. Such application features were supposed to reduce the amount of paper documentation and data collection time. Time losses, production defects, and production downtime—all these factors directly affect the profit of the enterprise and its place in the market [34]. Based on this information, it follows that there is an important factor in providing sustainable HR management and adaptation [35] to new technologies. The mobile application in this way should be tested on the biggest part of employees, welders, because they create product quality in the enterprise. Managers provide support to this group of employees and create "motivation", which is why it was decided to test the application in two steps: on managers and welders.

As "motivation" pros for managers, the following were selected:

- Corporate E-learning or M-learning [36] takes from 40% to 60% less time than the traditional learning process [25],
- 67% of people use mobile devices for access to E-learning courses and all active mobile phone users are 43 percent more productive than non-mobile devices users [34,35],
- Effectiveness of M-learning is proven by research by Globe News Wire, where it was written that in 2015, the mobile learning market was worth just 7.98 billion US dollars but by the end of 2019, that number had risen to 27.32 billion US dollars which proves that this market is promising. During the COVID-19 pandemic, there has been a steady worldwide increase in mobile users. Experts predict that the mobile and E-learning market will rise and the sector will maintain a Compound Annual Growth Rate (CAGR) of 36.3 percent down to 2025 [29,30,36],
- The growth of financial investments in M-education indicates that this method corresponds to modern trends in the technology market and makes possible a quick and affordable data exchange, which is especially important in manufacturing [5],
- Market research of employee engagement software has shown that the top 10 applications [37] and all these programs have some features that are very helpful in manufacturing enterprises such as:
    - customizable features,
    - easy to use,
    - feedback,
    - real-time score [37].

- The price of such applications ranges from 4 US dollars per month for one user to 20 US dollars [37]. As the proposed research was aimed at studying SMEs, it is reasonable to calculate the possible costs of implementing such an application. It is reasonable to take for calculating the maximum number of employees for a small enterprise (50 people) and the maximum number of employees for a medium enterprise (250 people) [32]. Paying 200 or 5000 US dollars per month for an application that is not tailored to the needs of a particular manufacturing enterprise is not a reasonable investment [35] which is why important to realize the research on the feasibility of a such decision.

The next stage was to determine the attractiveness of practical use of the mobile application for both groups of employees. The age group of workers under study and regular Slovak manufacturing enterprises is 25–70 years old, but most employees (about 90%) are in the age group of 25–55 years old. It is also necessary to study the group of the seasonal labor force, which mainly consists of people waiting for work in the labor market [38].

The biggest part of unemployed people in the Slovak republic is in the age group of 25–54 years.

According to information from Figures 3 and 4 and the information below, it can be said that the studied group of people consists of active smartphone users. Moreover, based on the above theoretical information, it is possible to hypothesize that the mobile application might be easy for using for the selected group of people.

	Age groups					
	less than 25 years		25-54 years		55 years and more	
People pending employment	Less than 20 year	5,024	25-29 years	18,912	55-59 years	21,042
			30-34 years	19,678		
			35-39 years	20,555		
	20-24 years	16,389	40-44 years	20,626	60 years and more	10,646
			45-49 years	19,131		
			50-54 years	18,864		
Total amount	214,113		117,766		31,688	

Figure 3. Structure of unemployed people of Slovak Republic in 2020 [38].

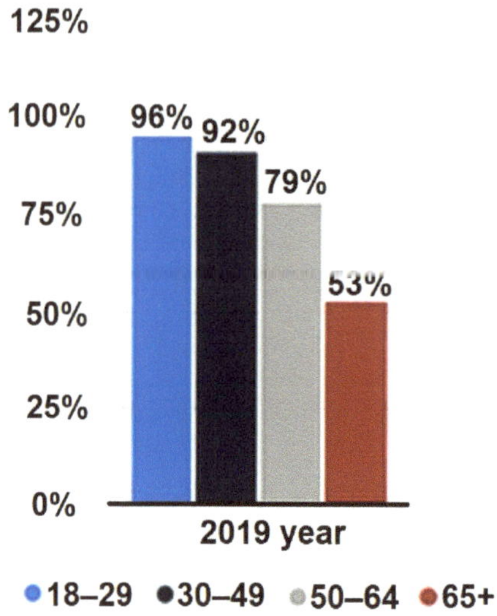

Figure 4. Smartphone users by age groups [34].

Management of the enterprise accepted the advantages of the proposed method of adaptation mentioned above and the mobile application went into the "Testing" stage (Figure 2) by the groups of users.

3. Results

Background of "Testing" and "Feedback" Stages

The proposed study is primarily aimed at evaluating the effectiveness of the mobile applications as a tool for adapting to new technologies in manufacturing SMEs. The production process in SMEs is based on workers and production managers and both are responsible for product and staff quality, which is why it was important to get feedback from both groups of users.

The first stage of "Testing" was to give the mobile application group of specialists (top management and stakeholders of the enterprise) and then was realized "Feedback" collecting process by evaluating its efficiency. This group was not chosen by chance, since in DEMATEL, MCDM, and INRD [7] theories and other assessment methods, these people are an important component of decision-making on automation.

The assessment was carried out using a questionnaire, where each criterion was assigned by a weighting factor as well as a scale of grades from 0 to 5, where "5" is the maximum value and 1 is the minimum one. To simplify the evaluation process, the original teaching method was called "mentoring" and the proposed was called "M-learning".

The consistency of expert opinions on each indicator is assessed by scattering estimates around the mean (variances) using Fisher's method.

The average score for each indicator "x" was calculated by the formula:

$$\overline{\mathrm{x}} = \frac{\sum_{i=1}^{m} \mathrm{x_i}}{m} \quad (1)$$

where $\mathrm{x_i}$—mark is given by the i-th expert on this indicator;

m—number of experts.

There were 4 experts (project manager, welder's team leader, and two specialists from the university) and 5 criteria (Table 1).

Table 1. F-test of expert assessment of training.

Criterion:	Value	Mentoring	Mentoring (X)	M-Learning	M-Learning (x)	S_1^2	S_2^2	F
The convenience of perception and understanding	0.2	4	0.8	4	0.8	0	0.01	0
		4	0.8	4	0.8			
		4	0.8	5	1			
		4	0.8	4	0.8			
	Average	4	0.8	4.25	0.85			
Easiness of study	0.2	5	1	4	0.8	0.01	0.013	1.33
		5	1	3	0.6			
		5	1	3	0.6			
		4	0.8	4	0.8			
	Average	4.75	0.95	3.5	0.7			
Need for additional advice	0.2	1	0.2	2	0.4	0.03	0.013	2
		0	0	1	0.2			
		2	0.4	1	0.2			
		1	0.2	2	0.4			
	Average	1	0.2	1.5	0.3			
The convenience of displaying information	0.1	4	0.4	5	0.5	0.0025	0.0025	1
		4	0.4	4	0.4			
		5	0.5	4	0.4			
		4	0.4	4	0.4			
	Average	4.25	0.425	4.25	0.425			
The quality of staff training	0.3	4	1.2	5	1.5	0.0225	0	0
		3	0.9	5	1.5			
		3	0.9	5	1.5			
		3	0.9	5	1.5			
	Average	3.25	0.975	5	1.5			

The consistency of expert opinions is assessed by the scatter of xi around the mean. To do this, it is necessary to calculate the variance of the estimate and compare the significance of its difference from the minimum possible variance according to Fisher's criterion.

The variance is calculated by the formula:

$$S_1^2 = \frac{\sum_{i=1}^{m}(x_i - \overline{x})^2}{m-1} \quad (2)$$

The variance for mentoring (S_1^2) was 0.08 and 0.05 for M-learning (S_2^2).

The next step was the F-Test: hypothesis tests for the variances of 2 normal distributions.

$$F = \frac{S_1^2}{S_2^2} \quad (3)$$

To assess the quality of the proposed teaching methodology, it was hypothesized that M-learning is equivalent in quality to E-learning. To confirm or refute the theory, calculations were made according to the above formulas for each criterion.

The average X for mentoring is 3.4 and for M-learning is 3.7 (Table 1.). The average score differs by only 0.3 which is not significant, given the assessment step of 1 point.

Provided that ($X \sim F(m,n)$) : $P(X \leq F_{\propto}(m,n)) = \propto$, where $\propto = 0.9$, $n = 3$, and $m = 4$ the hypothesis is confirmed, since Φ of each parameter is less than its tabular value (5.34).

Implementing training using mobile devices, the employees' level of skills in working with modern gadgets has to be considered, since not every employee can be an active user of mobile devices. The need for additional advice can be seen on the right side of the graph, line "3" (Figure 5) corresponding to "The quality of staff training" from Table 1.

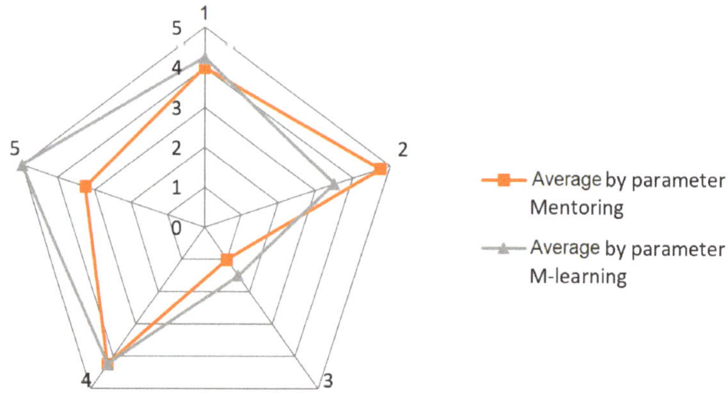

Figure 5. Comparison of mentoring and M-learning.

According to the experts evaluating (Figure 5), M-learning could replace hands-on learning (mentoring) because the mobile application is inherently multi-functional and able to expand the range of tasks to be performed.

After an expert evaluation of the suitability of the offered application and especially its content for the investigated company, the mobile application was handed over for practical testing to a group of 50 employees in the next part of "Testing".

After using the mobile application in the work process, selected welders evaluated the application by the Usability test [39], which is a standard test of the usability of the mobile application in practice. The system Usability Scale (SUS) is a base for the Usability test [39–41] and serves as a tool for getting feedback from users. This test is aimed at determining the level of usability, learnability, comprehensibility, and attractiveness of the target audience of product users [40].

SUS consists of 10 statements with 5 options for respondents (from strongly agree to strongly disagree) (Figure 6). The SUS evaluates a wide range of products and services, including hardware, software, mobile devices, websites, and applications [39]. The user's score for each statement is converted to a new number, added, and then multiplied by 2.5 to convert the original scores from 0–40 to 0–100. Although the scores are 0–100, they are not percentages and should only be considered in terms of their percentile ranking. According to the research, a SUS score of 68 and higher is considered to be above average, and a score that is less than 68 is considered to be below average.

The SUS questionnaire was completed by the group of target users. The average score of 50 SUS tests was 28 points. Then, this value was converted by multiplying by 2.5 and a score of 70 was obtained. Since 70 is more than 68, it means that the result is above average and satisfactory. The next step was to illustrate the conversion results (Figure 7).

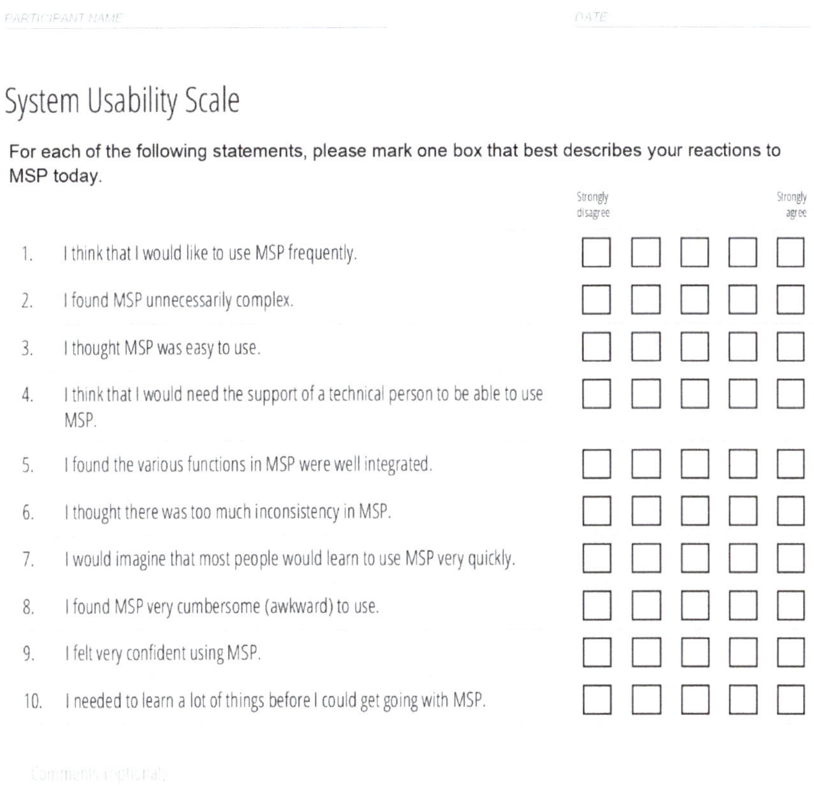

PARTICIPANT NAME DATE

System Usability Scale

For each of the following statements, please mark one box that best describes your reactions to MSP today.

		Strongly disagree				Strongly agree
1.	I think that I would like to use MSP frequently.	☐	☐	☐	☐	☐
2.	I found MSP unnecessarily complex.	☐	☐	☐	☐	☐
3.	I thought MSP was easy to use.	☐	☐	☐	☐	☐
4.	I think that I would need the support of a technical person to be able to use MSP.	☐	☐	☐	☐	☐
5.	I found the various functions in MSP were well integrated.	☐	☐	☐	☐	☐
6.	I thought there was too much inconsistency in MSP.	☐	☐	☐	☐	☐
7.	I would imagine that most people would learn to use MSP very quickly.	☐	☐	☐	☐	☐
8.	I found MSP very cumbersome (awkward) to use.	☐	☐	☐	☐	☐
9.	I felt very confident using MSP.	☐	☐	☐	☐	☐
10.	I needed to learn a lot of things before I could get going with MSP.	☐	☐	☐	☐	☐

Comments (optional)

Figure 6. Template of SUS test of application.

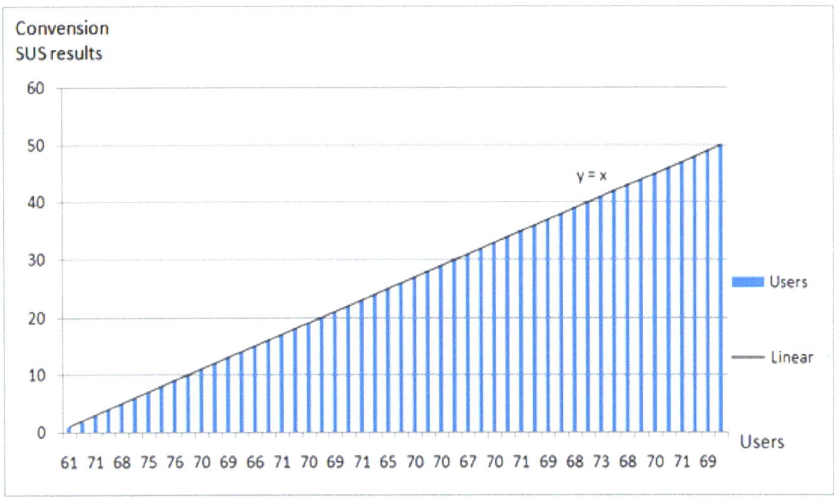

Figure 7. Conversion of SUS test.

According to the results of the SUS test, it can be stated that the experts and the target group of welders evaluate the application as effective. The next step was evaluation of nine g aspects (Figure 8), that can affect the process of the mobile application implementation.

Aspects/Element	Descriptions
Technology aspect (G1)	
Technology readiness (g1)	Technology needs to be organized to make procedure changes, and latent sites must make changes for Industry 4.0 technology if benefits are to accrue.
Technology security (g2)	This refers to internet platforms being considered secure for conducting online transactions and exchanging data, including isolated data security and protection in using Industry 4.0 technology.
Technology integration (g3)	This refers to improving the responsiveness of information systems and decreasing incompatibility between legacy systems.
Organization aspect (G2)	
Financial commitment (g4)	This refers to the company that can offer the financial resources , and is especially committed to investing in employee training, software, hardware, system development, and system integration.
Organizational readiness (g5)	This refers to potential sites having to make decisions and businesses needing to be prepared to make business procedure changes for Industry 4.0 technology if benefits are to accrue.
Top management support (g6)	This refers to the top managers offering the support, and a promise to ensure a positive influence on this Industry 4.0 technology application procedure.
Environment aspect (G3)	
Competitive pressure (g7)	This refers to the adoption of Industry 4.0 technology, so that companies can benefit from more accurate data collection and a greater operational efficiency.
Regulatory support (g8)	This refers to elements of conception such as the policies of the government, which impact the diffusion of IT.
Environmental uncertainty (g9)	This refers to the environmental uncertainty, and how managers and entrepreneurs tend to act proactively according to well-informed conjectures about the strategic path ahead.

Figure 8. Clarification of aspects [7].

Evaluation of mentioned aspects will show the strongest aspects and management will be able to provide sustainable modernization and development.

Mobile devices and mobile applications are components of Industry 4.0 and their implementation is associated with many factors that affect the conditions of their use, as well as the process of their implementation. Based on this, it was decided to turn to the research of Chang and colleagues and use elements for assessing the Industry 4.0 technology application effectiveness. This method is based on MCDM and DEMATEL combination.

For building the final graph of the effect of the provided implementation of technology on the manufacturing process using the expert method and evaluating of g factors affect [7], g is a set of factors related to the Industry 4.0 technology application within each aspect that is subsequently evaluated by 5 experts (stakeholders and top management) (Figure 8). The normalized average matrix of the evaluation is shown below (Table 2).

Table 2. Matrix of normalized average values of expert ratings.

Average Matrix	g1	g2	g3	g4	g5	g6	g7	g8	g9
g1	0	2	1.4	0.8	1.6	0.4	0.2	0.4	0.2
g2	0.6	0	1.6	0.8	1.8	0.2	2	0.4	0.8
g3	1.6	1.8	0	1.2	0.6	0.6	0.2	0.2	0.2
g4	1	0.2	0.8	0	1.2	2.2	0.6	0.8	1.4
g5	1.6	0.6	1.4	0.8	0	2.6	0.4	1.2	1
g6	1.4	0	1.8	0.2	1.6	0	0.4	0.6	0.4
g7	0.2	0.2	0.2	0.8	1.8	0.6	0	0.6	1.2
g8	0.4	0.2	2.6	1.8	0.4	0.4	0	0	0.2
g9	1.2	2.2	1.2	1	1.4	2.6	1.8	0.4	0

The final DEMATEL MCDM analysis can be represented in Table 3.

Table 3. Results of the DEMADEL MCDM analysis.

	Ri	Ci	Ri + Ci	Ri − Ci	Identify
g1	−1.18592	−1.41569	−2.60162	0.229772197	Cause
g2	−1.24655	−1.41866	−2.6652	0.172108986	Cause
g3	−1.13962	−1.40314	−2.54276	0.263520535	Cause
g4	−1.16896	−1.01881	−2.18778	−0.150149811	Effect
g5	−1.17654	−1.40442	−2.58096	0.227883863	Cause
g6	−0.99864	−1.0799	−2.07853	0.081262193	Cause
g7	−1.06374	−0.92671	−1.99045	−0.137021864	Effect
g8	−0.94888	−0.85661	−1.80548	−0.092269376	Effect
g9	−1.4399	−0.84479	−2.28468	−0.595106724	Effect

Table 3 represents four factors that can affect the final implementation of the proposed solution. g-factors that affect the implementation process are illustrated in Figure 9 by red dots.

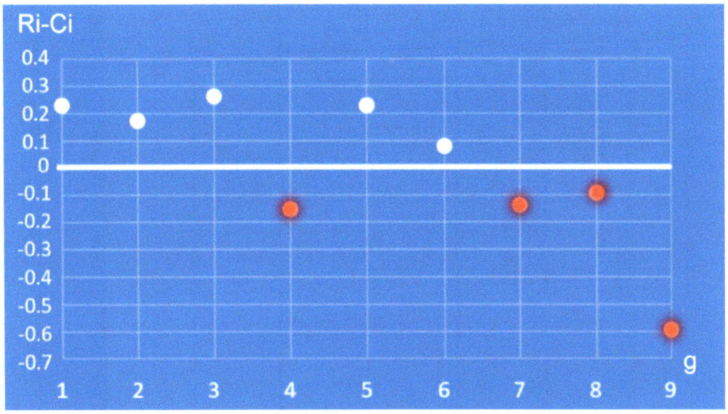

Figure 9. Graphical illustration of the DEMADEL MCDM analysis results.

The red dots are g4 (financial commitment), g7 (competitive pressure), g8 (regulatory support), and g9 (environmental uncertainty) which can be seen in Figure 9. According to the graph, it can be seen that factor g9 has the greatest impact on the implementation process. This factor is integral during the implementation of the components of Industry 4.0, this is explained by the fact that technologies are developing very quickly and so is the software. DEMATEL MCDM method is based on the assessment of the influence of factors on one another and therefore, it is important to find not only the most influential factors but also to emphasize their relationship.

To draw up a diagram of the influence of factors on each other, it is worth referring to the Table 4.

Table 4. Total reversion matrix of DEMADEL MCDM analysis results.

T Total Reversion Matrix	g1	g2	g3	g4	g5	g6	g7	g8	g9
g1	**−0.71202**	0.12849	−0.16953	−0.08183	0.01537	**−0.26235**	0.02985	−0.07672	−0.05718
g2	**−0.34181**	−0.85330	**−0.27959**	−0.15632	0.01736	−0.04786	**0.30742**	−0.02816	0.13572
g3	−0.08796	0.10679	−0.79430	−0.02976	−0.08048	−0.19448	0.08999	−0.13870	−0.01072
g4	0.02394	−0.21387	−0.07675	**−0.60679**	−0.15609	0.15783	**−0.24592**	−0,02674	−0.02458
g5	0.08845	−0.01721	0.06542	−0.05823	**−0.75647**	−0.05175	**−0.28315**	0,01490	−0.17849
g6	0.15497	0.04506	0.15685	−0.09297	−0.01575	−0.77666	**−0.20290**	−0.05466	**−0.21257**
g7	−0.15796	−0.36523	−0.19757	−0.06894	0.00859	0.14514	**−0.60098**	0.07296	0.10026
g8	−0.09838	−0.05444	**0.28359**	**0.43427**	**−0.39739**	**−0.23115**	**−0.27824**	**−0.48739**	−0.11975
g9	**−0.28493**	−0.19494	−0.39125	**−0.35824**	−0.03955	0.18137	**0.25721**	−0.13209	**−0.47748**
Ci	−1.41569	−1.41866	−1.40314	**−1.01881**	−1.40442	−1.0799	**−0.92671**	**−0.85661**	**−0.84479**
Theresold (alpha) value					0.20262				

Figure 10 shows that not only g9 factor is important (Figure 9) but g8 is important too. This can be explained by the fact that with the development of technology, the area of collected and processed data is constantly expanding, which entails changes in legislation, as well as the facts of the adoption of legal restrictions on the use of powerful computers, which have increased power consumption [42], etc. Laws affect the very development of technology and technology, as well as the emergence of the Industry 5.0 trend. In the case of the studied enterprise and SMEs in general, it can be said that mobile applications and devices are the keys to the implementation of Industry 4.0, but one should not forget to take into account external factors of influence during vertical and horizontal integration.

According to the results of the study, it can be stated that mobile applications can be used as a tool for adapting to modern technologies and offered cycles (Figure 1) and methods of evaluation could be useful for other enterprises in the manufacturing branch. The proposed cycle of innovations will ensure sustainable development of the enterprise, since even highly automated manufacturing plants involve people in one way or another, and it is especially important to ensure that they adapt to the proposed changes.

Mobile applications, due to their wide functionality, can be used as a tool for vertical and horizontal integration of Industry 4.0 components into the manufacturing process.

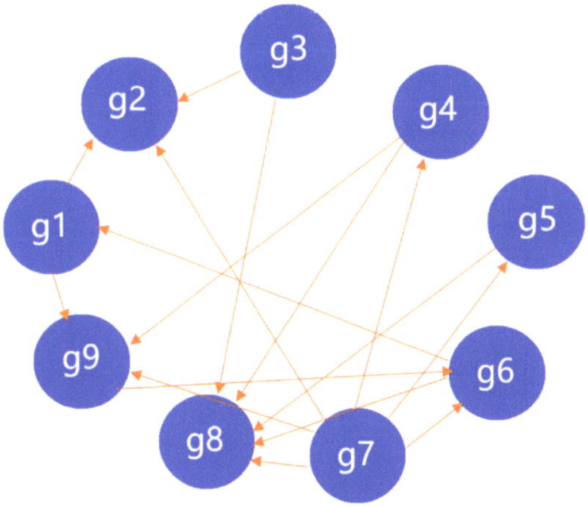

Figure 10. Graphic scheme diagram of the influence of factors on each other.

4. Conclusions

At the stage of developing the application content, the company's management took the idea positively and actively entered the stage of employee motivation. At the stage of application testing, it became obvious that the implementation of work instructions and basic functionality, such as registering an employee in the system, is just a small part of the functions that can be integrated into the application.

The introduced mobile application, only during the month of active use has shown a positive effect on the manufacturing process:

- the number of mistakes made by welders decreased by 18%,
- staff training time has been reduced from 1 h 30 min to 30 min,
- reduced production downtime associated with additional consultations on instructions from 1 h per work shift to 30 min,
- reduced the number of paper carriers of instructions,
- paperwork has been accelerated,
- data exchange become faster because the shift manager stopped spending 40 min of the shift collecting data from each workplace.

Workers noted that it was especially convenient that they were able to open necessary instructions and put them in front of them as a sample, according to which they made work and compared the quality of work performed. Visual control of the completed work is an integral part of the workflow of each employee, and the mobile application was useful for these purposes as well. The positive assessment of the application is especially emphasized by the results of the Usability test given in the section above.

Taking into account the fact that mobile application is primarily a program, it can be changed at any time according to the requirements of modernity.

A lot of manufacturing enterprises try to introduce the newest technologies in manufacturing processes, but they are every day faced with the problem of employee adaptation. The proposed study shows that mobile applications and mobile devices, as a component of Industry 4.0, can be accepted by employees of different ages, education, and professions. Feedback collected from specialists, management, and employees clearly showed the attractiveness of the proposed solution. Methods for analyzing the effectiveness of the mobile application used in the article can be used by other enterprises in the industry. It is worth emphasizing that to ensure the sustainable development of the enterprise, the proposed assessment methods should be applied according to the innovation cycle to ensure the adaptation of the personnel who will be involved in this.

Mobile devices are distinguished by their autonomy, and therefore, by expanding the functionality of the mobile application, it is possible to replace some devices for collecting and processing data in the enterprise, thereby reducing electricity costs and increasing the mobility of employees in the enterprise. The price of such a device will not be high because such a mobile application does not require high technical requirements, which is why the proposed mobile application can be introduced by many SMEs.

Author Contributions: Conceptualization, A.I. and M.B.; methodology, A.I. and K.Ž.; validation, K.Ž. and A.I.; formal analysis, M.B.; investigation, A.I.; resources, K.Ž.; data curation, M.B.; writing—original draft preparation, A.I.; writing—review and editing, A.I.; visualization, A.I.; supervision, M.B.; project administration, K.Ž.; funding acquisition, K.Ž. All authors have read and agreed to the published version of the manuscript.

Funding: This work was supported by the Slovak Research and Development Agency under contract No. APVV-19-0590, and by the projects VEGA 1/0700/20, KEGA 055TUKE-4/2020 granted by the Ministry of Education, Science, Research, and Sport of the Slovak Republic.

Data Availability Statement: Related works can be found here: https://doi.org/10.1007/978-3-030-90462-3_4 and DOI:10.1007/978-3-030-67241-6_6.

Acknowledgments: This work was supported by the Slovak Research and Development Agency under contract No. APVV-19-0590, and by the projects VEGA 1/0700/20, KEGA 055TUKE-4/2020 granted by the Ministry of Education, Science, Research, and Sport of the Slovak Republic.

Conflicts of Interest: The authors declare no conflict of interest.

References

1. Pavlák, M.; Písař, P. Strategic management controlling system and its importance for SMEs in the EU. *Probl. Perspect. Manag.* **2020**, *18*, 362–372. [CrossRef]
2. Haleem, F.; Jehangir, M.; Ullah, Z. Barriers to SMEs growth: An Exploratory Study of Khyber Pakhtunkhwa Manufacturing Sector. *Bus. Econ. Rev.* **2019**, *11*, 89–112. [CrossRef]
3. Bayraktar, M.; Algan, N. The Importance Of SMEs On World Economies. In Proceedings of the International Conference on Eurasian Economies 2019, Fammmagusta, Turkey, 11–13 June 2019. [CrossRef]
4. Andrei, J.; Chivu, L.; Gheorghe, I.; Grubor, A.; Sedlarski, T.; Sima, V.; Subić, J.; Vasic, M. Small and Medium-Sized Enterprises, Business Demography and European Socio-Economic Model: Does the Paradigm Really Converge? *J. Risk Financ. Manag.* **2021**, *14*, 64. [CrossRef]
5. Parfenov, A.; Shamina, L.; Niu, J.; Yadykin, V. Transformation of Distribution Logistics Management in the Digitalization of the Economy. *J. Open Innov. Technol. Mark. Complex.* **2021**, *7*, 58. [CrossRef]
6. Davidescu, A.; Apostu, S.-A.; Paul, A.; Casuneanu, I. Work Flexibility, Job Satisfaction, and Job Performance among Romanian Employees—Implications for Sustainable Human Resource Management. *Sustainability* **2020**, *12*, 6086. [CrossRef]
7. Chang, S.-C.; Chang, H.-H.; Lu, M.-T. Evaluating Industry 4.0 Technology Application in SMEs: Using a Hybrid MCDM Approach. *Mathematics* **2021**, *9*, 414. [CrossRef]
8. Ha, S.T.; Lo, M.C.; Suaidi, M.K.; Mohamad, A.A.; Bin Razak, Z. Knowledge Management Process, Entrepreneurial Orientation, and Performance in SMEs: Evidence from an Emerging Economy. *Sustainability* **2021**, *13*, 9791. [CrossRef]
9. Kamariotou, M.; Kitsios, F.; Madas, M. E-Business Strategy for Logistics Companies: Achieving Success through Information Systems Planning. *Logistics* **2021**, *5*, 73. [CrossRef]
10. Roles of Individual Perception in Technology Adoption at Organization Level: Behavioral Model versus TOE Framework. *J. Syst. Manag. Sci.* **2020**, *10*, 97–118. [CrossRef]
11. Martinez, M.C.; Fischer, F.M. Work Ability and Job Survival: Four-Year Follow-Up. *Int. J. Environ. Res. Public Health* **2019**, *16*, 3143. [CrossRef]
12. Verma, A.; Bansal, M.; Verma, J. Industry 4.0: Reshaping the future of HR. *Strat. Dir.* **2020**, *36*, 9–11. [CrossRef]
13. Sivathanu, B.; Pillai, R. Smart HR 4.0—How industry 4.0 is disrupting HR. *Hum. Resour. Manag. Int. Dig.* **2018**, *26*, 7–11. [CrossRef]
14. Ninan, N.; Chacko, J.T. Training the workpiece for industry 4.0. *Int. J. Res. Soc. Sci.* **2019**, *9*, 782–790.
15. Dhanpat, N.; Buthelezi, Z.P.; Joe, M.; Maphela, T.V.; Shongwe, N. Industry 4.0: The role of human resource professionals. *SA J. Hum. Resour. Manag.* **2020**, *18*, a1302. [CrossRef]
16. Bayraktar, O.; Ataç, C. The Effects of Industry 4.0 on Human Resources Management. In *Globalization, Institutions and Socio-Economic Performance*; Yıldırım, E., Çeştepe, H., Eds.; Peter Lang GmbH: Berlin, Germany, 2018; pp. 337–360.
17. Pikhart, M.; Klimova, B. Information and Communication Technology-Enhanced Business and Managerial Communication in SMEs in the Czech Republic. *Information* **2020**, *11*, 336. [CrossRef]

18. Choudhary, R. The Mobile Application to Track a Remote Employee. *Int. J. Res. Appl. Sci. Eng. Technol.* **2020**, *8*, 1197–1201. [CrossRef]
19. Alemão, D.; Rocha, A.; Barata, J. Smart Manufacturing Scheduling Approaches—Systematic Review and Future Directions. *Appl. Sci.* **2021**, *11*, 2186. [CrossRef]
20. Kearney, M.; Burden, K.; Schuck, S.; Kearney, M.; Burden, K.; Schuck, S. Mobile Learning and Ubiquitous Learning. In *Theorising and Implementing Mobile Learning*; Springer: Singapore, 2020; pp. 25–37.
21. Clarizia, F.; De Santo, M.; Lombardi, M.; Santaniello, D. E-Learning and Industry 4.0: A Chatbot for Training Employees. In Proceedings of the Fifth International Congress on Information and Communication Technology, London, UK, 20–21 February 2020. [CrossRef]
22. Ehimwenma, K.E.; Crowther, P.; Beer, M.; Al-Sharji, S. An SQL Domain Ontology Learning for Analyzing Hierarchies of Structures in Pre-Learning Assessment Agents. *SN Comput. Sci.* **2020**, *1*, 335. [CrossRef]
23. Richardson, J.T. Approaches to studying across the adult life span: Evidence from distance education. *Learn. Individ. Differ.* **2013**, *26*, 74–80. [CrossRef]
24. Sweetman, D.S. Making virtual learning engaging and interactive. *FASEB BioAdvances* **2020**, 3, 11–19. [CrossRef]
25. Wang, M.; Novak, D.; Shen, R. Assessing the Effectiveness of Mobile Learning in Large Hybrid/Blended Classrooms. In Proceedings of the International Conference on Hybrid Learning and Education, Hong Kong, China, 13–15 August 2008. [CrossRef]
26. Yousuf, M.I. Effectiveness of mobile learning in distance education. *Turk. Online J. Distance Educ.* **2007**, *8*, 114–124. [CrossRef]
27. Mileva, N. The Effectiveness of Mobile Learning in the Form of Performance Support System in Higher Education. *Int. J. Interact. Mob. Technol. (iJIM)* **2011**, *5*, 17–21. [CrossRef]
28. Coculov, J. The Analysis of the Selected Factors Influencing the Selection of Employee Training Methods. *J. Hum. Resour. Manag. Labor Stud.* **2017**, *5*. [CrossRef]
29. Leligou, H.C.; Zacharioudakis, E.; Bouta, L.; Niokos, E. 5G technologies boosting efficient mobile learning. *MATEC Web Conf.* **2017**, *125*, 3004. [CrossRef]
30. Woods, D. Interactive eLearning statistics modules for design of experiments and regression methods. *MSOR Connect.* **2007**, *7*, 12–17. [CrossRef]
31. Welbourne, T.M.; Rolf, S.; Schlachter, S. Employee Resource Groups: An Introduction, Review and Research Agenda. *Acad. Manag. Proc.* **2015**, *2015*, 15661. [CrossRef]
32. Urad Pre Vydávanie Publikácií Európskej Únie. Available online: https://op.europa.eu/sk/publication-detail/-/publication/79c0ce87-f4dc-11e6-8a35-01aa75ed71a1 (accessed on 3 March 2021).
33. Eurostat. Available online: https://ec.europa.eu/eurostat/web/national-accounts/methodology/european-accounts/productivity-indicators (accessed on 7 December 2022).
34. Çetinkaya, A.; Niavand, A.; Rashid, M. ORGANIZATIONAL CHANGE AND COMPETITIVE ADVANTAGE: BUSINESS SIZE MATTERS. *Bus. Manag. Stud. Int. J.* **2019**, *7*, 40–67. [CrossRef]
35. Santhoshkumar, G.; Jayanthy, S.; Velanganni, R. Employee Engagement. *J. Adv. Res. Dyn. Control Syst.* **2019**, *11*, 1100–1104. [CrossRef]
36. Cheng, P.-Y.; Huang, Y.-M.; Shadiev, R.; Hsu, C.-W.; Chu, S.-T. Investigating the Effectiveness of Video Segmentation on Decreasing Learners' Cognitive Load in Mobile Learning. In Proceedings of the International Conference on Web-Based Learning, Tallinn, Estonia, 14–17 August 2014.
37. Best Employee Engagement Software for 2022. Available online: https://techjury.net/best/employee-engagement-software/ (accessed on 7 December 2022).
38. Office of Labour, Social Affairs and Family. Current Problems on the Slovak Labor Market. Available online: https://www.upsvr.gov.sk/ (accessed on 8 November 2019).
39. Wario, R.; Ngari, B. A Framework for Mobile Learning Technology Usability Testing BT. In Proceedings of the 20th International Conference on Human-Computer Interaction, Las Vegas, NV, USA, 15–20 July 2018.
40. Zhang, D.; Adipat, B. Challenges, Methodologies, and Issues in the Usability Testing of Mobile Applications. *Int. J. Human–Computer Interact.* **2005**, *18*, 293–308. [CrossRef]
41. Prata, W.; Alvão, C.R.M.; Quaresma, M. Usability Testing of Mobile Applications Store: Purchase, Search and Reviews. In Proceedings of the International Conference of Design, User Experience, and Usability, Las Vegas, NV, USA, 21–26 July 2013.
42. General Data Protection Regulation. Available online: https://gdpr-info.eu/ (accessed on 27 December 2022).

applied sciences

Article

Tire Bubble Defect Detection Using Incremental Learning

Chuan-Yu Chang *, You-Da Su and Wei-Yi Li

Department of Computer Science and Information Engineering, National Yunlin University of Science and Technology, Douliou 64002, Taiwan
* Correspondence: chuanyu@yuntech.edu.tw

Abstract: Digital shearography is a technique that has recently been applied to material inspections that cannot be performed by the naked eyes, including the detection of air bubble defects in tires. Although digital shearography detects bubbles that are not visible to the naked eyes, the process of determining tire defects still relies on field operators, with inconsistent results depending on the experiences of the field operator personnel. New or different types of bubble defects that AI models have not previously recognized are often missed, resulting in an inadequate quality detection model. In this paper, we propose a bubble defect detection method based on an incremental YOLO architecture. The data for this research was provided by the largest tire manufacturer in Taiwan. In our research, we classify the defects into six distinct categories, pre-process the images to allow better detections of less-noticeable defects, increase the amount of training data used, and generate an initial training model with the YOLO framework. We also propose an incremental YOLO method using small-model training for previously unobserved defects to improve the model detection rate. We have observed detection accuracy and sensitivity of 98% and 90% in the experimental results, respectively. The methods proposed in this paper can assist tire manufacturers in achieving semi-automatic quality inspections and labor cost reductions.

Keywords: tire bubble defect detection; digital shearography; YOLO; deep learning

Citation: Chang, C.-Y.; Su, Y.-D.; Li, W.-Y. Tire Bubble Defect Detection Using Incremental Learning. *Appl. Sci.* **2022**, *12*, 12186. https://doi.org/10.3390/app122312186

Academic Editors: Tibor Krenicky, Maros Korenko and Juraj Ruzbarsky

Received: 10 November 2022
Accepted: 24 November 2022
Published: 28 November 2022

1. Introduction

Industry 4.0 enables data communication, collection and analysis by machines, enabling faster, more agile and efficient processes to manufacture quality products with minimal expenditure. Since 2012, researches on computer vision tasks based on convolutional neural networks have progressed fairly rapidly [1,2].

During the tire manufacturing process, air may be trapped in the tire to form bubbles. The bubbles can compromise the tire's structural integrity, leading to the risk of a punctured tire if the pressure is not distributed evenly. Therefore, tire manufacturers inspect the tires for bubble defects before they are sold to reduce the risk of punctures. The tire bubbles are difficult to detect by the naked eye since they are internal defects. The operator puts the tire into the machine, and air bubbles in the tire cause the tire surface to bulge in the vacuum. Digital shearography [3] is used to obtain speckle images of the tire at different angles for manual inspections. However, tired or inexperienced inspectors can lead to inaccuracies and inconsistency in the inspection outcomes.

The tire defect detection process is extremely labor and time-intensive, even though it is only a small part of the quality control process in the tire manufacturing industry. Many difficulties arise during the inspection due to the inconsistent and subjective assessments by inspectors with different experiences. To ensure product quality and reduce labor costs, deep learning has been gradually incorporated in the industry by many manufacturers to reduce labor-intensive work and improve production efficiency effectively. With the help of artificial intelligence, operators only need to confirm the bubble defects detected by machine, which can reduce the workload of labors. Manufacturers do not need to recruit a lot of labor to detect bubble defects one by one; developments in automation technology can also address the problems caused by inconsistent inspection standards.

In our previous research [4–6], we divided the entire speckle image into several blocks, and a convolutional neural network is used to detect the tread bubbles and the sidewall separately. However, there is still room for improvement regarding the detection speed and detection rate. In this paper, we use an object detection method to improve the bubble defect detection speed and detection rate. In addition, to address the situation where the model developed in the laboratory is actually introduced into the field with a reduced defect detection rate, we propose an incremental YOLO (You Only Look Once) architecture to add on-site misidentified samples to model training to improve the model's flaw detection accuracy on the actual site. The YOLO machine learning algorithm uses features learned by a deep convolutional neural network to detect objects in real time. YOLO has the advantage of being much faster than other networks and still maintaining high accuracy. Experimental results show that the improved YOLOv3 SPP outperforms other YOLO series architectures (Tiny YOLO, YOLOv3 and YOLOv3 SPP) in bubble defect detection [7–10]. Furthermore, the proposed incremental learning also demonstrates the effectiveness of improving the detection accuracy of bubble defects in real field.

The rest of this paper is organized as follows. First, the proposed method is described in detail in Section 2. Then, we present the experimental results and discuss the results in Section 3 and, finally, the conclusions are delivered in Section 4.

2. Proposed Method

Since the tire factory cannot provide enough defect images for model training, in practice, it often happens that the recognition rate of the model trained in the laboratory is greatly reduced when it is actually imported into the field. In order to shorten the time for importing the model into the field, this paper proposes an incremental learning method for bubble detection on tire tread and sidewalls. We test the lab-trained model with images from the tire production site and collect misidentified samples, incrementally train by freezing the Yolo Backbone, and only re-train the Neck and Prediction network layers to improve defect detection rates [11]. The main steps are described as follows:

1. Manually mark the position and type of the bubble and classify it into one of the six types;
2. Perform image enhancement and increase the number of training samples;
3. After resizing, use the YOLO architecture proposed in this paper for model training;
4. Overlay the detected defects on the original image for visualization.

We apply incremental learning to the inspection process to account for potentially overlooked defects. Incremental learning allows small models to be trained on the missed defects and improves the model's accuracy. The process is as shown in Figure 1.

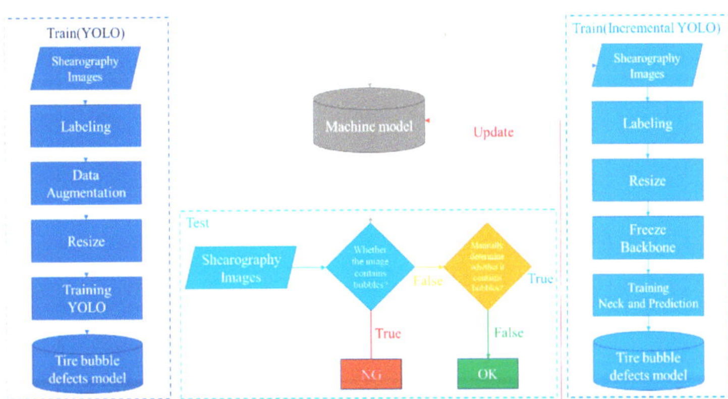

Figure 1. The bubble detection process.

2.1. Data Sample Categorization and Enhancement

Speckle images with dimensions of 1360 × 1024 pixels are used as input for inspection. Defects can be classified into different categories based on their location, size and shape. Using only a binary classification to separate images with and without bubbles would likely result in higher misjudgment rates. In this research, we divided the bubble defects into six categories, as listed below in Figure 2.

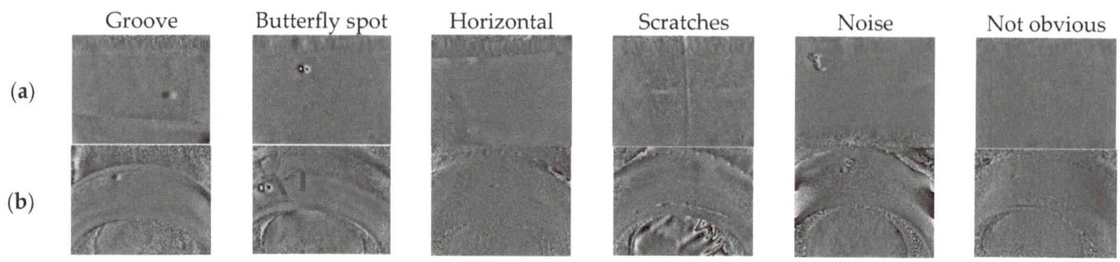

Figure 2. Six types of bubble images (**a**) Tire tread defects and (**b**) Tire sidewall defects.

To improve the accuracy for the hard-to-spot defects that are likely to be overlooked during the inspection, we used an image enhancement technique called Contrast Limited Adaptive Histogram Equalization (CLAHE) [12] to increase the visibility of the subtle bubbles present in the image. As shown in Figure 3, the bubbles in the red frame are much more obvious than the original image.

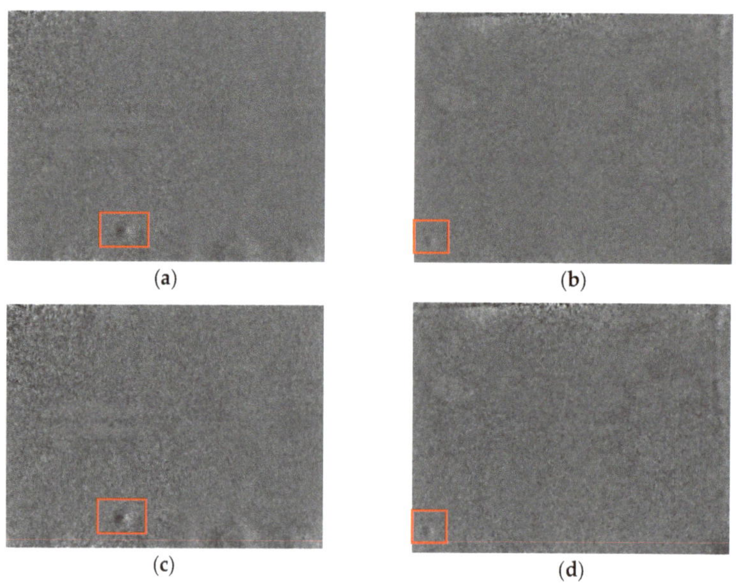

Figure 3. Image enhancement (**a**,**b**) original speckle images, (**c**,**d**) Contrast enhanced images.

As deep learning network training requires a large amount of sample data to attain a higher detection rate, insufficient defect samples in the dataset provided by the tire manufacturer may affect the effectiveness of the model. To address this issue, we combined the original and enhanced images in the model training set to increase sample data [13].

2.2. Image Resize

Since the sizes of the grids in the YOLOv3 architecture [8] are 32 × 32, 16 × 16 and 8 × 8, we resize the 1360 × 1024 input images to 1344 × 1024 so that the dimensions are multiples of 32. For computational convenience, we further divided the entire speckle image into multiple 480 × 480 sub-images. Figure 4 shows what the bubbles look like at three different scales (32 × 32, 16 × 16 and 8 × 8). Obviously, the area of the bubbles is smaller than the 32 × 32 grid. The finer grids allow for more features in the bubble defects to be learned.

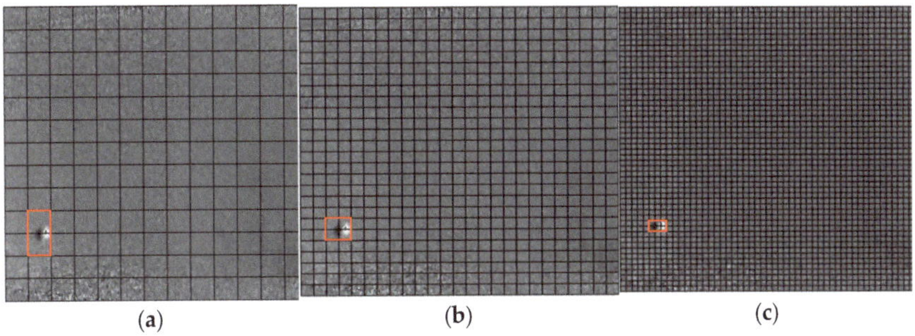

Figure 4. Visualization of different grid sizes: (**a**) Large scale (32 × 32), (**b**) Medium scale (16 × 16), and (**c**) Small scale (8 × 8).

2.3. Framework: Modify the YOLOv3 SPP Architecture

The YOLOv3 architecture [8] is mainly divided into four sections: the Input terminal, Backbone, Neck and Prediction, as shown in Figure 5. The functions of each section are given below,

1. Input terminal: Image pre-processing and data augmentation;
2. Backbone: Feature extraction for lower-level, shallow features which are relatively similar, including edges, colors and textures;
3. Neck: Feature enhancement conducted by processing and enhancing the shallow features extracted by the backbone and allowing the model to learn the characteristics of tire bubbles.
4. Prediction: Outputting a final prediction containing the loss function and the Non-Max Suppression Algorithm of the bounding and anchor boxes.

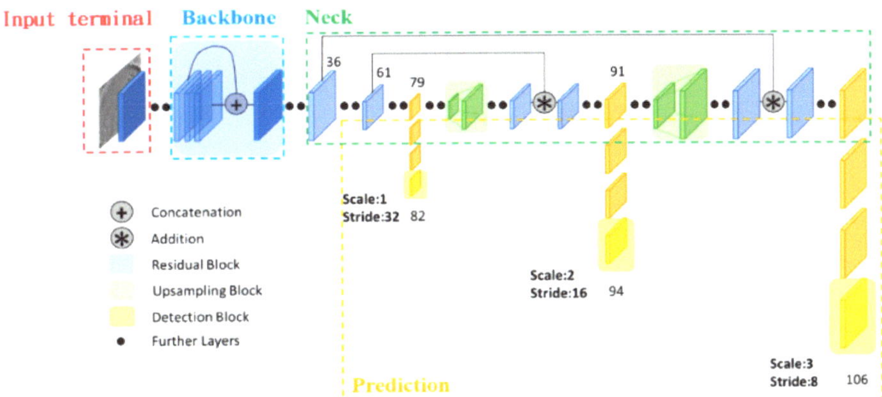

Figure 5. Four main sections of the YOLOv3 architecture [8].

The original YOLOv3 SPP [14] uses Darknet 53 as the backbone with a three-scale detection network. Since the bubble defects only occupy a small portion of the speckle image, therefore, we modify the YOLOv3 SPP architecture, adjust it to a two-scale output and modify the backbone network. As shown in Figure 6, we removed the large-scale network layers in the Darknet 53 architecture, adding the same number of layers back to the small-scale network layer to improve small-scale feature acquisition. As the large-scale sections consist of many channels, removing the layers from the backbone network reduces the number of parameters by nearly 1/3, significantly reducing the required computing resources. The FPN [11] structure is referenced for re-joining the backbone network and the prediction layer. Furthermore, we have found that the addition of the SPP method effectively increases the range of backbone features and significantly separates the most important context features. The addition of a layer of SPP structure after up-sampling achieves two-scale local and global feature fusion.

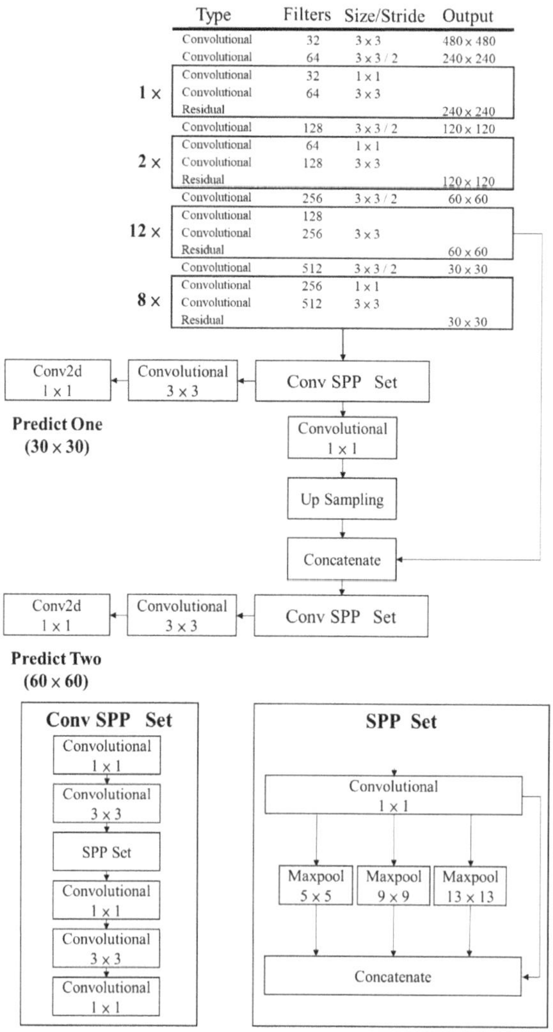

Figure 6. The YOLOv3 SPP two-scale output network architecture.

2.4. The Incremental YOLO Architecture

Previous methods merge new data into the training datasets before retrain to reduce missed detections due to incorrect identification of defective images by the original model. Although this method improves the detection rate of the model, it is a time-consuming process. We propose an incremental YOLO approach with the concept of transfer learning. Since we can only train the object detection model on images with objects, we have manually relabeled and added the misidentified images to the incremental YOLO training dataset. For incremental training, we retained the shallow features in the original model while training the deeper features of the network layer. As a result, the bubble defects missed during inspection accumulate and are used to rapidly update the model to improve the detection rate.

We froze the Backbone section for incremental training and only trained the Neck and Prediction network layers to improve the detection rate. Since the backbone network mainly extracts low-level features such as edges, colors, and textures of images. The neck layer is mainly to process and enhance the low-level features extracted by the backbone, so that the model can learn features suitable for tire bubbles. To reduce the computational load of feature extraction and efficiently learn the image features of the field, we froze the backbone layer and retrained the neck layer in incremental learning to strengthen the bubble features of the field. As shown in Figure 7, the blue box indicates the frozen Backbone section, while the red box indicates the sections where the incremental YOLO architecture performs parameter adjustments.

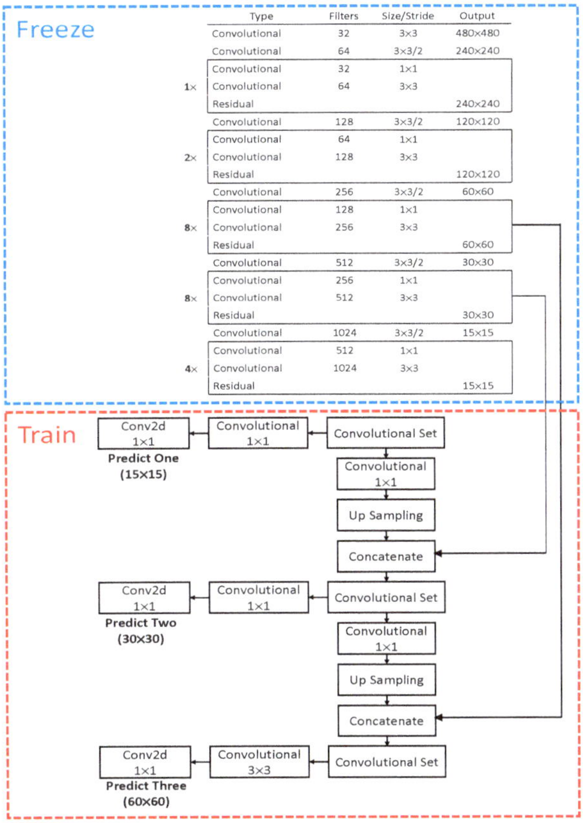

Figure 7. Steps of the incremental learning method.

3. Experimental Results

3.1. Dataset and Experimental Environment

The tire speckle images used in this paper were provided by Taiwan's largest tire manufacturer. The dataset consists of tire tread and sidewall images taken from different angles. Each set contains 24 to 26 images of size 1360 × 1024 pixels. One tire speckle image may include more than one bubble. The training and the testing dataset are independent. The training dataset contains 2450 images with 3116 bubble defects. The dataset is classified into six defect categories, as listed in Table 1. The sample tire tread images consist mostly of grooved bubbles, with 1421 images in the category, followed by butterfly shaped bubbles. The remaining categories only account for a small part of the overall training data. The dataset consists of 1763 tread samples, containing a total of 2247 bubbles. A similar pattern exists for the sidewall images, with 321 grooved images containing 377 grooved bubbles, followed by butterfly-shaped bubbles and images from the remaining categories. There are altogether 687 sidewall samples containing 869 bubbles. Altogether, the dataset contains 2450 training images and a total of 3116 bubbles.

Table 1. Training Dataset.

Data Class	Samples		Bubbles	
	Tread	Sidewall	Tread	Sidewall
Groove	1156	321	1421	377
Butterfly spot	322	265	431	347
Horizontal	91	22	134	49
Scratches	88	10	137	17
Noise	45	61	48	69
Not obvious	61	8	76	10
Total	1763	687	2247	869
Training	2450		3116	

The testing dataset is described in Table 2. Binary classification of normal and defective images is used in this study, as the primary purpose is to detect defective bubbles without ensuring the correct categorization. The test data contains 500 normal tread sample images and 810 tread images with bubble defects, as well as 500 normal sidewall sample images and 505 sidewall images with bubble defects.

Table 2. Testing Dataset.

	Data Class	# of Image
Tread	Non-bubbles	500
	Bubbles	810
Sidewall	Non-bubbles	500
	Bubbles	505

The experiments were performed in a Windows environment, using an Intel Core i7-8700K processor and a GeForce GTX 1080 Ti GPU with the Python Keras framework.

We used four evaluation methods to verify the results of this experiment: Accuracy, Sensitivity, Specificity and Precision. The equations are as follows;

$$Accuracy = (TP + TN)/(P + N) \tag{1}$$

$$Sensitivity = TP/P \tag{2}$$

$$Specificity = TN/N \quad (3)$$

$$Precision = TP/(TP + FP) \quad (4)$$

In the above equations, P represents the total number of images containing bubble defects, N represents the total number of images without defects. TP represents the number of true positives, which are the correctly identified images with defects, and TN represents the number of true negatives, which are the correctly identified images without defects. FP represents the number of false positives, which are images without defects and incorrectly marked as defective, and FN represents the number of false negatives, which are images with defects and incorrectly marked as normal. As the purpose of quality control in tire manufacturing is to detect all bubble defects, we use sensitivity as the main evaluation standard for this paper since false positives have a lesser effect compared to false negatives.

In order to verify that the proposed modified YOLOv3 SPP architecture is superior to other YOLO series architectures in bubble defect detection, this experiment will compare Tiny YOLO, YOLOv3, YOLOv3 SPP and other methods. In addition, we will also verify through experiments that the incremental learning method proposed in this paper can effectively improve the bubble defects detection accuracy on the actual site.

3.2. Results Obtained by the Tiny YOLO Architecture

The Tiny-YOLO architecture is a YOLO-based object detector [7]. It is designed to contain a simpler network structure and utilize fewer computing resources. From Table 3, the sensitivity of the Tiny YOLO architecture is 79.77%. The incremental Tiny YOLO architecture achieves a better bubble detection rate by an additional 5%.

Table 3. Results Obtained by (a) the Tiny YOLO Architecture, (b) the incremental Tiny YOLO Architecture.

Result	(a)		(b)	
	Tread	Sidewall	Tread	Sidewall
TP	638	411	690	424
TN	496	486	500	2
FP	4	14	0	8
FN	172	94	120	81
Sensitivity	79.77%		84.71%	

3.3. Results Obtained by the YOLOv3 Architecture

Experimental results in Table 4 show that the detection rates of the YOLOv3 architecture. For tire sidewall images, a larger input size has better recognition. We have noticed that the increased small-scale output of the YOLOv3 architecture improves the detection rate by an average of 2% compared to the Tiny YOLO architecture. This also proves that the model is more sensitive to bubble features at a smaller scale during the feature extraction stage. Adding the incremental YOLO architecture can further increase the detection rate by about 6%.

Table 4. Results Obtained by (a) the YOLOv3 Architecture, (b) the Incremental YOLOv3 Architecture.

Result	(a)		(b)	
	Tread	Sidewall	Tread	Sidewall
TP	650	419	722	435
TN	499	496	497	490
FP	1	4	3	10
FN	160	86	88	70
Sensitivity	81.29%		87.98%	

3.4. Results Obtained by the YOLOv3 SPP Architecture

Spatial Pyramid Pooling (SPP) [14] is a pooling layer, using $1 \times 1, 5 \times 5, 9 \times 9$, and 13×13 maximum pooling methods to achieve multi-scale fusions verification including local and global features. In this experiment, we added an SPP layer to the YOLOv3 backbone network to improve the detection accuracy at the cost of higher computing resource requirements.

Table 5 shows the detection results after adding an SPP structure to the YOLOv3 architecture to achieve local and global feature fusions. The results show that both tread and sidewall samples have higher detection accuracies than those obtained by the YOLOv3 architecture. For both the sidewall and the tread, after adding the SPP structure, the detector extracts more feature information than the YOLOv3 architecture in the feature extraction stage, which effectively increases the reception range of the backbone feature and significantly improves the detection rate. Adding the incremental YOLO architecture further increases the detection rate by approximately 6%.

Table 5. Results Obtained by (a) the YOLOv3 SPP Architecture, (b) the Incremental YOLOv3 SPP Architecture.

Result	(a)		(b)	
	Tread	**Sidewall**	**Tread**	**Sidewall**
TP	714	431	770	459
TN	497	492	489	477
FP	3	8	11	23
FN	96	74	40	46
Sensitivity	87.07%		93.46%	

3.5. Results Obtained by the Modified YOLOv3 Architecture

Although the results obtained by the modified YOLOv3 architecture are slightly lower when compared to those obtained by the YOLOv3 SPP architecture, we have observed a slightly improved detection rate relative to the YOLOv3 architecture in both the tread and sidewall defect detections, as seen by comparing Tables 4 and 6. The modified scale prediction network layer improves the detection rates and decreases the number of outputs, since reducing the large-scale feature extraction and strengthening the small-scale feature extraction increase the extracted bubble features. For the tire bubble detection, similarly, the two-output modification is more suitable for verification. The incremental YOLO architecture further increases the detection rate by about 4%.

Table 6. Results Obtained by (a) the Modified YOLOv3 Architecture, (b) the Incremental YOLOv3 Architecture.

Result	(a)		(b)	
	Tread	**Sidewall**	**Tread**	**Sidewall**
TP	668	442	736	431
TN	500	490	499	485
FP	0	10	1	15
FN	142	63	74	74
Sensitivity	84.81%		88.75%	

3.6. Results Obtained by the Modified YOLOv3 SPP Architecture

Table 7 shows the experimental results of the proposed architecture. The proposed architecture has achieved higher accuracies than the YOLOv3 SPP architecture in the tire tread and sidewall detections due to the backbone network modifications, where large-scale feature extraction layers were removed, and small-scale layers were strengthened to allow the network to learn detailed defect features. Verifying results using two outputs instead of three outputs also resulted in better accuracy. By adding the SPP structure before the two outputs, the detection rate of the architecture is increased significantly while reducing nearly 40 million parameters. Adding the incremental YOLO architecture further increases the detection rate by about 6%.

Table 7. Results Obtained by (a) the Modified YOLOv3 SPP Architecture, (b) the Incremental YOLOv3 SPP Architecture.

Result	(a)		(b)	
	Tread	**Sidewall**	**Tread**	**Sidewall**
TP	758	459	801	495
TN	490	487	481	432
FP	10	13	19	68
FN	52	46	9	10
Sensitivity	92.54%		98.56%	

3.7. Comparision Results of Different YOLO Architectures

Table 8 summarizes the results of the five architectures. Compared to the Tiny YOLO architecture, the YOLOv3 architecture has performed better in detection. Similarly, the modified YOLOv3 architecture performed slightly better than the original YOLOv3 detector. The SPP structure in the YOLOv3 SPP architecture has also significantly improved the detection rates of the YOLOv3 architecture. Finally, the modified YOLOv3 SPP architecture proposed in this paper has achieved the highest accuracy, with a 5.5% increase in accuracy compared to the unmodified YOLOv3 SPP model.

Table 8. Results Obtained by Different Architecture (a) Tiny YOLO (b) YOLOv3 (c) YOLOv3 SPP (d) Modified YOLOv3 (e) Proposed modified YOLOv3 SPP Model.

Result	Tread				Sidewall				Sensitivity
	TP	**TN**	**FP**	**FN**	**TP**	**TN**	**FP**	**FN**	
(a)	638	496	4	172	411	486	14	94	79.77%
(b)	650	499	1	160	419	496	4	86	81.29%
(c)	714	497	3	96	431	492	8	74	87.07%
(d)	668	500	0	142	442	490	10	63	84.81%
(e)	**758**	**490**	**10**	**52**	**459**	**487**	**13**	**46**	**92.54%**

3.8. Comparision of the Different Incremental YOLO Architectures

Table 9 lists the comparison results of the above five architectures in conjunction with the incremental learning. The proposed improved YOLOv3 SPP method achieves the highest sensitivity among the results obtained using the incremental learning. The proposed model's detection rate is 6% higher than the modified YOLOv3 SPP method without the incremental YOLO architecture (see Table 8), significantly reducing the number of false negatives in defect detection. The proposed method's detection time is shorter than all other methods except the Tiny YOLO architecture. Nevertheless, despite the Tiny YOLO architecture's simple, easy-to-use network structure and low computing resource requirements, it is unsuitable for the detection task due to its poor detection rates. Hence, our proposed method provides a semi-automated defect detection approach for tire manufacturers, effectively reducing labour costs and improving inspection accuracy.

Table 9. Results and detection times of the incremental YOLO architecture with different architectures (a) Tiny YOLO (b) YOLOv3 (c) YOLOv3 SPP (d) Modified YOLOv3 (e) Proposed Model.

Result	Tread				Sidewall				Sensitivity	Time
	TP	TN	FP	FN	TP	TN	FP	FN		
(a)	690	500	0	120	424	2	8	81	84.71%	70(s)
(b)	722	497	3	88	435	490	10	70	87.98%	132(s)
(c)	770	489	11	40	459	477	23	46	93.46%	134(s)
(d)	736	499	1	74	431	485	15	74	88.75%	133(s)
(e)	**801**	**481**	**19**	**9**	**495**	**432**	**68**	**10**	**98.56%**	**125(s)**

4. Conclusions

This paper proposes a semi-automated tire defect detection process using a dataset of speckle images provided by the largest tire manufacturer in Taiwan, and labeled by professionals into six defect categories. To supplement the dataset, we have enhanced images of defects that are difficult to detect and added them to the dataset during training to improve accuracy. Using the training sample, we obtained a detection rate of 92.54% with the initial model in the first stage. During the second stage, we implemented an incremental YOLO architecture upon the architecture in the first stage, which improved the detection rate to 98.56%, reduced the detection speed, and provided better detection rates than previous studies for the hard-to-detect bubbles. During the quality control process in tire manufacturing, the model is likely to encounter new defects that were not present in the training data. We can incrementally train the model to recognize new defects with the proposed incremental YOLO architecture, which significantly improves the accuracy compared to the original model. For defect recognition, the original and incremental models had misjudged rates of 2.3% and 7.7%, respectively. The misjudgments are primarily due to less visible defects in the speckle images. As a result, more research may be required to further improve the detection rate.

Author Contributions: Conceptualization, C.-Y.C. and Y.-D.S.; methodology, C.-Y.C.; software, W.-Y.L.; validation, C.-Y.C. and Y.-D.S.; investigation, W.-Y.L.; writing—original draft preparation, W.-Y.L. and Y.-D.S.; writing—review and editing, C.-Y.C. and Y.-D.S.; visualization, W.-Y.L.; supervision, C.-Y.C. and Y.-D.S.; project administration, C.-Y.C.; funding acquisition, C.-Y.C. All authors have read and agreed to the published version of the manuscript.

Funding: This research was funded by Intelligent Recognition Industry Service Center, a Featured Research Center of the Taiwan Ministry of Education's Higher Education SPROUT Project. And the APC was funded by Ministry of Education, Taiwan.

Institutional Review Board Statement: Not applicable.

Informed Consent Statement: Not applicable.

Conflicts of Interest: The authors declare no conflict of interest.

References

1. Bhat, S.S.; Selvam, V.; Ansari, G.A.; Ansari, M.D.; Rahman, M.H. Prevalence and Early Prediction of Diabetes Using Machine Learning in North Kashmir: A Case Study of District Bandipora. *Comput. Intell. Neurosci.* **2022**, *2020*, 2789760. [CrossRef] [PubMed]
2. Gaddam, D.K.R.; Ansari, M.D.; Vuppala, S.; Gunjan, V.K.; Sati, M.M. Human Facial Emotion Detection Using Deep Learning. In *ICDSMLA 2020*; Springer: Singapore, 2021; pp. 1417–1427.
3. Steinchen, W.; Yang, L. *Digital Shearography: Theory and Application of Digital Speckle Pattern Shearing Interferometry*; SPIE Press: Bellingham, DC, USA, 2003.
4. Chang, C.-Y.; Huang, J.-K. Tires Defects Detection Using Convolutional Neural Networks. In Proceedings of the Conference on Computer Vision, Graphics, and Image Processing, Taipei, Taiwan, 24–27 June 2017.
5. Chang, C.-Y.; Srinivasan, K.; Wang, W.-C.; Ganapathy, G.P.; Vincent, D.R.; Deepa, N. Quality Assessment of Tire Shearography Images via Ensemble Hybrid Faster Region-Based ConvNets. *Electronics* **2020**, *9*, 45. [CrossRef]

6. Chang, C.-Y.; Wang, F.-C. Tire Bubble Defects Detection Using ResNet. In Proceedings of the National Conference on Web Intelligence and Applications, Yunlin, Taiwan, 18–19 October 2019.
7. Redmon, J.; Divvala, S.; Girshick, R.; Farhadi, A. You Only Look Once: Unified, Real-Time Object Detection. In Proceedings of the 2016 IEEE Conference on Computer Vision and Pattern Recognition (CVPR), Las Vegas, NV, USA, 27–30 June 2016; pp. 779–788.
8. Redmon, J.; Farhadi, A. Yolov3: An incremental improvement. *arXiv* **2018**, arXiv:1804.02767.
9. Wang, Q.; Bi, S.; Sun, M.; Wang, Y.; Wang, D.; Yang, S. YOLOv3 Architecture. *PLoS ONE* **2019**. [CrossRef]
10. He, K.; Zhang, X.; Ren, S.; Sun, J. Spatial Pyramid Pooling in Deep Convolutional Networks for Visual Recognition. In Proceedings of the European Conference on Computer Vision, Zurich, Switzerland, 6–12 September 2014; pp. 346–361.
11. Cai, Z.; Fan, Q.; Feris, R.S.; Vasconcelos, N. A unified multi-scale deep convolutional neural network for fast object detection. In Proceedings of the European Conference on Computer Vision, Amsterdam, The Netherlands, 11–14 October 2016.
12. Zuiderveld, K. Contrast Limited Adaptive Histogram Equalization. In *Graphics Gems IV*; Elsevier: Amsterdam, The Netherlands, 1994; pp. 474–485.
13. Russakovsky, O.; Deng, J.; Su, H.; Krause, J.; Satheesh, S.; Ma, S.; Huang, Z.; Karpathy, A.; Khosla, A.; Bernstein, M.; et al. ImageNet Large Scale Visual Recognition Challenge. *Int. J. Comput. Vis.* **2015**, *115*, 211–252. [CrossRef]
14. Lin, T.-Y.; Dollar, P.; Girshick, R.; He, K.; Hariharan, B.; Belongie, S. Feature pyramid networks for object detection. In Proceedings of the IEEE Conference on Computer Vision and Pattern Recognition, Honolulu, HI, USA, 21–26 July 2017; pp. 2117–2125.

applied sciences

MDPI

Article

Method of Sequential Approximation in Modelling the Processes of Heat Transfer and Gas Dynamics in Combustion Equipment

Miroslav Rimar [1], Oleksandr Yeromin [2], Grigoriy Larionov [3], Andrii Kulikov [1], Marcel Fedak [1], Tibor Krenicky [4,*], Olena Gupalo [2] and Yana Myanovskaya [2]

[1] Department of Process Technique, Faculty of Manufacturing Technologies, Technical University of Košice, Bayerova 1, 08001 Prešov, Slovakia

[2] Institute of Industrial and Business Technologies, Ukrainian State University of Science and Technology, Lazaryana Str., 2, 49010 Dnipro, Ukraine

[3] M.S. Poliakov Institute of Geotechnical Mechanics of the National Academy of Sciences of Ukraine, Simferopolska Str., 2a, 49005 Dnipro, Ukraine

[4] Department of Technical Systems Design and Monitoring, Faculty of Manufacturing Technologies, Technical University of Košice, Bayerova 1, 08001 Prešov, Slovakia

* Correspondence: tibor.krenicky@tuke.sk

Abstract: The behavior of the processes taking place in furnaces determines the efficiency of fuel chemical energy utilization, the quality of the final products and the environmental safety of the production. Mathematical models of the processes of gas dynamics and heat transfer in the working space of heating equipment are quite complex, and do not allow the establishment of a direct analytical relationship between the quality indicator of the process (F) and the influencing parameters (x_i). To simplify the procedure for obtaining the values of the function F depending on the change of parameters x_i, a method of successive approximation is presented in the article. The main idea of the method is that the representation of the function around a point from the domain of the function can be extended to the entire domain for many problems of mechanics. The relative error in the definition of the function acquires its maximum value at the border of the area, and a reasonable narrowing of it allows control of the size of the error. Thus, the advantages of using the method are obvious; it is able to provide approximation of the function in a multiplicative form with a controlled error. The distribution of the method to the field of heat transfer problems is presented in this paper. The successful implementation of this method for solving problems of this kind shows that the solution of practical problems may be generalized for the entire domain of the function, despite the fact that the errors of such a representation increase to 5–7% when approaching its limit, which, however, may be considered acceptable for engineering calculations.

Keywords: gas dynamics; heat transfer; mathematical modelling; method of successive approximation; heating; standard of heating

Citation: Rimar, M.; Yeromin, O.; Larionov, G.; Kulikov, A.; Fedak, M.; Krenicky, T.; Gupalo, O.; Myanovskaya, Y. Method of Sequential Approximation in Modelling the Processes of Heat Transfer and Gas Dynamics in Combustion Equipment. *Appl. Sci.* **2022**, *12*, 11948. https://doi.org/10.3390/app122311948

Academic Editor: Satoru Okamoto

Received: 24 October 2022
Accepted: 19 November 2022
Published: 23 November 2022

1. Introduction

Material heating and heat treatment belong to fundamental processes that take place in furnaces heated by gaseous fuel. The processes of fuel combustion, movement of gases in the working chamber of the furnace, and external and internal heat exchange during heating have a high level of mutual interaction. Some parameters of these processes are uncertain and interdependent, and they impose numerous conditions of mutual application.

The main tool for examination of thermophysical processes in energetics, especially in metallurgy, is mathematical modelling. Owing of the high level of computer technology, the complexity of the software, and the ability to simulate the influence of different parameters, mathematical modelling prevails over real measurements. Modern mathematical models that use numerical methods allow implementation of an algorithm of any complexity, as well as considering a large number of non-linear parameters affecting processes. However, the increase in the number of parameters that is considered and studied in the mathematical model significantly increases the amount of preparatory work for modelling, complicates software development and handling, requires the introduction of additional databases and data arrays, and increases the calculation time. Thus, for example, the calculation time of one of the options for heating a cage of twelve multi-ton ingots in a regenerative heating well using a mathematical model may result in more than 5 h of computation time [1]. It should be noted that the accuracy of such a calculation usually depends on the reliability of the accepted values of the properties of the materials, the environment influence, and other coefficients used in the model. As practice has shown, the results of the calculations may fully correspond to the real data collected during 5-year operation of the simulation object—a regenerative heating well [1].

Depending on the required accuracy of calculations, complex mathematical models of gas-dynamic, mass- and heat-transfer processes or more simplified engineering models are used as an engineering tool, that for particular cases provide sufficient accuracy of calculations. A similarly important tool provides the physical modelling of processes and aggregates, which enables a clear and easy-to-understand representation of the studied phenomena [2].

To ensure the accuracy of the calculations sufficient for the development of engineering solutions and the evaluation of proposed modernization measures, or to improve the design of the furnace and its equipment, it is quite sufficient to obtain the dependence of the change of the selected final criterion of the process (the so-called quality function F) on the factors (independent variable parameters x_i) that is influenced. This, of course, does not exclude the need to create a mathematical model (MM) of the processes under study as a perfect mechanism for a calculation experiment, but it allows significantly reduction of the number of necessary calculations without significantly losing the quality of the obtained result.

Multiplicative models are widely used to process the results of calculation experiments (obtaining the function $F = f(x_i)$).

However, the problem of choosing a multiplier that precedes the product of functions has not been fully clarified. This multiplier is also called the coefficient of ignorance, and the evaluation of the obtained data depends on its choice. Initially, the multiplier was chosen as a function of independent parameters from the domain. It turned out that despite the high accuracy and significant costs for conducting the experiment, the quality function of the process had satisfactory accuracy only around the reference point. Attempts to determine the quality function at other points did not yield a positive result, despite the use of various multiplier selection algorithms [3]. It became clear that an experiment carried out at a point simply cannot satisfy the accuracy of the definition of the quality function at other points and should not be a function of coordinates, but a number for each point in the domain.

On the other hand, any research process that is conducted using experiments, simulations, or mathematical models can be represented as a black box (Figure 1). Any process that can be represented in the form of a black box model can be applied to the method of successive approximation (SAP) used in this paper. Unlike many methods with a similar name, it does not use any derivatives of functions, which makes it attractive for use in a wide range of tasks.

The black box model may be represented here by physical devices, systems of differential equations, mathematical formulas, and computer programs or application software packages (ASPs) that implement it. The results of research using the MM, as a rule, represent matrices of numbers or tables of numerical data. Functions are reproduced using their values on the parameter grid. The finer the mesh of a model, the more accurate approximations of the function are achievable. For complex problems, obtaining the values of functions on the parameters mesh requires significant expenditure of machine time, which makes the problem of reproduction of functions practically impossible. Therefore, the idea of using the method of obtaining MM according to a simplified procedure arose.

According to this approach, the MM is obtained not on the parameters mesh, but at a certain point of the area of their definition. The algorithm proposed by the author of [4] makes it possible to significantly reduce the computational costs of reproducing the quality function, but only in the vicinity of a certain point. Thus, instead of calculating the values of the function on the parameter mesh, they are calculated only on the coordinate lines, which significantly reduces the number of calculations of the resulting function (Figure 2).

Figure 1. Model of a black box: $X(x_1, x_2, \ldots x_n)$, control parameters or model variables. Y is the resulting factor or process quality function $Y = Y(x_1, x_2, \ldots x_n)$, the form of which must be reproduced from the data obtained after computer processing.

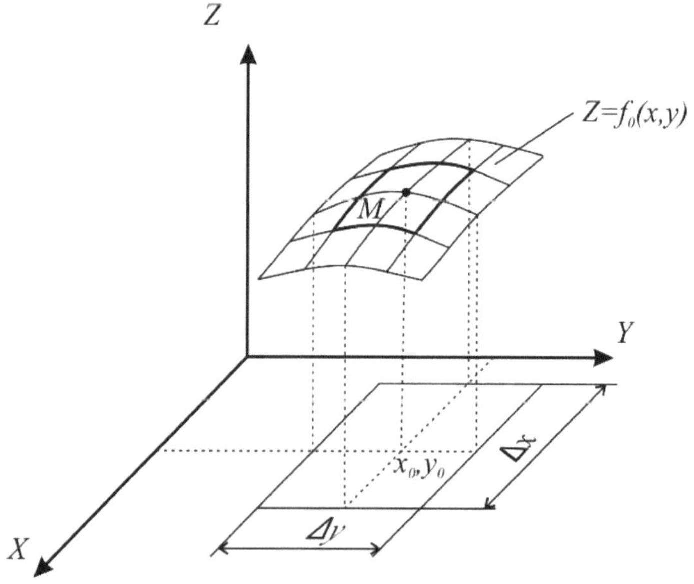

Figure 2. Selection of the point M vicinity with coordinates (x_0, y_0): $Z = f_0(x,y)$—the general view of the surface of the quality function, Δx, Δy—the step along the x and y coordinates.

The method of successive approximation (MSA) is a method that can be used in all fields from physical calculation to statistic and computer technology. It is possible to build an efficient and flexible computational method by using MSA [5]. This may be applied to the study of the fixed-point problems in the more abstract setting of Banach spaces (e.g., differential and integral equations, dynamical systems) [6]. The efficiency of such a method was also proven by comparing it to a variation iteration method under equivalent conditions [7].

MSA allows representation of functions in an analytical form (namely, in the form of a product of functions, each of which depends on one variable) when their values exist in tabular form on lines formed by the intersection of the functional surface with coordinate planes passing through a selected point from the definition area (Figure 2) [8].

The examples of MSA successful application on the problems of geotechnical mechanics [4] show that the solution of practical problems may be extended to the entire domain of the function, despite the fact that the errors of such a representation increase when approaching its boundary, and do not exceed the value of 5–7%. It turns out that such accuracy is satisfactory for engineering calculations in the field of geotechnical mechanics since the initial data for this area are determined with the same accuracy.

The main aim of this work was to determine the possibility of applying the method of successive approximation for the study of gas dynamics and heat transfer processes occurring in furnaces during the heating of metal products, and to estimate the error of approximation of the quality function from independent process parameters.

2. Materials and Methods

2.1. Mathematical Model

A heating furnace with a U-shaped trajectory of flue gas movement was used to solve the problem, the mathematical model of gas dynamics and heat transfer processes [9]. The scheme of the furnace is shown in Figure 3. The furnace is designed for heating metal ingots (3) before pressure treatment. The working space (2) of the furnace has the shape of a rectangular parallelepiped. To burn fuel, the furnace is equipped with a burner (4) located in the upper part of the left side wall of the stove. In the lower part of the same wall, there is a window (5) for the exhaust of a flue gas from the working space of the furnace into the flue gas channel, equipped with regenerators (1) for heating the air used for burning fuel.

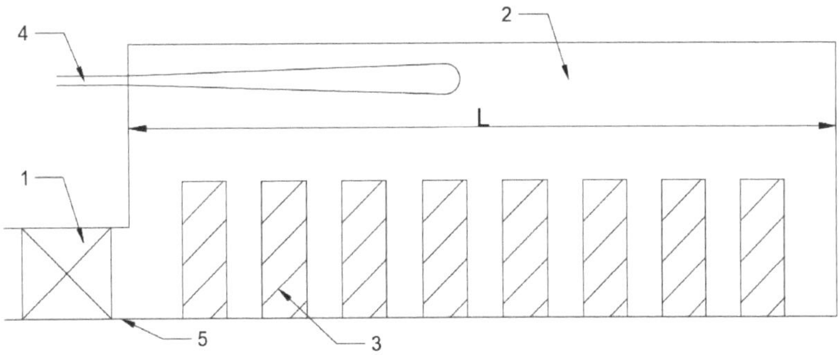

Figure 3. Scheme of a reheating furnace with regenerators and with distributed volumetric combustion: 1—regenerator; 2—working chamber; 3—heated material; 4—gas nozzle; 5—flue gas ducts; L—furnace length.

The furnace with a U-shaped trajectory of flue gas is among widely used equipment for material heating, so it was chosen to increase its parameters and, as a result, reduce of production costs and influence on the environment.

The main disadvantage of furnaces with a U-shaped trajectory of flue gas is the non-uniformity of gas temperatures in the working chamber, which does not allow obtaining the same quality of heating of all heated products [10]. Unevenness of heating in furnaces leads not only to an undesirable increase of the temperature of some product parts, but also to additional heat and material loss [11]. Therefore, the main tasks set during simulation of the processes taking place in such furnaces are:

- determination of gas temperature distribution along the trajectory of their movement in the working chamber of the furnace;
- calculation of the parameter of uniformity of gas temperature distribution in the working chamber ($F = \Delta t = t_{max} - t_{min}$);
- selection of independent parameters that affect uniformity;

- determining the necessary values of these parameters to increase the uniformity of gas temperature distribution in the working chamber.

The design parameters of the furnace and auxiliary equipment of the furnace are selected according to the results of mathematical modelling. For existing furnaces, the simulation results are used to develop measures to change the operating parameters of the unit and its equipment in order to improve the uniformity of gas and temperature distribution in the furnace.

During simulation, the working chamber of the furnace is divided into $i = 2n$ calculation zones (see Figure 4). It is assumed that the heating material is located in each calculated zone of the furnace, and there is an ideal mixing of gases; that is, the temperature of the gases within the volume is the same. A burner is located in the first (by gas flow) calculation zone $i = 1$, and a smoke window is located in the last zone $i = 2n$. The location of the burner and the smoke window on the same end wall contributes to the occurrence of large-scale gas recirculation, which significantly affects the uniformity of the smoke temperature distribution in the working chamber of the furnace.

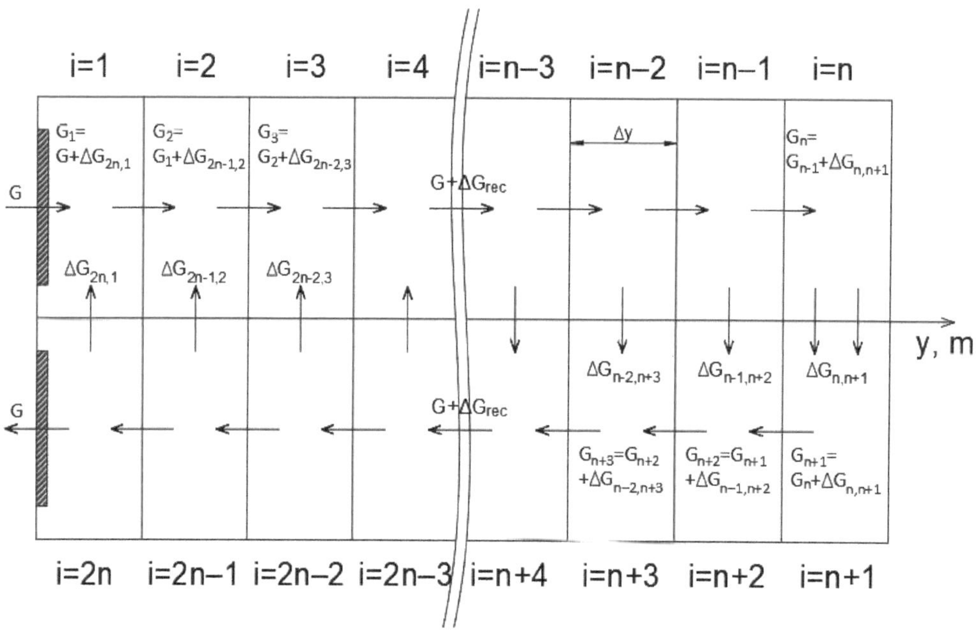

Figure 4. Calculation scheme of the furnace and the movement of flue gases.

In the central part of the furnace (Figure 4), on the border of the conditional division of the gas flow into forward and reverse trajectories of movement, local turbulent vortices appear, moving in forward and reverse directions. The mixing of gases in local zones through the flow separation boundary is a local recirculation of gases, which also affects the uniformity of the temperature distribution in the working chamber of the furnace.

The equation of heat balance of the U-shaped flue gas trajectory furnace is:

$$Q_{in\ i} + Q_{rec\ i} + Q_{bur\ i} = Q_{Me\ i} + Q_{lost\ i} + Q_{exg\ i} \quad (1)$$

where

$Q_{in\ i}$—heat entered into the 1st zone with fuel, air and fuel combustion products, (W);
$Q_{rec\ i}$—heat of combustion products passing through the furnace gas flow interface on the straight and reverse path, (W);

$Q_{bur\ i}$—heat released by the fuel combustion, (W);
$Q_{Me\ i}$—heat consumed to heat up the material in the zone, (W);
$Q_{lost\ i}$—heat loss through furnace walls, (W);
$Q_{exg\ i}$—heat loss of furnace gases leaving the ith zone, (W).

Heating components of the heat balance calculation are calculated according to Equations (2)–(4).

$$Q_{in\quad i} = \begin{cases} B_f \cdot t_f \cdot c_f + L_n \cdot B_f \cdot t_{air} \cdot c_{air} \quad \text{for} \quad i = 1 \\ \left(1 - \sum\limits_{k=1}^{i-1} m_k\right) \cdot \left(B_f \cdot t_{i-1} \cdot c_f + L_n \cdot B_f \cdot t_{i-1} \cdot c_{air}\right) + \\ + \sum\limits_{k=1}^{i-1} m_k \cdot v_{exg} \cdot B_f \cdot t_{i-1} \cdot c_{exg} + \sum\limits_{k=1}^{i-1} (\Delta G_{2n-k+1} - \Delta G_k) \cdot t_{i-1} \cdot c_{exg} \quad \text{for} \quad i = 2, \quad \ldots \quad, \quad 2n \end{cases} \tag{2}$$

$$Q_{rec\quad i} = \begin{cases} \Delta G_{2n-i+1} \cdot t_{2n-i+1} \cdot c_{exg} \quad \text{for} \quad i = 1, \quad \ldots \quad, \quad n/2 \quad \text{and} \quad i = n+1, \quad \ldots \quad, \quad 3n/2 \\ -\Delta G_{2n-i+1} \cdot t_{2n-i+1} \cdot c_{exg} \quad \text{for} \quad i = n/2+1, \quad \ldots \quad, \quad n \quad \text{and} \quad i = 3n/2+1, \quad \ldots \quad, \quad 2n \end{cases} \tag{3}$$

$$Q_{bur\ i} = m_i \cdot B_f \cdot Q_{low} \tag{4}$$

where
B_f—consumption, (kg/s);
t_f, t_{air}—temperature of heated fuel and air, (°C);
c_f, c_{air}, c_{exg}¯heat capacity of fuel, air and flue gases, (J/(m^3 K));
L_n—actual combustion air consumption, (kg air/kg fuel);
t—temperature of the gases in combustion chamber, (°C);
m_i—fraction of fuel burned in the ith zone;
v_{exg}—specific exhaust gases output during combustion, (kg/kg).
ΔG_i—consumption of gases that move between local recirculation zones, (kg/s);
Q_{low}—lower operating heat of combustion of fuel, (J/kg);
n—number of calculated zones along the length of the furnace, (pcs).

The value of ΔG_{ij} is determined by the law of the distribution of mass flow density of combustion products through the interface between the forward and reverse trajectories of their movement. For this purpose, the following formula can be used:

$$\Delta G_i = \begin{cases} 0 \quad \text{for} \quad 1 \le i \le n/2 \quad \text{and} \quad n < i \le 3n/2 \\ A \cdot \left(\frac{2 \cdot y_i}{L_{tr}}\right)^b \quad \text{for} \quad n/2 < i \le n \\ A \cdot \left(\frac{2 \cdot y_{i-n}}{L_{tr}}\right)^b \quad \text{for} \quad 3n/2 < i \le 2n \end{cases} \tag{5}$$

where
$y_i = \Delta y \cdot (i - 1/2)$—coordinate of zone i centre, (m);
$\Delta y = L_{tr}/(2 \cdot n)$—length of the ith zone, (m);
L_{tr}—total length gas movement trajectory, (m);
b—distribution function exponent;
A—the regularity coefficient of the distribution of the specific mass flow of combustion products across the interface of the forward and reverse trajectories of their movement, which is determined by the Equation (6):

$$B_f \cdot v_{exg} \cdot (k_p - 1) - A \cdot \left[\sum_{i=\frac{n}{2}}^{n} \left(\frac{2 \cdot y_{i+1}}{L_{tr}}\right)^b\right] = 0 \tag{6}$$

where
k_p—coefficient of the large scale (recirculation through all zones of the combustion chamber) recirculation.

The share of fuel that burns in the ith calculation zone of the furnace is determined as:

$$m_i = \begin{cases} 1 - C(L_i) & \text{for} \quad i = 1 \\ C(L_{i-1}) - C(L_i) & \text{for} \quad i = 2, \quad \ldots \quad , \quad 2n \end{cases} \tag{7}$$

where

$C(L_i) = exp\left(-\frac{K_g \cdot \Delta y \cdot i}{L_f}\right)$—the regularity of changes in the relative concentration of fuel depending on the length of the torch;

K_g—parameter of the fuel burnout model, which depends on the completeness of fuel combustion in the flare and the type of burner;

L_f—the length of the combustion zone, (m).

Output parts of the heat balance of the ith zone are calculated according to (8)–(10).

$$Q_{Me\ i} = C_{Me} \cdot \left[(t_i + 273)^4 - (t_{Me} + 273)^4\right] \cdot \Delta F_{Me_i} \tag{8}$$

$$Q_{lost\ i} = k_{lost} \cdot t_i \cdot \Delta F_{Me_i} \cdot \xi_{gr} \tag{9}$$

$$\begin{aligned} Q_{exg\ i} &= \left(1 - \sum_{k=1}^{i} m_k\right) \cdot \left(B_f \cdot t_i \cdot c_f + L_n \cdot B_f \cdot t_i \cdot c_{air}\right) + \\ &+ \sum_{k=1}^{i} m_k \cdot v_{exg} \cdot B_f \cdot t_i \cdot c_{exg} + \sum_{k=1}^{i} (\Delta G_{2n-k+1} - \Delta G_k) \cdot t_i \cdot c_{exg} \end{aligned} \tag{10}$$

where

C_{Me}—equivalent coefficient of radiation in the zone with respect to the conduction, $(W/(m^2\ K^4))$

t_{Me}—temperature of the material surface, (°C);

$\Delta F_{Me_i} = \frac{F_{Me}}{2 \cdot n}$—metal surface area in the ith zone, (m^2);

F_{Me}—total heat exchange area of the material, (m^2);

k_{lost}—coefficient of heat transfer to the environment, $(W/(m^2 \cdot K))$;

$\varsigma_{gr} = F_{lin}/F_{Me}$—gradation of furnace laying;

F_{lin}—total area of the furnace laying, (m^2);

To determine the temperature change of the gases in the working chamber of the furnace, a system of heat balance equations was compiled for each i-th calculation zone. The total number of equations was $2n$. Next, by solving the system, the temperature distribution of the gases along the trajectory of their movement was relatively determined. The value $\Delta t = t_{max} - t_{min}$ was calculated as a parameter of the uniformity of gas temperature distribution in the working chamber of the furnace, where t_{max} and t_{min} are the maximum and minimum temperatures of gases in the furnace).

Article [9] presents a comparison of the temperature distribution in the furnace chamber with a U-shaped trajectory of flue gas movement obtained by using a mathematical model with the results of temperature distribution measurements in a physical model- laboratory furnace. It was shown that the accuracy of the calculated results was approximately 1.4–2.45%.

The mathematical model of the processes of gas dynamics and heat exchange in the heating furnace is quite complex, and does not allow establishing the analytical dependence of the uniformity of the gas temperature distribution Δt on the parameters that affect it. The use of the method of successive approximation [3] allows establishing dependence and eliminating the need to solve a system of $2n$ nonlinear equations each time to determine the value of Δt.

2.2. Method of Successive Approximation

To extend the scope of the MSA application, a hypothesis was formulated, as follows.

Let there exist a scalar function $F(X) = F(x_1, x_2, x_3, \ldots x_n)$, that is bounded, defined and continuous in the closed region $\overline{D}$ of the scalar field P. Then for any point $M \subset \overline{D} \forall M \in D; \forall \varepsilon \geq$

$0\exists U_\varepsilon(M) \subset \overline{D}$ in the vicinity of the point $M_0(x_1^0, x_2^0, x_3^0, \ldots x_n^0)$ (Figure 2) the function $F(X)$ can be represented in the form:

$$|F(X) - \phi(X)| \leq \varepsilon \forall M_0 \in U_\varepsilon(M_0) \tag{11}$$

where

$U_\varepsilon(M_0)$—vicinity of the point $M_0(x_1^0, x_2^0, x_3^0, \ldots x_n^0)$;
$\phi(X) = \alpha \prod_{i=1}^n g_i(x_i)$;
$g_i(x_i)$—approximation functions for $f_1, f_2, f_3, \ldots, f_n$, which are given as follows:

$$f_1(x_1) = F(x_1^0, x_2^0, x_3^0, \ldots, x_n^0),\ f_2(x_2) = F(x_1^0, x_2^0, x_3^0, \ldots, x_n^0),\ f_3(x_3) = F(x_1^0, x_2^0, x_3^0, \ldots, x_n^0), \ldots, f_n(x_n) = F(x_1^0, x_2^0, x_3^0, \ldots, x_n^0) \tag{12}$$

α—the approximation coefficient is determined in accordance with the formula:

$$\alpha = \frac{F(M_0)}{g_1(x_1^0) \cdot g_2(x_2^0) \cdot g_3(x_3^0) \cdot \ldots \cdot g_n(x_n^0)} \tag{13}$$

Using the specified approach, the representation of the function $F(X) = F(x_1, x_2, x_3, \ldots x_n)$ around the point $M_0(x_1^0, x_2^0, x_3^0, \ldots x_n^0)$ has sufficient accuracy for engineering calculations on the entire definition area $\overline{D}$.

The MSA application algorithm can be represented by a sequence of the following steps:

1. Selecting a point from the function definition area

$$M = M_0\left(x_1^0, x_2^0, x_3^0, \ldots x_n^0\right),\ M \in \overline{D} \tag{14}$$

2. Creating the function $f_1(x_1) = F(x_1^0, x_2^0, x_3^0, \ldots, x_n^0)$.
3. Finding the type of function $g_1(x_1)$, that is an approximation for the function $f_1(x_1)$;
4. Finding $\varphi_1(x_1)$ in accordance with step 1: $\varphi_1(x_1) = \alpha_1 g_1(x_1)$, where α_1 is the approximation coefficient.
5. Defining the function around the point M from equality $F(x_1) \approx \varphi_1(x_1)$.
6. Repeating the steps 2–5 consecutively for the variables $x_j(j = \overline{2, n})$ and obtaining the representation as follows:

$$F(x_1, x_2, x_3, \ldots x_n) \approx \varphi_1(x_1, x_2, x_3, \ldots x_n) = \alpha \cdot g_1(x_1) \cdot g_2(x_2) \cdot g_3(x_3) \cdot \ldots \cdot g_n(x_n) \tag{15}$$

The location of a point $M = M(x_1^0, x_2^0, \ldots, x_n^0), M \in \overline{D}$ in the definition area significantly depends on its topology, and therefore affects the way it is represented. The choice of a point on the definition area is determined by prior knowledge of its features and is determined by the researcher's experience. In the case of complex functions and lack of prior knowledge about behavior of the resulting function, it is convenient to choose it in the center of the definition area, i.e., to determine the coordinates according to the formula:

$$x_j = \frac{b_j - a_j}{2} \tag{16}$$

where a_j, b_j represent the beginning and end proves (respectively) of the interval of changes of parameter x_j.

It should be noted that the surface of the desired function is represented using a generalized hyperbolic hyperboloid, and it is not worth hoping for a uniform behavior of the relative error.

It should also be noted that the problem of choosing classes of approximation functions is one of the most critical problems, not only in applied mathematics but also in technical applications. As research has shown [12], the coefficient of variation of approximating functions cannot act as a criterion for selecting functions. As a criterion that limits the choice of the class of approximating functions, it is suggested to choose the dimension of the original function, if possible. Using the limitations of the class of approximating functions is extremely important for evaluating the influence of parameters on the resulting function.

Wide application of MSA in practice has shown that the use of a class of power functions is particularly effective for evaluating the influence of parameters on the quality function. The impact assessment method is a reproduction of the problem solution (when it is constructed in the form of described tables of numbers) in the form of a product of power functions and a comparison of their indicators. The higher the exponent, the stronger the influence of the parameter on the function.

3. Initial Data, Quality Function, Independent Parameters and Their Limits

Using a mathematical model of gas dynamics and heat exchange processes, calculations of gas temperature distribution along the length of the trajectory of their movement in the working chamber of the furnace were performed.

The initial data for modeling are presented in Table 1. The most characteristic parameters of the work intended for heating steel ingots were taken as initial data. The length of the working chamber of the furnace (8 m) and height (43 m) are typical for the intended design and do not change depending on the aisles. As a rule, such furnaces are heated with a mixture of consumption of coke and blast-furnace gas with a calorific value of 5.5–812 MJ/m^3 [2]. At the same time, the higher the temperature of heating the fuel and air before their absorption, the lower the calorific value of the mixture of coke oven and blast-furnace gas that is used. In the assumed heating of air and visible gases, up to 800 °C and 300 °C, respectively, which is the most widely used heat leaving the flue gas furnace, the metal surface temperature of 1000 °C corresponds to the range of revolutions of its metal processing. The thermophysical properties of metal, gas, and air correspond to their temperature [10].

Table 1. Initial data.

B_f, kg/s	Q_f, J/kg	L_n, kg/kg	v_{exg}, kg/kg	t_f, °C	t_{air}, °C	t_{Me}, °C	c_f, J/(kg·K)
1.721	6.164×10^6	1.943	2.943	300	800	1000	1240
L_{tr}, m	C_{Me}, W/(m^2·K^4)	F_{Me}, m^2	ξ_{gr}	k_{lost}, W/(m^2·K)	n	c_{air}, J/(kg·K)	c_{exg}, J/(kg·K)
16	1.731×10^{-8}	122	1.344	2	8	1071	1238.2

As a quality function, an indicator that characterizes the uniformity of the gas temperature in the working chamber $F = t_{max} - t_{min}$ was chosen, where t_{max} and t_{min} are the maximum and minimum gas temperatures along the trajectory of their movement, respectively. According to the results of previous studies, six independent parameters that affect the quality function were determined, and the range of their possible change was also established.

(a) Fuel temperature $t_f = 0, \ldots, 300$ °C. It is known from literature sources [13] that it is not safe to increase the heating temperature of gaseous fuel in the existing heat exchangers of furnaces above 300 °C, for the reasons of explosive safety.

(b) The temperature of the air used for fuel combustion is $t_{air} = 0, \ldots, 1000$ °C. Due to the heating of air to high temperatures, significant fuel savings are achieved in methodical [14], ring [15], roller and shaft [16] furnaces, as well as in heating wells [17] of various designs. The maximum air heating temperature depends on the method of utilization of the heat of the flue gases leaving the furnace and the design of the heat exchangers, but usually this temperature does not exceed 1000 °C. At the same time, in outdated equipment that is still in operation at some enterprises, heating of the combustion air before burning is not provided.

(c) Multiplicity of internal large-scale recirculation of furnace gases $k_p = 1, \ldots, 6$.

The choice of such limits for varying the multiplicity of recirculation of furnace gases is determined by the practical research data obtained by numerous experimenters. So, when modeling a recirculation circuit with a built-in fan, V.D. Brook [18] discovered that increasing the recirculation ratio above 5 was impractical and led to an increase in fan power consumption. A shown in research conducted at the Gas Institute of the National Academy of Sciences [19], when the recirculation rate was increased above $k_p = 6$, the intensity of temperature equalization in the furnace decreased sharply, so further intensification of recirculation is impractical.

The indicator of the degree of the distribution function, which characterizes small-scale recirculation, was chosen on the base of considerations outlined in the paper [4] within $b = 0, \ldots, 15$.

The parameter of the fuel burnout model adopted by V.Y. Gubynskyi et al. [2] varies within the limits $K_g = 3, \ldots, 4.61$;

The length of the combustion zone $L_f = 8, \ldots, 16$ m was taken in metres from the conditions that the torch occupies the entire length of the furnace space, and according to the operation data of the operating furnaces.

Acting in accordance with the MPA algorithm, the range of change for each independent parameter was divided into 10 intervals with equal steps and a reference point was chosen $M\left(t_f^0, t_{air}^0, k_p^0, b^0, K_g^0, L_f^0\right)$, where $t_f^0 = 150$ °C; $t_{air}^0 = 500$ °C; $k_p^0 = 3.5$; $K_g^0 = 3.805$; $L_f^0 = 12$ m.

4. Results

Figure 5 shows the results of calculating the gas temperature distribution in the working chamber of the furnace. As can be seen from the figure, a gas minimal temperature value of 1053.9 °C was observed at the entrance to the working chamber of the furnace. The maximal gas temperature was observed in the sixth calculation zone, reaching 1268.4 °C. The quality function value at the reference point was $F(t_f^0, t_{air}^0, k_p^0, b^0, K_g^0, L_f^0) = 214.5$ °C.

Figures 6–11 show combined graphs of the function F values (marked by dots in the figures) and their approximations functions g_i (solid continuous lines).

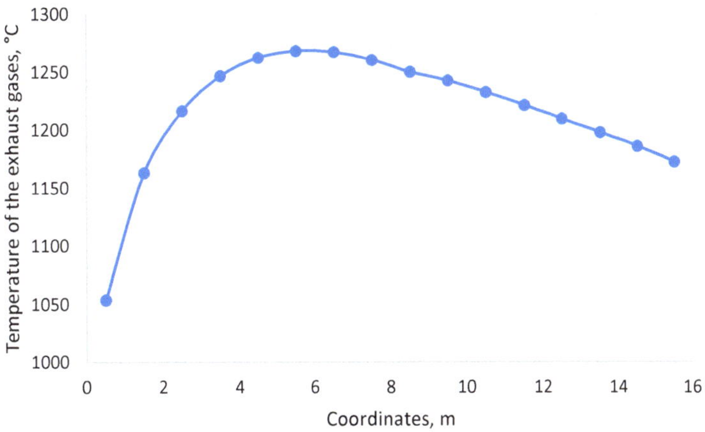

Figure 5. Distribution of gas temperature along the trajectory of movement.

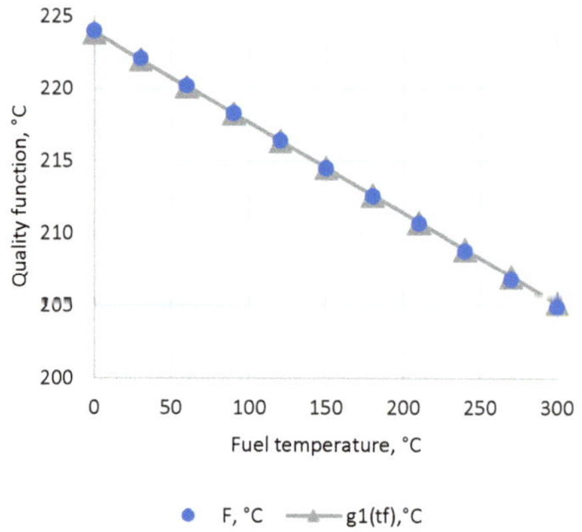

Figure 6. Combined graph of the values of the function F from the parameter t_f.

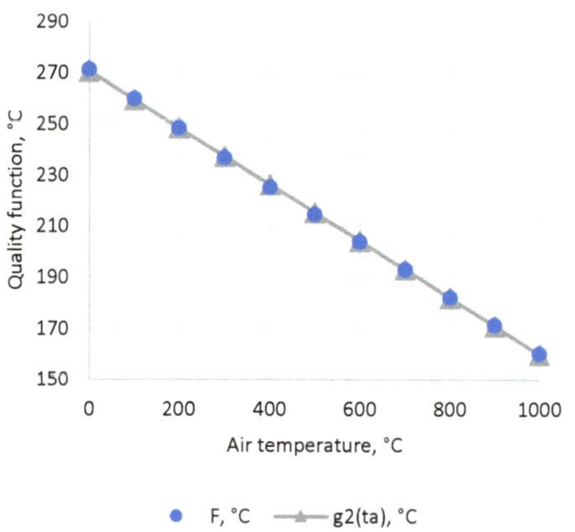

Figure 7. Combined graph of the values of the function F from the parameter t_{air}.

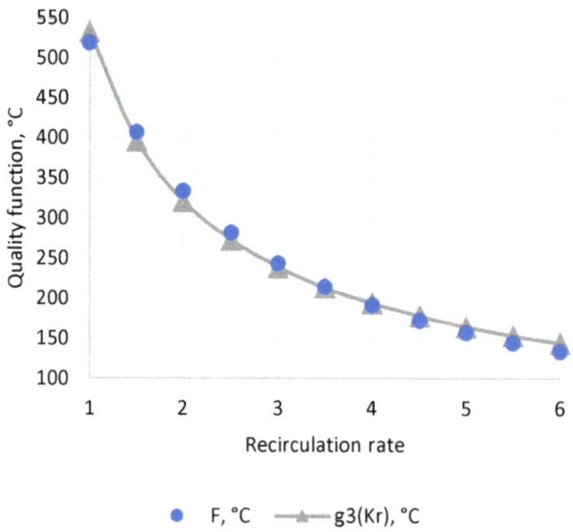

Figure 8. Combined graph of the values of the function F from the parameter k_r.

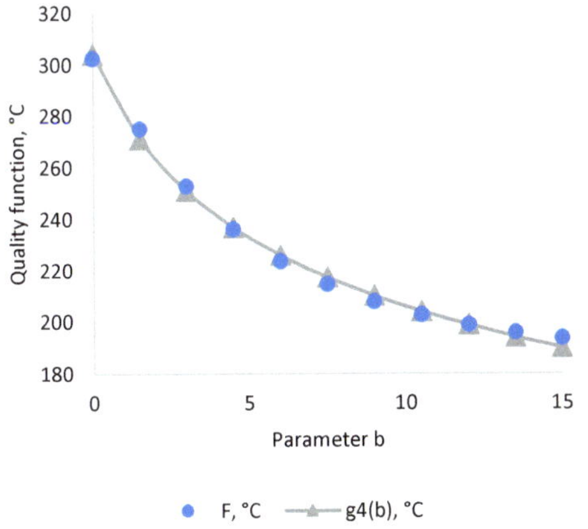

Figure 9. Combined graph of the values of the function *F* from the parameter *b*.

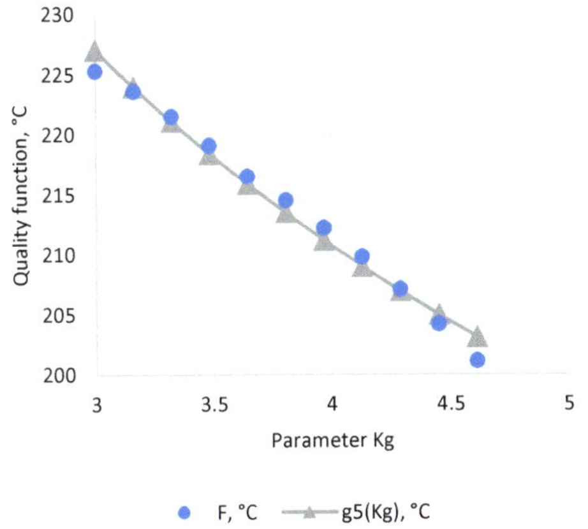

Figure 10. Combined graph of the values of the function *F* from the parameter K_g.

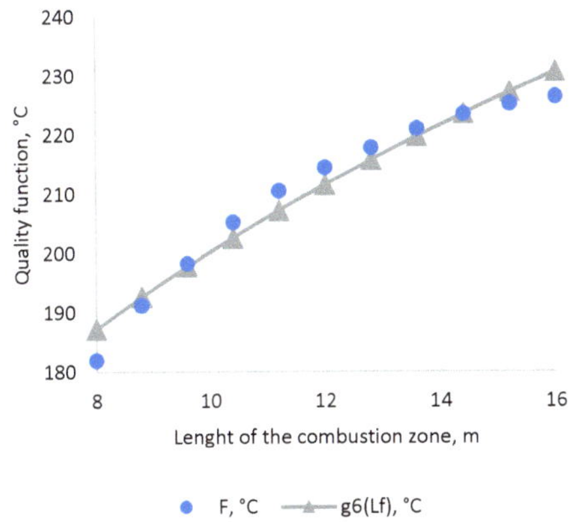

Figure 11. Combined graph of the values of the function *F* from the parameter L_f.

The approximation functions reached the following values:

$$\begin{gathered} g_1 = \left(223.75 - 0.0623939\ t_f\right) \quad g_2 = (270.441 - 0.110591\ t_{air}) \\ g_3 = 532.222/k_p^{0.729813} \quad g_4 = \frac{376.11426000}{(b+2.43204800)^{0.23907803}} \\ g_5 = 301.811/K_g^{0.259232} \quad g_6 = 100.019L_f^{0.301486} \end{gathered} \tag{17}$$

The choice of approximating functions g_i was carried out based on the best fit to the data obtained. Despite the above-mentioned advantages of approximating the obtained data by power functions, their use is not always possible. So, according to Figures 5–7, the dependencies of the quality function on air and fuel temperatures cannot be approximated by power functions, since they are linear dependencies [2]. Therefore, an exception was made for g_1 and g_2. The approximation function g_4 took on this form solely because of physical considerations and conditions of existence in the denominator of the expression close to zero, which leads to uncertainty. All other functions were chosen in the form of exponents for the convenience of considering the sensitivity of the function to the parameters.

According to the above hypothesis and algorithm, the calculated function F takes the form:

$$F = \frac{0.0134281171 \cdot L_f^{0.301486}\left(223.75 - 0.0623939 \cdot t_f\right)(270.441 - 0.110591 \cdot t_{air})}{(b + 2.432048)^{0.23907803} \cdot K_g^{0.259232} \cdot k_r^{0.729813}} \tag{18}$$

The relative errors of the approximating functions $g_1, g_2, g_3, g_4, g_5, g_6$ are characterized by graphical dependencies presented in Figure 12.

Figure 12 presents the distribution of relative deviations of the function from the one obtained by the formula on the lines of intersection of the functional space with coordinate planes. The accuracy obtained by the choice of approximating functions is at a satisfactory level for most of engineering calculations because the obtained relative errors are commensurate with the measurement errors of the corresponding quantities in industrial conditions. The behavior of errors on the hyper diagonal of the domain of definition is of interest. According to the authors, these deviations are decisive for evaluating the behavior of the function and adjusting the scope of its definition, that is, in fact, the range of the parameter variations.

As can be seen from the above graph, the last error value (32.8%) is not desirable for engineering calculations. So, for example, with the temperature distribution shown in Figure 6, the value of the quality function is 214.5 °C. With an error value of 32.8%, the error in determining the temperature unevenness in the working chamber of the furnace reaches 70.4 °C, which significantly exceeds the error in measuring temperatures in industrial conditions (10–15 °C).

To reduce the relative error when using the given formula, it is necessary to reduce the domain of the function by reducing the right-hand interval by the step of dividing the range of variations. The reduction is not significant but allows the range of errors in the definition domain of the function to be limited to 6%, which is a satisfactory result for most engineering calculations.

Figure 14 shows the graph of the relative error for the adjusted range of parameters: fuel temperature $t_f = 0, \ldots, 270$ °C; the temperature of the air used for fuel combustion $t_{air} = 0, \ldots, 900$ °C; multiplicity of internal large-scale recirculation of furnace gases $k_r = 1, \ldots, 5.5$; an indicator of the degree of distribution function characterizing small-scale recirculation $b = 0, \ldots, 13.5$; a parameter of the fuel burnout model $K_g = 3, \ldots, 4.449$, and the length of the combustion zone $L_f = 8, \ldots, 15.2$ m. In the specified range of parameter variations, the relative error of the quality function does not exceed 6%.

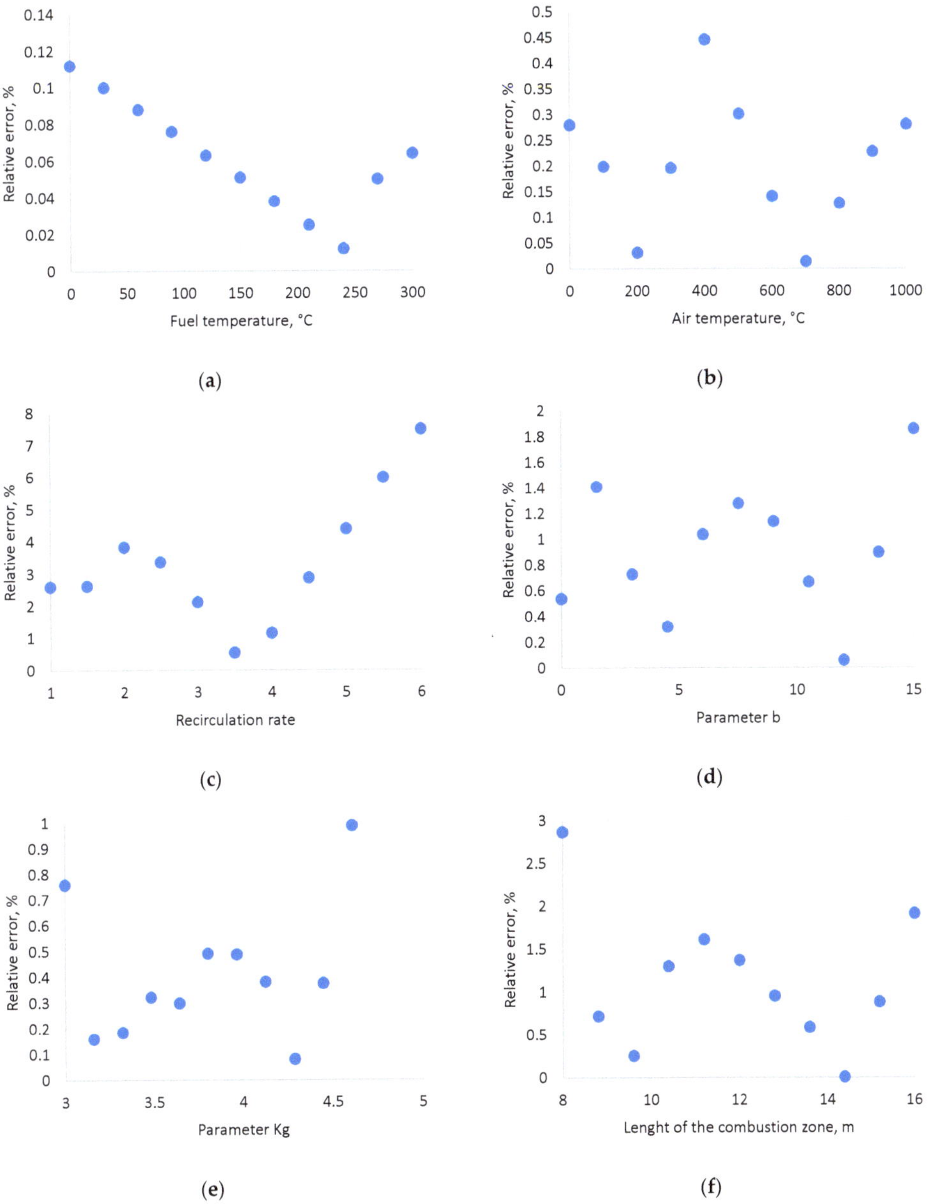

Figure 12. Relative error of approximating functions: (**a**) g_1, (**b**) g_2, (**c**) g_3, (**d**) g_4, (**e**) g_5, (**f**) g_6.

The distribution of the relative error on the generalized diagonal of the hypercube of the definition area is presented in Figure 13.

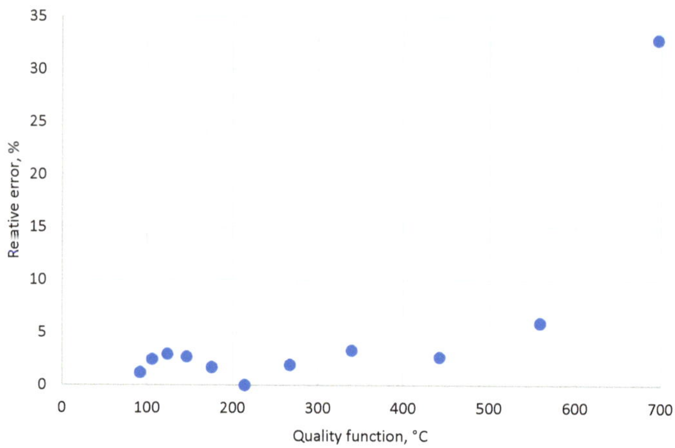

Figure 13. Relative error on the diagonal of the definition area.

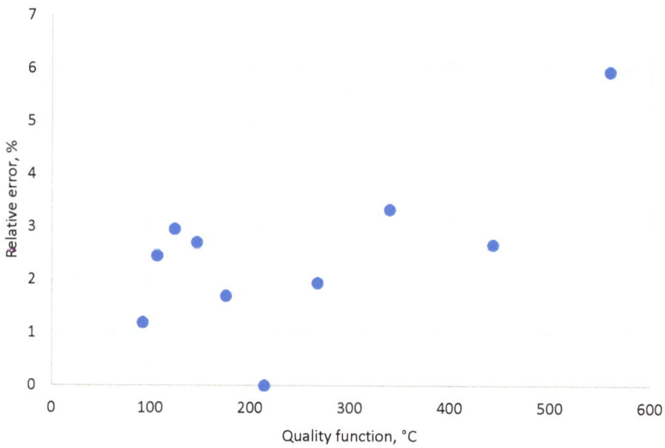

Figure 14. Relative error on the diagonal of the definition area of the parameters t_f, t_{air}, k_r, b, K_g, L_f.

5. Discussion

Based on the hypothesis formulated in previous section, the results can be written as follows:

$$f_1^*(x_1, x_2) = F^*\left(x_1, x_2, x_3^0, x_4^0, \ldots, x_n^0\right) = F(M_0)/\left(g_1\left(x_1^0\right) g_2\left(x_2^0\right)\right)$$
$$f_2^*(x_1, x_2, x_3) = F^*\left(x_1, x_2, x_3, x_4^0, \ldots, x_n^0\right) = F(M_0)/\left(g_1\left(x_1^0\right) g_2\left(x_2^0\right) g_3\left(x_3^0\right)\right)$$
$$f_3^*(x_1, x_2, x_3, x_4) = F^*\left(x_1, x_2, x_3, x_4, x_5^0, \ldots, x_n^0\right) = F(M_0)/\left(g_1\left(x_1^0\right) g_2\left(x_2^0\right) g_3\left(x_3^0\right) g_4\left(x_4^0\right)\right),$$
$$f_4^*(x_1, x_2, x_3, x_4, \ldots, x_n) = F^*(x_1, x_2, x_3, x_4, x_5, \ldots, x_n) = F(M_0)/\left(g_1\left(x_1^0\right) g_2\left(x_2^0\right) g_3\left(x_3^0\right) \ldots g_n\left(x_n^0\right)\right).$$

That means that moving sequentially, starting from the first step, one can obtain the entire sequence of representations of the function from these parameters. It all works when all other points are at the point $M_0\left(x_1^0, x_2^0, x_3^0, \ldots x_n^0\right)$.

The simulation results given above fully reflect qualitative ideas about furnace processes, which are known from their operation. Thus, among the selected independent parameters, the most significant influence on the uniformity of the temperature field in furnaces with a U-shaped trajectory of gas movement is characterized by the multiplicity of large-scale recirculation k_r. Changing this indicator value from $k_r = 1$ to $k_r = 6$ results in a tenfold decrease of the quality function value $F = t_{max} - t_{min}$ meaning the uniformity of the gas temperature in the working chamber of the furnace. The simulation results confirmed the thesis of a sharp drop in the intensity of temperature equalization with an increase in the recirculation rate. The processing of the approximation curve of the quality function F shows that the uniformity of the gas temperature increases by almost 40 % when the recirculation ratio increases from $k_r = 1$ to $k_r = 2$ and by 12.5% only when the same parameter increases from $k_r = 5$ to $k_r = 6$.

Small-scale recirculation has a similar quality, but smaller influence on function F. Such an effect on temperature uniformity in the working chamber of the furnace is quite natural. After all, both small-scale and large-scale recirculation characterize gas-dynamic and mass exchange processes, which are described by the same equations and have the same effect on the heat exchange. Analysis of the data on the impact of small-scale recirculation shows that with an increase in its share, the unevenness of the gas temperature distribution in the furnace can increase by an additional 15%. The reduction of small-scale local recirculation is associated with the elimination of obstacles and local resistances in the path of movement of furnace gases, and a sharp change in the direction of their movement.

The gas heating temperature and the parameters of the fuel combustion model have the least influence on the uniformity of the gas temperature out of the six selected parameters. The change in fuel temperature from the minimum to the maximum level (i.e., from 0 °C to 300 °C) leads to an insignificant improvement of the uniformity of gas temperature distribution in the working chamber of the furnace, as it ranges from 225 °C to 205 °C. The parameter of the model of gas combustion in the torch has almost the same effect.

The length of the flare affects the quality function F in direct proportion. Studies have confirmed the growth of heating unevenness with increasing torch length. In our case, such an increase value was 45 °C, which means an almost 25% increase.

As follows from Figure 12, the relative errors of the approximating functions g_1, g_2, g_3, g_4, g_5, g_6 are not significant and do not exceed 2–3% except for the error of the approximating function g_3 (multiplicity of recirculation) which reaches 7.5% at the limit point equal to 6. Under the conditions where such a value of the recirculation multiplicity for furnaces is too large, it may be considered that the accuracy of the approximation functions obtained in research is quite acceptable for engineering calculations.

The same conclusion can be made about the value of the relative error on the diagonal of the detection area on the base of Figure 13. Except for the extreme point, the relative error does not exceed 6%, which is within the limits for most of engineering calculations and measurement errors when using industrial metrological devices.

According to the literature [20,21], MSA is a suitable and effective method of mathematical modeling and calculation. The results of the studies carried out in this work indicate the possibility of expanding the scope of MSA to solve gas dynamics and heat transfer problems.

6. Conclusions

This paper presents the results of applying the successful approximation method for mathematical modeling of gas dynamics and heat transfer processes in a heating furnace with a U-shaped trajectory of flue gas movement. The influence of a number of independent parameters on the selected quality function, i.e., the uniformity of the temperature field of gases in the working chamber of the furnace, was studied. The following parameters were selected as independent parameters: fuel and air heating temperatures before combustion; the multiplicity of internal large-scale recirculation of furnace gases; the exponent of the distribution function, which characterizes the small-scale recirculation of gases in the working chamber; the flame burnup model parameter, and the length of the burning zone. It was established that the greatest influence on the quality function is exerted by the multiplicity of the internal large-scale recirculation of furnace gases, the change of which from 1 to 6 leads to a tenfold increase in the uniformity of the gas temperature distribution in the working chamber. Moreover, the main increase in the uniformity of temperature distribution (up to 40%) occurs with an increase in the recirculation ratio from 1 to 2. With a further increase in the recirculation ratio, the effect of this parameter on the quality function weakens and does not exceed 12.5% with a change in the recirculation ratio from 5 to 6. The fuel heating temperature before combustion has the least effect on the quality function (up to 8.9%). The remaining parameters considered are characterized by an average influence on the quality function. Their change in the considered interval causes a change in the uniformity of the temperature field of gases by 15–25%.

MSA, proposed for the approximate determination of regularities between system parameters, was confirmed for its efficiency in the calculation of gas dynamics and heat transfer problems. Relative errors for most parameters did not exceed 6%, which is a confirmation of the efficiency of the algorithm, on the one hand, and the possibility of using simplified formulas for practical engineering calculations.

The simplified method proposed by the authors, in contrast to that used in the aforementioned works, does not require continuity and smoothness of functions and does not use derivatives of the quality function, which widely extends the scope of its application in various areas of science including mathematical modeling of gas dynamic and heat transfer processes in industrial heating and thermal equipment.

The representation of the product functions by power functions allows swift evaluation of their influence on the quality function. Our research also indicates the possibility of using MSA as a tool for a digital twin design for thermal processes. In the future, the authors aim to continue with mathematical and experimental verification of the presented hypothesis.

Author Contributions: Conceptualization, M.R. and O.Y.; methodology, O.G.; software, A.K.; validation, G.L., O.G., Y.M. and A.K.; formal analysis, M.F.; investigation, O.Y.; resources, M.R.; data curation, A.K. and T.K.; writing—original draft preparation, O.G. and A.K.; writing—review and editing, A.K. and T.K.; visualization, A.K.; supervision, O.Y.; project administration, M.R.; funding acquisition, M.R. All authors have read and agreed to the published version of the manuscript.

Funding: The work is supported by KEGA grant agency under Grant KEGA 023TUKE-4/2021.

Institutional Review Board Statement: Not applicable.

Informed Consent Statement: Not applicable.

Data Availability Statement: Not applicable.

Conflicts of Interest: The authors declare no conflict of interest.

References

1. Gubinskiy, V.I.; Yeremin, A.O.; Tryapichkin, M.G. Experience of using volumetric-regenerative method of fuel combustion in soaking-pit furnaces. *Metall. Min. Ind.* **2010**, *2*, 164–167.
2. Gubinskiy, V.I.; Lu, Z.-U. *The Theory of Flame Furnaces*; Mashynostroyeniye: Moscow, Russia, 1995; p. 256.
3. Fedorets, V.A. The method of multifactorial study of the parameters of the heat transfer process. *Digatelestroyeniye* **1982**, *11*, 54–66.
4. Larionov, H.; Larionov, M. Chapter 4. On the One Parameters Influence Evaluating Method Employed to Evaluate the Support Capacity of a Metal-Resin Anchor. In *Modeling of the Soil-Structure Interaction: Selected Topics*; Zhelyazov, T., Ed.; Nova Science Publisher, Inc.: Hauppauge, NY, USA; Technical University of Sofia: Sofia, Bulgaria, 2020; pp. 87–101.
5. Torokhti, A.; Howlett, P.; Pearce, C. Method of Best Successive Approximations for Nonlinear Operators. *J. Comput. Anal. Appl.* **2003**, *5*, 299–312.
6. Potra, F.A. On the Superlinear Convergence of the Successive Approximations Method. *J. Optim. Theory Appl.* **2002**, *113*, 473–485.
7. Jaffari, H. A comparison between the variational iteration method and the successive approximations method. *Appl. Math. Lett.* **2014**, *32*, 1–5. [CrossRef]
8. da Silva, E.A.B.; Dutra, A.J.S. FIR Filter Design Based on Successive Approximation of Vectors. *IEEE Trans. Signal Process.* **2014**, *62*, 3833–3848. [CrossRef]
9. Rimar, M.; Kulikov, A.; Fedak, M.; Yeromin, O.; Sukhyy, K.; Gupalo, O.; Belyanovskaya, E.; Berta, R.; Smajda, M.; Ratnayake, M.R. Mathematical Model of a Heating Furnace Implemented with Volumetric Fuel Combustion. *Processes* **2020**, *8*, 469. [CrossRef]
10. Gubinskiy, V.I.; Timoshpolsky, V.I.; Olshansky, V.M. *Metallurgical Furnaces. Theory and Calculations: A Textbook in 2 Volumes*; Nauka: Minsk, Belorus, 2007; p. 832.
11. Revun, M.P.; Barishenko, E.N.; Cheprasov, A.I. New schemes for pulsed heating of heating and thermal furnaces. *Metal. Hornorudnaya Promyshlenost* **2005**, *3*, 97–100.

12. Ivakhnenko, A.G.; Iurachkovskii, I.P. *Modelirovanie Slozhnykh Sistem po Eksperimentadnym Dannym*; Radio i Sviaz: Moscow, Russia, 1987; p. 118.
13. Tebenkov, B.P. *Heat Exchangers for Industrial Furnaces*; Metalurgiya: Moscow, Russia, 1975; p. 295.
14. Rich, L.V. Regenerative burners in reheat furnaces. *Iron Steel Eng.* **1989**, *66*, 46–52.
15. Drehkammerofen mit Regenerativbrenner. *Stahl Und Eisen* **2006**, *126*, 19.
16. Newby, J.N. High-performance heat recovery with regenerative burners. *Iron Steel Eng.* **1987**, *5*, 20–24.
17. Chen, H.L. Current state and prospects for the development of high-performance regenerative furnaces (HRF) in China. In *Metalurgicheskaya Teplotekhnika: Collection of Scientific Papers*; GMetAU (Energy. Metallurgy): Dnipropetrovsk, Ukraine, 1999; pp. 195–200.
18. Bruk, V.D.; Neimark, L.A.; Brun, B.P. Study of a thermal furnace equipped with built-in fans. *Pechi Sushila Mashinostroitelnoi Promzshlennosti* **1977**, *42*, 44–51.
19. Pilipenko, R.A. Chamber thermal furnaces for precision metal heating. In Proceedings of the International Symposium "OTTOM-2": Sat. Report.—NNCKhFTI, Kharkiv, Ukraine, 3–5 October 2001.
20. Chen, X.; Dai1, H.; Wei1, W. Successive mth approximation method for the nonlinear eigenvalue problem. *Comp. Appl. Math.* **2017**, *36*, 1009–1021. [CrossRef]
21. Ghosh, P.; Ghosh, S.; Bera, N. Method of successive approximation and its application in chemistry. *Indian J. Phys.* **2021**, *95*, 1407–1422. [CrossRef]

applied sciences

MDPI

Communication

Study of Anisotropic Friction in Gears of Mechatronic Systems

Zuzana Ságová [1], Valerii Vasilevich Tarasov [2], Ivana Klačková [1,*], Alexander Ivanovich Korshunov [2] and Milan Sága [3]

1 Department of Automation and Production Systems, Faculty of Mechanical Engineering, University of Žilina, Univerzitná 8215/1, 010 26 Žilina, Slovakia
2 Federal State Budgetary Institution of Science, Udmurt Federal Research Center of the Ural Branch of the Russian Academy of Sciences, Institute of Mechanics, Tatyany Baramzinoy Str., 34, 426067 Izhevsk, Russia
3 Department of Applied Mechanics, Faculty of Mechanical Engineering, University of Žilina, Univerzitná 8215/1, 010 26 Žilina, Slovakia
* Correspondence: ivana.klackova@fstroj.uniza.sk; Tel.: +421-41-513-2812

Abstract: The article discusses the features of anisotropic friction, which can be used to refine the calculation of the efficiency in various friction and gear drives and transmissions in mechatronic systems. Friction processes are considered that determine the level of losses in friction and gear drives, which are complex and heterogeneous in a number of parameters: the contact patch, which depends on the quality of the contacting surfaces; the direction and intensity of sliding; load distribution, etc. A more complete understanding of the features of these processes requires the use of the concept of friction anisotropy, which is well known in tribology of mechatronics systems. The anisotropy effect is caused by the difference in the characteristics of the surface microgeometry and its physical and mechanical properties in relation to the direction of the tool marks remaining on the surface after machining. In the presence of anisotropic friction, in contrast to isotropic, the body moves at a certain angle to the direction of application of the perturbing (external) force. The situation is considered in detail within the framework of the tensor model of anisotropic friction. The model and methodological approaches considered in the paper to the estimation of friction anisotropy can be used to refine the calculations of friction losses. The aim of the work is to create mechanical and analytical models of frictional anisotropy for a more complete understanding of this phenomenon in relation to various friction pairs. This article may be of interest to specialists in the field of friction gears for solving problems related to improving the accuracy of calculations and quantifying friction losses.

Keywords: friction anisotropy; gear transmissions; friction; cylindrical surfaces; mechatronic systems

Citation: Ságová, Z.; Tarasov, V.V.; Klačková, I.; Korshunov, A.I.; Sága, M. Study of Anisotropic Friction in Gears of Mechatronic Systems. *Appl. Sci.* **2022**, *12*, 11021. https://doi.org/10.3390/app122111021

Academic Editor: Alessandro Gasparetto

Received: 12 September 2022
Accepted: 28 October 2022
Published: 31 October 2022

1. Introduction

The processes of friction, which eventually define the level of losses during friction and gear transmissions used in manufacturing machines and robots [1], have complicated and non-uniform characters over the contact spot, depending on the quality of contact surfaces, on the direction and the intensity of sliding, the distribution of loading, etc. [2,3]. There are known works aiming to measure the coefficient of, e.g., the sliding friction of pairs of materials by means of a variable inclination tribometer [4].

For a deeper understanding of the peculiarities of these processes, it is useful to introduce the notion of friction anisotropy, widely known in tribology in mechatronics systems. The anisotropy effect [5] is caused by the difference in the microgeometry characteristics of the surface and its physio-mechanical properties, depending on the direction of tracks of a tool, left after mechanical processing of the surface, typical for cutting metals with a defined cutting edge, such as turning, drilling, milling, etc. [6–8]. Orthotropic friction (a special case of anisotropic friction) is usually known as the unequal frictional characteristic in two mutually perpendicular directions in the contact plane.

The aim of the work is to create a mechanical and analytical models of frictional anisotropy for a more complete understanding of this phenomenon in relation to various friction pairs.

2. The Phenomenon of Anisotropic Friction

In the common case in the presence of anisotropic friction, in contrast to isotropic friction, the body moves at a certain angle to the direction of the application of the disturbing (imposed) force. We shall consider this situation within the framework of the tensor model of the anisotropic friction [9,10].

At the orthotropic friction there are two directions on the surface of the body, when the isotropic body is moving in these directions it is possible to know the maximal and minimal values of friction coefficient, respectively. We place the axes X and Y in these directions (Figure 1) and designate the corresponding friction coefficients as K and k. Let $\vec{e}$ be the unit vector in the direction of the body motion, disposed at the angle α (the angle of motion) with the axis X. In order for the body, pressed to the surface by normal force N, to move in the direction $\vec{e}$, it is necessary to apply to the body certain active force P along the line OO_1 at the angle β (the angle of the application of the force) with the axis X [11–13]. The active force P compensates the originating friction force $F_{tr} = -P$. We shall designate $\theta = \alpha - \beta$ (the angle of deviation).

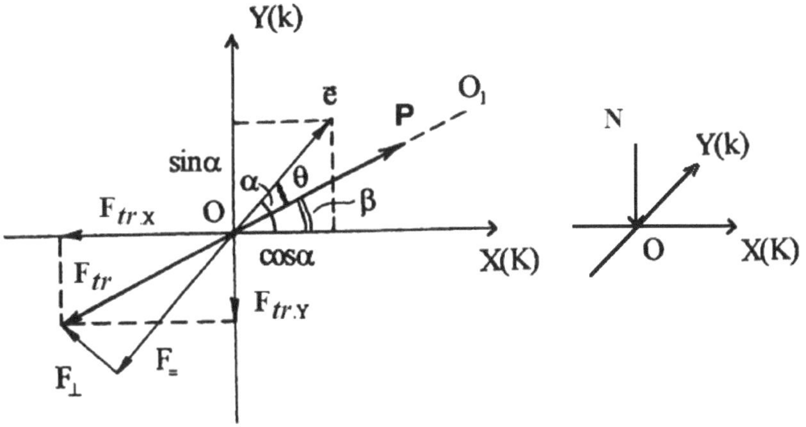

Figure 1. Scheme for the analysis of the process of anisotropic friction.

A normal force N is applied to a point O in a direction orthogonal to the drawing plane. Strength not shown.

In the accepted model of anisotropic friction for the constituents of the friction force along the axes X and Y we have:

$$\begin{cases} F_{trX} = -NK\cos\alpha \\ F_{trY} = -NK\sin\alpha \end{cases} \quad (1)$$

$$\text{Hence } tg = \frac{F_{trY}}{F_{trX}} = \frac{k}{K} tg\alpha, \quad (2)$$

and

$$F_{tr} = \sqrt{F_{trX}^2 + F_{trY}^2} = N\sqrt{K^2\cos^2\alpha + k^2\sin^2\alpha}\,. \quad (3)$$

The constituents of the friction force F_{tr}, projected in the direction of the $\vec{e}$ motion and on the direction normal to $\vec{e}$ [13–15], have the form

$$\begin{cases} F_{=} = F_{trX}\cos\alpha + F_{trY}\sin\alpha = -N(K\cos^2\alpha + k\sin^2\alpha) \\ F_{\perp} = -F_{trX}\sin\alpha + F_{trY}\cos\alpha = N(K-k)\cos\alpha\sin\alpha \end{cases} \quad (4)$$

For the angle of deviation θ we obtain

$$tg\theta = -\frac{F_{\perp}}{F_{=}} = \frac{(K-k)tg\alpha}{K + k\, tg^2\alpha}, \quad (5)$$

in this case, the angle θ lies within the limits from $-\theta_{max}$ to θ_{max}, where

$$tg\theta_{max} = 0.5\left|\sqrt{\frac{K}{k}} - \sqrt{\frac{k}{K}}\right|, 0 \le \theta_{max} < \frac{\pi}{2}. \quad (6)$$

In the first quadrant (at $0 < \alpha < \pi/2$) for the angle of deviation θ, we have $0 < \theta \le \theta_{max}$ if $K > k$ and $-\theta_{max} < \theta \le 0$ if $K < k$. In both cases the extreme value of the angle of deviation (θ_{max} or $-\theta_{max}$) is obtained at $tg\alpha = \sqrt{K/k}$. If $K = k$, then $\theta = 0$ for all the angles of motion α. Here at $k/K \Rightarrow 1$, friction degenerates into the isotropic one and at $k/K \Rightarrow 0$ it tends towards hypothetically absolute anisotropy.

In Figure 2, a mechanical simulation model is schematically presented for visual study of the peculiarities of friction and motion of the couple of bodies "anisotropic-isotropic surface". On the carrying plate of the model the supports of anisotropic friction are mounted as rolling bearings, having different resistance to the motion along to orthogonal directions [16–19]. Each support is given the angle of discrepancy of orientations among the supports φ_i (where $i = 1 \ldots n$; n = number of supports) with the direction of motion, it allows modeling different classes of indicatrix at unequal φ_i. The orientation of the model is rigidly fixed and its torque about its axis is compensated (the mode of plane-parallel motion).

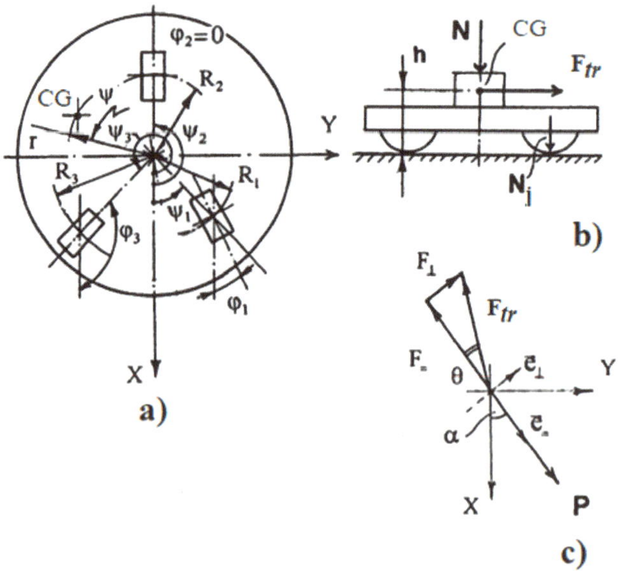

Figure 2. Calculation scheme of the mechanical model of anisotropic friction: (**a**) geometric parameters for calculation; (**b**) the scheme of application of forces; (**c**) power scheme, reduced to the center of gravity of the CG.

For the model, moving uniformly and linearly on three supports, the set of equilibrium equations has the following form:

$$\begin{cases} F_{=} = -\Sigma N_j\left[K\cos^2(\alpha-\varphi_j)+k\sin^2(\alpha-\varphi_j)\right] \\ F_{\perp} = \Sigma N_j(K-k)\sin(\alpha-\varphi_j)\cos(\alpha-\varphi_j) \\ N = \Sigma N_j \\ -F_{\perp}h + Nr\sin(\Psi-\alpha) = \Sigma N_j R_j \sin(\Psi_j-\alpha) \\ -F_{=}h + Nr\cos(\Psi-\alpha) = \Sigma N_j R_j \cos(\Psi_j-\alpha) \end{cases}, \quad (7)$$

where N, N_j is general normal force and constituent normal force, respectively, per the j, which is the support; K, k are friction coefficients on a support in the orthogonal directions ($k << K; R_j, r$, which is a radius of disposition of the j, i.e., the support and the center of gravity of the model; Ψ_j, Ψ is the angle of disposition of the j, which is the support and the center of gravity of the model; φ_j is the angle of discrepancy of orientation of the j, which is the support with the direction of motion; and h is the height of motive force application [20,21].

The set of Equation (7) was realized through the programming language BASIC, and the integration Gauss method was used. The main graphic dependences of change of force and angular parameters of the model are represented in Figure 3 [22,23].

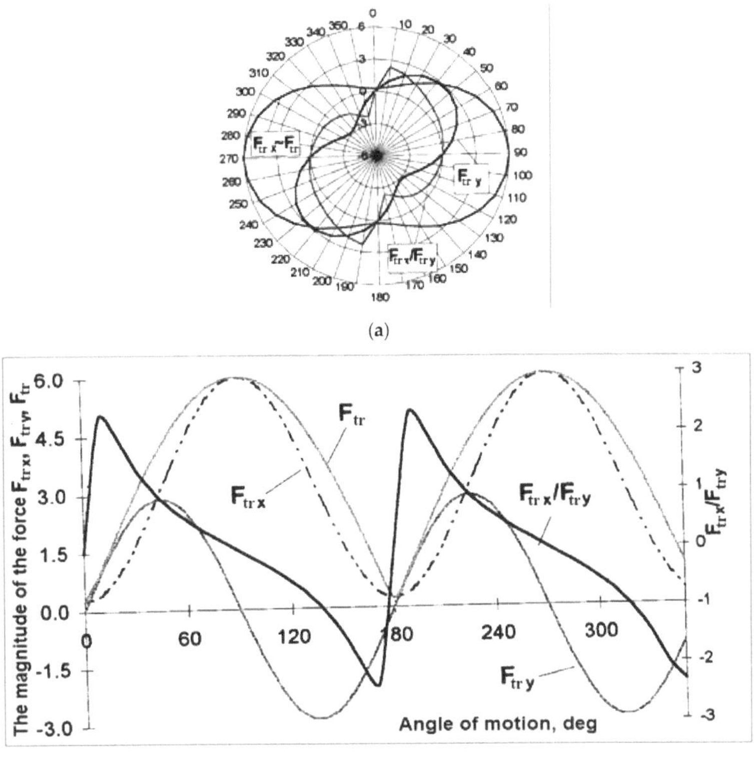

Figure 3. Change in the power characteristics of the mechanical model depending on the angle of motion in polar (**a**) and decretal (**b**) coordinates. Values F_{trX} and F_{trY} are projections of the friction force on the coordinate axes (Figure 1).

The changes of extreme values of the angular parameters of conjunction, depending on the factor k/K, are defined from the nomogram (Figure 4).

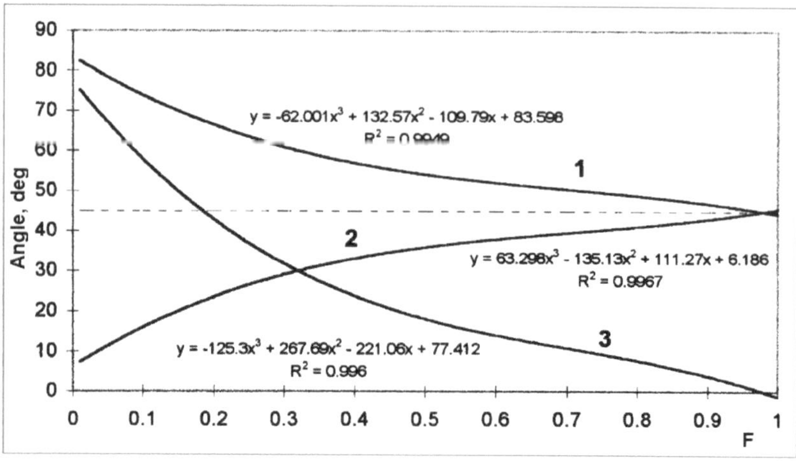

Figure 4. Nomogram for determining the maximum values of the angular parameters of the mechanical model from the friction anisotropy factor k/K.

In Figure 4, 1 is the angle of application of the driving force; 2 is angle of movement; and 3 is the slip angle.

For real pairs of friction $(k/K)_{max} > 0.6$, and technologically, it is difficult to achieve its further decrease [24–26]. The appearance of the materials with controlled friction characteristics allows the better use of the properties of anisotropic friction in various tribo-conjuctions. For friction transmissions (variators), which are kinematically similar to the simulation model (Figure 2), the anisotropy factor reaches the level of $(K_x/K_y)_{min} \approx 0.04$, and its influence becomes rather considerable; therefore, the revealed regularities can be used for designing the mechanisms with such transmissions and for computing their efficiency.

The comparison of the model representations according to Equation (7) with the experimental ones was carried out with the help of the mechanical model and special samples [27].

The experiments, conducted with the use of the mechanical model, have confirmed both the qualitative and quantitative correspondence of experimental and calculated results for the deviation angle θ and the force of resistance to the motion depending on the angle of motion α. The actual values of friction coefficients have been determined [28]. Thus, when the steel plate was used for the experiment, the following limiting coefficients were determined: $K_{max} = 0.123$ and $K_{min} = 0.05$ and $K_{min}/K_{max} = 0.041$, respectively.

The experiments with the use of special samples were carried out for the pairs "isotropic–anisotropic surface" without lubricant.

For the pair "steel 45–brass B62" under the loading Fn = 1.7 N, σ_K = 2.4 kPa, and the velocity V = 45–50 mm/min, the experimentally found friction coefficients were: $K_{max} = 0.25$ and $K_{min}/K_{max} = 0.816$. The deviation angle at different angles of the application of force changed within the range $\pm 3^{\circ}$. In the preliminary experiments with the pairs "anisotropic–anisotropic surface", in some cases the deviation angle was up to 20°.

The comparison of experimental and calculated data shows qualitative coincidence of the model representations with the real ones in both the deviation angle (Figure 5) and the character of change of the motive force depending on the angle of motion. The exact approximating functions are the following:

- on the deviation angle $\theta = -0.25 + 2.60\sin(2\alpha_{rad})$;
- on the motive force $F = 60.99 + 13.18\sin^2(\alpha_{rad})$.

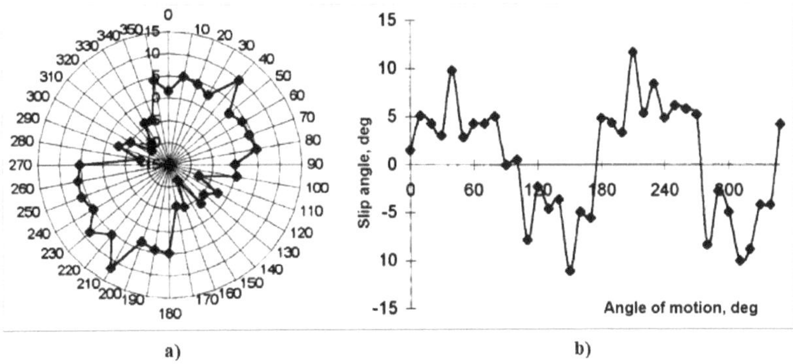

Figure 5. Change in the value of the slip angle from the angle of motion for full-scale samples in polar (**a**) and Cartesian (**b**) coordinates.

The quantitative coincidence of the model representations together with the real representations regarding the change of the components F_x and F_y of the friction force, depending on motion angle, has also been found. Experimental dependence is approximated by the curve of $F_x = 0.28 + 0.05\sin^2(\alpha_{rad})$.

It has been shown that the increase of anisotropy factor leads to the increase of the deviation angle, which deviated in the range ±12° in the last case.

3. Estimation of Anisotropy of Friction Properties on Flat and Cylindrical Surfaces

For the estimation of anisotropy of the friction coefficient on the contact surfaces of the elements of friction and gear transmissions, the following technique can be used [28]: on the control surface, the indenter is moved along the circular trajectory and the components of the total vector of resistance forces are measured, these measurements are used for building up indicatrix. In tribometry, for the pairs "anisotropic–anisotropic body", which include the majority of elements of friction and gear transmissions, the motion of the indenter over the surface should be arranged in a way that the direction of anisotropy would never change its position when the moving along the given trajectory on the control body.

For the pair of samples from steel 45 with different roughness and with use of the indenter 2H, the friction coefficients $K_{max} = 0.162$ and $K_{min} = 0.134$ were obtained. The components of the friction force, acting upon the indenter, were determined in two perpendicular directions with the help of tensometric yoke. In the experiments the indenter was moved, while the mutual position of the selected axes of anisotropy was kept constant in a special device [29,30].

The accuracy of measurements in this technique would be better if the radius of the indenter itself were as small as possible and the radius of rotation of the indenter over the control surface were as large as possible. This technique can be used for flat surfaces or the surfaces with a small curvature; in the latter case, the indenter should be spring-loaded. In this way, it is possible to estimate the anisotropy of friction properties over the working surface, for instance, over the surface of teeth of gears or the non-cylindrical elements of friction variators [19–21].

The estimation of the anisotropy of friction on cylindrical surfaces should be considered more thoroughly.

Let the cylindrical body 1 with the anisotropy of frictional surface properties move along the axis **A–A** with the velocity V_c (Figure 6). The isotropic strip 2 is pressed to the surface of this body with force P in the area of the element **a–a** at an angle ξ; the isotropic strip only has the possibility of moving in the direction of the axis **B–B**. In this case, as the result of the contact interaction, the strip moves along the axis **B–B** with the velocity V_S.

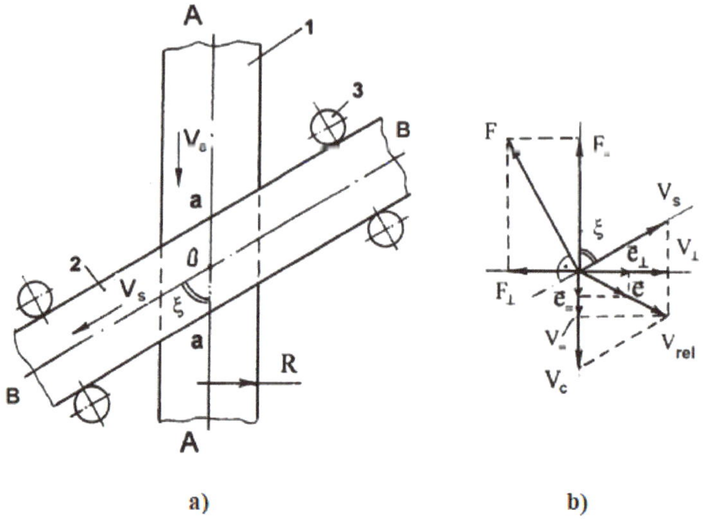

Figure 6. Calculation scheme of device for estimation of friction anisotropy on cylindrical surface (**a**) The design scheme of the device (Figure 7) and (**b**) the scheme of the application of forces.

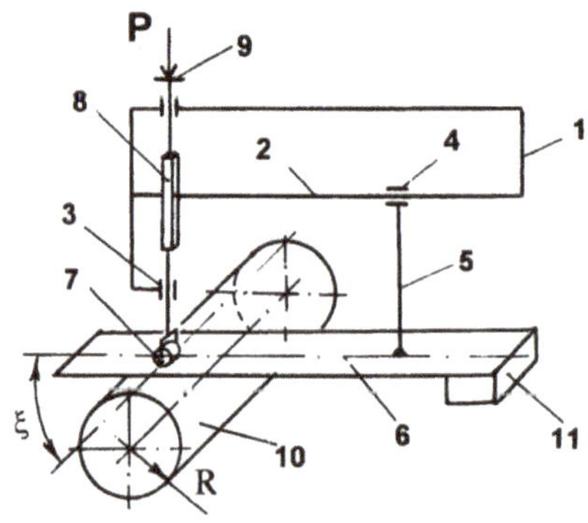

Figure 7. Kinematic scheme of device for the estimation of friction anisotropy on the cylindrical surface.

Following is the equilibrium equation for this case on the plane of tangency for the point **O**:

$$F_= \cos\xi - F_\perp \sin\xi = 0, \quad (8)$$

The values of $F_=$ and $F_\perp$ can be presented in the form

$$\begin{cases} F_= kNe_= \\ F_\perp = kNe_\perp \end{cases}', \quad (9)$$

where $e_=$ and $e_\perp$ are the projections of the unit vector e on the direction along the axis of the cylinder and perpendicular to it. They are determined as

$$\begin{cases} e_= = V_= / \sqrt{V_=^2 + V_\perp^2} \\ e_\perp = V_\perp / \sqrt{V_=^2 + V_\perp^2} \end{cases}', \quad (10)$$

Here, the velocities $V_{=}$ and $V_{\perp}$ are the constituents of the relative velocity V_{rel} of the motion of the cylinder and the strip [31], which can be found from the relationship:

$$\begin{cases} V_{=} = V_c - V_s \cos \xi \\ V_{\perp} = -V_s \sin \xi \end{cases}, \quad (11)$$

Substituting (8) into (9), taking into account (10) and (11), after transformation we obtain:

$$\frac{K}{k} = [V_C/V_S \sin \xi, \mathrm{ctg}\xi]\mathrm{ctg}\xi, \quad (12)$$

Thus, there is the possibility of determining the anisotropy factor experimentally with the help of the measured velocities V_C and V_S.

With the rotation of the cylindrical body with the velocity ωR (R-cylinder radius), the component $V_{\perp}$ in the expression (13) should be calculated according to the formula

$$V_{\perp} = -V_S \sin \xi, \pm \omega R, \quad (13)$$

In expression (13), the plus (+) sign is used when there is a coincidence in directions of the vector of rotation velocity and the vector of the velocity of the motion of the strip [31], otherwise the minus (−) sign is taken. In this case, Formula (12) assumes the following form:

$$\frac{K}{k} = [(V_C/ - V_S \sin \xi, \pm \omega R) - \mathrm{ctg}\xi]\mathrm{ctg}\xi, \quad (14)$$

In Figure 7, the device for the realization of the offered technique is shown.

The device in Figure 1 consists of a frame (1) with the axis (2) and guider (3). Regarding the axis (2) on the bearing block (4) through suspension (5), the standard strip (6) with isotropic roughness of the surface is attached; this strip is pressed with force P by the pressure roller (7) through the fork (8) to the control cylindrical sample (10) from the load center (9). To provide stability of motion, the iron rule (6) is fitted with a balance beam (11).

4. Conclusions

A simulation of a mechanical model has been developed in the form of a platform with oriented-controlled cylindrical rolling bearings, which makes it possible to simulate the effect of anisotropic friction when moving in different directions.

Based on the model and its mathematical representation, the relationships of the main characteristic parameters of tribosystems with anisotropic friction are obtained. Modeling the anisotropy of varying complexity (symmetries) is provided through the mutual misorientation of the supports.

The considered features of anisotropic friction can be used to improve the efficiency of the calculations of various friction and gear drives in mechatronic systems.

The proposed model and methodological approaches to estimating friction anisotropy can be used to improve the calculation of friction losses and develop control paths [31].

The authors suggest the continuation of experimental research in relation to gears in mechatronic systems.

The considered features of anisotropic friction, the proposed model, and methodological approaches to estimating the anisotropy of friction can be used for more accurate and reasonable calculations of the efficiency in various friction and gear drives and transmissions. This is also true for extremely loaded gears with a high coefficient of friction as presented by the authors in paper [32]. The application of the proposed method of calculation makes it possible to take into account the influence of a number of parameters that are omitted from the reasoning when conducting a significant number of calculations. The use of the tensor model of anisotropic friction has made it possible to refine a number of calculation formulas and provide a more accurate estimate.

Author Contributions: Conceptualization, V.V.T. and A.I.K.; methodology, Z.S. and I.K.; software, M.S.; validation, Z.S., I.K. and M.S.; formal analysis, V.V.T. and A.I.K.; investigation, M.S.; resources, I.K.; data curation, Z.S.; writing—original draft preparation, Z.S.; writing—review and editing, I.K.; visualization, I.K.; supervision, M.S.; project administration, I.K.; funding acquisition, I.K. All authors have read and agreed to the published version of the manuscript.

Funding: This work was supported through the project VEGA 1/0073/19. Project title: Computer modeling and development of algorithms for evaluating the reliability of composite structural elements produced by the additive technology based on Onyx for the purpose of optimal design and under the project KEGA 001ZU 1/2020. Project title: Implementing of additive technologies into the education process of engineers' study programs.

Conflicts of Interest: The authors declare no conflict of interest. The funders had no role in the design of the study; in the collection, analyses, or interpretation of data; in the writing of the manuscript, or in the decision to publish the results.

References

1. Peterka, J.; Nikitin, J.R.; Bozek, P. Diagnostics of automated technological devices. *MM Sci. J.* **2020**, *2020*, 4027–4034. [CrossRef]
2. Bragin, V.V.; Reshetov, D.N. *Designing of Cylindrical Gear Transmissions with High Tension*; Mashinostroenie: Moscow, Russia, 1991; p. 224.
3. Hebda, M.; Chichinadze, A.V. (Eds.) Techniques and means of tribotechnological testing. In *Tribotechnology Reference Book: V.3: Tribotechnology of Antifriction, Friction and Transmission Devices*; Mashinostroenie: Moscow, Russia, 1992; p. 730.
4. Kelemenova, T.; Dovica, M.; Bozek, P.; Kolarikova, I.; Benedik, O.; Virgala, I.; Prada, E.; Mikova, L.; Kot, T.; Kelemen, M. Specific Problems in Measurement of Coefficient of Friction Using Variable Incidence Tribometer. *Symmetry* **2020**, *12*, 1235. [CrossRef]
5. Opeyko, F.A. Experimental study of anisotropic friction. In *Collection of Scientific and Technical Works of MIMESH*; Soviet Science: Moscow, Russia, 1952; pp. 235–243.
6. Vopat, T.; Peterka, J.; Kovac, M.; Buransky, I. The Wear Measurement Process of Ball Nose End Mill in the Copy Milling Operations. In Proceedings of the 24th DAAAM International Symposium on Intelligent Manufacturing and Automation, Zadar, Croatia, 23–26 October 2013; Volume 69, pp. 1038–1047. [CrossRef]
7. Frankovský, P.; Delyová, I.; Trebuňová, M.; Čarák, P.; Kicko, M.; Kurylo, P. Motion Analysis of the Hydraulic Ladder. *Int. J. Appl. Mech. Eng.* **2019**, *24*, 230–240. [CrossRef]
8. Peterka, J.; Pokorny, P. Influence of the Lead Angle from the Vertical Axis Milling on Effective Radius of the Cutter. In Proceedings of the 7th International Congress of Precision Machining (ICPM 2013), Miskolc, Hungary, 3–5 October 2013; Volume 581, pp. 44–49.
9. Vopat, T.; Peterka, J.; Simna, V.; Kuruc, M. The Influence of Different Types of Copy Milling on the Surface Roughness and Tool Life of End Mills. In Proceedings of the 25th DAAAM International Symposium on Intelligent Manufacturing and Automation, Vienna, Austria, 26–29 November 2014.
10. Alexandrovich, A.I.; Vekshin, B.S.; Potapov, I.N. Tensor of friction coefficients of anisotropic surfaces. *Trenie Iznos* **1985**, *6*, 996–1004.
11. Tarasov, V.V. *Friction Anisotropy*; IM UB RAS: Izhevsk, Russia, 1998; p. 54.
12. Frankovsky, P.; Ostertag, O.; Ostertagova, E.; Trebuna, F.; Kostka, J.; Vyrostek, M. Experimental analysis of stress fields of rotating structural elements by means of reflection photoelasticity. *Appl. Opt.* **2017**, *56*, 3064–3070. [CrossRef]
13. Sharapin, Y.F. Elements of the theory of metal processing by pressure. *Metallurgizdat* **1961**, 208. (In Russian)
14. Kuric, I.; Klačková, I.; Domnina, K.; Stenchlák, V.; Sága, M., Jr. Implementation of Predictive Models in Industrial Machines with Proposed Automatic Adaptation Algorithm. *Appl. Sci.* **2022**, *12*, 1853. [CrossRef]
15. Tarasov, V.V. Method for Determination of Physico-Mechanical Properties of Material, Friction for the Most Part. A. s. USSR # 1388769, MKI G01N 19/02.
16. Saga, M.; Vasko, M.; Pechac, P. Chosen numerical algorithms for interval finite element analysis. *Procedia Eng.* **2014**, *96*, 400–409. [CrossRef]
17. Burduk, A.; Wiecek, D.; Tlach, V.; Ságová, Z.; Kochanska, J. Risk assessment of horizontal transport system in a copper mine. *Acta Montan. Slovaca* **2021**, *26*, 303–314. [CrossRef]
18. Kuric, I.; Tlach, V.; Císar, M.; Ságová, Z.; Zajačko, I. Examination of industrial robot performance parameters utilizing machine tool diagnostic methods. *Int. J. Adv. Robot. Syst.* **2020**, *17*, 1729881420905723. [CrossRef]
19. Sapietova, A.; Saga, M.; Novak, P.; Bednar, R.; Dizo, J. Design and Application of Multi-Software Platform for Solving of Mechanical Multi-Body System Problems. In Proceedings of the 9th International Conference on Mechatronics, Mechatronics: Recent Technological and Scientific Advances, Warsaw, Poland, 21–24 September 2011; pp. 345–354.
20. Maláková, S.; Puškár, M.; Frankovský, P.; Sivák, S.; Palko, M.; Palko, M. Meshing Stiffness—A Parameter Affecting the Emission of Gearboxes. *Appl. Sci.* **2020**, *10*, 8678. [CrossRef]
21. Muravev, V.V.; Muraveva, O.; Volkova, L.; Sága, M.; Ságová, Z. Measurement of Residual Stresses of Locomotive Wheel Treads During the Manufacturing Technological Cycle. *Manag. Syst. Prod. Eng.* **2019**, *27*, 236–241. [CrossRef]
22. Klarák, J.; Andok, R.; Hricko, J.; Klačková, I.; Tsai, H.-Y. Design of the Automated Calibration Process for an Experimental Laser Inspection Stand. *Sensors* **2022**, *22*, 5306. [CrossRef]

23. Segota, S.B.; Andelic, N.; Lorencin, I.; Saga, M.; Car, Z. Path planning optimization of six-degree-of-freedom robotic manipulators using evolutionary algorithms. *Int. J. Adv. Robot. Syst.* **2020**, *17*, 1729881420908076. [CrossRef]
24. Sága, M.; Blatnická, M.; Blatnický, M.; Dižo, J.; Gerlici, J. Research of the Fatigue Life of Welded Joints of High Strength Steel S960 QL Created Using Laser and Electron Beams. *Materials* **2020**, *13*, 2539. [CrossRef]
25. Puškár, M.; Jahnátek, A.; Kuric, I.; Kádárová, J.; Kopas, M.; Šoltésová, M. Complex analysis focused on influence of biodiesel and its mixture on regulated and unregulated emissions of motor vehicles with the aim to protect air quality and environment. *Air Qual. Atmos. Health* **2019**, *12*, 855–864. [CrossRef]
26. Kuric, I.; Klačková, I.; Nikitin, Y.R.; Zajačko, I.; Císar, M.; Tucki, K. Analysis of diagnostic methods and energy of production systems drives. *Processes* **2021**, *9*, 843. [CrossRef]
27. Ali, L.M.; Ali, Q.A.; Klačková, I.; Issa, H.A.; Yakimovich, B.A.; Kuvshimov, V.V. Developing a thermal design for steam power plants by using concentrating solar power technologies for a clean environment. *Acta Montan. Slovaca* **2021**, *26*, 773–783. [CrossRef]
28. Fiegel, A.; Klačková, I. Safety requirements for mining complexes controlled in automatic mode. *Acta Montan. Slovaca* **2020**, *25*, 417–426. [CrossRef]
29. Maláková, S.; Puškár, M.; Frankovský, P.; Sivák, S.; Harachová, D. Influence of the Shape of Gear Wheel Bodies in Marine Engines on the Gearing Deformation and Meshing Stiffness. *J. Mar. Sci. Eng.* **2021**, *9*, 1060. [CrossRef]
30. Handrik, M.; Kopas, P.; Baniari, V.; Vasko, M.; Saga, M. Analysis of stress and strain of fatigue specimens localised in the cross-sectional area of the gauge section testing on bi-axial fatigue machine loaded in the high-cycle fatigue region. In Proceedings of the XXI Polish-Slovak Scientific Conference Machine Modeling and Simulations MMS 2016, Hucisko, Poland, 6–8 September 2016; Volume 177, pp. 516–519. [CrossRef]
31. Abramov, I.V.; Abramov, A.I.; Nikitin, Y.; Sosnovich, E.; Božek, P.; Stollmann, V. Diagnostics of electrical drives. In Proceedings of the 18th International Conference on Electrical Drives and Power Electronics, Kosice, Slovakia, 21–23 September 2015. [CrossRef]
32. Virgala, I.; Kelemen, M.; Bozek, P.; Borovský, Z.; Hagara, M.; Prada, E.; Oscadal, P.; Varga, M. Investigation of snake robot locomotion possibilities in a pipe. *Symmetry* **2020**, *12*, 939. [CrossRef]

applied sciences

MDPI

Article

The Difficulty of Measuring the Roughness of Glossy Surfaces Using the Triangulation Principle

Juraj Ružbarský

Faculty of Manufacturing Technologies, Technical University of Košice, Štúrova 31, 080 01 Prešov, Slovakia; juraj.ruzbarsky@tuke.sk

Abstract: In the experiment, the roughness was measured on a machined surface with high gloss, which was also the main requirement for the test samples. For this reason, the samples made of stainless steel A304 and aluminum alloy AW 2017 were created by a progressive laser using material-cutting technology. This article explains a contact-free measurement method that uses the triangulation principle, which constitutes the basis on which the device used in the experiment, i.e., the laser profilometry, works. The surface roughness of the cut surfaces was examined on the manufactured samples through the selected roughness parameters of Ra and Rz. These parameters are commonly used in industry to quantify the roughness of a surface. The values measured in a contact-free manner were then compared with the reference values measured in a contact manner. Data from individual experimental measurements were graphed as dependencies based on which problem areas of measuring the roughness of glossy material surfaces with laser profilometry were described. Laser profilometry is a non-contact method for measuring the roughness of surfaces, and given the presented results of the experimental measurements and selected roughness parameters of the cut surface using a laser, we do not recommend using it for materials that have a glossy surface.

Keywords: profilometry; laser; glossy surface; roughness; stainless steel; aluminum alloy

Citation: Ružbarský, J. The Difficulty of Measuring the Roughness of Glossy Surfaces Using the Triangulation Principle. *Appl. Sci.* **2023**, *13*, 5155. https://doi.org/10.3390/app13085155

Academic Editor: Cem Selcuk

Received: 14 March 2023
Revised: 6 April 2023
Accepted: 18 April 2023
Published: 20 April 2023

1. Introduction

Increasing the service life and functional reliability of components depends on the high quality of their functional surfaces, i.e., dimensional and shape accuracy and surface roughness. With the increasing use of progressive construction materials (generally with impaired machinability), the development of cutting materials and machine tools requires, among other things, the development of measurement technology and evaluation methods for the quality of component surfaces [1].

Roughness parameters are only one of many factors affecting machining quality, which has a direct impact on productivity. However, higher roughness values are undesirable in terms of mechanical vibrations that can cause noise and dynamic loads. Subsequently, these vibrations can cause fatigue, malfunctions of the machine structure or its functional parts, energy losses, and reduced performance. Therefore, it is important to control and minimize roughness to maintain machine integrity and achieve optimal performance. By eliminating unwanted roughness, the machine operator can minimize the risk of these negative effects and increase the overall efficiency and effectiveness of the machining process [2,3].

The measurement and analysis of surface roughness in the machining process have several important aspects. Undesirable surface roughness can signal damage to functional parts of the machine tool, wear of the machining tool, workpiece, cutting head, or other machine parts. These processes can cause damage to the machine structure, its rigidity, bearing components, or other parts used in the machining process. Therefore, it is important to understand and control the surface roughness in the machining process to maintain the efficiency and durability of the machine tool and the quality of the machined parts [4,5].

Through the evaluation of roughness parameters, it becomes possible to predict the necessary sequence of operations to achieve the desired surface quality and to optimize the process of surface creation [6].

The value of surface roughness is often a critical factor for products that come into contact with each other [7]. It also has a significant impact on the durability, reliability, and operation of technical equipment [8,9].

The emerging surface roughness is not only a carrier of partial information but above all an image of the technology used to create it [10].

The clear trend of automating production processes also affects the automation of surface structure control. Nowadays, a common solution is to compare the measured surface with the sample surface and then evaluate it for technological modifications that should ensure the desired surface of the product. The ideal solution in fully automated production is to find out the resulting deviation of the surface from the defined standard and then adapt the process by changing the machining corrections to achieve the required surface structure quality [5].

When evaluating surface quality, not only roughness parameters are taken into account, but also dimensional uncertainty, geometric shape and position, and surface hardness of components. These surface characteristics are studied in the field of topography, which deals with the description of the surface of the studied object [11].

The 3D measurement and evaluation of surface topography (Figure 1) are new possibilities for how to express and present the surface structure. Due to the additional dimension and increasing amount of data, there is a possibility of more objective surface evaluation, possible prediction of functional properties of the surface, and their gradual modification during operation. Subsequent spatial analyses of the measured surface structure provide a graphical representation of the profile in various views in the form of a topographic map or a record of the coordinate's intensity. This form of measurement yields a large amount of information that an experienced observer can use [12].

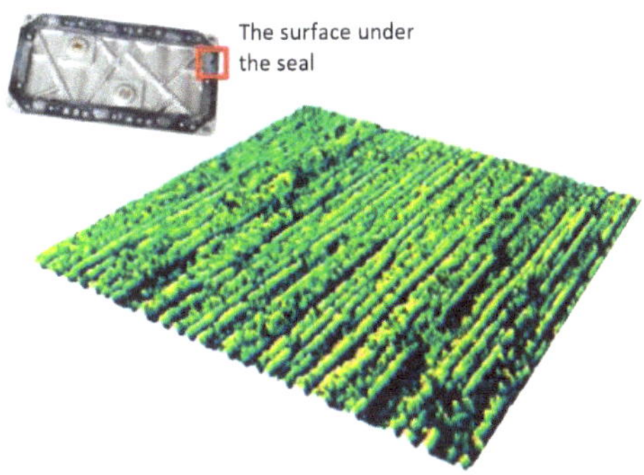

Figure 1. 3D structure of the gearbox cover.

3D surface topography uses spatial characteristics of the surface, which provide a large amount of information for a truly realistic representation of the surface under examination. Surface characteristics, e.g., the count, size, shape, and distribution of protrusions and profile recesses, not only increase the objectivity of surface evaluation but can also be used to predict the properties of functionally loaded surfaces and their changes during operation and solve several problems in tribology and production technology. Expanding the possibilities of the quantitative evaluation of geometric and dimensional changes in the surface profile at various stages of the components' production processes or upon their completion and even during their function is one of the main advantages of the spatial evaluation of the surface structure. Thus, there is a clear shift from the original contact measurement to a contact-free measurement of a range of profiles, providing fast, touch-free, and accurate measurement of the surface topography of common machine components, where the advantage is to display the surface roughness over the component's entire functional area [13].

The measurement and processing of a greater amount of data in spatial surface evaluation are made possible by a new generation of measuring devices that simultaneously stimulate research in many technical and other areas. The real image of the surface obtained contains a large amount of information that can be interpreted by an experienced observer. A quantified description of the surface is provided by precisely defined parameters. It should be remembered that the parameters cannot describe the surface comprehensively. Each parameter brings information only about certain properties of the surface structure and requires correct interpretation [14].

When comparing the spatial measurement of the surface texture with the evaluation of a single cut (profile) of the surface, it can be concluded that 3D surface texture measurement is more objective and gives more accurate information about the state of the surface with a greater statistical significance of the characteristic evaluated. The evaluated spatial parameters of the texture are based on a significantly larger amount of measured data and have a much higher reliability. Thus, the obtained value of the parameter is much more reliable. Using the spatial evaluation of the surface structure also significantly reduces the dangerous neglect of the influence of one of the essential properties of the surface [15].

Most of the spatial surface parameters are, by analogy, dependent on the surface roughness parameters in 2D. The classic 2D roughness measurement is still prevalent due to the standardized and unambiguous surface roughness parameters and because it constitutes the foundation of the entire surface metrology. The parameters such as the spatial density of the protrusion are often estimated from 2D measurements. The resulting value is usually 20% higher than in reality, but it is enough to convey an idea of a functional area. Scanning the surface for spatial evaluation is mostly carried out on parallel profiles [16].

Although several instruments for measuring and evaluating the spatial characteristics of the surface are currently available, it is clear that the practical transition from evaluating a single profile to evaluating a surface is a long-term process. The use of a qualitatively new method of surface evaluation will require solving or specifying the number of problems associated not only with the actual measurement of the surface, but also with the processing of the results and, above all, with the effective use of the results obtained. The success of the enforcement of spatial surface quality control in metrological practice will undoubtedly be determined not only by technical but also by economic aspects [17].

The advantages of the spatial assessment of the surface structure show that these are the metrological methods of the future. Their wider practical use will undoubtedly be supported not only by increasing demands on the accuracy and quality of the existing production technologies but also by the development of new materials and technological methods, such as nanotechnology, etc. [18].

The use of spatial surface inspection offers a demonstrably significant improvement in the properties of functional surfaces, as well as progress in production processes, e.g., energy savings, etc. Therefore, it is in the interest of the technical development of production and inspection technologies to create all the prerequisites for the comprehensive use of spatial surface evaluation in regular production and metrological practice [19].

Experimental setups for optical topography methods have a similar basic structure, which consists of an optical source (for easier setup and calibration of the experimental setup, a light source is often used, that is, an optical source shining at a wavelength of 380–760 nm), an optical detector and a measured object. The optical source generates a light beam that is projected onto the surface of the measured object through the so-called optical probe, which can be point, linear, area, or intense, and subsequently reflects and refracts. An optical detector, which usually consists of a camera with a CCD or CMOS sensor, captures the reflected light beam and generates an image that is subsequently processed and analyzed. Different surface analysis methods vary in the type of optical probe used and how it interacts with the surface being measured [20].

In various laboratory or industrial measurement applications, measuring and assessing objects' spatial form by the contactless measurement of their height or thickness of a 3D profile is essential. Laser profilometry is designed to carry out these tasks quickly and efficiently [21].

Contact-free measurement methods replace contact measurements of surfaces sensitive to mechanical damage, such as soft materials, etc. Contact-free measurement avoids damaging the measured surface, and the inspected surface is monitored by a focused measuring head. The program-controlled adjustment of the measuring head is quick and easy [1].

3D profilometry using a laser is one of the most recently developed approaches to measure surface roughness, while pulse thermography is used to combine the effect of surface roughness with thermal radiometry [22].

Laser profilometry uses a well-known triangulation principle (Figure 2). The principle consists of the projection of a linear laser beam onto the measured object and the detection of its deformed image on the measured surface by a camera on the computer. The light source, the measured object, and the camera form a triangle whose parameters (distance of the camera and light source from the reference plane, their angle to the reference plane) are known. In this way, each scanned image can be used to create a real 3D profile [23].

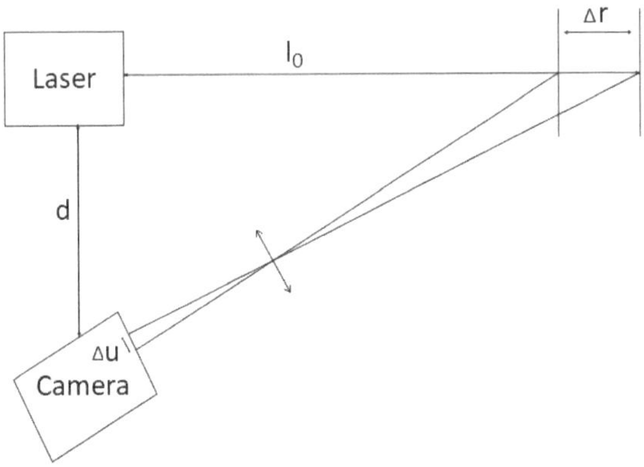

Figure 2. Scheme of the laser triangulation experimental setup.

The sought topographical deviation Δr is the distance of each point of the measured surface whose coordinates correspond to the pixels of the camera chip matrix from the reference plane. This topographic distance is determined by the relation

$$\Delta r = \frac{\Delta u}{b + a\Delta u}, \tag{1}$$

where:

Δr—sought topographic deviation,
Δu—the number of pixels on the matrix of the camera chip by which the display of the point is shifted during projection onto the reference plane and the measured point,
a, b—parameters of the mapping algorithm, which can be calculated from the known parameters of the experimental setup or obtained from the calibration of this setup.

The possibilities of profilometry do not end with the basic laser line–camera configuration. For example, a substantial increase in accuracy can be achieved in certain cases by using a projection grid instead of a line [11].

Laser profilometry is a non-contact optical method of measuring the profile of a scanned object surface, which then allows the creation of a 3D model of the object. This method utilizes a laser beam emitted by an optical source of electromagnetic radiation and a sensing device in the form of a camera, the basis of which is a CCD or CMOS chip, to capture the profile of the object's surface [5].

Just like other methods of roughness measurement, laser profilometry has its pros and cons. One of the advantages of contact-free laser profilometry is that there is no wear on the measuring tool or damage to the surface of the measured object during measurement. It is capable of detecting cracks and micro-unevenness, which is a problem when using contact roughness meters that rely on tip geometry. Laser profilometry provides better repeatability of measurements and allows for measuring even highly curved surfaces. It can also evaluate multiple parameters of roughness and undulation and measure the entire surface area of the sample rather than just one line. Problems often associated with non-contact laser profilometry include difficulties in measuring glossy surfaces, problems in measuring multi-layered surfaces, the high cost of equipment acquisition, and the need for skilled operators [24,25].

After conducting a comprehensive study of the existing literature, the objectives of the paper were established. When choosing a specific topographical method, several criteria should be considered, and these criteria depend on the nature of the experiment. Some of the most important criteria to consider include the measurement speed, size of the measurement area, sensitivity of the method, accuracy of results, and financial difficulty. Part of the experiment was to assess whether laser profilometry was a suitable method for measuring the roughness of materials with a glossy surface or not. The test samples were made from stainless steel material A304 and aluminum alloy AW 2017 by cutting the respective material using laser technology. The main requirement for the samples was that their cleavage surface showed a very glossy surface when laser cleaved, and that was the reason why this cleaving procedure was used. The experiment of measuring the selected roughness parameters was carried out with a contact-free laser profilometer, and to verify the accuracy of the measured values and compare the measured values, the Mitutoyo Surftest SJ-400 contact roughness meter was used as well. At the end of the experiment, the experimental data from individual measurements were presented in graphs and analyzed. Another objective of the study was to compare two methods of measuring roughness: contact-free laser profilometry and the Surftest SJ-400 contact roughness meter. The experiment provided practical insights into the suitability of laser profilometry for measuring the roughness of glossy material surfaces based on the analysis of the sample surfaces.

2. Materials and Methods

The conducted experiments were aimed at monitoring the suitability of the contact-free laser profilometer for measuring the roughness parameters of glossy surfaces. The tested samples were made of stainless steel and aluminum alloy. The roughness on the cut surface was measured in a contact-free and contact way. Both methods were applied to the same cut areas measured. A Mitutoyo contact roughness meter (Surftest SJ-400, Tokyo, Japan) and a new commercial system on the market, a non-contact laser profilometer (KVANT, Bratislava, Slovakia), were used to measure and evaluate the samples.

In evaluating the surface roughness, the parameter Ra, the mean arithmetic deviation of the profile, and the parameter Rz, the tallest height of the profile unevenness, were monitored. In practice, the Ra and Rz parameters are those most monitored, capable of providing sufficient information about the quality of the measured surface [26]. The measured values of the sample materials were graphed as dependencies for both the Ra and Rz quality parameters of roughness.

2.1. Sample Preparation and Description of the Materials Tested

The sample production consisted of cutting the material with laser technology, which was implemented directly in the production plant of DRC s.r.o, Prešov, Slovakia where the TruLaser 3040 cutting machine from TRUMPF was used for cutting the material (Figure 3). The cut surfaces of the samples were very glossy once the samples were made.

Figure 3. TruLaser 3040 material-cutting device from TRUMPF.

However, on glossy surfaces, there is an intense reflection of light that can cause some details on the surface not to be captured accurately. To verify the difficulty of measuring the roughness of glossy surfaces of materials with laser profilometry, samples made of stainless steel and aluminum alloy were used. These materials have different physical properties and may have different levels of glossiness and roughness. Using these samples allowed for comparing the measurement results with the expected results and identifying any issues in measuring the roughness of glossy surfaces.

Technical parameters of the cutting machine used to prepare the experimental samples are presented in Table 1.

Table 1. Technical parameters of the TruLaser 3040 cutting machine.

Parameter	Value
Laser power	3.200 W
Cutting speed	2.1 m × min^{-1}
Maximum cutting depth	12 mm
Slice-gap width	0.2 mm
Focal length	4.5 mm
The diameter of the focused beam on the nozzle	2.3 mm
The type and pressure of auxiliary gas used	Nitrogen (17 Bar)

The analyzed materials were selected based on essential requirements such as high usability, availability of materials in the industry, as well as the various properties of the studied material (high gloss), which were taken into consideration during the experiment.

Material designation:

- stainless steel A304—material thickness 5 mm, material-cutting speed 2.1 m × min^{-1}
- aluminum alloy AW 2017—material thickness 5 mm, material-cutting speed 2.1 m × min^{-1}

The test samples were made by cutting the material using laser technology (Figure 4). The reason was that the material cut with laser technology manifests a highly glossy surface of the cut area.

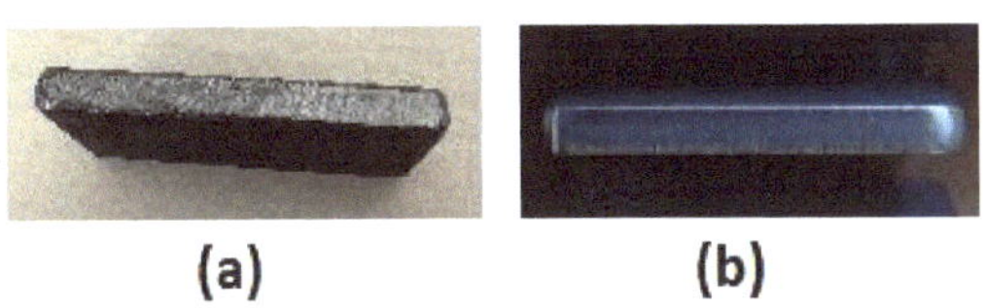

Figure 4. Experimental samples were used. (**a**) Stainless steel; (**b**) Aluminum alloy.

2.2. Measuring the Surface in a Contact-Free Way

Figure 5 shows the setup of the laser profilometer, which consists of basic and additional parts. The basic components included a sturdy aluminum support structure comprising components for the vertical positioning of the measuring tube, a worktop mounted on stepper motors for movement along the X and Y axes, a laser beam source, a lens, and a camera equipped with a CMOS (complementary metal–oxide semiconductor) sensor and backlight. In our experiment, the light source was a laser blue diode StingRay with a wavelength of 445 nm. Additional parts of the laser profilometer setup included a computer with laser profilometry View control and evaluation software, as well as an image splitter.

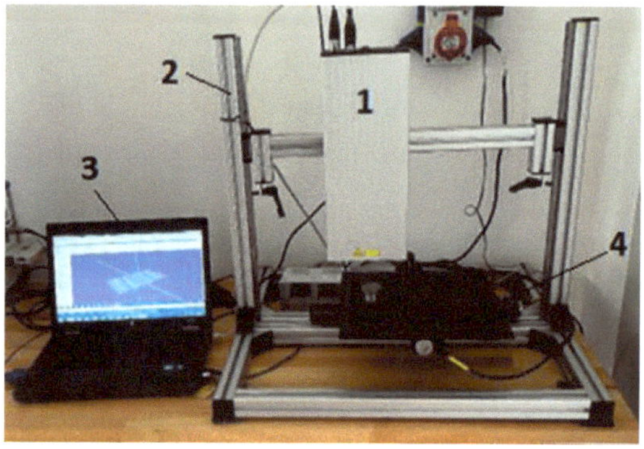

Figure 5. The assembly of the laser profilometer. (1) tube with laser and camera; (2) frame; (3) PC with operating and evaluation software; (4) worktop with step motors movable in X and Y axes.

When measured by the contact-free laser profilometry system, the experimental samples on the worktop were first placed in an anti-vibration mass (Figure 6). The first reason was that the machined surface of the sample was imperfectly fitted to the laser profilometry worktop and the sample on the worktop shifted during the movement of the stepper motors in the course of measurement. The second reason was to eliminate the influence of ambient vibrations, which can affect the quality of the measured data.

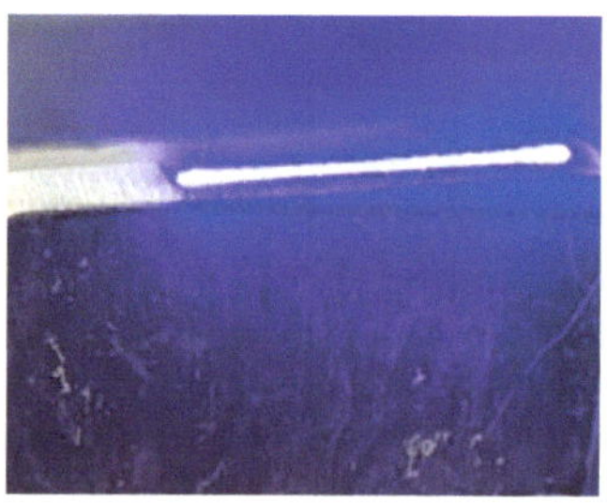

Figure 6. Measurement of the sample with the laser profilometer.

The input parameters for setting up the laser profilometry for measuring surface roughness are shown in Table 2.

Table 2. Input Parameters for Sample Measurement with the Laser Profilometer Setting.

Axis Y Scan Distance	Axis X Scan Line
Scan lines count 45 Scan lines step 110 μm	4000 μm

The gain mode (image signal amplification in the camera) was set to 1. The shutter time (time of measurement) for the laser profilometer was set to 19.520 ms after previous test measurements. This shutter time was found to produce the clearest image with the least amount of noise. By optimizing the shutter time, the laser profilometer was able to capture a clear and accurate representation of the surface of the sample. This allowed for precise measurements of roughness and other surface features to be obtained. This setup was consistent with the observations made by Srivastava et al. [27], Ružbarský [5], and Mital' et al. [26].

The experimental laser profilometry enables the measurement and evaluation of surface roughness parameters of samples by ISO 4287 standard (Rq, Rv, Rz, Ra, Rp). The measured roughness data can be exported in a CSV format, which is suitable for the further processing of experiments in the form of raw data. The surface roughness parameters evaluated using laser profilometry were Ra (arithmetic mean deviation of the surface profile), and Rz (the tallest protrusion in the profile's unevenness).

By following the recommended parameters, the laser profilometer was able to accurately measure the roughness of the sample surfaces. The parameters included in Table 2 may include the measurement range, the sampling frequency, and the scanning speed, among others, which can vary depending on the specific requirements of the measurement.

2.3. Measuring the Surface in a Contact Way

As mentioned in the introduction, to verify the correctness of the values measured with the contactless laser profilometer (Figure 7) and to compare them to those measured in a contact way, the measurement of samples was also carried out with the Mitutoyo Surftest SJ-400 roughness meter (Figure 8). Table 3 shows the individual parameters of the contact roughness meter.

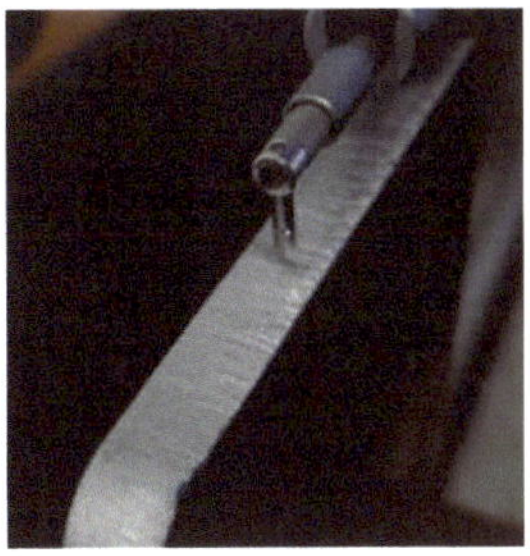

Figure 7. Measurement with the Surftest SJ-400 contact roughness meter.

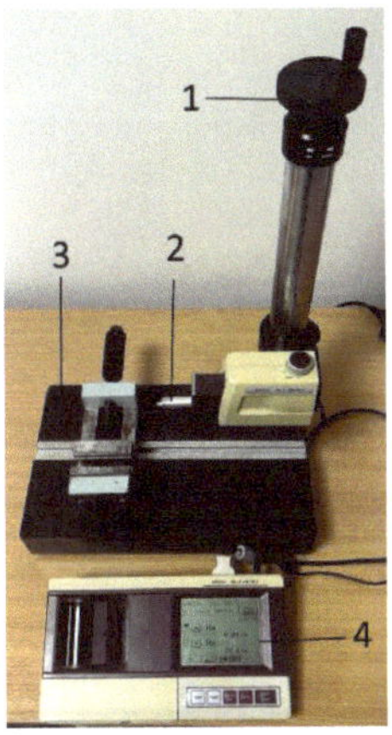

Figure 8. Surftest SJ-400 contact roughness meter. (1) Adjusting the standoff distance of the measuring tip; (2) measuring tip; (3) worktop; (4) touch display with the printer.

Table 3. Individual parameters of the Surftest SJ-400 roughness meter.

Parameters	Value
Measuring speed (mm $\times$ s^{-1})	0.05; 0.1; 0.5; 1.0
Speed of return (mm $\times$ s^{-1})	0.5; 1.0; 2.0
Positioning	±1.5° (inclination); 10 mm (up/down)
Measurement range/resolution	P (primary); R (roughness); W (filtered waviness)
Evaluated parameters	0.889
Digital filter	2CR; PC75; Gauss
Cutoff length (mm)	0.08; 0.25; 1.8; 2.5; 8

The same part of the measured surface was selected for each sample for both measurement methods. The materials of the samples were defined according to the relevant standard as stainless steel A304 and aluminum alloy EN AW 2017.

2.4. Experimental Samples Measurement

Roughness measurement of the material surfaces cut with laser technology was carried out on samples of two types of material. The samples made from the two types of materials were of the same thickness and the cutting speed when splitting both materials was 2.1 m $\times$ min^{-1}. Each material was made from a single sample.

The measurement of the roughness parameters of Ra and Rz of the cut surface of the sample when measured with the laser profilometer and the Surftest SJ-400 was determined on each sample at three levels of measuring the cut surface, i.e., at the smooth, the medium and the coarse zone. In addition, the measured surface was transversally divided into parts A, B, and C (Figure 9). Individual measurement lines (1, 2, 3) were set to 1, 2.5, and 4 mm beginning in the smooth zone of the sample. The thickness of the sample material measured in both ways was the same.

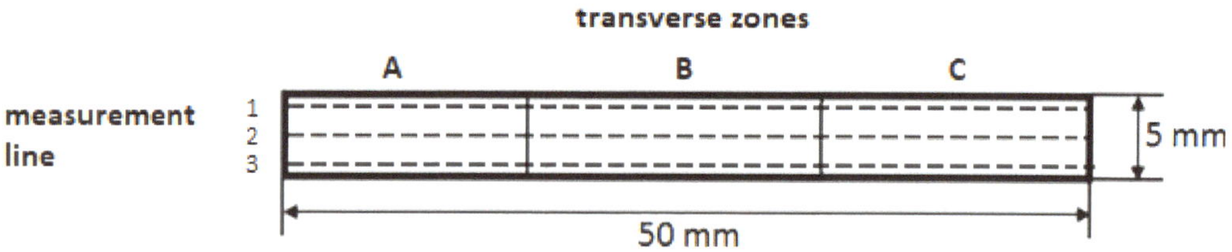

Figure 9. Samples with designated measurement lines and transverse zones.

The measurement itself was performed at several measurement lines, namely at a smooth zone of 1 mm, a medium zone of 2.5 mm, and a rough zone of 4 mm. Each of these measurement lines was divided into parts labeled A, B, and C, and three repetitions of roughness measurements were performed on each part.

3. Results

Due to the characteristics of glossy surfaces, measuring their roughness using contact-free laser profilometry can be challenging. The samples were made using laser-cutting technology. The roughness parameters values measured using the laser profilometry were then compared with the roughness parameters values measured with the Surftest SJ-400 contact roughness meter.

The graphical representation of the Ra parameter (Figure 10) illustrates the average surface roughness at different measurement lines (1, 2, 3) and parts of samples A, B, and C (length of the measuring section) for stainless steel and aluminum alloy, measured using contact-free and contact methods. These results were consistent with the conclusions reached by Jurko et al. [21], Foldyna et al. [28], and Azhari et al. [29].

Figure 10. Graphed dependence of the average Ra values on the length of the measurement lines (1, 2, 3) and section (A, B, C) for stainless steel A304 and aluminum alloy AW 2017.

A comparison of the graphed dependencies involving the Ra parameter, which was measured in individual samples in a contact-free and contact manner, shows an obvious difference if we do not take into account the material used for the production of the samples (Figure 10). The measured Ra parameter values in the A304 material using the contact-free laser profilometer ranged from 12 to 18 μm, while the reference values measured with the contact roughness meter ranged from 1.6 to 2.5 μm. Similarly, the measured Ra parameter values in the AW 2017 material using the contact-free laser profilometer ranged from 8 to 13 μm, while the reference values ranged from 5.7 to 6.7 μm. This difference between the values measured with the contact-free method and the reference values are caused by the unwanted glare of the camera sensor, which leads to measurement error. However, this error can be partially eliminated by deleting the outlier values from the exported tables. These conclusions are supported by the findings of Stojanovic et al. [30] and Mital' et al. [31].

The 3D model shown in Figure 11 was generated by combining multiple profiles measured by the laser profilometer using the laser profilometry View software. In the image, the blue arrow indicates areas on the sample where the laser light reflected most strongly onto the CMOS camera. The software interpreted these areas as having the greatest surface roughness and reported them as sharp increases in the Ra parameter values.

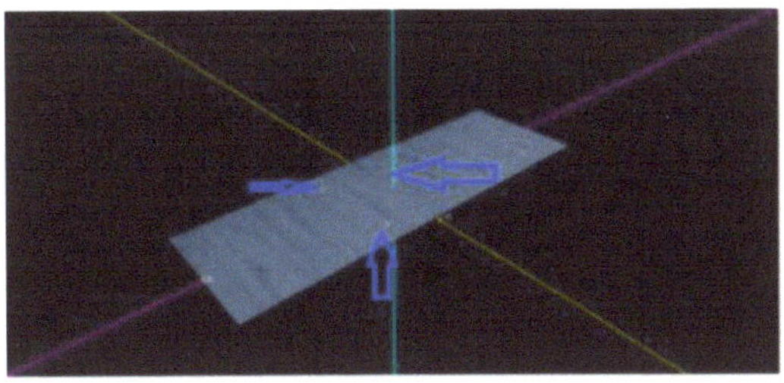

Figure 11. 3D model of the scanned sample made of AW 2017 (section A, 5 × 22 mm). *X*-axis—red line; *Y*-axis—yellow line; *Z*-axis—blue line.

There is also a visible difference in the Ra parameter in the materials of the samples used that were produced by laser technology of material cutting. In the contact method, the value of the roughness parameter Ra of the AW 2017 aluminum alloy material is in the order of 4 μm greater than that of the stainless steel material sample. In general, we can conclude that the stainless steel used is significantly harder than the aluminum alloy, which means that the roughness we achieve on the surfaces cut by laser-cutting technology is better for harder materials.

In the contact-free method, the measured values of the roughness parameter Ra are higher for a sample made from stainless steel than from the aluminum alloy. This is due to the higher gloss of the cut surface of the stainless steel material than that of the aluminum alloy.

The graphed dependence of the Rz parameter (Figure 12) shows the average surface roughness at different measurement lines (1, 2, 3) and parts of samples A, B, and C (length of the measuring section) for stainless steel and aluminum alloy measured in contact free and contact methods. These results are also confirmed by the conclusions reached by Akkurt [32], Hreha et al. [33], and Boud et al. [34].

Figure 12. Graphed dependence of the average Rz values on the length of the measurement lines (1, 2, 3) and section (A, B, C) for stainless steel A304 and aluminum alloy AW 2017.

A comparison of the graphed dependencies involving the Rz parameter, which was measured on individual samples in a contact-free and contact manner, shows an obvious difference if we do not take into account the material used for the production of the samples (Figure 12). The measured Rz parameter values in the A304 material using the contact-free laser profilometer ranged from 65 to 95 μm, while the reference values measured with the contact roughness meter ranged from 55 to 67 μm. Similarly, the measured Rz parameter values in the AW 2017 material using the contact-free laser profilometer ranged from 64 to 89 μm, while the reference values ranged from 54 to 75 μm. We assume that the increase in Rz values measured by the laser profilometry view software compared to the contact roughness meter is due to the reflection of laser light from the surface of the sample to the CMOS camera. They also suggest that the Rz values measured by the contact roughness meter may be reduced because the diameter of the measuring contact tip could not traverse each surface depression. This is because the diameter of the measuring tip is larger than that of the surface depression. These results are consistent with the conclusions reached by Azhari et al. [35], Mital' et al. [31], and Ružbarský et al. [5].

There is also a visible difference in the Rz parameter in the materials of the samples used, produced by the laser-cutting technology. In the contact method, the value of the Rz roughness parameter of the AW 2017 aluminum alloy and that of the stainless steel is approximately the same. In general, we can state that the stainless steel used is significantly harder than the aluminum alloy, which means that the Rz parameter, obtained on the surfaces cut with the laser material-cutting technology, is better in the smooth and medium areas of softer materials.

So, in the contact-free method, the measured Rz values were similar in both materials, with only a slight difference observed in the medium area of the cut surface. This difference may be due to the reflection of the laser light from the surface back to the CMOS camera, which is influenced by the material's glossiness. These findings are consistent with the results obtained by Mitaľ et al. [31] and Krenický [36].

4. Conclusions

Overall, the goal of the experiment was to determine whether contact-free laser profilometry is an accurate and reliable method for measuring the roughness of glossy surfaces. The results of the experiment can have important implications for industries that rely on precise measurements of surface roughness, such as aerospace, automotive, medical device manufacturing, and others. The obtained measurement results can be summarized as follows:

- The principle of the penetration of reflections into the measured data is based on the penetration of the reflected laser beam from the glossy surface of the sample into the CMOS camera, which causes an increase in the measured data compared to reality.
- Glossy surfaces can cause reflections of laser light, which can affect the accuracy of contact-free measurements of surface roughness parameters such as Rz. These reflections are captured by the CMOS camera and can be erroneously evaluated by the laser profilometry software as increased surface roughness. Therefore, it can be challenging to obtain accurate measurements of surface roughness parameters using a contact-free method for surfaces with high gloss. In such cases, contact methods may be more appropriate for obtaining accurate measurements (Figures 10 and 12).
- The stainless steel used is significantly harder than the aluminum alloy, which means that the Ra parameter obtained from laser-cut surfaces is better in harder materials (Figure 10).
- The 3D model created using the laser profilometry view software allows for a more detailed examination of the surface of the sample and can pinpoint areas where the laser light is reflecting the CMOS camera, leading to increased surface roughness values. The software evaluates these reflections as the greatest surface unevenness and records them as sharp increases in the Ra parameter (Figure 11).
- One way to partially limit this phenomenon is to coat the sample surface with a special transparent coating that absorbs some of the reflections, or the surface can be smudged or made matte. However, any such "finish" on the surface can distort the measured surface roughness results.
- We assume that another option to reduce the gloss of the material is to use a different color of laser light. However, it is necessary to compare individual colors of lasers as they affect the accuracy of surface roughness measurements.
- Previously used spatial (3D) surface assessment (x, y, z) can be supplemented with the fourth "dimension", variable (t), which can be time, temperature, pressure, or other physical parameters, the so-called 4D progressive analysis of surface changes. In this way, it is possible to evaluate not only the static properties of the examined surface but above all to monitor changes in surface texture, the course of action in the functional process, the impact of wear, deformation, cracking, erosion, and other structural changes.

The disadvantage of the SJ-400 contact roughness meter compared to the laser profilometry device is that it does not provide a complete picture of the roughness on the evaluated surface or the roughness in individual parts of the evaluated area. The disadvantage of the contact-free laser profilometer is that it is not able to reliably scan parameters from a glossy surface, from which laser light is reflected into the scanning camera, and subsequently, the laser profilometer software evaluates the reflection points as surface roughness problem areas. Laser profilometric measurement is a modern and reliable method of obtaining information about the surface of the measured object. This is thanks to the possibility of obtaining detailed data on the surface profile. One of the advantages of the laser profilometer is that it can obtain information about the largest depression and the largest protrusion of the surface. This allows for obtaining more detailed information about the surface of the measured object, which is useful in various applications. This is made possible with the help of high-precision targeting of the laser beam and subsequent processing of the obtained data. Another advantage of laser profilometry is the ability to capture a series of surface profiles at different points, which allows it to generate a 3D image of the surface that allows a better view of the surface character of the object. This 3D image provides a more detailed view of the surface, including any problem areas caused by reflections of the laser beam.

A more detailed evaluation of the individual roughness parameters that were selected depending on the increasing feed rate could not be carried out precisely because of the very small number of samples, and thus the small volume of necessary data. As can be seen in Figures 10 and 12, this area deserves attention not only because of the changing roughness parameters Ra, and Rz, the changes in which are noticeable at first glance but also because of the changing composition of the surface roughness profile.

Funding: This research was funded by the Cultural and educational grant agency of the Ministry of Education, Science, Research and Sport of the Slovak Republic (KEGA), grant number 015TUKE-4/2022.

Institutional Review Board Statement: Not applicable.

Informed Consent Statement: Not applicable.

Data Availability Statement: Not applicable.

Acknowledgments: The author would like to thank the KEGA grant agency for supporting research work within project 015TUKE-4/2022.

Conflicts of Interest: The author declares no conflict of interest. The funders had no role in the design of the study; in the collection, analyses, or interpretation of data; in the writing of the manuscript; or in the decision to publish the results.

References

1. Ružbarský, J. *Contactless Systems for Measurement and Evaluation of Machined Surface*; Springer Nature: Cham, Switzerland, 2022; 97p, ISBN 978-3-031-08980-0.
2. Božek, P.; Nikitin, Y.; Krenický, T. Model Systems for Diagnosticing of Mechatronic. In *Diagnostics of Mechatronic Systems*; Series: Studies in Systems, Decision, and Control; Springer: Berlin/Heidelberg, Germany, 2021; Volume 345, pp. 27–61; ISBN 978-3-030-67055-9.
3. Gajdos, I.; Spišák, E.; Kaščák, L.; Krasinskyi, V. Surface finish techniques for FDM parts. *Mater. Sci. Forum* **2015**, *818*, 45–48. [CrossRef]
4. Panda, A.; Olejarova, S.; Valicek, J.; Harnicarova, M. Monitoring of the condition of turning machine bearing housing through vibrations. *Int. J. Adv. Manuf. Technol.* **2018**, *97*, 401–411. [CrossRef]
5. Ružbarský, J.; Mitaľ, G. Diagnostics of selected surface characteristics with laser profilometry. *MM Sci. J.* **2018**, 2140–2143. [CrossRef]
6. Vikram, G.; Babu, N.R. Modelling and Analysis of Abrasive Water Jet Cut Surface Topography. *Int. J. Mach. Tools Manuf.* **2002**, *42*, 1345–1354. [CrossRef]
7. Yashchuk, V.V.; Tyurin, Y.N.; Tyurina, A.Y. Modeling of Surface Metrology of State-of-the-Art X-ray Mirrors as a Result of Stochastic Polishing Process. *Opt. Eng.* **2016**, *55*, 74106. [CrossRef]
8. Wang, W.P.; Zhang, S.L.; Li, Y. Surface Microstructure Profilometry Based on Laser Confocal Feedback. *Rev. Sci. Instrum.* **2015**, *86*, 3108. [CrossRef]
9. Chen, L.C.; Nguyen, X.L.; Huang, H.S.; Chen, J.L. Dynamic Surface Profilometry and Resonant-Mode Detection for Microstructure Characterization Using Nonconventional Stroboscopic Interferometry. *IEEE Trans. Ind. Electron.* **2010**, *57*, 1120–1126. [CrossRef]
10. Hlavac, L.M.; Kocich, R.; Gembalova, L.; Jonsta, P.; Hlavacova, I.M. AWJ Cutting of Copper Processed by ECAP. *Int. J. Adv. Manuf. Technol.* **2016**, *86*, 885–894. [CrossRef]
11. Das, T.; Bhattacharya, K. Refractive index profilometry using the total internally reflected light field. *Appl. Opt.* **2017**, *56*, 9241–9246. [CrossRef]
12. Drbúl, M.; Šajgalík, M.; Šemcer, J.; Czánová, T.; Petřkovská, L.; Čepová, L. Strojárska Metrológia a Kvalita Povrchov Vytvorených Technológiami Obrábania. ŽU Žilina a VŠB-TU Ostrava. 2014. Available online: https://projekty.fs.vsb.cz/064/docs/metrologie.pdf (accessed on 8 February 2023).

13. Murav'ev, V.I.; Fizulakov, R.A.; Mel'nichuk, A.F.; Semashko, N.A.; Matveenko, D.V. Effect of Gas-Laser Cutting on Formation of Texture, Structure, and Chemical Composition of Cut Surfaces of Titanium Preforms. *Met. Sci. Heat Treat. Titan. Its Alloys* **2005**, *47*, 62–65. [CrossRef]
14. Fizulakov, R.A.; Murav'ev, V.I.; Pitsyk, V.S. Estimation of the Quality of Gas Laser Sawing of Sheets of Aluminum Alloys. *Met. Sci. Heat Treat.* **2016**, *58*, 147–152. [CrossRef]
15. Afanasieva, L.E.; Barabonova, I.A.; Ratkevich, G.V.; Novoselova, M.V.; Grechishkin, R.M. Surface Micromorphology and Abrasive Wear Resistance of Tool Steel after Gas-Laser Cutting. *J. Surf. Investig.* **2016**, *10*, 1231–1238. [CrossRef]
16. Kulisz, M.; Zagorski, I.; Jozwik, J. Modeling and Testing of 2D Surface Roughness Parameters After Milling AZ91D Magnesium Alloy with the HSS Tool. *IEEE Metroaerospace* **2022**, *2022*, 70–75.
17. Grabchenko, A.I.; Dobroskok, V.L.; Garashchenko, Y.N.; Abdurajimov, L.N. Integral characteristics of triangulation 3d models of products. *Precis. Mach. VII* **2014**, *581*, 1013–9826. [CrossRef]
18. Bernal, C.; Camacho, A.M.; Marin, M.M.; de Agustina, B. Methodology for the evaluation of 3D surface topography in multiple indentation processes. *Int. J. Adv. Manuf. Technol.* **2013**, *69*, 2091–2098. [CrossRef]
19. Bharathi, S.; Ratnam, M.M. Evaluation of 3D Surface Roughness of Milled Surfaces using Laser Speckle Pattern. In Proceedings of the International Conference on Recent Advances in Industrial Engineering and Manufacturing, Penang, Malaysia, 12–13 December 2018; pp. 1757–8981.
20. Jin, G.C.; Bao, N.K. Surface Detection and 3D Profilometry for Microstructure Using Optical Metrology. *Opt. Lasers Eng.* **2001**, *36*, 1–9. [CrossRef]
21. Jurko, J.; Panda, A.; Valicek, J.; Harnicarova, M.; Pandova, I. Study on Cone Roller Bearing Surface Roughness Improvement and the Effect of Surface Roughness on Tapered Roller Bearing Service Life. *Int. J. Adv. Manuf. Technol.* **2016**, *82*, 1099–1106. [CrossRef]
22. Stemp, W.J.; Stemp, M. UBM Laser Profilometry and Lithic Use-Wear Analysis: A Variable Length Scale Investigation of Surface Topography. *J. Archaeol. Sci.* **2001**, *28*, 81–88. [CrossRef]
23. Salvi, J.; Fernandez, S.; Pribanic, T.; Llado, X. A State of the Art in Structured Light Patterns for Surface Profilometry. *Pattern Recognit.* **2010**, *43*, 2666–2680. [CrossRef]
24. Kumar, B.M.; Ratnam, M.M. Machine Vision Method for Non-Contact Measurement of Surface Roughness of a Rotating Workpiece. *Sens. Rev.* **2015**, *35*, 10–19. [CrossRef]
25. Lu, C. Study on Prediction of Surface Quality in Machining Process. *J. Mater. Process. Technol.* **2008**, *205*, 439–450. [CrossRef]
26. Mitaľ, G.; Dobránsky, J.; Ružbarský, J.; Olejárová, Š. Application of Laser Profilometry to Evaluation of the Surface of the Workpiece Machined by Abrasive Waterjet Technology. *Appl. Sci.* **2019**, *9*, 2134. [CrossRef]
27. Srivastava, M.; Hloch, S.; Krejci, L.; Chattopadhyaya, S.; Dixit, A.M.; Foldyna, J. Residual Stress and Surface Properties of Stainless-Steel Welded Joints Induced by Ultrasonic Pulsed Water Jet Peening. *Measurement* **2018**, *127*, 453–462. [CrossRef]
28. Foldyna, J.; Sitek, L.; Scucka, J.; Martinec, P.; Valicek, J.; Palenikova, K. Effects of Pulsating Water Jet Impact on Aluminium Surface. *J. Mater. Process. Technol.* **2009**, *209*, 6174–6180. [CrossRef]
29. Azhari, A.; Schindler, C.; Bo, L. Effect of Waterjet Peening on Aluminum Alloy 5005. *Int. J. Adv. Manuf. Technol.* **2013**, *67*, 785–795. [CrossRef]
30. Stojanovic, R.; Koubias, S.; Stojanovic, S.; Georgoudakis, M. A measuring method for laser-based profilometry and its applications in non-destructive testing and quality control. In Proceedings of the Fifth International Conference on Vibration Measurements by Laser Techniques: Advances and Applications, Ancona, Italy, 18–21 June 2002; Volume 4827, pp. 553–563.
31. Mitaľ, G.; Spišák, E. 3D hodnotenie geometrických vlastností povrchu pomocou laserovej triangulačnej metódy. *Transf. Inovácií* **2021**, *44*, 1337–7094.
32. Akkurt, A. Surface Properties of the Cut Face Obtained by Di_erent Cutting Methods from AISI 304 Stainless Steel Materials. *Indian J. Eng. Mater. Sci.* **2009**, *16*, 373–384.
33. Hreha, P.; Radvanska, A.; Knapcikova, L.; Krolczyk, G.M.; Legutko, S.; Krolczyk, J.B.; Hloch, S.; Monka, P. Roughness Parameters Calculation by Means of On-LineVibration Monitoring Emerging from AWJ Interaction with Material. *Metrol. Meas. Syst.* **2015**, *22*, 315–326. [CrossRef]
34. Boud, F.; Carpenter, C.; Folkes, J.; Shipway, P.H. AbrasiveWaterjet Cutting of a Titanium Alloy: The Influence of Abrasive Morphology and Mechanical Properties onWorkpiece Grit Embedment and Cut Quality. *J. Mater. Process. Technol.* **2010**, *210*, 2197–2205. [CrossRef]
35. Azhari, A.; Schindler, C.; Kerscher, P.; Grad, P. Improving Surface Hardness of Austenitic Stainless-Steel Using Waterjet Peening Process. *Int. J. Adv. Manuf. Technol.* **2012**, *63*, 1035–1046. [CrossRef]
36. Krenický, T. Non-contact study of surfaces created using the AWJ technology. *Manuf. Technol.* **2015**, *15*, 61–64. [CrossRef]

applied sciences

MDPI

Article

Large-Scale 3D Multi-Story Enterprise Layout Design in a New Type of Industrial Park in China

Xuemin Liu [1], Guozhong Huang [1], Xuehong Gao [1,*], Haoxuan Li [1], Shengnan Ou [1] and Ibrahim M. Hezam [2]

1 Research Institute of Macro-Safety Science, School of Civil & Resource Engineering, University of Science and Technology Beijing, Beijing 100083, China
2 Department of Statistics & Operations Research, College of Sciences, King Saud University, Riyadh 11451, Saudi Arabia
* Correspondence: gaoxh2020@ustb.edu.cn

Abstract: In the new-type clustered industrial park, the closer distance between enterprises leads to risk aggregation, and the layout of enterprises affects the safety and economy of the park. However, previous studies have often paid insufficient attention to safety, and few studies have considered park profits. To address this issue, a bi-level three-dimensional layout optimization model was proposed to minimize the overall association risk of the park and maximize the rental profit. In particular, this article explained the enterprise association risks and provided calculation formulas, considering multiple risk types. To solve the proposed nonlinear model, a specific variable conversion method was presented to reduce the problem scale. Subsequently, an improved genetic algorithm was developed and applied to obtain the layout results. Furthermore, a case study of an industrial park was conducted, and the computational results indicated the validity of the model and methods. Finally, two different scenarios were implemented, and critical parameters were tested to provide valuable management insights.

Keywords: industrial park layout; association risk; rental profit; multi-story layout design

Citation: Liu, X.; Huang, G.; Gao, X.; Li, H.; Ou, S.; Hezam, I.M. Large-Scale 3D Multi-Story Enterprise Layout Design in a New Type of Industrial Park in China. *Appl. Sci.* **2022**, *12*, 8110. https://doi.org/10.3390/app12168110

Academic Editors: Tibor Krenicky, Juraj Ruzbarsky and Maros Korenko

Received: 20 July 2022
Accepted: 8 August 2022
Published: 12 August 2022

1. Introduction

The layout problem in new-type industrial parks is the large-scale 3D multi-story enterprise layout design (3DMSELD). According to certain principles, the enterprises will be reasonably distributed on all floors of each building in the park. Traditional industrial park layout problems are mainly carried out on a two-dimensional plane [1]. However, intensive production models have become increasingly common with economic development. Many new-type clustered industrial parks have emerged in China and are distributed in the eastern coastal region. These parks generally have multiple multistory buildings leased to dozens of enterprises, which involves the large-scale 3DMSELD problem. Factory intensification can significantly improve land and infrastructure utilization rates. However, it also creates a concentration of risks. The shortening distance between enterprises has led to a greater possibility of risk interactions. An accident in one enterprise can lead to injury or property damage to others and may even trigger a domino effect [2]. Therefore, it is necessary to consider the risk factors in the park layout problems.

With social development, people attach greater importance to safety, and how to reduce risk through layout has become the research focus. Most studies involving safety have considered safety factors as safety costs or constraints. For instance, Caputo [3] demonstrated that safety-related costs are relevant for defining optimal layout; Wang [4] treated multiple hazard sources as risk costs in the objective function. However, the safety status of enterprises in clustered industrial parks is closely related, and it is worthwhile to focus on safety factors separately. In particular, Wang [5] argued that it would be inappropriate to turn safety issues into economic numbers in objective functions. The layout problem has been around for a long time; many previous studies have focused on the single-story layout of a facility [6]. Rosenblatt [7] proposed a combined quantitative and qualitative approach to the facilities layout problem on a two-dimensional plane. Penteado [8] constructed a model to provide a chemical plant's optimal layout of process equipment. These studies did not solve the three-dimensional layout problem. Research on multi-story layouts has become a hot topic with an increase in multi-story workshops. Kia [9] and Latifi [10] presented two 3D models to describe the facility layout problem. However, there are few studies on the three-dimensional optimization layouts at the industrial park level with enterprises as the smallest layout unit. In addition, the park management party's main profit comes from rent affected by the layout. Nevertheless, this has rarely been considered in previous studies.

Although significant progress has been made in related fields, we can get some research gaps from the above discussion. First, there is a lack of theoretical research on the large-scale 3DMSELD problem; Second, researchers have not treated the risk impact between enterprises in much detail; Third, there is a lack of consideration for rental profit in the layout problem. To address the gaps mentioned above, this study used a bi-level integer nonlinear programming (BLINLP) model considering enterprise association risks and rental profit to make hierarchical decisions at the layout stage. First, settled enterprises were divided into risk and non/low risk enterprises. Priority was given to reducing the risk level of the parks. Because the enterprise's inherent risks do not change with location, this study only considered the associated risk between enterprises. For non/low risk enterprises, rents were considered to create enterprise layouts to maximize the park's income. Subsequently, a BLINLP model was presented to solve the large-scale 3DMSELD problem. To reduce problem size, a specific variable conversion method was proposed. Then, an improved evolutionary algorithm was applied. In addition, a case study was conducted using data from an industrial park in Shunde. Consequently, this study focuses on addressing the large-scale 3DMSELD problem considering enterprise association risk and rental profit. The following key questions are required to address the main concerns of this study:

(1) *How to define the enterprise association risk?*
(2) *How to formulate a bi-level optimization model for the large-scale 3DMSELD problem?*
(3) *How to solve the proposed nonlinear bi-level model?*

To solve the large-scale 3DMSELD problem with the above questions, a BLINLP model was developed in this study. Compared with the existing studies, this work makes the following main contributions: (a) this study considers a three-dimensional layout optimization problem with multi-building, multi-floor, and multi-enterprises factors; (b) the enterprises association risk calculation formulas are proposed and used as the layout goal; (c) our study constructs a BLINLP model and proposes a variable conversion strategy to reduce the problem size. The remainder of this article is organized as follows. Section 2 presents a literature review of the main concerns of this study, in which research gaps are identified. Section 3 describes the enterprise layout problem with two goals, where the association risks are explained and calculated. To address this problem, the BLINLP model is presented in Section 4. Then, an improved genetic algorithm (GA) is developed and presented after the decision variables are converted in Section 5. Section 6 presents a case study in which the obtained results validate the effectiveness of the proposed model. Finally, Section 7 concludes the study by summarizing the findings and highlighting future directions.

2. Literature Review

Our work aims to develop a computable BLINLP model to solve the large-scale 3DMSELD problem. This section reviews the literature by discussing dozens of studies on the influencing factors in the layout, formulation of association risks, mathematical programming, and solving algorithms.

2.1. Influencing Factors in Layout

The facility layout in a single plant or plant layout in an industrial area significantly impacts manufacturing costs, energy losses [11], enterprise performance, and safety [12,13]. Economy and safety are the two main factors considered in the layout design. Previous research has focused on minimizing the layout costs [14], including pipelines [15], land, and internal costs (e.g., cost of moving materials). Wang [16] proposed a model to minimize the piping investment, pump power, land, and floor construction costs. Derhami [17] focused on reducing the operational costs of warehouses by layout. The park leases plants to the enterprise, and rent is the primary profit source of the park. Unlike previous studies that considered costs, this study sets maximizing rental profits as an objective function.

In addition, some studies have considered safety factors in the layout [18]. de Lira-Flores [19] proposed a mixed integer nonlinear programming (MINLP) approach to solve the plant layout problem with the objective of reducing the risk of the process plant. Medina-Herrera [20] used quantitative risk analysis (QRA) to design a plant layout. Factors of multiple hazard sources in explosion accidents were considered in Wang [4]. Latifi [10] considered the toxic release risk and possible scenarios of fire, explosion, and domino effects in their layout model. To address safety issues, this study considered the risk association between enterprises in the optimization model.

2.2. Formulation of Association Risks

The industrial park acts as a complex system in which enterprises interact with each other. Enterprise association risk reflects the degree of impact between enterprise risks. A critical issue in studying association risk is its measurement. Wang [5] adopted the trinitrotoluene (TNT) equivalent and Pasquill-Gifford models to evaluate workers' death probabilities and structural damage to buildings. Groth [21] used a software toolkit to conduct a QRA and consequence analysis relevant to assessing the safety of hydrogen fueling and storage infrastructure. Medina-Herrera [20] used consequence analyses and probit models to quantify the effects of catastrophic outcomes. Brunoro Ahumada [18] developed a framework integrating layout formulation with a QRA method.

In addition, domino accidents [22,23] can occur in the new-type clustered industrial parks, and studies have been conducted to describe them quantitatively. Hazardous events such as pool fires, jet fires, flash fires, fireballs, and blast waves resulting from explosions were considered by Ejeh [24]. An optimization-based approach for obtaining safe plant layout designs using the domino hazard index was presented in this article. Liu [25] established a consequence and probability model for the domino effect initiated by fires and explosions. Ding [26] modeled the spatial-temporal evolution of domino effects, and the risk was assessed by the time to failure and escalation probability. Ni [27] proposed a relative risk model to assess the domino effect in the chemical process industry. Furthermore, field theory and Monte Carlo simulations have been adopted for the risk assessment of domino accidents [28]. Although many relevant studies exist, the association risks between enterprises in a three-dimensional space still need to be further classified and studied.

2.3. Bi-Level Optimization Model

Both practical and theoretical studies have demonstrated the importance of properly constructing mathematical models for layouts involving multiple buildings, multiple stories, and multiple enterprises. Layout issues often involve multiple goals, and these goals are not necessarily considered to have the same priority. Multilayer optimization models can reflect hierarchical relationships. Lu [29] introduced bi-level and tri-level decision making. Cao [30] proposed a fuzzy bi-level optimization model for multi-period post-disaster relief distribution. Yang [31] constructed a bi-level optimization model for large-scale electric vehicle charging. Although bi-level optimization theory is relatively mature, the current research on layout problems basically use single-level programming. For instance, Che [32] proposed a bi-objective MINLP model to minimize the total material handling cost and total occupied room area. Martinez-Gomez [33] considered the economic and safety issues in a single-level programming model. Because our study has two goals and is more focused on minimizing risk, a hierarchical optimization model is used that differs from previous ones.

2.4. Model Solving Algorithm

Optimization model solving strategies can be divided into exact and heuristic algorithms. For exact algorithms, the simplex algorithm [34], branch-and-bound algorithm [35], and Benders decomposition algorithms are commonly used. For instance, Ahmadi [36] used the CPLEX software to precisely solve the proposed multiple-stage mathematical programming model. However, exact algorithms are more suitable for solving small-scale linear models and more challenging for large-scale problems. Therefore, many studies have used heuristic algorithms to solve these problems [37,38]. Caputo [3] proposed a method based on a GA to optimize the process plant layout. Chen [39] presented a heuristic approach combining an improved adaptive GA with scatter search. In the present study, an improved GA was used.

2.5. Summary

In addition to the discussion above, Table 1 provides a comprehensive literature review. Some gaps in the literature can be addressed as follows. First, many recent studies have focused on single-story facility layouts in plants. However, only a few studies have addressed multi-story layout problems. Second, most existing studies do not pay sufficient attention to risk, often only adding risk as a risk cost to the objective function of minimizing layout costs. Third, previous studies scarcely considered parks' profits. Therefore, this study focuses on the large-scale 3DMSELD problem considering risk and profit. In addition, a BLINLP model is proposed to formulate the above problem, which has two goals: (a) minimization of association risk and (b) maximization of rental profit. To solve this problem, a special decision variable conversion method is first presented to reduce the problem scale. This study also applied an improved GA method to handle large-scale problems. Finally, a case study with different scenario simulations and a sensitivity analysis was implemented to validate the proposed nonlinear model and solution methods.

Table 1. Summary of the literature on the layout operations.

Article	Main Problem	Multi-Story Layout	Risk	Profit	Cost *	Hierarchy	Approach
Balakrishnan [40]	Plant layout	No	No	No	Yes	Single	Heuristic
Patsiatzis [41]	Process plant layout	No	Yes	No	Yes	Single	Exact
Vázquez-Román [42]	Facility layout	No	Yes	No	Yes	Single	Exact
Cheng [43]	Facility layout	No	No	No	No	Single	Heuristic
Xu [44]	Chemical Plant Layout	No	Yes	No	Yes	Single	Heuristic
Emami [45]	Dynamic facility layout	No	No	No	Yes	Single	Heuristic
Medina-Herrera [20]	Plant layout	Yes	Yes	No	Yes	Single	Exact
Kia [9]	Dynamic cellular manufacturing systems layout	Yes	No	No	Yes	Single	Heuristic
Izadinia [46]	Multi-floor layout problem	Yes	No	No	Yes	Single	Exact
Caputo [3]	Process plant layout	No	Yes	No	Yes	Single	Heuristic
Alves [47]	Chemical plant layout	No	Yes	No	No	Single	Heuristic
Ahmadi [36]	Facility layout	Yes	No	No	Yes	Single	Exact
Wang [1]	Industrial area layout	No	Yes	No	Yes	Single	Heuristic
Che [32]	Facility layout	Yes	No	No	Yes	Single	Exact
Latifi [10]	Process plant layout	No	Yes	No	Yes	Single	Heuristic
Zhang [48]	Warehouse layout	No	No	No	Yes	Single	Heuristic
Arnaout [49]	Warehouse layout	No	No	No	Yes	Multiple	Heuristic

Table 1. *Cont.*

Article	Main Problem	Multi-Story Layout	Risk	Profit	Cost *	Hierarchy	Approach
Wang [4]	Industrial park layout	No	Yes	No	Yes	Single	Heuristic
Şahin [6]	Facility layout	No	No	No	Yes	Single	Heuristic
Ejeh [24]	Process plant layout	Yes	Yes	No	Yes	Single	Exact
This study	**Industrial park layout**	**Yes**	**Yes**	**Yes**	**No**	**Multiple**	**Heuristic**

* does not include costs related to safety and risk but only the actual economic outlay.

3. Problem Description

The new-type clustered industrial park layout involves multiple buildings and floors, considering the enterprises association risk and rental profit. This section describes the main problem in the following four subsections.

3.1. New-Type Clustered Industrial Park

The new-type clustered industrial parks are currently numerous in China and have the following characteristics: (a) they contain several multi-story buildings, (b) their factories are intensive and standard. The enterprise leases one or more factories for production and operation, and (c) complete supporting services. The park management party manages the entire park and provides complementary services. Each building in the park may consist of multiple enterprises of different types, such as packaging and printing, machinery and equipment, e-commerce, and furniture manufacturing. These enterprises are geographically located in the park and share park infrastructure, creating a concentration of safety risks.

3.2. Association Safety Risks

The new-type clustered industrial park is a complex system in which enterprises are risk-related. This risk correlation can partially be explained by the domino effect. The occurrence of the domino effect is only possible when the energy of the initial accident is sufficiently large to affect other enterprises within the range. It is generally believed that the spread of an initial accident must at least cause a secondary accident to happen [25]. The association safety risk of Enterprise A to Enterprise B is the size of the risk faced by Enterprise B from Enterprise A. Unlike the domino effect risk, the association safety risk also includes the possibility that Enterprise A did not cause a secondary accident in Enterprise B but only led to harm Enterprise B. The impact of an accident in Enterprise A on Enterprise B is illustrated in Figure 1.

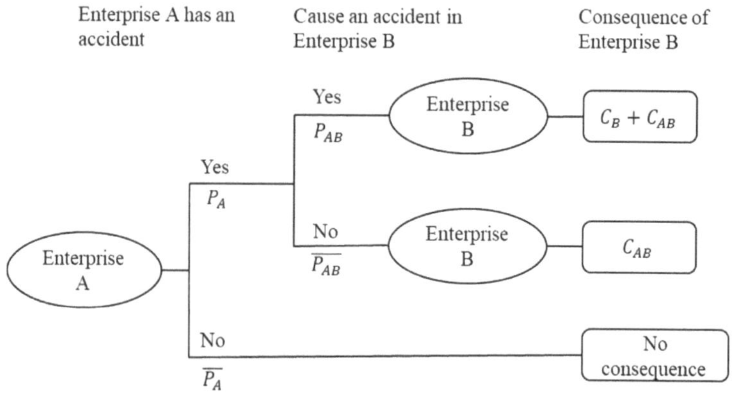

Figure 1. Impact of an accident in Enterprise A on Enterprise B.

Where P_A is the probability of an accident in Enterprise A, P_{AB} is the probability that an accident in Enterprise A causes a secondary accident in Enterprise B, C_B is the loss caused by the secondary accident in Enterprise B, and C_{AB} is the loss in Enterprise B caused by an accident in Enterprise A. It should be noted that C_{AB} may be 0. That is, an accident at Enterprise A will not affect Enterprise B. Risk can be expressed as a function of probability and consequence [50]. The association safety risk R_{AB} of Enterprises A to B can be calculated by

$$R_{AB} = P_A P_{AB}(C_B + C_{AB}) + P_A \overline{P_{AB}} C_{AB} \quad (1)$$

From engineering experience and previous research, it has been found that the farther away locations are from the flashpoint of risk events, the less they are affected. However, the distance between multiple enterprises is unknown before the layout is implemented, and the specific location of the dangerous equipment is unknown. This article takes the enterprise as the smallest layout unit. The actual association risk assessment steps between enterprises were simplified to reduce the calculation scale and improve the feasibility of the layout model. The specific steps are as follows: (a) classify enterprises according to the category of the most severe risk event; (b) consider only the most severe risk event and assess the association risk when two enterprises are adjacent; (c) determine the diffusion coefficient functions for different risk events; (d) determine the association risk change between enterprises when the distance changes according to the risk diffusion coefficient function. The specific steps are as follows:

(1) Step 1: Risk enterprise classification

There are many types of enterprises in the new-type industrial parks, and accidents that may occur include fires, explosions, object strikes, electric shocks, and mechanical injuries. This study examines the interactions between enterprises. The domino effect is caused by the following physical effects: fire heat radiation, overpressure of an explosion shock wave, and propellant fragments. Accidents causing the domino effect can be divided into fires and explosions [25]. These two types of accidents are also the focus of our work. New-type industrial parks typically contain multiple multistory buildings, and fires spread rapidly in high-rise buildings. Furthermore, some enterprises have pressure vessels, which increases the risk of explosion. Therefore, risk enterprises were classified according to the most severe dangerous events categories: (1) fire risk enterprises, (2) explosion risk enterprises, and (3) other types of risk enterprises. This study only calculates the fire and explosion risks separately and does not consider situations in which both occur simultaneously.

(2) Step 2: Assessment of association risk with neighboring enterprises

Many studies have conducted quantitative assessments of risk for fixed-location enterprises. Luo [51] proposed a QRA method that combines a risk matrix, fault tree, and fishbone diagram model. First, the two enterprises to be evaluated are considered adjacent. The probability of a primary risk event occurring can be calculated based on a fault tree. The domino effect extension probability model [25] can be used to calculate the probability of accidents caused by other enterprises. The consequences of an accident can be analyzed using the Dao chemical index method. The risk matrix method can determine the probability and severity of the consequences. Finally, based on Equation (1), the associated safety risk when enterprises are adjacent is obtained. This method determines accident probability among enterprises by merely considering the occurrence of the most severe risk event. This moderately simplified approach aims to reduce the magnitude of computing complexity without losing the rationality of the enterprise layouts.

(3) Step 3: Risk diffusion coefficient function

The risk diffusion coefficient function identifies the degree to which the risk value decreases with an increase in distance from the risk event flashpoint and is a function of distance. The layout points used in this study were fixed. Thus, the distances are discrete and can be approximated as floor differences.

(i) Fire risk enterprises

Accidents that cause fires can generally be divided into pool fires, jet fires, fireballs, and flash fires, and the main hazard is thermal radiation [28]. The point source model is simple, easy to manage, and widely used to calculate fire thermal radiation. May [52] first hypothesized that the fire source was a point heat source and used the square ratio theorem to calculate the thermal radiation intensity at any target point around the flame. However, they did not construct a perfect mathematical calculation model. Modak [53] set the location of the point heat source at half the height of the flame above the flame centerline based on their predecessors and derived a mathematical calculation model of the single point heat source model through theoretical derivation. Sivathanu [54] proposed a method for estimating the total radiant output of turbulent jet flames by measuring the radiative heat flux at a single location. This study examined the impact of enterprises that are distant and suitable for applying point source models. The general formula for the application of the single-point source model to thermal radiation in pool fires is

$$I_p = \frac{Qt_c}{4\pi x^2} \tag{2}$$

where I_p is the intensity of the thermal radiation at the target point, Q is the total thermal radiation of the pool fire, t_c is the thermal conductivity, and x is the distance from the target point to the center of the flame.

The point source model reduces the flame to a radiant emission point in a spherical shape that radiates heat into the surrounding space based on the inverse ratio of the intensity of the thermal radiation to the square of the distance to the radiation source. The point source model is shown in Figure 2. Referring to the point source model, the fire risk diffusion coefficient function is defined as follows:

$$\gamma^f(D_{ij}) = \begin{cases} 1 & if\ D_{ij} = 0 \\ \frac{\beta_1}{(bD_{ij})^2} & if\ D_{ij} > 0 \\ \frac{\beta_2}{(bD_{ij})^2} & if\ D_{ij} < 0 \end{cases} \tag{3}$$

where D_{ij} is the floor difference between enterprise j and enterprise i, b is the floor-to-floor height, and β_1 and β_2 are the correction parameters.

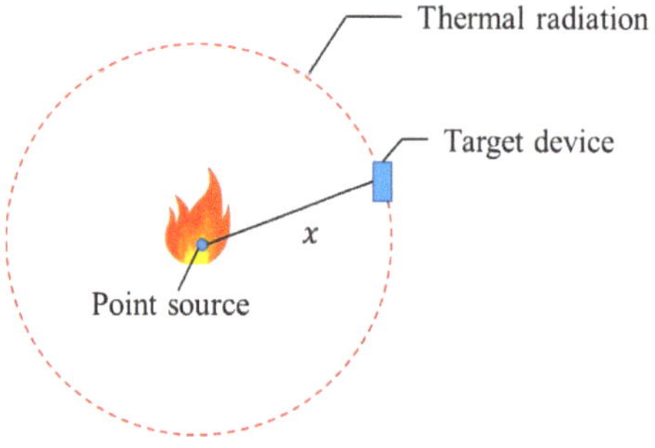

Figure 2. Schematic of the point source model.

(ii) Explosion risk enterprises

The physical effects corresponding to an explosion are the overpressure of the explosion shock waves, propellant fragments, and thermal radiation. One of the most representative types of damage is shockwave damage, which spreads widely and quickly and may cause chain damage. Shockwave injuries and destructive effects can be assessed using the overpressure criterion. When the shock wave overpressure reaches a specific value, it causes inevitable damage. In the shockwave propagation method, the energy propagates outward from the center of the explosion source. Its propagation characteristic is that the farther away from the center it is, the smaller the effect of the overpressure state. The shockwave peak overpressure is generally considered a function of the distance. Many experimental studies have been conducted, and various empirical formulas have been summarized [55,56]. Yang [57] compared and analyzed the prediction results of multiple empirical formulas and summarized the following formulas:

$$Z = \frac{R}{W^{1/3}} \tag{4}$$

$$P_{so} = \begin{cases} \frac{0.084}{Z} + \frac{0.27}{Z^2} + \frac{0.7}{Z^3} & if\ Z \leq 1 \\ \frac{0.084}{Z} + \frac{0.255}{Z^2} + \frac{0.65}{Z^3} & if\ 1 < Z \leq 15 \end{cases} \tag{5}$$

where Z is the proportional distance, R is the distance between the target point and burst center, W is the equivalent TNT dosage, and P_{so} is the shock wave peak overpressure. The explosion risk diffusion coefficient function is defined as follows:

$$\gamma^e(D_{ij}) = \begin{cases} 1 & if\ D_{ij} = 0 \\ \frac{\alpha_1}{bD_{ij}} + \frac{\alpha_2}{(bD_{ij})^2} + \frac{\alpha_3}{(bD_{ij})^3} & if\ D_{ij} > 0 \\ \frac{\alpha_4}{bD_{ij}} + \frac{\alpha_5}{(bD_{ij})^2} + \frac{\alpha_6}{(bD_{ij})^3} & if\ D_{ij} < 0 \end{cases} \tag{6}$$

where α_1, α_2, α_3, α_4, α_5, and α_6 are correction parameters. When considering other environmental impact factors, such as wind direction, wall thickness, and safety barriers, the risk diffusion coefficient can be adjusted using correction parameters based on actual engineering experience. Furthermore, it can be combined with expert evaluations to increase the model's flexibility.

(iii) Other types of risk enterprises

In addition to fires and explosions, enterprises can also have other accidents. The model only considers the impact on the same floor for accidents that are less likely to occur or have a smaller scope of impact. The diffusion coefficient function is defined as follows:

$$\gamma^o(D_{ij}) = \begin{cases} 1 & if\ D_{ij} = 0 \\ 0 & \text{otherwise} \end{cases} \tag{7}$$

(4) Step 4: Determination of actual association safety risks

Based on the above steps, the actual association risk of enterprises is determined by:

$$R_{ij}^a = \begin{cases} R_{ij} * \gamma^f(D_{ij}) & if\ i \in F \\ R_{ij} * \gamma^e(D_{ij}) & if\ i \in E \\ R_{ij} * \gamma^o(D_{ij}) & if\ i \in O \end{cases} \tag{8}$$

where R_{ij}^a is the actual association risk with enterprises i and j, R_{ij} is the association safety risk where enterprises i and j are adjacent, and F, E, and O represent fire types, explosion types, and other types of enterprises, respectively.

3.3. Profit

In addition to safety issues, park management needs to consider profit. The profit of the new-type industrial park is affected by many factors, such as factory rent, management costs, and government support. Among these, rent is an important source of profit for the park. Standardized factories and supporting service facilities have led to certain increases in costs and rent. Most parks begin to attract enterprises during the construction process. The new-type industrial park generally consists of several multistory buildings. To obtain greater rental profits, the park is more inclined to choose negotiable pricing to achieve park management and enterprise consistency. In particular, different enterprises have different affordability values for each floor. For example, enterprises that need to move large goods frequently are more willing to pay higher rents for the first floor. Therefore, the layout results of enterprises can have an impact on rental profits, which is worth noting. This study considers only rental profits for non/low-risk enterprises.

3.4. Goals

This study considers two levels: the goal of the upper level is to achieve minimum association safety risks, and the lower-level goal is to maximize the park rental profit. At a certain point in time, the risk enterprises that need to be prioritized are identified. In addition to the enterprise's inherent risks, there is also a risk correlation between enterprises. The enterprise's inherent risks do not change with location and cannot be reduced by optimizing the park layout. From Section 3.2, it can be seen that the enterprise association risk affected by distance reflects the degree of impact between enterprise risks. By optimizing the facility layout, the distance between risky enterprises can be changed, thereby reducing the overall associated risk of the park. Consequently, this study takes the minimization of the total association risk as the upper-level goal.

Based on achieving the first goal, the rent is considered, and the layout of the remaining enterprises is determined. In particular, the decision variables for the two goals are different. In this study, improving the safety level of parks is a priority. Therefore, the upper optimal solution is first calculated, and the upper-level result is used as the constraint of the lower level.

4. Bi-Level Programming Model Formulation

4.1. Notations

Sets

I	Set of risk enterprises, indexed by $i \in I$.
F	Set of fire risk enterprises, indexed by $f \in F$.
E	Set of explosion risk enterprises, indexed by $e \in E$.
O	Set of other types of risk enterprises, indexed by $o \in O$.
J	Set of non/low - risk enterprises, indexed by $j \in J$.
B	Set of buildings, indexed by $b \in B$.
S	Set of floors, indexed by $s \in S$.
Area parameters	
Q_{bs}	Size of $s \in S$ in $b \in B$.
A_i	Area required by enterprise $i \in I$.
A_j	Area required by enterprise $j \in J$.
Risk parameters	
R_{fi}	Association risk where enterprises $f \in F$ and $i \in I$ are adjacent.
R_{ei}	Association risk where enterprises $e \in E$ and $i \in I$ are adjacent.
R_{oi}	Association risk where enterprises $o \in O$ and $i \in I$ are adjacent.
$\gamma^f(d)$	Fire is the floor difference.
$\gamma^e(d)$	Explosion is the floor difference.
$\gamma^o(d)$	Other is the floor difference.
Rent parameter	
Z_{js}	Expected rent of $j \in J$ on $s \in S$.
Other parameter	
N_s	specific floor value of $s \in S$.

Particularly:

$$F \cup E \cup O = I \tag{9}$$

$$F \cap E, E \cap O, F \cap O = 0 \tag{10}$$

Decision variables:

$$x_{bsi} = \begin{cases} 1 \text{ Enterprise } i \in I \text{ on } s \in S \text{ in } b \in B \\ 0 \text{ otherwise} \end{cases}$$

$$x_{bsf} = \begin{cases} 1 \text{ Enterprise } f \in F \text{ on } s \in S \text{ in } b \in B \\ 0 \text{ otherwise} \end{cases}$$

$$x_{bse} = \begin{cases} 1 \text{ Enterprise } e \in E \text{ on } s \in S \text{ in } b \in B \\ 0 \text{ otherwise} \end{cases}$$

$$x_{bso} = \begin{cases} 1 \text{ Enterprise } o \in O \text{ on } s \in S \text{ in } b \in B \\ 0 \text{ otherwise} \end{cases}$$

$$x_{bsj} = \begin{cases} 1 \text{ Enterprise } j \in J \text{ on } s \in S \text{ in } b \in B \\ 0 \text{ otherwise} \end{cases}$$

$$x_{bfi} = \begin{cases} 1 \text{ Enterprises } f \in F \text{ and } i \in I \text{ in } b \in B \\ 0 \text{ otherwise} \end{cases}$$

$$x_{bei} = \begin{cases} 1 \text{ Enterprises } e \in E \text{ and } i \in I \text{ in } b \in B \\ 0 \text{ otherwise} \end{cases}$$

$$x_{boi} = \begin{cases} 1 \text{ Enterprises } o \in O \text{ and } i \in I \text{ in } b \in B \\ 0 \text{ otherwise} \end{cases}$$

4.2. Assumptions

Three assumptions were postulated for the BLINLP model.

(i) An enterprise can be located on only one floor of a building. A few enterprises that need to be located on multiple floors are treated as independent enterprises.
(ii) The associated safety risk can only occur between two enterprises when the two enterprises are located in the same building. This study does not consider the interactions between enterprises located in different buildings.
(iii) The types of enterprises and the association risks of adjacent enterprises are known. The specific determination process is described in Section 3.

4.3. Bi-Level Integer Programming Model

In this subsection, the BLINLP model is defined using Equations (11)–(25).

$$\begin{aligned} \Psi_1 = \text{Min} \sum_{b\in B}\sum_{f\in F}\sum_{i\in I} R_{fi} * \gamma^f \left(\sum_{s\in S} x_{bsf}N_s - \sum_{s\in S} x_{bsi}N_s\right) * x_{bfi} + \\ \sum_{b\in B}\sum_{e\in E}\sum_{i\in I} R_{ei} * \gamma^e \left(\sum_{s\in S} x_{bse}N_s - \sum_{s\in S} x_{bsi}N_s\right) * x_{bei} + \\ \sum_{b\in B}\sum_{o\in O}\sum_{i\in I} R_{oi} * \gamma^o \left(\sum_{s\in S} x_{bso}N_s - \sum_{s\in S} x_{bsi}N_s\right) * x_{boi} \end{aligned} \tag{11}$$

s.t.

$$\sum_{i\in I} A_i * x_{bsi} \le Q_{bs} \ \ \forall\, b \in B, s \in S \tag{12}$$

$$\sum_{b\in B}\sum_{s\in S} x_{bsf} = 1 \ \ \forall\, f \in F \tag{13}$$

$$\sum_{b\in B}\sum_{s\in S} x_{bse} = 1 \ \ \forall\, e \in E \tag{14}$$

$$\sum_{b\in B}\sum_{s\in S} x_{bso} = 1 \ \ \forall\, o \in O \tag{15}$$

$$\sum_{b\in B}\sum_{s\in S} x_{bsi} = 1 \ \ \forall\, i \in I \tag{16}$$

$$\sum_{s\in S} x_{bsf} = \sum_{s\in S} x_{bsi} \ \ \forall\, b \in B, f \in F, i \in I, f = i \tag{17}$$

$$\sum_{s\in S} x_{bse} = \sum_{s\in S} x_{bsi} \ \ \forall\, b \in B, e \in E, i \in I, e = i \tag{18}$$

$$\sum_{s\in S} x_{bso} = \sum_{s\in S} x_{bsi} \ \ \forall\, b \in B, o \in O, i \in I, o = i \tag{19}$$

$$x_{bfi} = \text{Min}\left\{\sum_{s\in S} x_{bsf}, \sum_{s\in S} x_{bsi}\right\} \ \forall\, b \in B, f \in F, i \in I \tag{20}$$

$$x_{bei} = \text{Min}\left\{\sum_{s\in S} x_{bse}, \sum_{s\in S} x_{bsi}\right\} \ \forall\, b \in B, e \in E, i \in I \tag{21}$$

$$x_{boi} = \text{Min}\left\{\sum_{s\in S} x_{bso}, \sum_{s\in S} x_{bsi}\right\} \ \forall\, b \in B, o \in O, i \in I \tag{22}$$

$$\Psi_2 = \text{Max} \sum_{b\in B}\sum_{s\in S}\sum_{j\in J} Z_{js} * x_{bsj} * A_j \tag{23}$$

s.t.

$$\sum_{j\in J} A_j * x_{bsj} \le Q_{bs} - \sum_{i\in I} A_i * x_{bsi} \ \forall\, b \in B, s \in S \tag{24}$$

$$\sum_{b\in B}\sum_{s\in S} x_{bsj} = 1 \ \forall\, j \in J \tag{25}$$

In this model, Equations (11)–(22) define the optimization model of the upper-level problem. Equation (11) minimizes the overall association risk of the park. Constraint (12) guarantees that the sum of the enterprise areas of each floor cannot exceed this monolayer area. Constraints (13)–(16) restrict an enterprise to having only one specific location and thus cannot be at multiple locations. Constraints (17)–(19) guarantee that each enterprise can be assigned to only one building. Constraints (20)–(22) represent the relationships between the decision variables. The optimization problem of the lower level is defined by Equations (23)–(25). Specifically, Equation (23) describes the objective function of the lower-level problem as the maximization of rental profit for the park. Constraint (24) prevents enterprises from overlapping. Constraint (25) guarantees that each enterprise is located in one location.

In addition, the proposed model allows some enterprises' locations that are given in advance, which is more realistic in the industry park. If the current enterprises' locations are provided, constraints can be added to the model to ensure the accuracy of layout results. Consequently, the current layout of the industry park is considered when applying the proposed model for designing the next stage layout.

5. Solution Strategy

The proposed mathematical model for the large-scale 3DMSELD problem contained a large number of decision variables. In particular, calculation efficiency can be enhanced by reducing the number of decision variables. Therefore, new parameters and variables were introduced to rewrite the original decision variables. These parameters and variables are given by

N_b specific building value of $b \in B$.

b_i The building in which the enterprise i is located.

s_i The floor on which the enterprise i is located.

b_j The building in which the enterprise j is located.

s_j The floor on which the enterprise j is located.

b_i, s_i, b_j, and s_j are nonnegative integer variables.

Then, the original decision variables are reformulated as

$$b_i = \sum_{b\in B}\sum_{s\in S} x_{bsi} N_b \ \forall\, i \in I \tag{26}$$

$$s_i = \sum_{b\in B}\sum_{s\in S} x_{bsi} N_s \ \forall\, i \in I \tag{27}$$

$$b_j = \sum_{b\in B}\sum_{s\in S} x_{bsj} N_b \ \forall\, j \in J \tag{28}$$

$$s_j = \sum_{b \in B} \sum_{s \in S} x_{bsj} N_s \ \forall j \in J \tag{29}$$

At a certain point in time, determine the input parameters and the limitations are imposed by the current layout of the park. Then, an improved GA was employed to solve the problem. A GA imitates the evolution of biology in nature to obtain a relatively optimal solution [58]. The selection, crossover, and mutation operations were implemented in the GA [1]. Our study adopts an improved GA with enhanced elite retention strategies to address the problem of slow convergence of the classical GA. The main process is as follows:

(1) Step 1: Determine the coding mechanism

How the variables in the actual problem are translated into genetic algorithms depends on the coding mechanism to be solved. Therefore, the feasible and initial solutions should be encoded to facilitate the identification and operation of the algorithm. Common coding mechanisms for genetic algorithms are binary coding, floating-point coding, symbolic coding, and gray coding. The new decision variables in the proposed model are integer variables, and the concept of floating-point coding is straightforward and convenient for subsequent evolutionary operations. In particular, these new decision variables were set as integers within a range.

(2) Step 2: Generate the initial population

Each new decision variable was numbered, and the chromosome length was defined as the size of the new decision variables. At the beginning of the algorithm, a population is randomly generated, the size of which is given in advance. As shown in Equation (30), the initial population contains multiple integer codes, each representing a particular layout situation. The population size affects the efficiency of the algorithm. This study establishes constraints based on area, etc., and substituting them can improve the algorithm operation efficiency and reduce the search scope of the algorithms.

$$\text{pop} = \begin{pmatrix} x_{1,1} & x_{1,2} & \cdots & x_{1,L} \\ x_{2,1} & x_{2,2} & \cdots & x_{2,L} \\ \vdots & \vdots & \ddots & \vdots \\ x_{N,1} & x_{N,2} & \cdots & x_{N,L} \end{pmatrix} \tag{30}$$

where x stands for the decision variables, L is the number of new decision variables, and N is the size of the population (i.e., the number of individuals in the population). Each row corresponds to a layout situation, and each column corresponds to a decision variable.

(3) Step 3: Fitness evaluation

Fitness refers to the ability of individuals in a population to adapt to their environment. The objective function was used to obtain the fitness function. It is necessary to establish a reasonable mapping relationship between the objective and fitness functions to ensure that the direction of the increase in adaptability is consistent with the optimization direction of the objective function. Linear fitness scaling can be used to calculate the fitness values. The calculation formula is as follows:

$$Fit' = m * Fit + n \tag{31}$$

where Fit' is the fitness value, Fit is the objective function value, and m and n are the transform parameters. In particular, Fit' needs to meet two conditions: (a) the average of the new fitness values after the linear scale transformation is equal to the original mean, and (b) after a linear scale transformation, the maximum fitness is equal to the specified multiple of the mean of the original fitness. The proposed layout model contains two objective functions that require separate fitness scaling. Through linear transformation, the degree of fitness difference between individuals can be reduced, which is conducive to maintaining the population diversity.

(4) Step 4: Evolutionary operations

The selection is generally based on individual fitness. The classic choice operators include roulette wheel selection, stochastic universal sampling, tournament selection, and truncation selection. The tournament selection mechanism was adopted in our study, and a specific theory can be found in Blickle [59]. The most common crossover methods are single-point, double-point, and multipoint crossing. This study used a double-point crossover, and the schematic diagram is shown in Figure 3. The mutation increases the population diversity and reduces the risk of algorithms falling into local optimal solutions. In our work, the mutation operation of the breeder GA was used, which can be found in Mühlenbein [60].

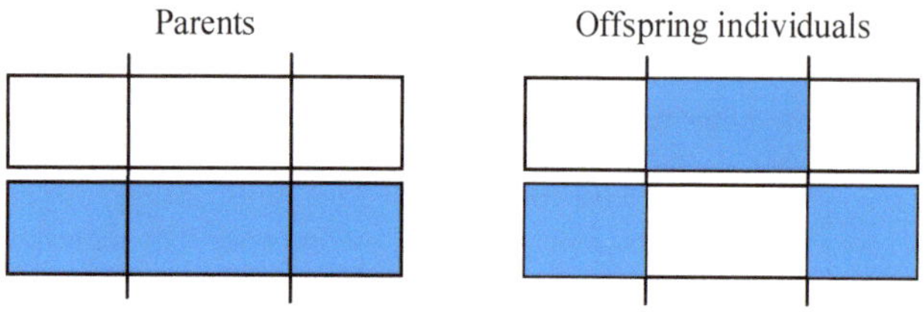

Figure 3. Double-point crossing.

Compared with the classical GA, the improved GA used in this study merges the parent population and the resulting cross-mutation population to obtain a population twice the size. From the merged population, half of the individuals with high fitness are selected to obtain the next generation population, which retains elite individuals.

6. Case Study

This study mainly focuses on two goals: (i) minimizing association risk at the upper level and (ii) maximizing rental profits at the lower level. The decision variables for the two processes were different. The upper-level layout is for risk enterprises, and the lower-level layout is for non/low-risk enterprises. The overall layout design considers multiple buildings, floors, and enterprises. In addition, our model prioritized safety. A case study of the Shunde industrial park was conducted to validate the effectiveness of the proposed model and approach, and the results are reported in this section. The proposed model was implemented using Geatpy [61] (version 2.7.0) and encoded using Python 3.7. All experiments were performed on a computer with an 11th Gen Intel(R) Core (TM) i5-1135G7 @ 2.40GHz and 16 GB memory, with a Windows 10 operating system.

6.1. Data Introduction

An industrial park in Shunde was chosen to obtain relevant data. The park's scale (see Table 2), area parameters, etc., were provided by the park and enterprises. The risk and rent parameters were assumed according to the type of settled enterprise, local environment, and economy, as it is difficult to collect complete and realistic data. The data we set up is general and is used to demonstrate the layout process and conduct case studies. The park contained five buildings, and each building had six floors, with a total of 45 enterprises to be laid out. There were 13 fire risk enterprises (i.e., I1–I13), 8 explosion risk enterprises (i.e., I14–I21), and 14 other types of risk enterprises (i.e., I22–I35), and the upper-level layout of these enterprises was carried out to minimize the association risk; there are 10 non/low-risk enterprises (i.e., J1–I10), and the lower-level layout of these enterprises was carried out to maximize the rental profit. At present, there is no unified statement regarding the population size of the GA. However, it is generally accepted that the larger the population size, the more likely it is to obtain an optimal solution. To obtain a better solution, a more extensive initial population (i.e., 300) and the number of evolutions (i.e., 3000) were chosen.

Table 2. Basic building information of the park.

Building ID	Number of Floors	Floor Area (m^2)
1	6	2332
2	6	3475
3	6	8211
4	6	2180
5	6	2934

6.2. Layout Result

In this subsection, the main results for the layout problem are presented. After variable conversion (Equations (26)–(29)), the scale of the decision variables in the upper-level layout was changed from 1050 to 70, and the size of the decision variables in the lower level was converted from 300 to 20, which significantly reduced the problem scale. In addition, the model was solved three times, and the optimal value was obtained to ensure that the global optimal solution was obtained as much as possible. This study first provided the approximate optimal layout results of 35 risk enterprises (see Figures 4a and 5). As shown in Figure 4a, as the population iterated, the association risk value gradually decreased and stabilized by 2100. The approximate optimal solution was 350.6. Then, 10 non/low risk enterprises were laid out in the remaining space to obtain the maximum rental profit (i.e., 420,592 yuan). The layout results for the 45 enterprises are shown in Figure 5.

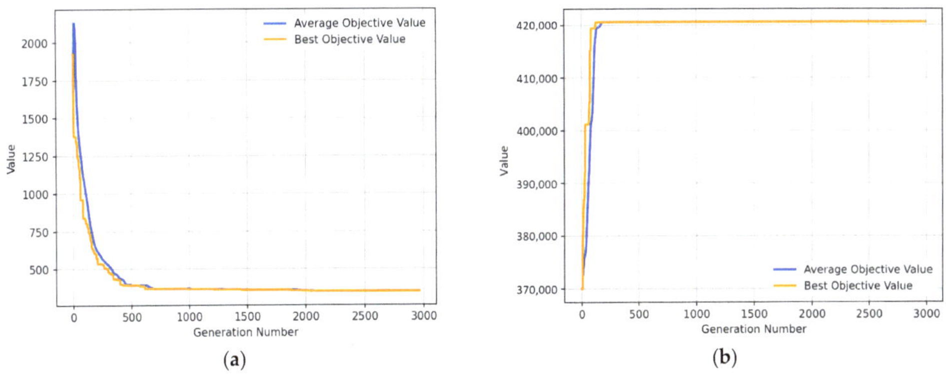

Figure 4. Trace plots of the improved GA solving: (**a**) Trace plot of upper-level layout; (**b**) Trace plot of lower-level layout.

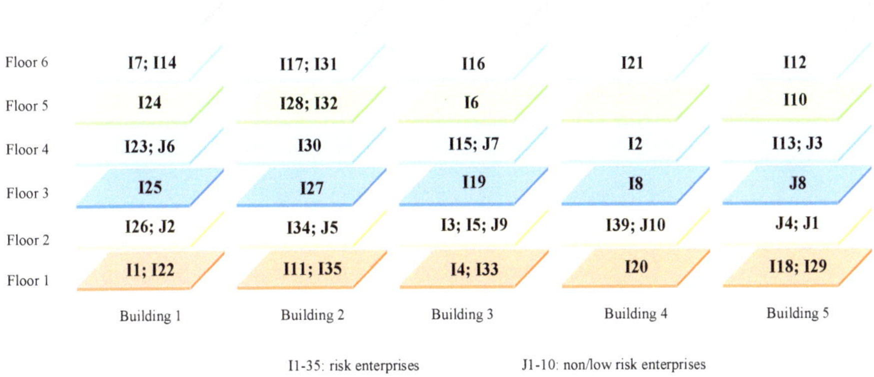

Figure 5. Multi-enterprises layout results.

6.3. Computational Results in Different Scenarios

6.3.1. Only the Profits Are Considered

The overall association risk of the park was measured when considering rent only. Owing to the convenience of moving and unloading goods on the first floor, many enterprises compete to lease lower floors, resulting in enterprises being more willing to pay relatively high rents for lower floors. For 45 enterprises, only the lower-level layout was implemented. In other words, only rental profit was considered, and the layout results are shown in Figures 6 and 7. When the improved GA iteration was approximately 1750 generations, the target value tended to stabilize, with a maximum rental profit of 2,476,679 yuan. As shown in Figure 7, enterprises tended to congregate on the lower floors. According to the calculation of the layout result and risk parameters, the total association risk value in this layout case was 2048.2, which is much higher than the result considering the risk (i.e., 350.6). This also shows that it is necessary to consider the risk factors.

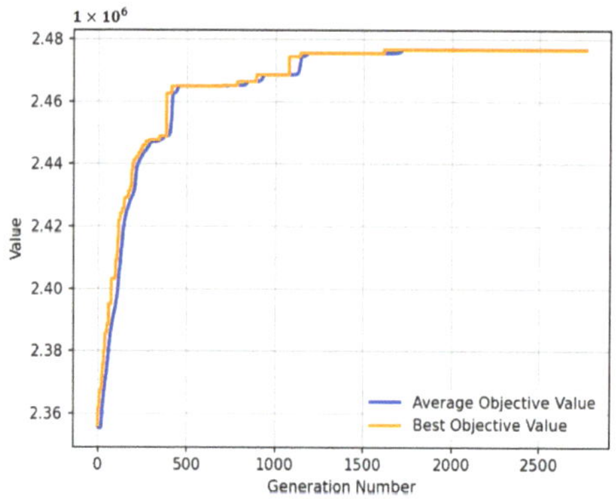

Figure 6. Trace plot considering only the rental profit.

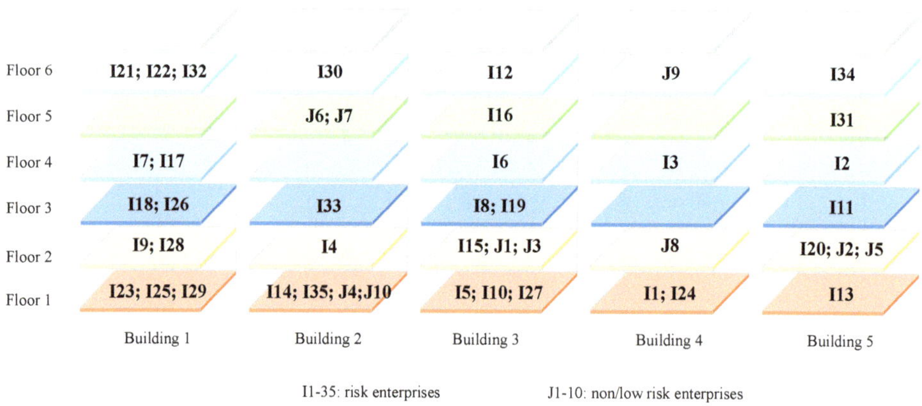

Figure 7. Layout results considering only rental profit.

6.3.2. Consider Current Layout

One of the problems that must be considered when determining a layout is that several enterprises in the park have already determined the advanced location. These enterprises will also affect the overall risk value of the park; therefore, the current layout is needed to consider. Six risk enterprises (I1, I2, I14, I15, I22, and I23) from three different risk enterprise categories were taken to determine the locations in advance. In this case, the location layout of the remaining 29 enterprises was determined.

First, considering the area constraint, the locations of these enterprises (see Figure 8) were assumed, and constraints were added to the solution algorithm. The model was then solved, and the results are shown in Figures 9 and 10. The overall association risk value of the park was 476.3, which was greater than the original layout result (i.e., 350.6). This also reminds us that, when optimizing the layout considering risks, try to conduct unified planning for all enterprises, which is more likely to obtain better results. The rental profit of 10 non/low-risk enterprises was 438,772 yuan, an increase. As can be seen from the layout results in Figure 10, most non/low-risk enterprises were located on the first floor, which may lead to higher rental profits. In the proposed model, rental profit is significantly affected by the upper-level layout. The advanced location of some risk enterprises may also positively impact rental profits.

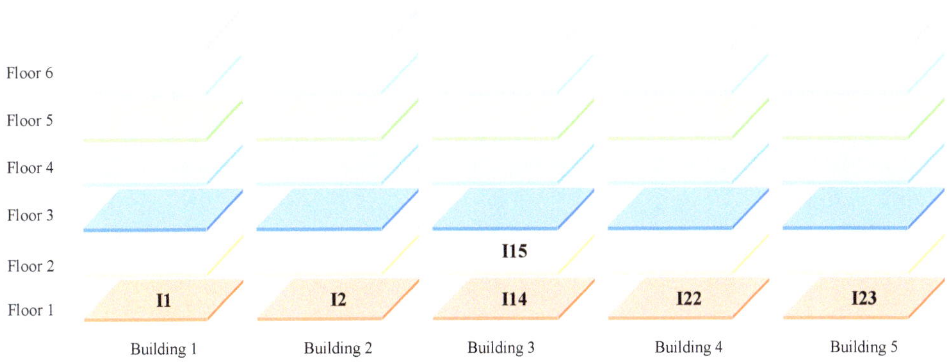

Figure 8. Current layout to consider.

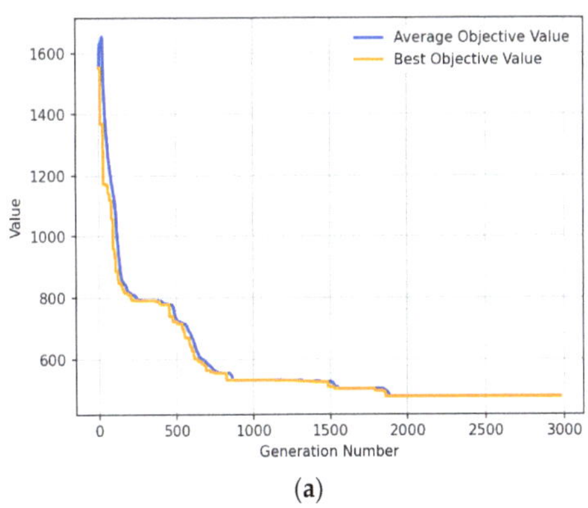

(**a**)

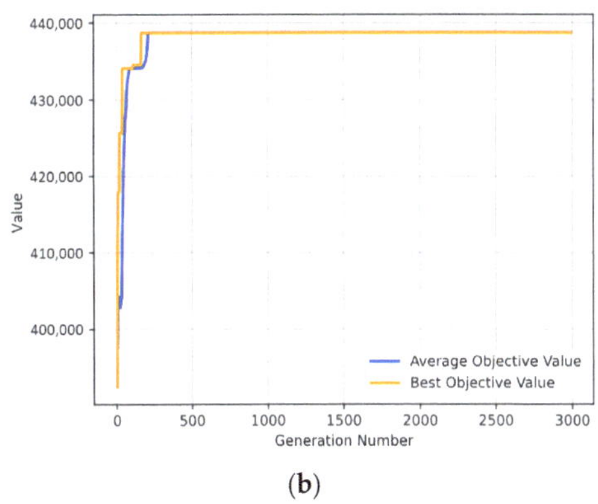

(**b**)

Figure 9. Trace plots of the improved GA solving considering current layout: (**a**) Trace plot of upper-level layout; (**b**) Trace plot of lower-level layout.

From the above scenario simulation, we can see that the proposed model can implement the function of dynamic layout. By changing the input parameters and adding corresponding constraints, the layout of the park at any specific moment can be conducted. Because the lease terms signed by enterprises and the park are generally longer and the frequency of enterprise changes is low, this study did not consider time variables. However, as time goes on and the number of enterprises increases, the total association risk value may change significantly. In addition, from the layout results, it is better to lay out multiple enterprises at once than multiple times. In this sense, it is advisable to re-plan the layout of all enterprises when the total associated risk value exceeds the acceptable level of the industrial park. Achieving enterprise reallocation is a complex issue that requires multi-party coordination. Consequently, this model is more suitable for solving the layout problem when a large number of enterprises settle in the early stages of the park.

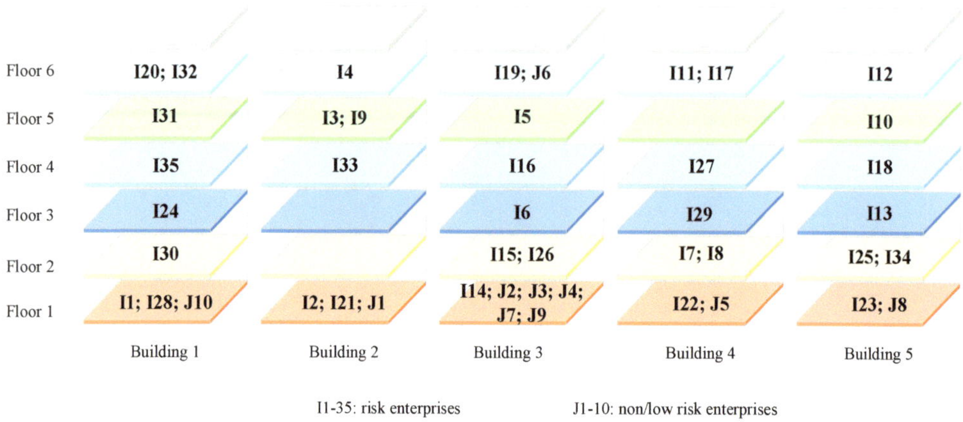

Figure 10. Layout results when some enterprise locations have been determined.

6.4. Sensitivity Analysis

To verify the proposed model and solution strategy, a sensitivity analysis was conducted to investigate the consequences of varying the critical parameters.

6.4.1. Sensitivity to the Number of Risk Enterprises

The number of risk enterprises had a significant influence on the overall risk level of the park. The numbers of risk enterprises in the three categories were changed separately, and the changes in the associated risk value and rental profit were observed. That is, the number of fire risk enterprises ranged from 9 to 17, the number of explosion risk enterprises ranged from 4 to 12, other risk enterprises ranged from 10 to 18, and the total number of enterprises ranged from 31 to 39. The simulation results are shown in Figure 11. In addition, when increasing the number of enterprises, there is a situation in which the areas of some enterprises are too large to be laid out, so the area parameters are appropriately adjusted.

Overall, the total association risk value increased with the number of the three types of risk enterprises (see Figure 11a). Explosion risk enterprises had the most significant change, from 138.5 to 681, while other risk enterprises had the least significant change, from 268.3 to 446.5. This phenomenon may occur because the risk parameter values of explosion-risk enterprises are larger than those of the other two types. In addition, the results remind park administrators to conduct a risk assessment of enterprises before they enter the park and to limit the number of risky enterprises as much as possible. Rental profit decreases with an increase in risk enterprises, but there are fluctuations. Overall, profit changes were minor. The third type of risk enterprise had the most significant risk change: the rental profit decreased from 427,392 yuan to 411,740 yuan, a decrease of 3.7%. The rent profit is insensitive to small changes in the number of risk enterprises.

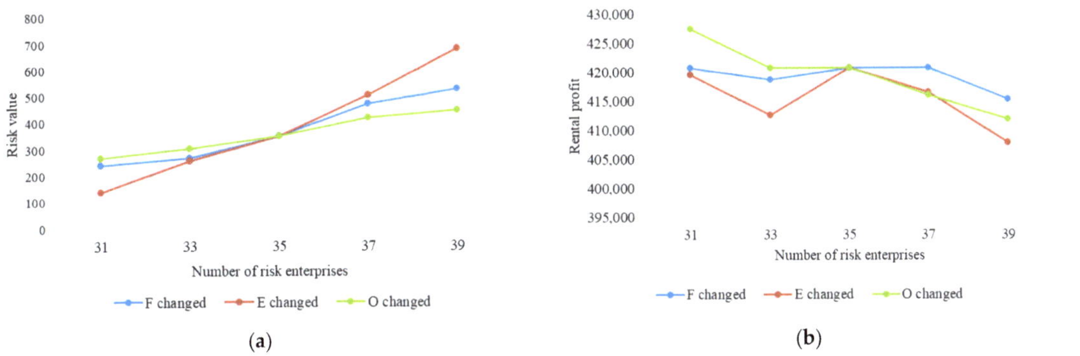

Figure 11. Results under different numbers of risk enterprises: (**a**) Association risk value; (**b**) Rental profit.

6.4.2. Sensitivity to the Size of the Industry Park

The changes in the number of buildings and floors were considered as examples to illustrate the impact of the size of the park on the layout results. The risk value and rental profit were simulated when the number of buildings was four, five, or six. As shown in Figure 12a, the risk value decreased with an increase in the number of buildings and varied significantly. This is because the association risk is greatly affected by distance. In addition, the model assumed that an association risk exists only between enterprises in the same building. When the number of buildings changed from four to five, the rental profit was greatly improved, whereas when the number changed from five to six, the rental profit did not change.

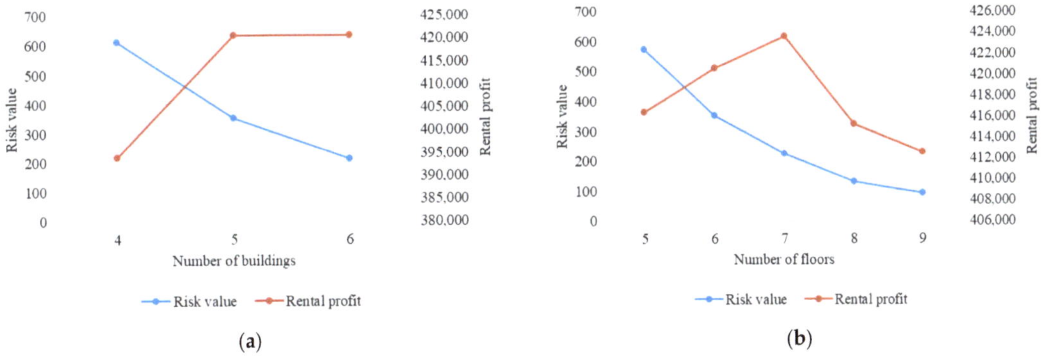

Figure 12. Results under different industry park sizes: (**a**) Number of buildings; (**b**) Number of floors.

Considering the area constraint, the number of floors in the industrial park was changed from five to nine, and the changes in risk value and rental profit were observed (see Figure 12b). The total association risk value decreased as the number of floors increased, which is consistent with our expectations because the risk diffusion coefficient decreases as the floor difference increases. The rate of change in rental profit was insignificant, and Figure 12b shows that the rental profit did not show a linear increasing trend. The rental profit was also insensitive to changes in the number of floors. The rental profit was affected by the results of the upper-level layout. When the number of floors exceeded a particular value, increasing the number of floors did not help rental profit. This may be because enterprises prefer to be located on the lower floors.

7. Conclusions

This study investigated the layout problem of enterprises in a new-type clustered industrial park. First, the association risk was defined, and the calculation formulas were given in combination with the QRA method and domino effect. Risk enterprises were divided into three types: fire risk enterprises, explosion risk enterprises, and other types. Subsequently, a BLINLP model was proposed to formulate the large-scale 3DMSELD problem. In the proposed model, minimizing the total association risk was the upper-level goal. The lower-level goal was to maximize the rental profit of the park. However, the proposed model was difficult to get a global optimum as the problem size was large. In this sense, the decision variables were first transformed to reduce the problem scale. Then, an improved GA was applied. Finally, a case study of the Shunde Industry Park was conducted to validate the proposed model and approach. Two different scenarios were implemented, and critical parameters were tested to provide valuable management insights.

Several insights on the theory and practice of the large-scale 3DMSELD problem considering risk and profit are obtained based on the results in Section 6. First, the bi-level optimization theory represents the hierarchical relationship between layout objectives, which is rarely considered in the existing literature. This model prioritizes enterprise association risk affected by distance: the farther the distance, the smaller the association risk. Secondly, the strategy we propose solves the problem that it is difficult to obtain an optimal solution for nonlinear models built for large-scale problems. It can be seen from the tracking plots (e.g., Figure 9) that the optimal objective value change curve converges well, and the change amplitude is large. In practice, the proposed model and approach are suitable for modeling such problems. Optimal layout results can be obtained that greatly dramatically reduces the total association risk and improves the rental profit. From the layout results, it is found that it is imperative to consider the association risks in the actual layout problem. When laying out considering rental profit only, the risk increased from 350.6 to 2048.2, which validates the effectiveness of the proposed model. In particular, the model considers the impact of existing layouts that often appear in real-world scenarios. It can be seen from the simulation results that a unified layout for all enterprises is more conducive to reducing the overall association risk of the park. In addition, the number of risk enterprises and the size of the park have a significant impact on the total association risk, which is in line with our engineering experience. Finally, the rental profit to the park manager from non/low-risk enterprises is significantly affected by the layout results of the risk enterprises. As can be seen from the sensitivity analysis, rental profit fluctuates but does not change significantly. However, the lower-level layout can still improve rental profit (see Figure 4b).

Despite these contributions and novelties, our study has several potential directions that can be further investigated. The layout results identify the floor on which the enterprise is located, but detailed explanations and constraints on single-floor layouts are lacking. Our work mainly focuses on the risks of fire and explosion, and there is a shortage of investigation on other risks. In addition, our study only considered the current risk level of the enterprise to be laid out; however, risk changes over its lifespan. Therefore, potential future work can be done to (a) add a single-floor specific location layout and form a multi-stage layout framework to solve the large-scale 3DMSELD problem; (b) include other factors in the analysis, such as emergency facilities of risk enterprises and toxic and harmful gas leakage; and (c) study the risk changes during the lease period of enterprises and improve the layout model.

Author Contributions: Conceptualization, X.L. and X.G.; methodology, X.L. and X.G.; validation, H.L.; formal analysis, H.L.; investigation, X.G.; data curation, G.H., S.O.; writing—original draft preparation, X.L.; writing—review and editing, H.L., I.M.H.; visualization, S.O.; supervision, G.H. and S.O.; project administration, G.H.; funding acquisition, X.G. All authors have read and agreed to the published version of the manuscript.

Funding: This work was supported by Interdisciplinary Research Project for Young Teachers of USTB (Fundamental Research Funds for the Central Universities with grant number FRF-IDRY-21-016) and researchers supporting project with grant number (RSP-2021/389) from King Saud University, Riyadh, Saudi Arabia.

Institutional Review Board Statement: Not applicable.

Informed Consent Statement: Not applicable.

Data Availability Statement: Not applicable.

Conflicts of Interest: The authors declare no conflict of interest.

References

1. Wang, R.; Wu, Y.; Wang, Y.; Feng, X. An industrial area layout design methodology considering piping and safety using genetic algorithm. *J. Clean. Prod.* **2017**, *167*, 23–31. [CrossRef]
2. Heikkilä, A.-M.; Malmén, Y.; Nissilä, M.; Kortelainen, H. Challenges in risk management in multi-company industrial parks. *Saf. Sci.* **2010**, *48*, 430–435. [CrossRef]
3. Caputo, A.C.; Pelagagge, P.M.; Palumbo, M.; Salini, P. Safety-based process plant layout using genetic algorithm. *J. Loss Prev. Process Ind.* **2015**, *34*, 139–150. [CrossRef]
4. Wang, R.; Wang, Y.; Gundersen, T.; Wu, Y.; Feng, X.; Liu, M. A layout design method for an industrial park based on a novel arrangement algorithm—Consideration of pipe network and multiple hazard sources. *Chem. Eng. Sci.* **2020**, *227*, 115929. [CrossRef]
5. Wang, R.; Wang, Y.; Gundersen, T.; Wu, Y.; Feng, X.; Liu, M. A multi-objective optimization method for industrial park layout design: The trade-off between economy and safety. *Chem. Eng. Sci.* **2021**, *235*, 116471. [CrossRef]
6. Şahin, R.; Niroomand, S.; Durmaz, E.D.; Molla-Alizadeh-Zavardehi, S. Mathematical formulation and hybrid meta-heuristic solution approaches for dynamic single row facility layout problem. *Ann. Oper. Res.* **2020**, *295*, 313–336. [CrossRef]
7. Rosenblatt, M.J. The facilities layout problem: A multi-goal approach. *Int. J. Prod. Res.* **1979**, *17*, 323–332. [CrossRef]
8. Penteado, F.D.; Ciric, A.R. An MINLP approach for safe process plant layout. *Ind. Eng. Chem. Res.* **1996**, *35*, 1354–1361. [CrossRef]
9. Kia, R.; Khaksar-Haghani, F.; Javadian, N.; Tavakkoli-Moghaddam, R. Solving a multi-floor layout design model of a dynamic cellular manufacturing system by an efficient genetic algorithm. *J. Manuf. Syst.* **2014**, *33*, 218–232. [CrossRef]
10. Latifi, S.E.; Mohammadi, E.; Khakzad, N. Process plant layout optimization with uncertainty and considering risk. *Comput. Chem. Eng.* **2017**, *106*, 224–242. [CrossRef]
11. Sonta, A.; Dougherty, T.R.; Jain, R.K. Data-driven optimization of building layouts for energy efficiency. *Energy Build.* **2021**, *238*, 110815. [CrossRef]
12. Wu, Y.; Wang, Y. A chemical industry area-wide layout design methodology for piping implementation. *Chem. Eng. Res. Des.* **2017**, *118*, 81–93. [CrossRef]
13. Al-Zubaidi, S.Q.D.; Fantoni, G.; Failli, F. Analysis of drivers for solving facility layout problems: A Literature review. *J. Ind. Inf. Integr.* **2021**, *21*, 100187. [CrossRef]
14. Ingole, S.; Singh, D. Unequal-area, fixed-shape facility layout problems using the firefly algorithm. *Eng. Optim.* **2017**, *49*, 1097–1115. [CrossRef]
15. Wang, R.; Wu, Y.; Wang, Y.; Feng, X.; Liu, M. An Industrial Park Layout Design Method Considering Pipeline Length Based on FLUTE Algorithm. In *Computer Aided Chemical Engineering*; Eden, M.R., Ierapetritou, M.G., Towler, G.P., Eds.; Elsevier: Amsterdam, The Netherlands, 2018; Volume 44, pp. 193–198.
16. Wang, R.; Zhao, H.; Wu, Y.; Wang, Y.; Feng, X.; Liu, M. An industrial facility layout design method considering energy saving based on surplus rectangle fill algorithm. *Energy* **2018**, *158*, 1038–1051. [CrossRef]
17. Derhami, S.; Smith, J.S.; Gue, K.R. A simulation-based optimization approach to design optimal layouts for block stacking warehouses. *Int. J. Prod. Econ.* **2020**, *223*, 107525. [CrossRef]
18. Brunoro Ahumada, C.; Quddus, N.; Mannan, M.S. A method for facility layout optimisation including stochastic risk assessment. *Process Saf. Environ. Prot.* **2018**, *117*, 616–628. [CrossRef]
19. de Lira-Flores, J.A.; López-Molina, A.; Gutiérrez-Antonio, C.; Vázquez-Román, R. Optimal plant layout considering the safety instrumented system design for hazardous equipment. *Process Saf. Environ. Prot.* **2019**, *124*, 97–120. [CrossRef]
20. Medina-Herrera, N.; Jiménez-Gutiérrez, A.; Grossmann, I.E. A mathematical programming model for optimal layout considering quantitative risk analysis. *Comput. Chem. Eng.* **2014**, *68*, 165–181. [CrossRef]
21. Groth, K.M.; Hecht, E.S. HyRAM: A methodology and toolkit for quantitative risk assessment of hydrogen systems. *Int. J. Hydrogen Energy* **2017**, *42*, 7485–7493. [CrossRef]
22. Cozzani, V.; Gubinelli, G.; Antonioni, G.; Spadoni, G.; Zanelli, S. The assessment of risk caused by domino effect in quantitative area risk analysis. *J. Hazard. Mater.* **2005**, *127*, 14–30. [CrossRef] [PubMed]
23. Zhou, J.; Reniers, G.; Cozzani, V. Improved probit models to assess equipment failure caused by domino effect accounting for dynamic and synergistic effects of multiple fires. *Process Saf. Environ. Prot.* **2021**, *154*, 306–314. [CrossRef]

24. Ejeh, J.O.; Liu, S.; Papageorgiou, L.G. An MILP model for safe multi-floor process plant layout using the domino hazard index. *Process Saf. Environ. Prot.* **2021**, *148*, 137–165. [CrossRef]
25. Liu, A.; Wu, C.; Peng, X. Research on Area Risk Assessment for Chemical Park based on Domino Effect Model. *Procedia Eng.* **2012**, *45*, 47–52. [CrossRef]
26. Ding, L.; Khan, F.; Ji, J. A novel approach for domino effects modeling and risk analysis based on synergistic effect and accident evidence. *Reliab. Eng. Syst. Saf.* **2020**, *203*, 107109. [CrossRef]
27. Ni, Z.-j.; Wang, Y.; Yin, Z. Relative risk model for assessing domino effect in chemical process industry. *Saf. Sci.* **2016**, *87*, 156–166. [CrossRef]
28. He, Z.; Weng, W. A dynamic and simulation-based method for quantitative risk assessment of the domino accident in chemical industry. *Process Saf. Environ. Prot.* **2020**, *144*, 79–92. [CrossRef]
29. Lu, J.; Han, J.; Hu, Y.; Zhang, G. Multilevel decision-making: A survey. *Inf. Sci.* **2016**, *346*, 463–487. [CrossRef]
30. Cao, C.; Liu, Y.; Tang, O.; Gao, X. A fuzzy bi-level optimization model for multi-period post-disaster relief distribution in sustainable humanitarian supply chains. *Int. J. Prod. Econ.* **2021**, *235*, 108081. [CrossRef]
31. Yang, X.; Niu, D.; Sun, L.; Ji, Z.; Zhou, J.; Wang, K.; Siqin, Z. A bi-level optimization model for electric vehicle charging strategy based on regional grid load following. *J. Clean. Prod.* **2021**, *325*, 129313. [CrossRef]
32. Che, A.; Zhang, Y.; Feng, J. Bi-objective optimization for multi-floor facility layout problem with fixed inner configuration and room adjacency constraints. *Comput. Ind. Eng.* **2017**, *105*, 265–276. [CrossRef]
33. Martinez-Gomez, J.; Nápoles-Rivera, F.; Ponce-Ortega, J.M.; Serna-González, M.; El-Halwagi, M.M. Optimization of facility location and reallocation in an industrial plant through a multi-annual framework accounting for economic and safety issues. *J. Loss Prev. Process Ind.* **2015**, *33*, 129–139. [CrossRef]
34. Nabli, H. An overview on the simplex algorithm. *Appl. Math. Comput.* **2009**, *210*, 479–489. [CrossRef]
35. Xie, W.; Sahinidis, N.V. A branch-and-bound algorithm for the continuous facility layout problem. *Comput. Chem. Eng.* **2008**, *32*, 1016–1028. [CrossRef]
36. Ahmadi, A.; Akbari Jokar, M.R. An efficient multiple-stage mathematical programming method for advanced single and multi-floor facility layout problems. *Appl. Math. Model.* **2016**, *40*, 5605–5620. [CrossRef]
37. Kulturel-Konak, S.; Konak, A. A large-scale hybrid simulated annealing algorithm for cyclic facility layout problems. *Eng. Optim.* **2015**, *47*, 963–978. [CrossRef]
38. Chalupa, D.; Nielsen, P. A simple and robust Monte Carlo hybrid local search algorithm for the facility location problem. *Eng. Optim.* **2019**, *51*, 832–845. [CrossRef]
39. Chen, Y.; Jiang, Y.; Wahab, M.I.M.; Long, X. The facility layout problem in non-rectangular logistics parks with split lines. *Expert Syst. Appl.* **2015**, *42*, 7768–7780. [CrossRef]
40. Balakrishnan, J.; Cheng, C.H.; Conway, D.G.; Lau, C.M. A hybrid genetic algorithm for the dynamic plant layout problem. *Int. J. Prod. Econ.* **2003**, *86*, 107–120. [CrossRef]
41. Patsiatzis, D.I.; Knight, G.; Papageorgiou, L.G. An MILP Approach to Safe Process Plant Layout. *Chem. Eng. Res. Des.* **2004**, *82*, 579–586. [CrossRef]
42. Vázquez-Román, R.; Lee, J.-H.; Jung, S.; Mannan, M.S. Optimal facility layout under toxic release in process facilities: A stochastic approach. *Comput. Chem. Eng.* **2010**, *34*, 122–133. [CrossRef]
43. Cheng, M.-Y.; Lien, L.-C. A hybrid AI-based particle bee algorithm for facility layout optimization. *Eng. Comput.* **2012**, *28*, 57–69. [CrossRef]
44. Xu, Y.; Wang, Z.; Zhu, Q. An Improved Hybrid Genetic Algorithm for Chemical Plant Layout Optimization with Novel Non-overlapping and Toxic Gas Dispersion Constraints. *Chin. J. Chem. Eng.* **2013**, *21*, 412–419. [CrossRef]
45. Emami, S.; Nookabadi, A.S. Managing a new multi-objective model for the dynamic facility layout problem. *Int. J. Adv. Manuf. Technol.* **2013**, *68*, 2215–2228. [CrossRef]
46. Izadinia, N.; Eshghi, K.; Salmani, M.H. A robust model for multi-floor layout problem. *Comput. Ind. Eng.* **2014**, *78*, 127–134. [CrossRef]
47. Alves, D.T.S.; de Medeiros, J.L.; Araújo, O.d.Q.F. Optimal determination of chemical plant layout via minimization of risk to general public using Monte Carlo and Simulated Annealing techniques. *J. Loss Prev. Process Ind.* **2016**, *41*, 202–214. [CrossRef]
48. Zhang, G.; Nishi, T.; Turner, S.D.O.; Oga, K.; Li, X. An integrated strategy for a production planning and warehouse layout problem: Modeling and solution approaches. *Omega* **2017**, *68*, 85–94. [CrossRef]
49. Arnaout, J.-P. Worm optimization for the multiple level warehouse layout problem. *Ann. Oper. Res.* **2018**, *269*, 29–51. [CrossRef]
50. Ikwan, F.; Sanders, D.; Hassan, M. Safety evaluation of leak in a storage tank using fault tree analysis and risk matrix analysis. *J. Loss Prev. Process Ind.* **2021**, *73*, 104597. [CrossRef]
51. Luo, T.; Wu, C.; Duan, L. Fishbone diagram and risk matrix analysis method and its application in safety assessment of natural gas spherical tank. *J. Clean. Prod.* **2018**, *174*, 296–304. [CrossRef]
52. May, W.G.; McQueen, W. Radiation from Large Liquefied Natural Gas Fires. *Combust. Sci. Technol.* **1973**, *7*, 51–56. [CrossRef]
53. Modak, A.T. Thermal radiation from pool fires. *Combust. Flame* **1977**, *29*, 177–192. [CrossRef]
54. Sivathanu, Y.R.; Gore, J.P. Total radiative heat loss in jet flames from single point radiative flux measurements. *Combust. Flame* **1993**, *94*, 265–270. [CrossRef]

55. Wu, C.; Hao, H. Modeling of simultaneous ground shock and airblast pressure on nearby structures from surface explosions. *Int. J. Impact Eng.* **2005**, *31*, 699–717. [CrossRef]
56. Brode, H.L. Blast Wave from a Spherical Charge. *Phys. Fluids* **1959**, *2*, 217–229. [CrossRef]
57. Yang, X.; Shi, S.; Cheng, P. Forecast and Simulation of Peak Overpressure of TNT Explosion Shock Wave in the air. *Blasting* **2008**, *25*, 15–18. (In Chinese)
58. Zhou, G.; Zhu, Z.; Luo, S. Location optimization of electric vehicle charging stations: Based on cost model and genetic algorithm. *Energy* **2022**, *247*, 123437. [CrossRef]
59. Blickle, T.; Thiele, L. A Comparison of Selection Schemes Used in Evolutionary Algorithms. *Evol. Comput.* **1996**, *4*, 361–394. [CrossRef]
60. Mühlenbein, H.; Schlierkamp-Voosen, D. Predictive Models for the Breeder Genetic Algorithm I. *Continuous Parameter Optimization. Evol. Comput.* **1993**, *1*, 25–49.
61. Jazzbin, e.a. Geatpy: The Genetic and Evolutionary Algorithm Toolbox with High Performance in Python. Available online: http://www.geatpy.com/ (accessed on 2 March 2022).

applied sciences

MDPI

Article

Model for Integrating the Electricity Cost Consumption and Power Demand into Aggregate Production Planning

Camila Matos [1,2], Antônio Vanderley Herrero Sola [1], Gustavo de Souza Matias [2,3], Fernando Henrique Lermen [4,*], José Luis Duarte Ribeiro [5] and Hugo Valadares Siqueira [1]

1 Graduate Program in Industrial Engineering, Universidade Tecnológica Federal do Paraná, Street Doutor Washington Subtil Chueire 330, Ponta Grossa 84017-220, Brazil; matoscamila@hotmail.com (C.M.); sola@utfpr.edu.br (A.V.H.S.); hugosiqueira@utfpr.edu.br (H.V.S.)
2 Industrial Engineering Department, Universidade Estadual do Paraná, Street Comendador Correia Júnior, 117, Paranaguá 83203-560, Brazil; gusmatias@gmail.com
3 Chemical Engineering Graduate Program, Chemical Engineering Department, Universidade Estadual de Maringá, Avenue Colombo 5790, Maringá 87020-900, Brazil
4 Industrial Engineering Department, Universidad Tecnológica del Perú, Avenue Arequipa 265, Lima 15046, Peru
5 Industrial Engineering Department, Federal University of Rio Grande do Sul, Avenue Osvaldo Aranha, 99, Porto Alegre 90035-190, Brazil; ribeiro@producao.ufrgs.br
* Correspondence: fernando-lermen@hotmail.com; Tel.: +55-(51)-98494-0121

Citation: Matos, C.; Sola, A.V.H.; Matias, G.d.S.; Lermen, F.H.; Ribeiro, J.L.D.; Siqueira, H.V. Model for Integrating the Electricity Cost Consumption and Power Demand into Aggregate Production Planning. *Appl. Sci.* **2022**, *12*, 7577. https://doi.org/10.3390/app12157577

Academic Editors: Tibor Krenicky, Maros Korenko and Juraj Ruzbarsky

Received: 26 June 2022
Accepted: 25 July 2022
Published: 27 July 2022

Abstract: The constant increases in electricity tax costs and the mandatory contracting of power demand in advance by companies connected to the high-voltage electrical system drive organizations to improve energy planning in their production processes. In addition, market uncertainties make only stochastic methods insufficient for forecasting future production demand. To fill this gap, this study proposes a model that integrates the cost with electricity consumption and power demand into the aggregate production planning, considering the market uncertainties. The model was empirically applied in the food industry, considering a family of potato chips products. From the collected data, a demand forecast was carried out for a later realization of the aggregate planning, using the Holt–Winters forecast model. Before modeling, the new energy demand was calculated, and finally, the model solution verification was performed. In the case study, after application, it was possible to reduce two workers and a cost reduction of R$ 14,288.00. Moreover, the proposal managed to define a power demand that minimized the costs of electric energy and the total costs of the aggregate production planning.

Keywords: planning and production control; aggregate production planning; electricity; power demand; production engineering

1. Introduction

The industrial sector is the largest energy consumer, representing more than a third of primary energy consumption worldwide [1]. With the increasing electricity costs in the recent years, it has become crucial to consider them during production planning [2]. From the producer's point of view, reducing the production process cost can make the product more competitive in the market. Furthermore, within a social context, reducing electricity costs in the productive sector makes it possible to offer more accessible products to the population.

Moreover, the increase in electricity consumption in developing countries follows the population growth and economic progress. In general, it is observed that the population needs to increase the energy use to sustain a better quality of life [3]. Thus, the increase in electricity consumption is one of the results of the economic recovery of emerging countries, but scientists also point it out as one of the factors with the greatest climatic and environmental impact [4]. It is estimated that if consumption in developing countries continues to follow the current trend, it may surpass that of developed countries [5].

The current challenge for developing countries is to ensure more efficient and sustainable resource management to minimize climate and environmental impacts related to energy production [6]. Unfortunately, Brazil's energy consumption data indicate that the country follows the same line as other developing countries, with increasing consumption over time.

Since the Brazilian energy matrix is heavily dependent on water resources, the low rainfall due to the seasons can become a bottleneck for the Brazilian society and its production chain. In the country, some seasons are systematically characterized by a low amount of rainfall, which influences water resources [5,7]. Given this, decision-makers in Brazil must know how to manage energy consumption better, reducing dependence on thermoelectric plants since they burn fossil fuel, pollute, and have higher generated energy costs, unlike hydroelectric plants.

In this scenario, it is essential that companies efficiently manage their energy consumption, minimize the impact that bad weather has on energy supply, guarantee the efficiency of their operations, and optimize their electricity costs. The efficient use of electrical energy is a way for companies to reduce operating costs and increase their economic performance [8,9].

Several studies in the literature present predictions about energy consumption. Chou et al. [10] employed Machine Learning models and Artificial Neural Networks to predict the demand and ensure the supply to customers of energy supply networks. Hu et al. [8] used prediction models to predict energy consumption; the application case was implemented in energy pipelines to meet the needs and optimize the management. Maaouane et al. [11] used regression models to predict industrial energy demand in Morocco. Mourtzis et al. [12] developed a model to manage energy demand and predicted prices in a way that allows companies to reduce energy costs. On the other hand, no works are found in the literature which differentiates demand forecasting and aggregate production planning.

Aggregate production planning is a way of managing production capacity. Usually, it is carried out considering a period of 2 to 18 months to meet the fluctuation in demand [13]. The aim is to determine the optimal amount of raw material, production, workforce, and inventory levels for each planning period considering limited resource capacities [14,15]. Given the identified gap, this study proposes and applies a model that integrates the cost of electricity consumption and power demand to aggregate production planning. The proposed model allows production planning to consider medium-term energy demand forecasts to reduce waste in procurement in a food production company in Brazil.

This study can assist in the energy planning of companies from different industrial segments since ANEEL (National Electric Energy Agency) Resolution 414/2010 determines that consumers of high voltage electricity must previously contract a power demand to be used throughout the contract period [16]. This requires companies to plan production to meet, as accurately as possible, the amount of demand needed.

To meet the demands of companies, this study aims to answer the research question: how can electric energy and power demand be considered for cost reduction in aggregate production planning? To fill this question, this study aims to propose a model that integrates the cost with electricity consumption and power demand into the aggregate production planning, considering the market uncertainties.

2. Theoretical Background

Traditional PAP problem models can be classified into six categories: (i) linear programming (LP); (ii) linear decision rule (RDL); (iii) method of transport; (iv) management coefficient approach; (v) search decision rule (RDB); and (vi) simulation [17]. Feng et al. [18] complemented with three more methods: (i) dynamic programming; (ii) transport tables, and (iii) fuzzy logic mathematical programming. When using any of the planning models, the objectives and inputs of the model are generally considered to be deterministic, and problems can only be solved if they have the sole objective of minimizing the cost during the planning period [15].

According to Arruda Junior [19], studies have been using models to support aggregate planning in natural systems since the 1980s. For example, Mehdizadeh et al. [20] developed a bi-objective optimization model for an aggregate planning problem with work learning effect and machine deterioration. The first objective maximizes profit, and the second minimizes costs associated with machine repairs and deterioration. The objective was to obtain the appropriate production rates at regular and overtime levels, inventory, and shortages, hiring and firing of workers, and the quantities of subcontractors. To validate the mathematical formulation of the model, it was converted into a single-objective model using the fuzzy objective programming method, based on which computational experiments are performed on a set of small-sized random instances solved by the software LINGO.

Izadpanahi and Modarres [21] performed a linear programming model with three objective functions, which minimize operational cost, energy cost, and carbon emission. It is assumed that several types of energy can be used. They differ in the amount of pollution, calorific value, and cost. The model was applied to the foundry factory, which produced a wide range of products such as aluminum, copper, and lead, and used different machines and manufacturing equipment capable of consuming different types of fossil fuels with different carbon emissions. The objective was to determine the most effective means of satisfying anticipated aggregate demand by adjusting production rates, inventory levels, backlogs, and other controllable variables. The model was solved as a single-objective model, applying a goal-setting technique, and to overcome the effects of uncertain input data, a robust optimization approach was applied to the model. These objective functions sought to minimize operating costs, energy, and carbon footprints. The authors concluded that by increasing the budget level of uncertainty, the amount of increase for electricity cost is much greater than other objective function costs.

Latifoglu et al. [22] presented an optimization model for aggregate production planning under the assumption that electricity supply is subject to uncertain interruptions caused by participation in uninterruptible load contracts. The objective was to minimize the cost of electricity used in production by providing a robust production plan to guarantee demand satisfaction in all possible outage scenarios. The model considered the operational level problem, aggregate production, and inventory planning with the electricity supply uncertainty and deterministic demand. Production decisions were separated from interruption decisions, as the former belonged to the industrial company, while the latter related to the electricity retailer. A heuristic was developed to find a viable solution to the model, in which it found an optimal solution for each instance in 7 periods. The outage uncertainty framework allows different contract and operational rules to be incorporated into the production planning problem simultaneously, such as limiting production in post-outage recovery or prohibiting production level increases in specific periods.

Zhang et al. [2] proposed a new mathematical model to determine efficient scheduling to minimize energy and electricity consumption for the flexible workshop scheduling problem in a time-of-use policy. In addition, a speed selection option has been added to the model, which represents the selection of variable operating speeds. It can be applied to industry and help decision-makers develop production schedules that can reduce time and electricity costs.

Choi and Xirouchakis [23] proposed a definition with a holistic approach to production planning in a reconfiguration manufacturing system with energy consumption and environmental effects. A methodology was defined based on reconfigurable manufacturing. The considered model was developed in linear programming with multiobjective functions to minimize energy consumption and maximize throughput, subject to linear constraints related to various resources and constraints for customer demands. The results showed the developed approach's efficient and practical applicability.

3. Materials and Methods

Figure 1 presents the model proposed in this work to integrate energy demand forecasting and aggregate planning, which is composed of four steps: (i) Data collection; (ii) Power Demand Definition; (iii) Production Demand Forecast; and (iv) Aggregate Planning Modeling and Solution.

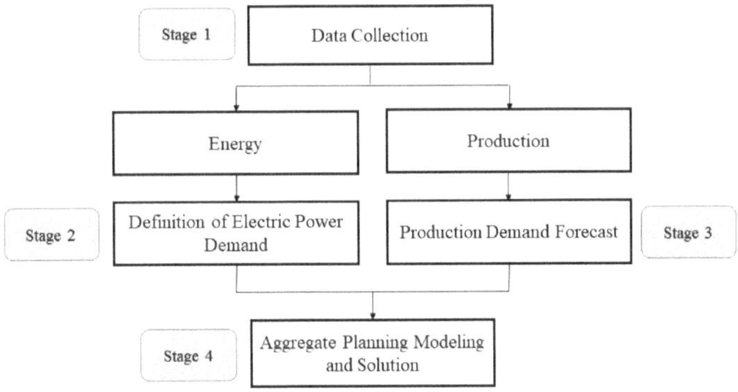

Figure 1. Stages of the proposed model.

The development and application of the model were based on a study carried out in a cooperative in the agro-industrial sector, located in southern Brazil, which produces a family of products, namely: straw potato; wavy potato chips; and plain potato chips (traditional and bacon with cheddar). In addition, the cooperative produces an average of 90 tons/month of French fries in a make-to-order environment.

3.1. Stage 1—Data Collection

Initially, the product family was identified based on data provided by the studied cooperative. Regarding electricity, it is necessary to collect: the type of contract (blue or green modes); the consumption and demand of power on and off the peak; the energy consumption taxes (R$/kWh), and the power demand (R$/kW) on and off-peak; and the demand for contracted and consumed power.

The peak time, defined by ANEEL—National Electric Energy Agency—in its Resolution 414/2010, occurs between 6 pm and 9 pm when electricity consumption is higher, except on Saturdays, Sundays, and national holidays. The off-peak time is the period composed of the set of consecutive daily hours and complementary to those defined during peak hours.

To discourage the use of electricity at peak hours by high voltage consumers, higher tariffs are imposed at that time. For this reason, the electricity bill has two modalities: blue and green. Consumers powered by 69 kV or more are charged only in the blue mode, while the others can choose blue or green, whichever is more convenient for the consumer. In the blue mode, there is a tariff differentiated by the contracted power demand at the peak and off-peak, and in the green mode, the power demand tariff is unique and can be consumed both at the peak and off [16].

Concerning production, the production demand forecast must be identified; the productive capacity of production; the cost of production; the cost of labor; the cost of hiring employees; the cost of dismissing employees; the forecast of energy consumption; the cost of electricity; the cost of power demand; the number of employees available in the period; and the cost of overtime.

3.2. Stage 2—Definition of Electric Power Demand

The charging of the electricity demand taxes, according to Normative Resolution No. 414, of 9 September 2010 [16], is calculated by Equation (1).

$$CDP = TDPC \times DPM \quad (1)$$

where *CDP* is the monthly cost of power demand (R$), *TDPC* is the contracted power demand taxes (R$/kW), and *DPM* is the Measured Power Demand (kW).

Overshoot occurs when an amount above the power demand contracted by the company is consumed. In this way, a different charge is made, obtained from Equation (2) [16].

$$CUDP = (DPM - DPC) \times (2 \times TDPC) \quad (2)$$

CUDP is the cost of exceeding power demand; *DPC* is the Contracted Power Demand; *TDPC* is the Contracted Power Demand Taxes.

For example, during the contract period of one year, the company pays monthly for the previously contracted power demand, even if it is not used. Overshoot occurs when the electric power utility measures a value above the power demand contracted by the unit. This tax is stipulated by ANEEL and costs twice as much as contracted. In this condition, the company pays the contracted value plus the value of exceeding demand.

As the company must contract only one demand value charged monthly during the contract period, the proposed model considers each value observed in recent years as a possible power demand to be contracted (DEC_j). This is compared with the other measured electrical demands (DEM_i) to check for overshoot.

The power demand to be contracted is mandatory and continuously made available by the distributor, at the point of delivery, according to the value and period of validity previously established in the contract, and which must be paid in full, whether or not used during the billing period, expressed in kilowatts (kW). During the entire contract period, for example, one year, the company pays monthly for the previously contracted power demand, even if it is not used. Overshoot occurs when the electric power utility measures a value above the power demand contracted by the unit. The rate for exceeding demand stipulated by ANEEL is double the contracted rate. In this condition, the company pays the contracted value plus the value of exceeding demand. In practice, companies usually contract the highest power demand value measured between the months of the previous year to avoid exceeding demand during the next contract period.

A contracted tax (TC_i) is applied to the contracted demand; if consumption is greater than the contracted demand, an overtaking tax (TU_i) is also applied. TC_i and TU_i are defined by ANEEL [17]. The cost during the period is the sum of the costs of the '*n*' months of the contract. The power demand to be contracted, then, is the one that results in the lowest cost of the period to be contracted (minCDE_j), obtained by Equation (3).

$$CDP_j = \sum_{i=1}^{n} \left(\left(DPC_j \times TDPC_i \right) + k \times \left(DPM_i - DPC_j \right) \cdot \left(2 \times TDPC_i \right) \right) \quad (3)$$

where

$$k = \begin{cases} 1, & DPM_i > DPC_j \\ 0, & otherwise \end{cases}$$

3.3. Stage 3—Production Demand Forecasting

This step consists of performing the production demand forecast. The Holt–Winters model was used because it has many uses in demands that present variations, seasonality, and trends. It contains triple exponential smoothing without recalculating seasonal factors from scratch [24,25]. The software R Project x64 (version 3.2.4, R Core Team, Vienna, Austria) and Excel developed the models. The R Core Team [26] developed the R Cran software to read and model the demand data. The TSA, MASS, tseries, and forecast packages were used for the modeling.

The first manipulation step is plotting the correlogram and performing an initial analysis to identify the components of randomness, trend, and seasonality. When the series is random, autocorrelations close to zero are observed. However, there are drops or positive peaks in the values when there is a trend or seasonality. Therefore, if the data show a trend, it is necessary to carry out the differentiation process to remove this component and then repeat the initial analysis of the differentiated series.

This repetition can take place until the trend component is removed. To prove the assumption of the removal of the trend component, it is necessary to carry out a hypothesis test at 0.05 of significance (Augmented Dickkey–Fuller) [27]. This test is based on the following hypotheses: H0: non-stationary process (null hypothesis); H1: stationary process (alternative hypothesis).

The series is broken down into trend and seasonality in the second step, and the graph is displayed for verification. After this procedure, the database is divided into testing and training, in which 87% train (fit) the model and 13% test the model's prediction quality.

The training process involves a series of steps. First, there is the adjustment of the Holt–Winters model, which can be additive or multiplicative, and, after that, an eight-step forward prediction is performed with these two approaches. Next, the forecasted models must be analyzed to choose the best among them. This analysis is performed from the calculation of forecast errors. Finally, with the results in hand, the prediction made by the model that presents the best performance is chosen.

Then, it is necessary to carry out graphical tests of normality and autocorrelation. The most commonly used test is the Shapiro–Wilk test [28]. The model's simplifying hypotheses are H0: residuals are not normal (null hypothesis); H1: residuals are normal (alternative hypothesis). If the *p*-value of the test is less than 5%, the null hypothesis is rejected so that the model meets the normality assumption.

3.4. Step 4—Aggregate Planning Modeling and Solution

The model developed to solve the Aggregate Production Planning (APP) consists of a linear programming model. The model equation was elaborated involving two dimensions, labor and financial. The first case involves hiring and subcontracting personnel. As for production, to carry out aggregate planning, it is necessary to collect the data presented in Section 3.1.

It is necessary to define its horizon to prepare the planning model, denoted in this equation by H. For aggregate production planning problems, the horizon usually varies from three to fifteen months ($3 \leq H \leq 15$). Thus, the H can be changed according to the problem. In this study, H is equal to 12 months since the idea is to develop an annual production plan.

After defining the horizon, it is necessary to determine the main variables and parameters that describe the problem. At this stage, the new reduced electric energy cost and power demand are allocated as variables and have their restrictions. With the literature support, it was possible to define those that will be used in the mathematical equation of the proposed model according to the nomenclature table presented in the Nomenclature list.

To obtain the cost of electricity, consider the electricity taxes (TE) in R$/kWh and an energy performance indicator (I) [29] that measures the relationship between energy consumption electricity by the amount produced (kWh/kg). This represents the average value from January monthly indicators 2014 to December 2018. Thus, the cost of electricity (R$) is obtained by: $C_t^{ee} = I \times TE$.

As this study focuses on integrating this power demand to reduce costs, the power demand cost in the period was considered a parameter, and the contracted power demand was considered a decision variable. In this way, from such definitions, it is possible to build the cost minimization function (Equation (4)) [30]:

$$Min \sum_{t=1}^{T} (C_t^u \cdot U_t + C_t^{ee} \cdot U_t + C_t^e \cdot U_t^e + C_t^m \cdot T_t + C_t^c \cdot T_t^c + C_t^D \cdot T_t^D + C_t^{He} \cdot H_t^e + C_t^S \cdot U_t^S + C_t^f \cdot U_t^f . + C_t^{de} \cdot D_t^e) \quad (4)$$

This function has the following restrictions, based on the nomenclature presented in the "Nomenclature list", such as the conservation of workforce, the limitation of production in the period, the limits of overtime, the limit of stock, the limits of backorders, the power limitation, and non-negativity constraints.

In the first constraint (Equation (5)), the number of workers in the current period (T_t) must be equal to the number of workers in the previous period (T_{t-1}) added to the hirings in the current period (T_t^c) and subtracted from layoffs in the current period (T_t^D), being [30]:

$$T_t = T_{t-1} + T_t^c - T_t^D \text{ para } 1 \leq t \leq H \quad (5)$$

The second constraint (Equation (6)) is characterized by the limitation of production in period t, in which the produced capacity cannot exceed the available capacity. This is internally defined based on available labor hours, regular or extra. Thus, it is determined by the amount that the company can produce in normal hours per month (H_t^N) and in overtime (H_t^H) [30]:

$$Ca_t = H_t^N + H_t^H \text{ para } 1 \leq t \leq H \quad (6)$$

In the third constraint (Equation (7)), there is a limit for overtime (established by the company itself); that is, if it is not desired that overtime production exceeds 25% of production capacity in each period (H_t), the constraint appears to be [30]:

$$H_t \leq 0.25 \text{ to } 1 \leq t \leq H \quad (7)$$

This variable is balanced at the end of each period in the stock restriction. The net demand is obtained from the sum of the demand D_t and the number of units missing U_{f-1} from the previous period. Thus, this can be met by current production (in normal U_t, subcontracting U_t^S, extra hour U_t^H and by previous stock E_{f-1}), or part of it is accumulated. This relationship is represented by Equation (8) [30]:

$$U_t^e = E_i + H_t^n + U_t^{He} + U_t^S - D_t - U_{f-1}, \text{ para } 1 \leq t \leq H \quad (8)$$

The backorder restriction (Equation (9)) is limited to the number of products allowed by the company that will be out of stock, being [30]:

$$U_t^f \leq X \text{ to } 1 \leq t \leq T \quad (9)$$

The power limitation constraint (Equation (10)) is limited to contracted demand power [30].

$$D_t^{pe} = X, \text{ to } 1 \leq t \leq H \quad (10)$$

The non-negativity constraints indicate that the number of units produced (U_t), in final stock (U_t^e), of available workers (T_t), hired (T_t^c), and fired (T_t^D), must be greater than or equal to zero, according to Equation (11) [30].

$$U_t \, ; \, U_t^e \, ; \, T_t \, ; \, T_t^c \, ; \, T_t^D ; \, U_t^s ; \, U_t^f ; \, H_t \geq 0 \text{ to } 1 \leq t \leq H \quad (11)$$

All parameters and variables that involve the production process were considered. However, according to the company's characteristics, they can be removed from the equation during the application or reset to zero. Thus, the resulting model determines the most effective means of satisfying the expected aggregate demand, adjusting production rates, the amount of electrical energy, and other controllable variables.

To solve the model, the equation is solved with the help of the Solver tool of the Excel software. Then, as planned, the production plan for the next 12 months was drawn up.

After this procedure, the current scenario of the cooperative is compared with the proposed scenario using the new power demand.

3.5. Model Verification (Sensitivity Analysis)

Sensitivity analysis consists of verifying the stability of the solution from possible variations of the parameters used in linear programming; that is, it determines how the solution can be modified from changes in its parameters [31]. For Butler et al. [32], it is the one who determines the robustness of the solutions obtained. In this analysis, some points are verified, in which Bazaraa et al. [33] reported that the shadow price is associated with the constraints of Linear Programming; if it increases, the objective function also changes in the same proportion, such as, for example, each more unit of raw material that is available for production also increases the profit in the objective function.

The reduced cost of variables measures the impact on the objective function caused by the entry of a variable unit in the solution. Thus, when a variable presents positive values, it does not participate in the optimal solution, and a penalty is to be paid if used. Sensitivity analysis was performed using the Solver tool of the Excel Software (version 3, Microsoft, Seatle, WA, USA).

4. Results

This section is separated into five subsections: (i) Overview of the collected data; (ii) Definition of New Power Demand; (iii) Production Demand Forecast; (iv) Modeling and Solution of Aggregate Planning in the Cooperative; (v) Production plan for the case studied; and (vi) Model Verification from Sensitivity Analysis. This research is also delimited in the economic sector based on the objectives outlined. It was developed in a company in a financial segment geographically located in the southern region of Brazil. As for the branch of activity, it is agro-industrial. The type of activity fits into a single family of products, with a specific profile to produce French fries.

Due to the cooperative opening and interest in the proposed study, the model was applied in an agro-industrial cooperative in southern Brazil. It comprises 877 members and 3153 employees, with a turnover of 2.91 billion reais, with business units divided into operations (agricultural, meat, milk, potato, beans, and administration) and industrial (meat, milk, potato, and beer). For example, the French fries unit has 36 employees (17 in the potato production line and the rest in administration and other sectors), producing an average of 90 tons/month of French fries in packages ranging from 80 g to 380 g.

This study was carried out in a make-to-order-rated environment. This system is dynamic and has flaws, such as a lack of operators and materials and low adherence to production orders due to pre-established delivery times [34,35].

The cooperative produces only a single family of products: potato straw (traditional, extra-fine, and very fine), wavy potato chips (traditional, meat, cheese and onion, and salsa), and smooth potato chips (traditional and bacon with cheddar). Its production process is represented in Figure 2.

Initially, the raw material arrives at the unit already processed (cleaned and peeled). Afterward, the potatoes are washed and directed by an infinite conveyor for a selection, in which the employees cut the defects presented. Next, they follow the conveyor belt and go through the slicing stage, utilizing a cutter according to the model of the potato produced (straw potatoes, wave potato chips, and plain potato chips). Then, the sliced potatoes are washed to remove the starch.

In the next step, the excess water is removed, and they go on to the oil frying step by submersion using mats. Afterward, the potatoes are cooled and seasoned, and a selection is made to remove those outside standards. Finally, the ready-made potatoes go to packaging, filling them into the packages together with nitrogen; the packages are placed in cardboard boxes and sent to the shipment as a finished product.

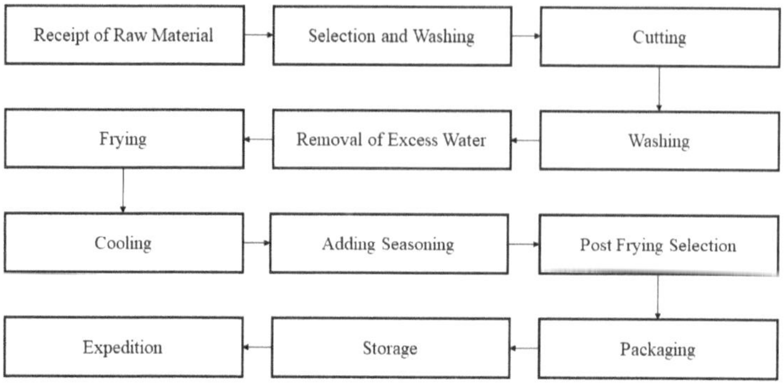

Figure 2. Potato production process.

4.1. Overview of the Collected Data

The power demand contracted by the industry in the last contract period was 255 kW. The consumer unit pays the exceeding part if more than 5% of the contracted demand is consumed. Because of this, the contracted power demand taxes have a value of R$ 13.14/kW, and the overtake taxes, R$ 26.28/kW.

From the history of electric power consumption over the last five years, available in Appendix A (Table A1), it was possible to plot a graph to analyze the behavior of this consumption, as shown in Figure 3.

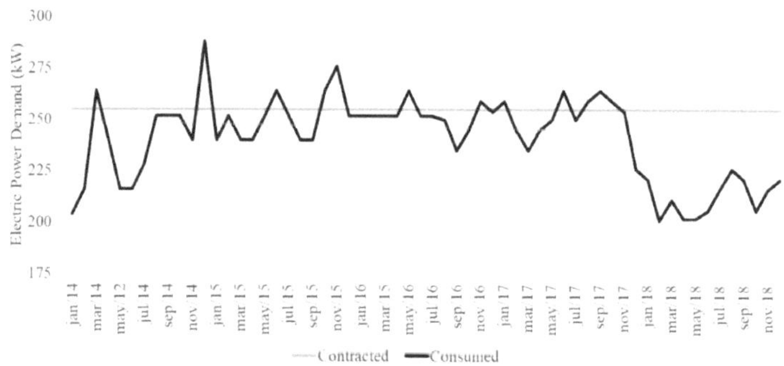

Figure 3. Demand for contracted and consumed electrical power over time.

From Figure 3, it is possible to see that, over the last five years, the consumer unit exceeded the contracted power demand in some months, and in most of them, it did not use the contracted value of 255 kW. In 2014 and 2016, consumption was exceeded in just two months, while in the other months, there was a loss of contracted power demand. In 2015, there were three overtaking. It was in 2017 when most overtaking occurred, totaling five months. On the other hand, during 2018, the consumer unit paid for a much higher demand than used. This shows a lack of reconciliation between production planning and energy planning. Two overruns in the contracted demand generated a fine, causing losses to the company, highlighting the need to carry out this study.

4.2. Definition of New Power Demand

From the data collected, all the power demands measured from January 2014 to December 2018 were organized in a table (Appendix A, Table A2). From this, and with the help of Equation (3), it was possible to obtain the total costs of each value and, consequently, define the best among them, as shown in Figure 4.

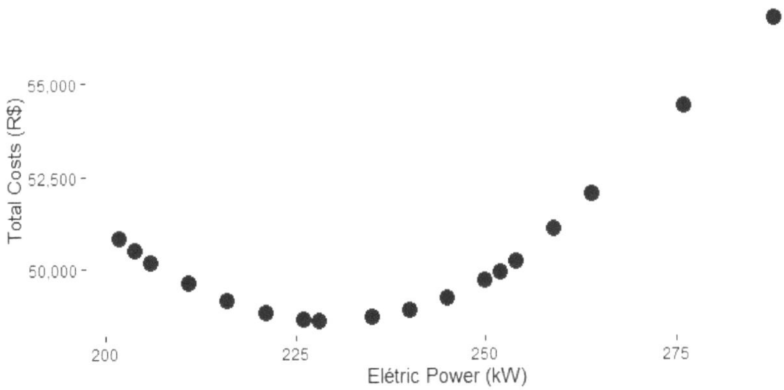

Figure 4. Power demand costs.

As shown in Figure 4, it is noted that the 240 W and 235 W demands had the exact value of R$ 67,881.24. Thus, the demand to be contracted by the cooperative is 235 W.

4.3. Production Demand Forecast

To perform aggregate planning, it is needed to identify demand forecast data. As the studied company did not have its demand forecast, it was necessary to perform it. Otherwise, only data would be collected.

To prepare the production demand forecast, the historical data of the production of kg of potato were considered in the period of 60 months, from January 2014 to December 2018, given in Appendix A (Table A3). Therefore, the first step was to analyze the data profile surveyed to identify the components present, as shown in Figure 5.

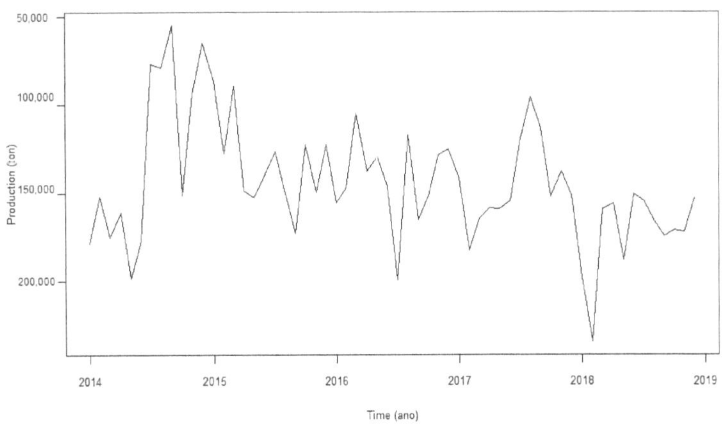

Figure 5. Potato production history from January 2014 to December 2018.

From Figure 5, it is possible to notice that potato production is a series that shows a slight downward trend with high and low peaks. However, it is also observed that it does not present a well-defined pattern of seasonality and cyclicality. Therefore, the correlogram was plotted to understand if the series was random or if it had any tendency or seasonality, as seen in Figure 6. When the series is random, it has autocorrelations close to zero. However, when there is a trend or a seasonality, the series shows a downward trend or positive peaks in the values.

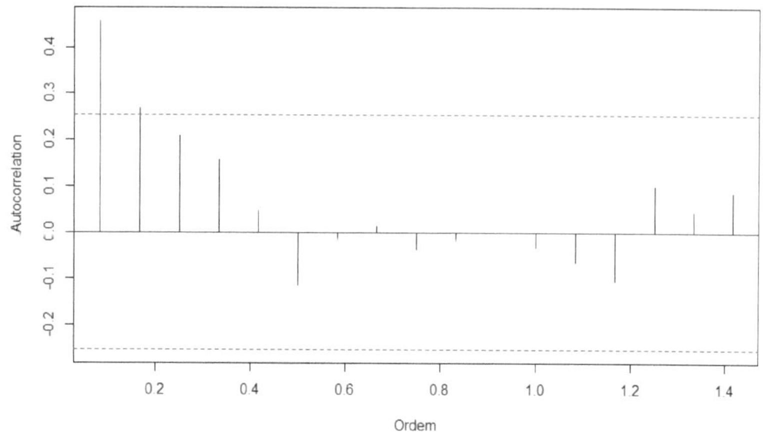

Figure 6. Correlogram of the potato production time series with normalized data.

In Figure 6, the vertical axis indicates autocorrelation, and the horizontal axis indicates lag. The blue dashed line indicates where it is significantly different from zero. Note that all values except the first are within the blue dashed line limit. This means zero autocorrelation, indicating that potato production can be random and supposedly stationary. On the other hand, if the values were above the dashed line, the series would not be random and would have to be treated with a moving average, which is not the case. However, to be sure of the statements, it was necessary to carry out a test to prove the assumptions.

The Augmented Dickey–Fuller (ADF) test resulted in −3.5467, Lag order = 3, and p-value = 0.0453, confirming the alternative hypothesis that the process is stationary. In this case, potato production can be random and stationary. From the ADF test, the series decomposition into trend and seasonality was performed, as shown in the graph in Figure 7.

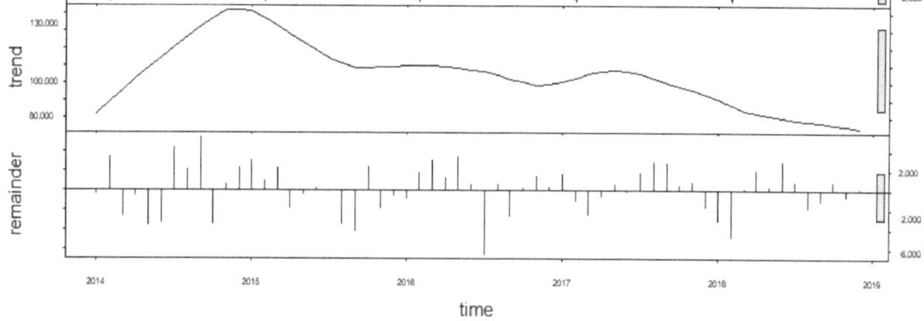

Figure 7. Components of trend and seasonality.

After differentiating the actual demand, the components' performance and behavior show seasonality in the production of potatoes, always in the same period over five years. As for the trend, it is possible to observe a slight growth in 2014; from 2015, it started to decrease. This can be explained by the crisis in mid-2014, with the drop in the Brazilian Gross Domestic Product (GDP) [36], which directly affected the cooperative's sales.

The STL algorithm performs time series smoothing using the LOESS algorithm [37] in two loops; the inner loop iterates between seasonal and trend smoothing, and the outer loop minimizes the effect of outliers. During the inner loop, the seasonal component is calculated first and removed to calculate the trend component. Then, the remainder is calculated by subtracting the seasonal and trend components from the time series. Decomposing seasonal trends using LOESS (STL) is a robust time-series decomposition method often used in economic and environmental analyses. The STL method uses locally adjusted regression models to decompose a time series into trend, seasonal, and remainder components [38–40].

The series was divided into two sets of data, the first consisting of samples from 1 to 52 and the second from 53 to 60, to check the quality and to plot the profile of the additive and multiplicative Holt–Winters models, as shown in Figure 8.

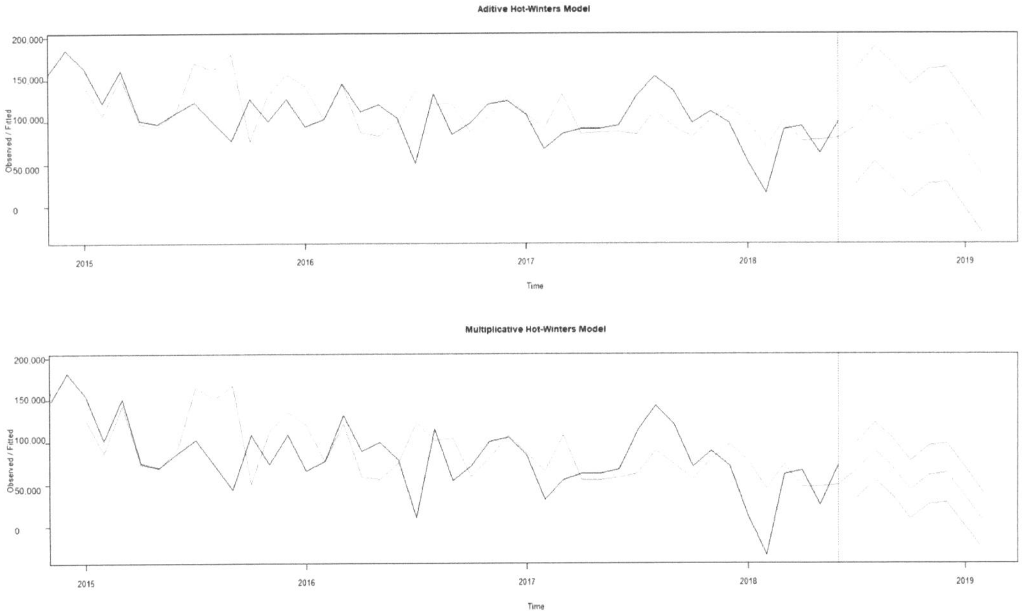

Figure 8. Holt–Winters additive and multiplicative model.

It can be noticed from Figure 8 that the profile behavior of the two models is close to the profile of the original data. However, this visual analysis does not define the best model to represent the demand forecast. For this reason, forecast errors were calculated. The calculation results are shown in Table 1, where REQM is the root mean square error, MAE is the mean absolute error, and MAPE is the mean percentage of the absolute error.

Table 1. Model performance measures.

Models	REQM	MAE	MAPE
Additive	18,928.18	14,549.97	62.30
Multiplicative	15,047.58	13,076.68	28.61

From Table 2, it is possible to verify that the multiplicative model has the lowest error considering all metrics; that is, its performance is better for forecasting production demand. Therefore, the errors of the multiplicative model were statistically tested in order to identify normalities. The Shapiro–Wilk test resulted in W = 0.929 and *p*-value = 0.0152, thus rejecting the null hypothesis. This indicates that the multiplicative model is normal. Afterward, the production demand forecast was carried out for the next 12 months, as shown in Table 2.

Table 2. Production demand forecast.

Month	Production Demand Forecast (kg)
January	69,579.00
February	45,862.00
March	84,995.00
April	74,710.46
May	69,592.76
June	71,464.08
July	82,127.11
August	97,713.42
September	84,283.04
October	66,475.00
November	77,758.62
December	79,578.84

It is noted that the production forecast presents a variation between 45,862.00 kg and 97,713.00 kg, representing 46.9%.

The demand forecast performed will be used in aggregate production planning. Because to do it, it is necessary to know the demand forecast.

4.4. Modeling and Solution of Aggregate Planning in the Cooperative

The strategy used to develop aggregate planning monitored demand with constant labor and overtime. This strategy provides continuity of labor and avoids emotional and tangible costs of hiring and firing personnel. The data used for the chosen strategy, as shown in Table 3, were collected with the cooperative understudy and calculated.

Table 3. Cooperative Costs.

Costs	Price (R$)
Production unit	15.00/Kg
Labor	2153.25/month
Regular time	13.45/hour
Extra hour	23.53/hour
Hiring	2100.00/worker
Resignation	18,120.48/worker
Electricity taxes	0.30/R$/kWh
Power demand	13.14/Watts

The company calculates the unit cost of production, encompassing all costs involved in the process. As for labor, during regular hours, the workers receive a salary of R$ 1485.00, but the real value that the company has with this cost is R$ 2153.25 (R$ 13.45 × 8 h/day × 20 days/month) due to charges (R$ 668.25) paid every month. As for the cost of overtime, 75% of regular hours is added.

As for the hiring and firing costs, the high value of both justifies why the cooperative does not use the hiring and firing strategy. The cost of hiring comprises recruitment, medical costs (doctor and nurse), registration and integration, and training costs. The dismissal involves even more costs, with a 40% fine on taxes (R$ 640.00), termination (R$ 475.00), prior notice (R$ 1485.00), vacation (R$ 1485.00), 1/3 vacation (R$ 495.00), 13th salary (R$ 1485.00), salary (R$ 1485.00), and charges (R$ 10,570.28). It is worth mentioning that these values are variable, depending on the employee's working time, which is at least one year of work in the cooperative.

To obtain the cost of electricity, the tax value (TE = 0.30 R$/kWh) from the contract with the concessionaire (Table 3) was multiplied by the value of the energy performance index (I), obtained by the average of the last five years. Correlation analysis of electricity consumption as a production function showed a coefficient of determination R^2 = 0.64. Some values of atypical months were then excluded (mainly as a result of the economic crisis in the country from 2014 onwards) as they are extreme values or values that are further away from the straight line (outliers), and R^2 = 0.77 was obtained. Thus, the average value of the energy performance indicator was I = 0.80 kWh/kg, with a Standard Deviation = 0.126 and error = ± 0.017. Thus, the unit cost of electricity was $C_t^{ee} = 0.24$ R$/kg.

The cost of production, electricity, and demand for electricity is proportional to the production volume. In each period, demand must be satisfied, as orders are made to order over time can be used to fulfill the plan. Physical resources are considered fixed during the 12-month planning horizon. The total cost verified during the planning horizon is the sum of all the mentioned costs, given by Equation (4), being presented with the data replaced by Equation (12).

$$Min \sum_{t=1}^{T} (18.37 \times U_t + 0.24 \times U_t + 2153.00 \times T_t + 2100.00 \times T_t^c + 18,120.48 \times T_t^D + 3.14 \times D_t^{pe} + 23.53 \times H_t^e) \quad (12)$$

The study aims to find an aggregate plan that minimizes the total cost incurred during the planning horizon. Therefore, the values are subject to a series of restrictions linked to the decision variables presented in Section 3.4.

As for the workforce restriction, hiring, and firing (Equation (5)), the cooperative does not carry out the hiring and firing strategy, so the number of workers in the current period will always be the same. As for the production limitation constraint (Equation (6)), the capacity is determined by the company's normal hours per month and overtime. According to information provided by the cooperative, the production capacity is 88,000.00 kg per month, and it is possible to produce 550 kg/h or 88,000.00 kg/month at normal hours (8 h/day).

The overtime (Equation (7)) is carried out with all workers on the line. The cooperative works with the minimum number of people necessary to meet demand, with only two overtime hours allowed per day. The cooperative is limited to 40 monthly overtime hours/worker in this case. For the power demand constraint (Equation 10), the power demand value that minimizes the total costs was found, so it must be maintained for all planning periods, 235 W.

4.5. Production Plan for the Case Studied

With the tool used in the research work, it was possible to prepare the production plan for the next 12 months and, subsequently, compare the current scenario with the proposed one, considering the new power demand.

4.5.1. Current Scenario

The company's current scenario is to keep the workforce constant, using overtime when necessary and contracting 255 watts of power demand (Table 4).

Table 4. Aggregate plan of the current scenario with the contracted power demand.

Period (month)	Contractors (Number of People)	Fired (Number of People)	Workforce (Number of People)	Production in Overtime (Hours)	Production (kg)	Power Demand (kW)	Production Demand (kg)
0	0	0	17	0	0	255	0
1	0	0	17	0	73,058.02	255	73,058.02
2	0	0	17	0	48,155.40	255	48,155.40
3	0	0	17	0	80,745.51	255	80,745.51
4	0	0	17	0	70,974.94	255	70,974.94
5	0	0	17	0	73,072.40	255	73,072.40
6	0	0	17	0	75,037.28	255	75,037.28
7	0	0	17	0	78,020.75	255	78,020.75
8	0	0	17	9	92,827.75	255	92,827.75
9	0	0	17	0	80,068.89	255	80,068.89
10	0	0	17	0	69,798.75	255	69,798.75
11	0	0	17	0	81,646.55	255	81,646.55
12	0	0	17	0	83,557.78	255	83,557.78
Total cost				14,301,554.61			

It can be seen from Table 4 that the production demand (maximum 88,000.00 kg/month) is met, except in the eighth period (in bold), in which it will be necessary to perform nine overtime hours to meet the production demand.

4.5.2. Innovative Scenario with the Proposed Power Demand

Keeping the strategy used by the company of constant workforce, it was possible to define a new plan with the new defined electric power demand, as shown in Table 5.

Table 5. Aggregate plan with the proposed power demand.

Period (Month)	Contractors (Number of People)	Fired (Number of People)	Workforce (Number of People)	Production in Overtime (Hours)	Production (kg)	Power Demand (kW)	Production Demand (kg)
0	0	0	17	0	0	235	0
1	0	0	17	0	73,058.02	235	73,058.02
2	0	0	17	0	48,155.40	235	48,155.40
3	0	0	17	0	80,745.51	235	80,745.51
4	0	0	17	0	70,974.94	235	70,974.94
5	0	0	17	0	73,072.40	235	73,072.40
6	0	0	17	0	75,037.28	235	75,037.28
7	0	0	17	0	78,020.75	235	78,020.75
8	0	0	17	9	92,827.75	235	92,827.75
9	0	0	17	0	80,068.89	235	80,068.89
10	0	0	17	0	69,798.75	235	69,798.75
11	0	0	17	0	81,646.55	235	81,646.55
12	0	0	17	0	83,557.78	235	83,557.78
Total cost				14,298,401.01			

Keeping the strategy, what differentiates this plan (Table 5) from the previous one (Table 4) is the reduced cost of R$ 3153.60 reais due to the new power demand contracted for this year. However, even with this reduction, another plan was prepared to minimize total costs if the company allowed the hiring and firing of employees, if necessary. The result is shown in Table 6.

Table 6. Aggregate plan as a proposed scenario considering power demand.

Period (Month)	Contractors (Number of People)	Fired (Number of People)	Workforce (Number of People)	Production in Overtime (Hours)	Production (kg)	Power Demand (kW)	Production Demand (kg)
0	0	0	17	0	0	235	0
1	0	2	15	0	73,058.02	235	73,058.02
2	0	0	15	0	48,155.40	235	48,155.40
3	0	0	15	0	80,745.51	235	80,745.51
4	0	0	15	0	70,974.94	235	70,974.94
5	0	0	15	0	73,072.40	235	73,072.40
6	0	0	15	0	75,037.28	235	75,037.28
7	0	0	15	1	78,020.75	235	78,020.75
8	0	0	15	28	92,827.75	235	92,827.75
9	0	0	15	4	80,068.89	235	80,068.89
10	0	0	15	0	69,798.75	235	69,798.75
11	0	0	15	7	81,646.55	235	81,646.55
12	0	0	15	11	83,557.78	235	83,557.78
Total cost				14,284,113.02			

Note that this plan (Table 6) allows for dismissal and hiring and presents a lower cost for the cooperative than the previous plan (Table 5). However, it is worth mentioning that it was only allowed to fire two people because the production line could not operate with less than 15 workers. Thus, in the months in which demand was not satisfied, only with the staff, it would be necessary to work overtime, as shown in Table 6. In this way, there would be a cost reduction of R$ 14,288.00.

4.6. Model Verification from Sensitivity Analysis

The sensitivity analysis is divided into sensitivity analysis on variable cells and sensitivity analysis on constraints; the results presented were collected from the report issued by Solver in the resolution of the aggregate planning.

The first scenario, with a constant workforce and new power demand, shows that the reduced cost of the variables is either positive or equal to zero; that is, the variables are part of the solution, and there is no slack. The exception is the dismissed workers variable, which has a negative reduced cost; it shows that it has extra workers for production due to the cooperative's capacity. As for the permission to increase or reduce, without modifying the optimal solution, the only variable that allows an increase in its cost is that of the workforce, presenting a different increase in each period, and the variables that enable the reduction are the production of hours overtime, production, and laid-off workers. The one for dismissed workers allows a different value for each period, while the others present the same value for all. Variables with a value equal to zero cannot be changed because the optimal solution is also changed.

As for restrictions, contractors, production, and overtime have a favorable shadow price; the objective function will increase when they are increased. For example, if the number of overtime hours increases, the cost of the objective function will also increase. The workforce, in turn, has a negative shadow price; if the workforce increases, the optimal solution is not changed since the Solver interprets the surplus of workers due to production capacity. This is because 17 workers have been fixed as a restriction of the cooperative. As for the permission to increase, it is possible to increase the cost of the workforce by up to R$ 0.85 in all periods, those hired by up to R$ 0.93 in the first eight periods, overtime by 671.22 h in the eighth period, and in all others in 680 h, the maximum limit that can be accomplished. The production constraint has a different increase limit for each period. As for the permission to reduce, the restriction of contractors does not allow reduction due to the limit set by the cooperative. The workforce can be reduced by 0.93 kg in the first eight periods, while it cannot be reduced in the others. Production can only be reduced in the eighth period by 4827.76 kg.

In the second scenario, allowing hiring and firing with the new power demand in the variables, the reduced cost remains the same as in the previous scenario. It can be noted that the dismissed workers, in this scenario, have zero cost, different from the last scenario, because now the restriction allowed to keep at least 15 workers, no more than necessary. As for the permission to increase or reduce, without modifying the optimal solution, the only variable that allows an increase in its cost is laid-off workers, as two of them are laid off at the beginning of the period. In the other periods, it is allowed to increase the cost in the variable workforce in all and the variable overtime in periods 3, 7, 8, 9, 11, and 12. All variables allow a reduction limit.

As for the restrictions and the workforce continue to have a negative price shadow in the first nine periods and the 12th period, the Solver continues to consider production less than capacity concerning workers. However, according to the cooperative manager, 15 workers present the minimum framework for the show. On the other hand, periods 10 and 11 have a positive price shadow; if there is an increase in the workforce in this period, the optimal solution may change. The other constraints also have a positive or zero value.

5. Discussion

The planning horizon defined for this study was 12 months. Entezaminia et al. [41] stated that it is usually performed from 3 to 18 months. Filho et al. [42], and Feng et al. [19] also planned with a horizon of 12 months. On the other hand, Kopanos et al. [43] used a horizon of only five days and, Bilgen and Dogan [44] of 7 days and three weeks.

The formulated model addressed the demand monitoring strategy, using overtime, hiring, and firing. The strategy depends on the reality of the place, as is the case of studies that considered the stock (i.e., [27–31]), subcontracting [25,29,32], and missing orders [21].

This work is similar to that of Mehdizadeh et al. [21] regarding aggregate planning since it determined the number of items to be produced in regular and overtime and the need to keep the workforce constant or not, as well as Kopanos et al. [43] and Bilgen and Dogan [44]. They also determined a regular production plan. In contrast, Feng et al. [19] used APP also to obtain the production lead time.

This presented model was formulated with linear programming, following Rajaram and Karmarkar [45], who stated that APP models are generally designed as linear programming that minimizes costs. The model formulated in this study sought to minimize production costs as demand was met. Therefore, it is possible to affirm that, in the literature discussed here, the planning models usually seek to minimize costs through strategies according to the reality of the study site. For example, Filho et al. [42] considered inventory levels, production rates, overtime, regular labor, and subcontracting. On the other hand, Feng et al. [18] used minimum inventory costs, ideal production rate, and production balance load rate, seeking minimization.

Kopanos et al. [43] focused on minimizing inventory, operating, batch recipe preparation, unit usage, family moving, and external production costs. On the other hand, Bilgen and Dogan [44] sought to maximize their production, as did Kadambur and Kotecha [46], who also aimed to maximize profit while determining the amount and type of product that should be produced. In this way, these authors go in the opposite direction to Wang and Liang [15], who claimed that the problems of any of the planning models can only be solved if they have the sole objective of minimizing the cost.

A similar work to this study is that of Izadpanahi and Modarres [21], which included energy cost and carbon emission in the model. The authors used three objective functions seeking to minimize such costs simultaneously. The main point was to find the ideal power demand value before including it in the aggregate planning. This contributes to the literature since even some studies considering electric energy in their costs do not consider the demand for power and, consequently, did not seek to reduce it before adding it. Mehdizadeh et al. [20] reported that aggregate planning costs involve inventory, ordering, and production costs, and yet, Stevenson [47] complemented with unit costs related to regular hours, overtime, subcontracting, layoffs, or other factors that affect the costs in a relevant way, that is, it does not consider the cost of electricity.

Choi and Xirouchakis [1] minimized energy consumption while maximizing production based on a multiobjective function. Mouzon et al. [48] sought to reduce the energy used by production equipment based on mathematical programming, similar to the study by Rager et al. [49], who reduced energy use in production by minimizing the final energy demand of the machines involved in the process. Finally, Santiago et al. [50] and Chaturvedi [31] aimed to reduce energy in the process facilities and shared resources of aggregate planning.

In contrast to these studies, this paper sought to include the power demand in the planning, considering minimizing its cost from an equation before adding it to the other existing ones. By having the cost of electricity in the aggregate planning, the proposed model contributes to the literature and companies connected to the high-voltage electrical system that need to contract power demand in advance. It is essential to highlight that the models found in the literature present a production schedule based on stochastic data.

6. Conclusions

This study aimed to propose an aggregate planning model integrating the cost with electrical energy and power demand. The model is innovative in three aspects: the cost of contracted power demand in aggregate planning, defining the power demand to be contracted based on the least cost principle, and the treatment of uncertainties in the production demand forecast with the participation of the decision-maker.

The application of the model was carried out in a cooperative of the small agro-industrial sector. However, this study provides insights for a company of any size that needs to contract power demand in advance and presents a high degree of uncertainty in the market. After application, it was possible to reduce two workers and a cost reduction of R$ 14,288.00. The main limitation of the proposed model is the single case study. Furthermore, although it has been applied in a make-to-order business, it can also be used in make-to-stock or project-to-order businesses.

A characteristic of the company used in the case study is that customers usually confirm their orders a few days before shipment; such orders can be changed. Thus, the suggestion for future study would be to generate a production plan at the end of the month, but as the power demand must be contracted in advance for an extended period, this can be considered impossible. Thus, for these cases, the suggestion is that the power demand forecast is revised, if the contract allows, to reduce the period and generate a plan with a smaller number of months, or even to integrate the power demand equation in the resolution of the aggregated planning so that, when solving it, the best option is presented.

Author Contributions: Conceptualization, C.M.; Data curation, G.d.S.M. and C.M.; Formal analysis, G.d.S.M. and C.M.; Investigation, C.M.; Methodology, C.M. and F.H.L.; Project administration, A.V.H.S.; Resources, C.M.; Supervision, A.V.H.S.; Writing—original draft, C.M.; Writing—review & editing, A.V.H.S., F.H.L., J.L.D.R. and H.V.S. All authors have read and agreed to the published version of the manuscript.

Funding: This research was funded by CAPES Foundation and The APC was funded by Universidad Tecnológica de Peru.

Institutional Review Board Statement: Not applicable.

Informed Consent Statement: Not applicable.

Data Availability Statement: The data will be provided by the corresponding author if asked.

Conflicts of Interest: The authors declare that they have no known competing financial interests or personal relationships that could have appeared to influence the work reported in this paper.

Nomenclature

(DEC_j)	**power to be contracted**
(DEM_i)	**Other measured demands**
(TC_i)	**Contracted taxes**
(TU_i)	**Overtaking taxes**
PARAMETERS	
H	Planning horizon in periods
t	Period index
D_t	Expected demand for period t
C_t^u	Unit cost of production in period t
C_t^{ee}	Unit cost of electricity in period t
C_t^{de}	Power demand cost in period t
C_t^e	Inventory cost of one unit in period t
C_t^m	Unit cost of labor in period t
C_t^c	cost of hiring an employee in period t
C_t^D	cost of dismissing an employee in period t
C_t^S	Subcontracting cost in period t
C_t^f	Out-of-stock cost in the period t
C_t^{He}	Overtime cost in period t
DECISION VARIABLES	
U_t	number of units produced in period t
U_t^S	number of units produced under subcontracting in period t
U_t^e	number of units in stock at the end of period t
U_t^f	number of units missing in the period t
T_t	number of workers available in the period t
T_t^c	number of workers hired in period t
T_t^D	number of workers laid off in period t
Ca_t	Production production capacity in period t
D_t^c	Contracted power demand in period t
E_f	inventory at the end of the total period
H_t^n	number of normal hours worked in period t
H_t^e	number of overtime hours worked in period t

Appendix A

Table A1. Power demand consumption history.

Month	Potency (kw)	Month	Potency (kw)	Month	Potency (kw)	Month	Potency (kw)	Month	Potency (kw)
Jan/2014	204	Jan/2015	240	Jan/2016	252	Jan/2017	259	Jan/2018	221
Feb/2014	216	Feb/2015	252	Feb/2016	252	Feb/2017	245	Feb/2018	201
Mar/2014	264	Mar/2015	240	Mar/2016	252	Mar/2017	235	Mar/2018	211
Apr/2014	240	Apr/2015	240	Apr/2016	252	Apr/2017	245	Apr/2018	202
May/2014	216	May/2015	252	May/2016	264	May/2017	250	May/2018	202

Table A1. *Cont.*

Month	Potency (kw)	Month	Potency (kw)	Month	Potency (kw)	Month	Potency (kw)	Month	Potency (kw)
Jun/2014	216	Jun/2015	264	Jun/2016	252	Jun/2017	264	Jun/2018	206
Jul/2014	228	Jul/2015	252	Jul/2016	252	Jul/2017	250	Jul/2018	216
Aug/2014	252	Aug/2015	240	Aug/2016	250	Aug/2017	259	Aug/2018	226
Sep/2014	252	Sep/2015	240	Sep/2016	235	Sep/2017	264	Sep/2018	221
Oct/2014	252	Oct/2015	264	Oct/2016	245	Oct/2017	259	Oct/2018	206
Nov/2014	240	Nov/2015	276	Nov/2016	259	Nov/2017	254	Nov/2018	216
Dec/2014	288	Dec/2015	252	Dec/2016	254	Dec/2017	226	Dec/2018	221

Table A2. Electricity consumption history.

Month	Energy (kwh)	Month	Energy (kwh)	Month	Energy (kwh)	Month	Energy (kwh)	Month	Energy (kwh)
Jan/2014	66,203	Jan/2015	83,532	Jan/2016	70,272	Jan/2017	86,027	Jan/2018	79,529
Feb/2014	75,792	Feb/2015	100,980	Feb/2016	89,592	Feb/2017	82,935	Feb/2018	34,769
Mar/2014	59,220	Mar/2015	89,136	Mar/2016	88,008	Mar/2017	68,050	Mar/2018	69,184
Apr/2014	74,364	Apr/2015	78,108	Apr/2016	85,044	Apr/2017	82,784	Apr/2018	76,301
May/2014	59,832	May/2015	75,420	May/2016	87,216	May/2017	75,050	May/2018	71,020
Jun/2014	65,304	Jun/2015	92,784	Jun/2016	87,636	Jun/2017	89,838	Jun/2018	70,393
Jul/2014	94,164	Jul/2015	91,308	Jul/2016	53,832	Jul/2017	93,682	Jul/2018	79,856
Aug/2014	102,504	Aug/2015	84,912	Aug/2016	89,736	Aug/2017	102,359	Aug/2018	73,906
Sep/2014	99,948	Sep/2015	70,188	Sep/2016	66,703	Sep/2017	102,745	Sep/2018	73,524
Oct/2014	52,356	Oct/2015	95,364	Oct/2016	77,791	Oct/2017	69,419	Oct/2018	72,326
Nov/2014	84,516	Nov/2015	81,852	Nov/2016	95,322	Nov/2017	88,272	Nov/2018	72,262
Dec/2014	107,928	Dec/2015	96,444	Dec/2016	88,956	Dec/2017	80,510	Dec/2018	72,413

Table A3. Potato production history.

Month	Production (kg)	Month	Production (kg)	Month	Production (kg)	Month	Production (kg)	Month	Production (kg)
Jan/2014	71,449.80	Jan/2015	163,406.92	Jan/2016	94,561.22	Jan/2017	108,315.34	Jan/2018	52,178.56
Feb/2014	98,108.06	Feb/2015	122,309.88	Feb/2016	102,871.56	Feb/2017	67,618.20	Feb/2018	15,807.6
Mar/2014	75,186.06	Mar/2015	160,586.38	Mar/2016	145,226.16	Mar/2017	85,486.90	Mar/2018	90,957.72
Apr/2014	89,184.04	Apr/2015	101,419.64	Apr/2016	112,505.82	Apr/2017	91,313.48	Apr/2018	94,088.215
May/2014	51,441.18	May/2015	97,440.4	May/2016	120,193.48	May/2017	91,153.62	May/2018	62,223.9
Jun/2014	73,104.84	Jun/2015	110,822.38	Jun/2016	104,055.4	Jun/2017	95,449.86	Jun/2018	99,377.09
Jul/2014	172,998.24	Jul/2015	123,069.82	Jul/2016	50,575.16	Jul/2017	130,865.92	Jul/2018	95,340.575
Aug/2014	170,871.62	Aug/2015	99,924.98	Aug/2016	132,796.44	Aug/2017	154,074.48	Aug/2018	83,872.86
Sep/2014	195,048.08	Sep/2015	77,072.10	Sep/2016	84,820.46	Sep/2017	136,504.22	Sep/2018	75,521.85
Oct/2014	98,406.44	Oct/2015	127,472.06	Oct/2016	98,265.96	Oct/2017	97,858.3	Oct/2018	78,968.54
Nov/2014	15,5987.16	Nov/2015	100,275.86	Nov/2016	121,060.3	Nov/2017	112,327.76	Nov/2018	78,199.93
Dec/2014	185,057.78	Dec/2015	127,309.30	Dec/2016	124,903.44	Dec/2017	98,389.48	Dec/2018	96,792.86

References

1. Choi, Y.C.; Xirouchakis, P. A production planning in highly automated manufacturing system considering multiple process plans with different energy requirements. *Int. J. Adv. Manuf. Technol.* **2014**, *70*, 853–867. [CrossRef]
2. Zhang, H.; Dai, Z.W.; Zhang, W.Y.; Zhang, S.; Wang, Y.; Liu, R.Y. A New Energy-Aware Flexible Job Shop Scheduling Method Using Modified Biogeography-Based Optimization. *Math. Probl. Eng.* **2017**, *2017*, 12. [CrossRef]
3. Rediske, G.; Siluk, J.C.M.; Gastaldo, N.G.; Rigo, P.D.; Rosa, C.B. Determinant factors in site selection for photovoltaic projects: A systematic review. *Int. J. Energy Res.* **2019**, *43*, 1689–1701. [CrossRef]
4. Lu, Y.; Khan, Z.A.; Alvarez-Alvarado, M.S.; Zhang, Y.; Huang, Z.; Imran, M. A critical review of sustainable energy policies for the promotion of renewable energy sources. *Sustainability* **2020**, *12*, 5078. [CrossRef]
5. Garlet, T.B.; Ribeiro, J.L.D.; de Savian, F.; Siluk, J.C.M. Value chain in distributed generation of photovoltaic energy and factors for competitiveness: A systematic review. *Sol. Energy* **2020**, *211*, 396–411. [CrossRef]
6. Sampaio, P.G.V.; González, M.O.A. Photovoltaic solar energy: Conceptual framework. *Renew. Sustain. Energy Rev.* **2017**, *74*, 590–601. [CrossRef]
7. De Andrade Guerra, J.B.S.O.; Dutra, L.; Schwinden, N.B.C.; De Andrade, S.F. Future scenarios and trends in energy generation in Brazil: Supply and demand and mitigation forecasts. *J. Clean. Prod.* **2015**, *103*, 197–210. [CrossRef]
8. Hu, G.; Xu, Z.; Wang, G.; Zeng, B.; Liu, Y.; Lei, Y. Forecasting energy consumption of long-distance oil products pipeline based on improved fruit fly optimization algorithm and support vector regression. *Energy* **2021**, *224*, 120153. [CrossRef]
9. Zeng, C.; Wu, C.; Zuo, L.; Zhang, B.; Hu, X. Predicting energy consumption of multiproduct pipeline using artificial neural networks. *Energy* **2014**, *66*, 791–798. [CrossRef]
10. Chou, J.S.; Truong, D.N.; Kuo, C.C. Imaging time-series with features to enable visual recognition of regional energy consumption by bio-inspired optimization of deep learning. *Energy* **2021**, *224*, 120100. [CrossRef]
11. Maaouane, M.; Zouggar, S.; Krajačić, G.; Zahboune, H. Modelling industry energy demand using multiple linear regression analysis based on consumed quantity of goods. *Energy* **2021**, *225*, 120270. [CrossRef]
12. Mourtzis, D.; Boli, N.; Xanthakis, E.; Alexopoulos, K. Energy trade market effect on production scheduling: An Industrial Product-Service System (IPSS) approach. *Int. J. Comput. Integr. Manuf.* **2021**, *34*, 76–94. [CrossRef]
13. Nugraha, I.; Hisjam, M.; Sutopo, W. Aggregate Planning Method as Production Quantity Planning and Control to Minimizing Cost. In *IOP Conference Series: Materials Science and Engineering*; IOP Publishing: Bristol, UK, 2020; p. 012045.
14. Rahmani, D.; Zandi, A.; Behdad, S.; Entezaminia, A. A light robust model for aggregate production planning with consideration of environmental impacts of machines. *Oper. Res.* **2021**, *21*, 273–297. [CrossRef]
15. Wang, R.C.; Liang, T.F. Application of fuzzy multiobjective linear programming to aggregate production planning. *Comput. Ind. Eng.* **2004**, *46*, 17–41. [CrossRef]
16. ANEEL—Agência Nacional de Energia Elétrica. Resolução Normativa nº 414, de 9 de Setembro de 2010. Estabelece as Condições Gerais de Fornecimento de Energia Elétrica de Forma Atualizada e Consolidada. Available online: http://www.aneel.gov.br/ren-414 (accessed on 29 May 2019).
17. SAAD, G. An overview of production planning model: Structure classification and empirical assessment. *Int. J. Prod. Res.* **1982**, *20*, 105–114. [CrossRef]
18. Feng, P.F.; Zhang, J.F.; Wu, Z.J.; Yu, D.W. An improved production planning method for process industries. *Int. J. Prod. Res.* **2011**, *49*, 4223–4243. [CrossRef]
19. Arruda, O.R., Jr. Planejamento Tático da Produção Agroindustrial com Fluxo Divergente e Produção em Dois Estágios. Master's Thesis, Departamento de Engenharia de Produção, Escola Politécnica da Universidade de São Paulo, São Paulo, Brazil, 2014.
20. Mehdizadeha, E.; Niaki, S.T.A.; Hemati, M. A bi-objective aggregate production planning problem with learning effect and machine deterioration: Modeling and solution. *Comput. Oper. Res.* **2017**, *91*, 21–36. [CrossRef]
21. Izadpanahi, E.; Modarres, M. Aggregate production planning by focusing on energy saving: A robust optimization approach. *J. Clean. Prod.* **2016**, *133*, 1074–1085.
22. Latifoglu, C.; Belotti, P.; Snyder, L.V. Models for Production Planning under Power Interruptions Naval. *Res. Logist.* **2013**, *60*, 413–431.
23. Choi, Y.C.; Xirouchakis, P. A holistic production planning approach in a reconfigurable manufacturing system with energy consumption and environmental effects. *Int. J. Comput. Integr. Manuf.* **2015**, *28*, 379–394. [CrossRef]
24. Lawton, R. How Should Additive Holt-Winters Estimates Be orrected? *Int. J. Forecast.* **1998**, *14*, 393–403. [CrossRef]
25. Nahmias, S. *Production and Operations Analysis*; Irwin: Homewood, IL, USA, 1995.
26. R Core Team. *R: A Language and Environment for Statistical Computing*; R Foundation for Statistical Computing: Vienna, Austria, 2021. Available online: https://www.R-project.org/ (accessed on 15 July 2021).
27. Dickey, D.; Fuller, W.A. Distribution of the Estimates for Autoregressive Time Series with a Unit Root. *J. Am. Stat. Assoc.* **1979**, *74*, 427–431.
28. González-Estrada, E.; Cosmes, W. Shapiro–Wilk test for skew normal distributions based on data transformations. *J. Stat. Comput. Simul.* **2019**, *89*, 3258–3272. [CrossRef]
29. Bakar, N.N.A.; Hassan, M.Y.; Abdullah, H.; Rahman, H.A.; Abdullah, M.P.; Hussin, F.; Bandi, M. Energy efficiency index as an indicator for measuring building energy performance: A review. *Renew. Sustain. Energy Rev.* **2015**, *44*, 1–11. [CrossRef]

30. Godinho Filho, M.F. Paradigmas Estratégicos de Gestão da Manufatura. Ph.D. Thesis, Universidade Federal de São Carlos (UFSC), São Paulo, Brazil, 2004.
31. Chaturvedi, N.D. Minimizing energy consumption via multiple installations aggregate production planning Clean Technologies and Environmental. *Policy* **2017**, *19*, 1977–1984.
32. Butler, J.; Jia, J.; Dyer, J. Simulation techniques for the sensitivity analysis of multi-criteria decision models. *Eur. J. Oper. Res.* **1997**, *103*, 531–546. [CrossRef]
33. Bazaraa, M.S.; Jarvis, J.J.; Sherali, H.D. *Linear Programming and Network Flows*, 4th ed.; Wiley: New York, NY, USA, 2010.
34. Land, M.; Gaalman, G. Workload control concepts in job shops: A critical assessment. *Int. J. Prod. Econ.* **1996**, *46*, 535–538. [CrossRef]
35. Corti, D.; Pozzetti, A.; Zornini, M. A capacity-driven approach to establish reliable due dats in a MTO environment. *Int. J. Prod. Econ.* **2006**, *104*, 536–554. [CrossRef]
36. Filho, F.H.B. A crise econômica de 2014/2017. Estudos avanChatuçados. vol.31 no.89 São Paulo Jan./Apr. 2017. Available online: https://doi.org/10.1590/s0103-40142017.31890006 (accessed on 15 January 2019).
37. Stone, C.J. Consistent nonparametric regression, with discussion. *Ann. Stat.* **1977**, *5*, 549–645. [CrossRef]
38. Cleveland, R.B.; Cleveland, W.S.; McRae, J.E.; Terpenning, I. STL: A seasonal-trend decomposition procedure based on LOESS. *J. Off. Stat.* **1990**, *6*, 3–73.
39. Watson, G.S. Smooth regression analysis. *Sankhya Ser. A* **1964**, *26*, 359–372.
40. Stone, C.J. Optimal rates of convergence for nonparametric estimators. *Ann. Stat.* **1980**, *8*, 1348–1360. [CrossRef]
41. Entezaminia, A.; Heydare, M.; Rahmani, D. A multiobjective model for multi-product multi-site aggregate Production planning In a green supply chain: Considering collection and recycling centers. *J. Manuf. Syst.* **2016**, *40*, 63–75. [CrossRef]
42. Filho, O.S.S.; Cezarino, W.; Ratto, J. Aggregate Production Planning: Modeling and Solution via Excel Spreadsheet and Solver. In *Management and Control of Production Logistics*; University of Coimbra: Coimbra, Portugal, 2010; Volume 43, pp. 89–94.
43. Kopanos, G.M.; Georgiadis, M.C.; Pistikopoulos, E.N. Energy production planning of a network of micro combined heat and power generators. *Appl. Energy* **2013**, *102*, 1522–1534. [CrossRef]
44. Bilgen, B.; Dogan, K. Multistage Production Planning in the Dairy Industry: A Mixed-Integer Programming Approach. *Ind. Eng. Chem. Res.* **2015**, *54*, 11709–11719. [CrossRef]
45. Rajaram, K.; Karmarkar, U.S. Aggregate production planning for process industries under oligopolistic competition. *Eur. J. Oper. Res.* **2012**, *223*, 680–689.
46. Kadambur, R.; Kotecha, P. Optimal Production Planning in a Petrochemical Industry using Multiple Levels. *Comput. Ind. Eng.* **2015**, *100*, 133–143. [CrossRef]
47. Stevenson, W.J. *Administração de Operações de Produção*, 6th ed.; LTC: Rio de Janeiro, Brazil, 2001.
48. Mouzon, G.; Yildirim, M.B.; Twomey, J. Operational methods for minimization of energy consumption of manufacturing equipment. *Int. J. Prod. Res.* **2007**, *45*, 4247–4271. [CrossRef]
49. Rager, M.; Gahm, C.; Denz, F. Energy-oriented scheduling based on Evolutionary Algorithms. *Comput. Oper. Res.* **2015**, *54*, 218–231. [CrossRef]
50. Santiago, G.C.A.; Del Ser, J.; Upton, C.; Quilligan, F.; Gil-Lopez, S.; Salcedo-Sanz, S. A random-key encoded harmony search approach for energy-efficient production scheduling with shared resources. *Eng. Optim.* **2015**, *47*, 1481–1496. [CrossRef]

applied sciences

MDPI

Article

Parametric Optimization of Combined Wind-Solar Energy Power Plants for Sustainable Smart City Development

Pavel Kuznetsov [1], Miroslav Rimar [2], Boris Yakimovich [1], Olha Kulikova [2,*], Martin Lopusniak [3], Dmitry Voronin [4] and Vladislav Evstigneev [4]

[1] Institute of Nuclear Energy and Industry, Sevastopol State University Sevastopol, 299053 Sevastopol, Russia; pavelnik2@gmail.com (P.K.); yakimovich52@gmail.com (B.Y.)
[2] Faculty of Manufacturing Technologies with the Seat in Presov, Technical University of Kosice, Bayerova 1, 080 01 Presov, Slovakia; miroslav.rimar@tuke.sk
[3] The Faculty of Civil Engineering, Technical University of Kosice, Vysokoskolska 4, 042 00 Kosice, Slovakia; martin.lopusniak@tuke.sk
[4] The National Technology Initiative Institute, Sevastopol State University Sevastopol, 299053 Sevastopol, Russia; voronin@sevsu.ru (D.V.); vald_e@rambler.ru (V.E.)
* Correspondence: olha.kulikova@tuke.sk

Abstract: To minimize the existing risks associated with the energy production, distribution, and development of urban areas, the concept of sustainable development of smart manufacturing is being introduced everywhere. However, most of existing digital transformation technologies do not sufficiently take into account environmental, economic, and social components. It requires the development of specialized solutions that, on the one hand, contribute to the development of the urban environment, and on the other, are focused on the use of innovative technologies that have a positive effect on the environment. However, in order to provide electric power to objects of a smart urban environment, we propose using autonomous power plants with renewable energy sources (wind and sun) instead of classical central power supply. The article describes the benefits of using combined wind-solar power plants to provide electricity to smart urban environments. The article considers advantages and disadvantages of this solution. We describe the design of the combined wind-solar installation with a vertical Darier rotor and photoelectric converters located on a common supporting structure. This solution allows getting a positive synergistic effect from the use of several renewable energy sources. The presented main advantages of this solution include increasing the energy efficiency of photovoltaic converters due to the intensification of heat removal from the surface of solar cells by the wind flow from the Darier rotor. In addition, this article considers the efficient use of the area and stability of electricity output. Digital aerodynamic modeling performed the parametric optimization of the Darier rotor having three blades with NACA 0021profile. The simulation demonstrated that the blades installed at an angle of 3.8° provide the maximum value of the wind utilization coefficient. This article also considers parametric optimization of a photovoltaic installation. In particular, the optimal installation angle of photovoltaic modules for Sevastopol is 34°.

Keywords: wind-solar installation; wind energy; photovoltaics; renewable energy; hybrid installation; energy smart city; sustainable development

Citation: Kuznetsov, P.; Rimar, M.; Yakimovich, B.; Kulikova, O.; Lopusniak, M.; Voronin, D.; Evstigneev, V. Parametric Optimization of Combined Wind-Solar Energy Power Plants for Sustainable Smart City Development. *Appl. Sci.* **2021**, *11*, 10351. https://doi.org/10.3390/app112110351

Academic Editor: Joachim Müller

Received: 7 October 2021
Accepted: 31 October 2021
Published: 4 November 2021

1. Introduction

The use of modern technologies to improve the efficiency of organizing the functioning of manufacturing and urban processes and the provision of innovative services forms the basis of smart sustainable city and society concept [1–5]. Emerging environmental problems and other risks associated with an increase in the proportion of the population living in cities [1,2] can be solved using assessing models [4,5]. Special attention should be paid to the set of strategies for the conceptual development of urban areas, which are called "smart" cities in the literature, proposed by many studies [6–8]. The concept of a "smart" city aims to find new ways to meet these challenges based on the integration of information and communication technologies, the introduction of modern advances in artificial intelligence, big data processing, decision support technologies, and IoT [7–15]. Thus, in practice, the implementation of smart sustainable city concept allows to meet the growing needs of the present and future generations while minimizing the negative impact of the implemented technical solutions on the economic, social, and environmental components of the city [6–10]. The technological base for ensuring sustainable development of smart sustainable city (SSC) cities is the so-called "smart" solutions in the field of digital transformation of urbanized territories [9–17]. Examples of such solutions are smart power grids [16–21], effective video surveillance systems [22–24], intelligent traffic management [25,26], and housing and communal services [27]. However, the difficulty of electricity providing using city power supply networks limits the widespread implementation of such solutions. This is due the need of significant financial costs and additional construction works in some cases many times higher than the cost of the installed equipment. In addition, meeting the increased energy needs only through the use of traditional types of energy will lead to a deterioration of the environmental situation.

These problems can be solved by using power plants that implement the concept of small-scale distributed energy, with generating plants located in the immediate vicinity of electricity consumers. This concept fits well with the use of renewable energy sources (RES) and, in particular solar and wind power plants. In addition, the advantages of such installations are environmental friendliness, which allows them to be used on almost any scale without causing damage to the environment, as well as their availability in almost every point of our planet, differing in radiation density by no more than two times [28,29], which is very important for the implementation of the concept of a smart urban environment. However, despite the stated advantages, the widespread use of PGR in Russia is constrained by a number of disadvantages. One of the main such disadvantages of most PGR-based power plants is the inconstancy of the output power [30]. This disadvantage can be leveled by using in one design several generating plants that convert different types of renewable energy sources and an energy storage system. This is because various types of renewable energy sources, as a rule, have a weak correlation, which makes it possible to significantly reduce their influence in combined wind-solar installations.

The authors analyzed the existing technical solutions, the simultaneous use of photoelectric converters (PV) and a wind power plant (WPP) [31]. This analysis shows that the technical implementation of such a combination does not always bring a noticeable synergistic effect and may lead to a decrease in the energy efficiency of each of the converters. In this regard, when designing these installations, it is necessary initially to take into account the design features of PV and wind turbines, as well as the features of their work.

2. Materials and Methods

The analysis showed that one of the most promising installations that allows to obtain a positive synergistic effect from the combined use of PV and wind turbines is the development of the Scientific and Production Center "Solar Energy". The design of which provides for the placement of photovoltaic modules on the supporting structure of a wind turbine with a vertical-axial rotor of the Darrieus type (Figure 1). The peculiarity of this solution is that the photovoltaic modules are located below the rotor of the wind turbine, the design of which allows the forced air supply to the rear surface of the photovoltaic converters, intensifying the heat transfer from them to the environment. This solution makes it possible to increase the energy efficiency of solar cells by lowering their temperature. To ensure the maximum extraction of electrical energy from the wind-solar plant, it includes an MPPT controller and an automatic voltage stabilizer, which allows the voltage matching of the wind and photovoltaic plants when they are connected in parallel [32]. To simulate the operating mode and calculate the output power of photovoltaic modules, in order to assess its energy parameters and efficiency, a system of Equation (1) was compiled. The determination of the values of the intensity of solar radiation entering the surface of the PVC was carried out according to the methods [33,34] based on data from [35] and PVGIS online service (Photovoltaic Geographical Information System https://re.jrc.ec.europa.eu/pvg_tools/en/tools.html, accessed on 1 September 2021).

$$\begin{cases} P_{PV} = I_{PV}^{st} U_{PV}^{st} = I_{PV}^{st} U_{WT} \\ I_{PV}^{st} = \frac{U_{PV} I_{PV}}{U_{WT}} \eta_{PV}^{st} \\ I_{PV} = I_{ph} - I_0 \left[\exp\left(\frac{q(U_{PV}+I_{PV}R_s)}{n_s(AkT)} \right) - 1 \right] \\ \frac{q(U_{PV}+I_{PV}R_s)}{n_{BX}(AkT)} - \ln\left[\frac{I_{ph}-I_0}{I_0\left(1+\frac{q(U_{PV}+I_{PV}R_s)}{n_s(AkT)}\right)} \right] = 0 \\ I_{ph} = \left[I_{sc(stc)} + K_i(T - T_{stc}) \right] \frac{IR}{IR_{stc}} \\ I_0 = \frac{I_{ph}}{\exp\left[\frac{q\left[U_{id(stc)}+K_u(T-T_{stc})\right]}{n_S(AkT)} \right] - 1} \\ T = T_{env} + K_t IR \end{cases}, \quad (1)$$

where P_{PV}—photovoltaic module power, W; I_{PV}^{st}, current at the output of the automatic voltage stabilizer of the photovoltaic module, A; U_{PV}^{st}—voltage at the output of the automatic voltage stabilizer of the photovoltaic module, V; U_{WT}—wind turbine output voltage, V; U_{PV}—voltage at the output of the MPPT controller of the photovoltaic module, V; I_{PV}—current at the output of the MPPT controller of the photovoltaic module, V; η_{PV}^{st}—the efficiency of the voltage stabilizer of the photovoltaic module; I_{ph}—photocurrent of solar cells, A; I_0—reverse saturation current of solar cells, A; q—electron charge, ($1.602 \cdot 10^{-19}$ C); R_s—series resistance of solar cells, Ohm; A—the ideality factor, depending on the thickness of the p-n-junction and the material, takes values for a silicon solar cell from 1.2 to 5 (Table 1); k—Boltzmann constant, ($1.381 \cdot 10^{-23}$ J/K); T—absolute solar cell temperature, K; $I_{sc(stc)}$—short-circuit current of the solar cell under standard conditions, A; K_i—temperature coefficient of the short-circuit current of the solar cell, A/K; K_u—temperature coefficient of no-load voltage, V/K; K_t—temperature coefficient of temperature rise, $(K \cdot m^2)/W$; T_{stc}—temperature standard conditions, (298 K); $n_s$number of solar cells connected in series; T_{env}—ambient temperature, K; $U_{id(stc)}$—open circuit voltage of the solar cell under standard conditions, V; IR—value of the intensity of solar radiation, W/m^2; and IR_{stc}—solar radiation intensity under standard conditions, (W/m^2).

The block diagram of the electrical part of the power plant is shown in Figure 2. The principle of operation is as follows: electrical energy from the alternating current generator of the wind turbine (1) through the charge controller (6) is transferred to the storage battery (7). In this case, the conversion of alternating voltage to direct voltage is carried out by means of a rectifier (2). The electrical energy from the photovoltaic system (5) is also transferred to the battery through the charge controller. At the same time, in order to ensure efficient extraction of electrical energy, using the MPPT controller (4), an energy extraction mode is set corresponding to the maximum power point of photovoltaic converters and voltage matching is performed by means of a stabilizer (3). The voltage regulator and the charge controller are controlled by a microcontroller device (9). Electric energy from the battery is transmitted to the consumer through the inverter (8), which converts DC voltage into AC voltage.

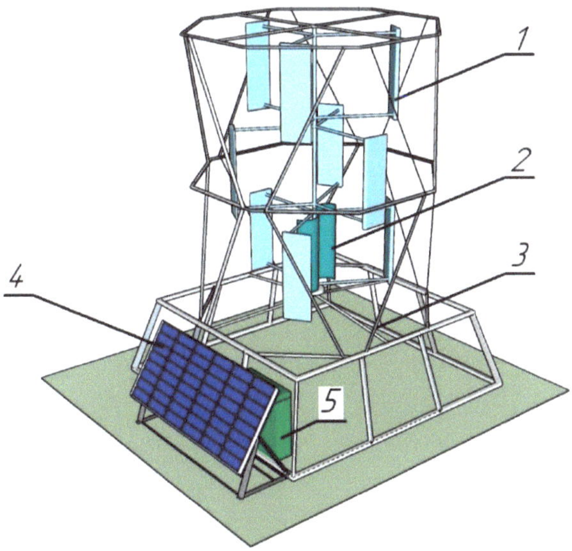

Figure 1. Combined wind-solar power station: (1) Darier rotor; (2) Savonius rotor; (3) supporting structure; (4) photoelectric converters; (5) control cabinet with rechargeable batteries.

Table 1. Coefficient of ideality solar cell from material [6].

Material	Coefficient A
Si—monocrystalline	1.2
Si—polycrystalline	1.3
a-Si:H—amorphous hydrogenated	1.8
a-Si:H double	3.3
a-Si:H triple	5

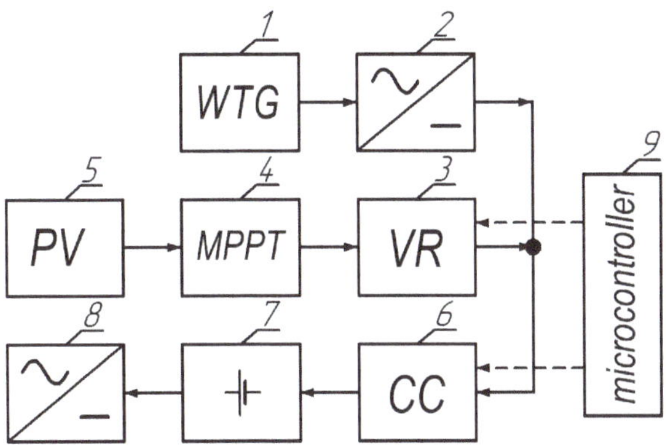

Figure 2. Block diagram of the electrical part of a wind-solar power station.

We carried out a digital aerodynamic simulation of the operation of the Darrieus rotor of a wind turbine with three blades with NACA 0021 profiles using the finite volume method [36]. A 2D computational domain with a length exceeding the rotor diameter by a factor of 5 has been developed, which makes it possible to obtain the parameters of the flow around the rotor with an accuracy sufficient for engineering calculations. To speed up the calculations without reducing the accuracy of their results, the computational domain was composed of three subdomains with different grid densities: the outer space around the rotor—the largest mesh size and the inner rotor space—the average mesh size, and the near-wall areas around each blade—the smallest mesh size. The main purpose of the simulation was to determine the angle of the rotor blades ϕ (the angle between the chord of the blade and the tangent to the trajectory of its motion on the fixing axis), at which the value of the coefficient of utilization of wind energy Cp will take the maximum value. The utilization factor of wind energy is a dimensionless value showing the efficiency of converting wind energy into mechanical energy of the rotor. Its value was calculated as the ratio of the mechanical power of the wind turbine rotor to the power of the wind flow:

$$C_p = \frac{P_M}{P_W} = \frac{\omega M}{0.5\rho S V^3}, \quad (2)$$

where P_M—mechanical power of the wind turbine rotor; P_W—wind power; ω—angular speed of rotation of the rotor; M—torque; ρ—air density; S—swept rotor area; and V—wind speed.

The torque value was calculated using the torque coefficients [37,38] calculated for each blade in one full revolution using the following formula:

$$M = 0.5(C_{M1} + C_{M2} + C_{M3})\rho S V^2 R, \quad (3)$$

where $C_{M1\ldots3}$—moment coefficients of rotor blades and R—rotor radius.

3. Results and Discussion

To simulate the operation of the Darrieus rotor and calculate its energy characteristics, a mathematical model was developed using the finite element method.

The geometry was created at the software CATIA P3V5 (Mechanical Design). The calculations and simulations were made at the software ANSYS 14.5, ANSYS Fluent.

At the first stage, the geometry of the NACA 0021 blade was created. To construct the model of the Darrieus rotor blades, in the CATIA program, the geometric points of the edges were calculated using the known models of the blade profiles The model was used as a base profile naca0021-il, which has the next parameters: Max thickness 21% at 30% chord, Max camber 0% at 0% chord. Profile characteristics are shown in the Figure 3.

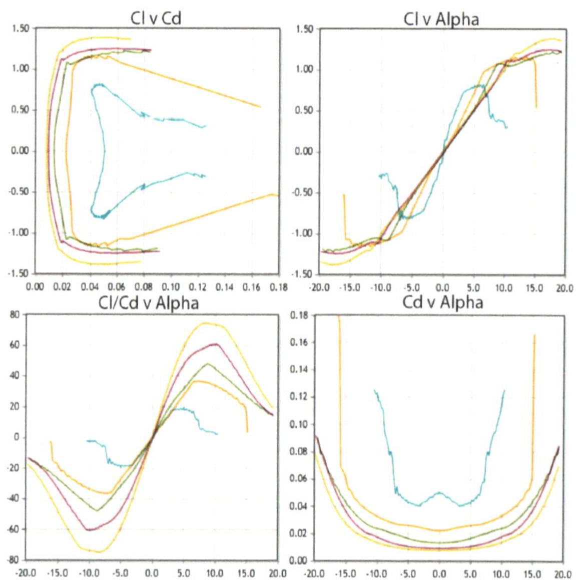

Figure 3. Profile characteristics of the naca0021-il.

The naca0021-il profile points were transferred to the CATIA program and a sketch of the Darrieus rotor blade was created (Figure 4).

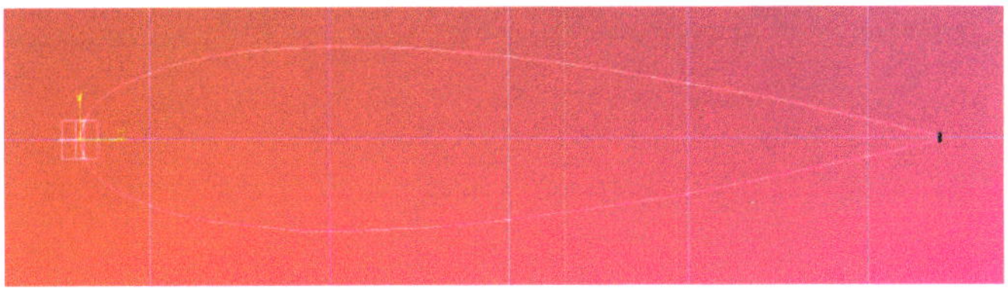

Figure 4. Sketch of the Darrieus rotor blade in CATIA.

Then, this blade was transformed into a full-fledged geometric copy of the rotor (Figure 5).

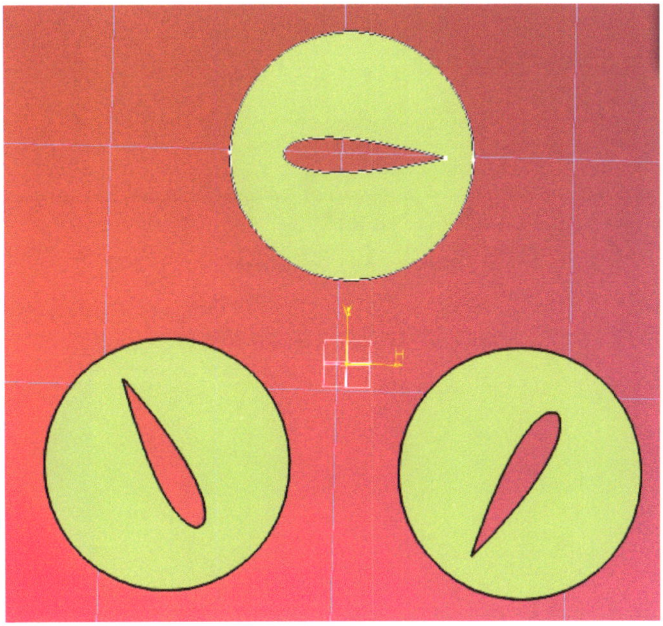

Figure 5. Sketch of the Darrieus rotor blades in CATIA.

The next step was to set additional geometry elements in the form of geometric contours of boundary conditions, which were necessary for performing subsequent calculations in ANSYS and ANSYS FLUENT software.

Figure 6 shows an image of the final version of the geometry of the simulated Darrieus rotor with a given geometry of all boundary conditions, including three main boundaries: the area remote from the rotor (for approximate modeling of the environment—meshes with large elements), the area near the rotor (for more accurate modeling of the environment around the rotor we used meshes with middle elements), and the area around the rotor blades (for accurate modeling of the environment and the flow of the wind flow along the geometry of the blade we used meshes with small elements).

Figure 6. Sketch of the Darrieus rotor blades in CATIA.

The mesh was generated in ANSYS Mesh using the proximity and curvature advanced size function. The mesh is shown on the Figure 7.

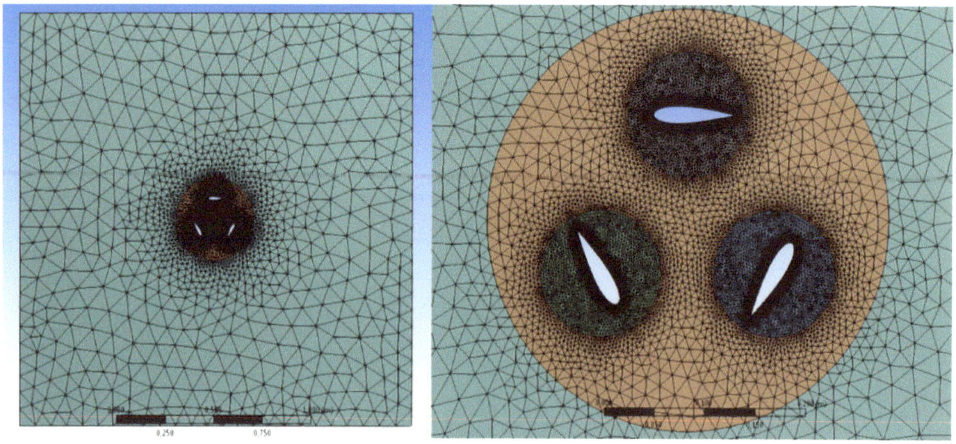

Figure 7. Mesh of the model.

The results of the simulation of the velocities are shown in Figure 8a,b.

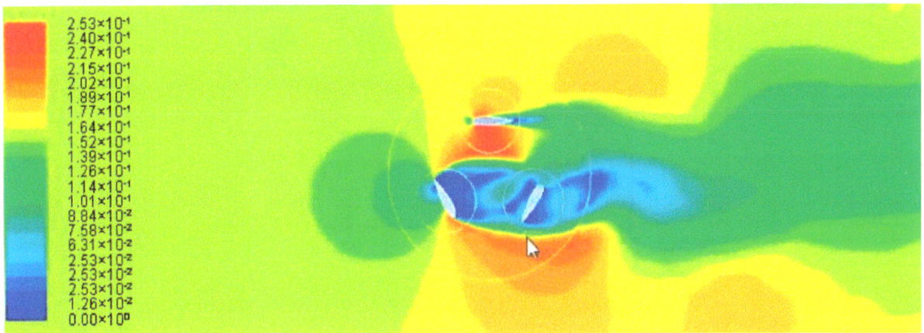

(**a**)

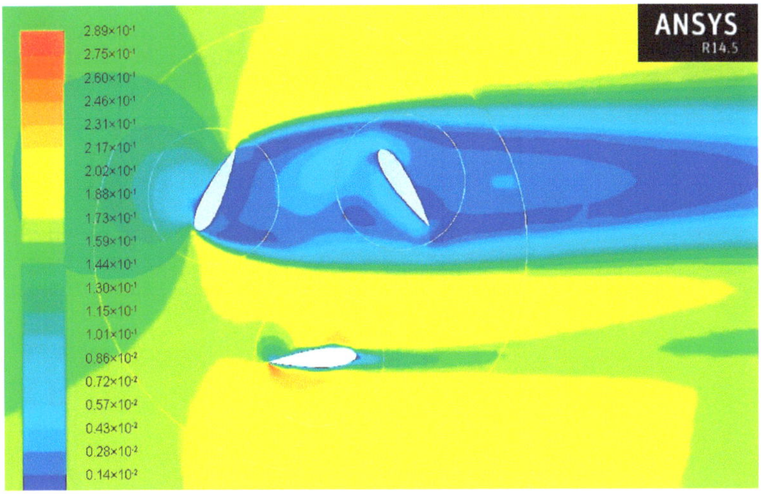

(**b**)

Figure 8. Results of the simulation (velocities): (**a**) simulation results and (**b**) enlarged view of the simulation model.

As a result of the simulation, the characteristic of the utilization factor (Cp) from the peripheral speed was obtained and presented in Figure 8. From the presented characteristics, it follows that the maximum value of the utilization factor is 0.345, which can be assumed as a small value with the respect to the number of model assumptions. As a result, it is necessary to carry out work aimed at increasing this indicator by changing the geometry parameters of the wind turbine rotor and the use of other technical design improvements. The work planned in the subsequent stages of work under this grant will be aimed at solving these problems.

Figure 9 shows the flow velocity diagram in the cross section of a wind turbine with a diameter of 2 m, obtained because of the simulation at a wind speed of 10 m/s using the finite volume method.

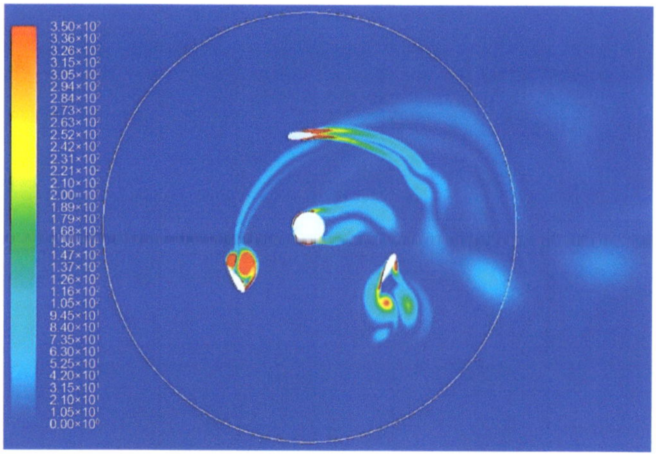

Figure 9. Velocity map in the turbine's cross section at 10 m/s of wind speed.

During modeling, the dependence of the coefficient of use of wind energy C_p at different angles of installation of the rotor blades was determined. The results obtained are shown in the summary graph (Figure 10), from which it follows that the maximum value of the wind energy utilization factor C_p is achieved when setting the blade angle $\varphi = 3.8°$, while the design value of this angle, regulated by the manufacturer, is 7.8°.

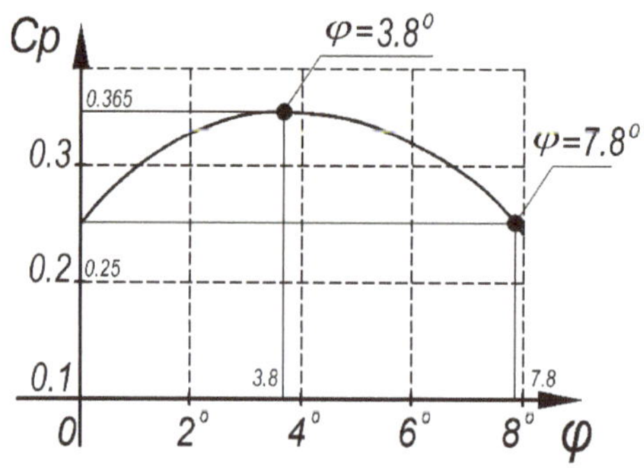

Figure 10. Power efficiency C_p as a function of the blade pitch angle φ.

Thus, the simulation showed that initially the wind turbine was operating at limited capacity. An experimental study was also carried out, the purpose of which was to verify the results obtained from digital simulations. To carry out this study, in the city of Sevastopol, on the territory of the student sports and health camp "Horizon", a physical model of a wind-solar installation was installed, a photograph of which is shown in Figure 11. The rated design power of the wind turbine is 3 kW at a wind speed of 12 m/s. The Darrieus rotor area is 12 m^2. Photovoltaic installation formed by two series-connected modules Delta SM 250-24 P and has a rated power of 500 watts.

Figure 11. Combined wind-solar power plant with a three-section Darier rotor and two photoelectric modules (Sevastopol, SRC «Horizont»).

The technique [39] was used to estimate the true value of the measured values from the results of the experiment. By the beginning of the mathematical processing of the measurement results, all gross and systematic errors were eliminated by introducing appropriate corrections to the measurement results and discarding obviously erroneous results [40]. These operations were eliminated in the first stage of mathematical processing of the measurement results. Thus, the measurement accuracy in the experiments was determined only by random errors.

When processing the experimental results, the "three sigma" rule was used. Thus, the confidence score, given that the sigma value is not known, was calculated from the expression

$$|a - \overline{x}| < \frac{3s}{\sqrt{n}}, \tag{4}$$

where a—true measured value; $\overline{x}$—arithmetic mean of measurement results; n—number of independent measurements; and s—an estimate of the standard deviation determined from the ratio:

$$s = \sqrt{\frac{1}{n-1}\sum\nolimits_{i=1}^{n}(x_i - \overline{x})^2}, \tag{5}$$

where x_i—i-th measurement result.

The accuracy of such an assessment depends significantly on the number of measurements, and therefore, during the experimental study, at least five measurements were carried out, which makes it possible to obtain the value of the reliability assessment $P > 0.96$.

Calculation of deviations of true values from the arithmetic mean values of measurement results shows that the errors of electrical parameters do not go beyond the limit ±0.2%, for non-electric—±5.2%.

Figure 12 shows the obtained power characteristics of the wind turbine (dependence of power on wind speed) at an angle of installation of the blades 3.8° and 7.8°. The power values obtained in the experimental study were divided by 3 due to the fact that the simulation was performed only for one of the three sections of the Darrieus rotor [41]. The theoretical characteristics were constructed using the wind turbine power formula

$$P = 0.5\rho S C_p \eta_G \eta_M \eta V^3, \tag{6}$$

where *P*—wind turbine power and $\eta_G \eta_M$Efficiency of the generator and mechanical transmission of the wind turbine.

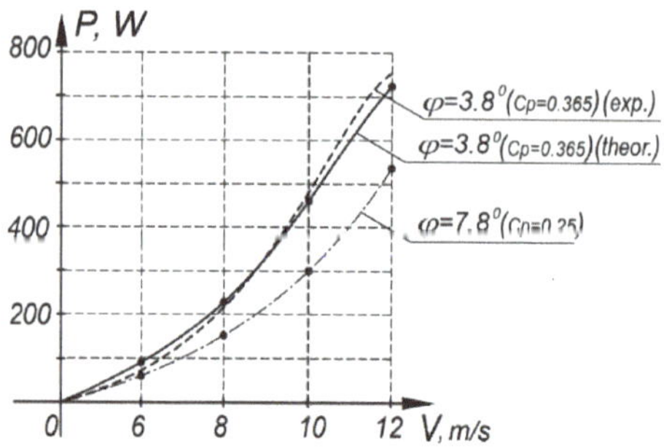

Figure 12. Power curves of the wind turbine.

It follows from the characteristics that the change in the angle of the blades made it possible to increase the energy efficiency of the operation of the wind turbine by more than 30% in almost the entire measured range of wind speed.

During the experimental study, the temperature of the rear surface of the photovoltaic modules was also measured by the pyrometric method. The measurement showed that the rotation of the rotor of the wind turbine makes it possible to reduce the temperature from 41.1 °C to 32 °C at wind speed 5 m/s, southbound, ambient temperature 25 °C and the intensity of solar radiation is about 650 W/m^2. This made it possible to increase the output power of photovoltaic modules by 3%. This result made it possible to obtain a qualitative result that confirms an increase in the energy efficiency of photovoltaic modules due to an increase in heat transfer from a solar cell by means of additional convective heat transfer.

Figure 13 shows the current-voltage characteristics at the output of photovoltaic modules and a matching stabilizer, obtained as a result of simulation. It can be seen from the characteristics that the matching stabilizer allows the transfer of electrical energy from photovoltaic modules at the point of maximum power at any voltage values at the output of the wind turbine generator.

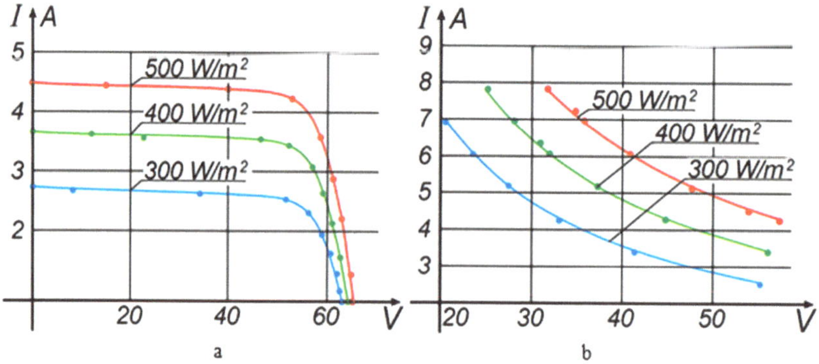

Figure 13. Current-voltage characteristics of photovoltaic modules (**a**) and a matching stabilizer (**b**) (dots indicate experimental).

The results of calculations of the annual power generation of photovoltaic modules showed that the optimal angle of installation of photovoltaic modules for the city of Sevastopol is 34°. From this it follows that they worked at limited power, as the module was fixed at a suboptimal angle 50°, set by the manufacturer. Figure 14 shows diagrams of annual energy production for optimal (34°) and suboptimal installation angle (50°) of photovoltaic modules.

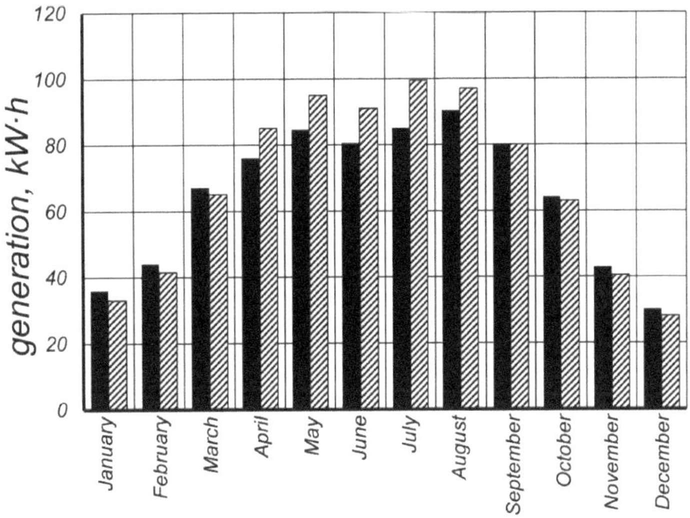

Figure 14. Diagram of monthly energy output of photovoltaic modules installed at an angle of 34° and 50° to the horizon.

It follows from the diagram that when the angle of photovoltaic modules is set to 34°, the annual power generation is 817 kW·h, and when the angle of installation is 50° it is 783 kW·h. Thus, changing the angle makes it possible to increase the power generation of the photovoltaic installation by 4.3%. The diagram also shows that for an even greater increase in efficiency, it is required to change the angle during operation: in the cold season, set the angle 50°, and in warm—34°. This event will allow to work out electrical energy on 6.2% more.

4. Conclusions

The processes of energy transformation from the RES have a direct impact on the environment. Implementation of the multivalent RES stations will decrease the dependency on the energy form the fossil fuel in the manufacturing and urban sectors primarily in the areas with the poor fossil fuel network.

Thus, the article shows that in order to minimize the existing risks associated with the development of urbanized areas, the concept of sustainable development of a smart urban environment is being introduced everywhere. However, existing digital transformation technologies do not sufficiently take into account environmental, economic, and social components. It requires the development of specialized solutions that, on the one hand, contribute to the development of the urban environment, and on the other, are focused on the use of innovative technologies that have a beneficial effect on the environment. In view of this, in order to provide electricity to objects of a smart urban environment, the authors propose, instead of the classical central power supply, to use autonomous power plants operating from renewable energy sources, in particular, wind and sun. The article showed the advantages of using combined wind-solar power plants, the main of which is the increased stability of electricity generation. An important advantage of this solution is also that it allows to increase the energy efficiency of photovoltaic converters by intensifying heat removal from the surface of the photocells, which significantly increases its total energy characteristics and increases the production of thermal and electrical energy. The presented results of digital aerodynamic modeling of the operation of the Darrieus wind turbine rotor using the finite volume method show that by setting the optimal blade angle, its energy efficiency can be significantly increased. The simulation results show that the maximum value of the wind utilization coefficient is achieved when the blades are installed at an angle 3.8°. The results of the parametric optimization of the photovoltaic installation in the conditions of Sevastopol showed that the highest efficiency is achieved when the modules are installed at an angle 50° to the horizon in the winter season, and in the summer—34°, which allows to generate 6.2% more electrical energy.

Estimated calculations of the energy parameters of the combined wind-solar plant with a rated power of 3.5 kW show that the potential annual energy production is 9817 kWh (9000 kWh is generated by the wind generator and 817 kWh by photovoltaic modules).

Author Contributions: Conceptualization, P.K. and M.R.; methodology, B.Y.; software, P.K. and D.V.; validation, V.E.; formal analysis, P.K.; investigation, B.Y.; resources, D.V.; data curation, V.E.; writing—original draft preparation, P.K.; writing—review and editing, P.K. and O.K.; visualization, O.K.; supervision, M.R. and M.L. All authors have read and agreed to the published version of the manuscript.

Funding: This research was funded by APVV-16-0192 and KEGA: 023TUKE-4/2021.

Institutional Review Board Statement: Not applicable.

Informed Consent Statement: Not applicable.

Data Availability Statement: Not applicable.

Acknowledgments: This work was supported by the Slovak Research and Development Agency under the contract No. APVV-16-0192. The autors would like to thank the KEGA grant agency for supporting research work and co-financing the project KEGA: 023TUKE-4/2021.

Conflicts of Interest: The authors declare no conflict of interest.

References

1. Höjer, M.; Wangel, J. Smart sustainable cities: Definition and challenges. *ICT Innov. Sustain.* **2015**, *310*, 333–349.
2. Martin, C.J.; Evans, J.; Karvonen, A. Smart and sustainable? Five tensions in the visions and practices of the smart-sustainable city in Europe and North America. *Technol. Forecast. Soc. Chang.* **2018**, *133*, 269–278. [CrossRef]
3. Hara, M. New key performance indicators for a smart sustainable city. *Sustainability* **2016**, *8*, 206. [CrossRef]
4. Bibri, S.E. A foundational framework for smart sustainable city development: Theoretical, disciplinary, and discursive dimensions and their synergies. *Sustain. Cities Soc.* **2018**, *38*, 758–794. [CrossRef]
5. Al-Nasrawi, S.; Adams, C.; El-Zaart, A. A conceptual multidimensional model for assessing smart sustainable cities. *J. Inf. Syst. Technol. Manag.* **2015**, *12*, 541–558.
6. Albino, V.; Berardi, U.; Dangelico, R.M. Smart cities: Definitions, dimensions, performance, and initiatives. *J. Urban Technol.* **2015**, *22*, 3–21. [CrossRef]
7. Kummitha, R.K.; Crutzen, N. How do we understand smart cities? An evolutionary perspective. *Cities* **2017**, *67*, 43–52. [CrossRef]
8. Batty, M.; Axhausen, K.W.; Giannotti, F.; Pozdnoukhov, A.; Bazzani, A.; Wachowicz, M.; Ouzounis, G.; Portugali, Y. Smart cities of the future. *Eur. Phys. J. Spec. Top.* **2012**, *214*, 481–518. [CrossRef]
9. Hader, M.; Rodzi, A. The smart city infrastructure development and monitoring. *Theor. Empir. Res. Urban Manag.* **2009**, *4*, 87–94.
10. Ramaprasad, A.; Sánchez-Ortiz, A.; Syn, T. A unified definition of a smart city. In *International Conference on Electronic Government*; Springer: Cham, Switzerland, 2017; pp. 13–24.
11. Ghosal, A.; Halder, S. Building Intelligent Systems for Smart Cities: Issues, Challenges and Approaches. *Comput. Commun. Netw.* **2018**, *18*, 107–125. [CrossRef]
12. Gupta, S.K.; Vanjale, S.B. Internet of things in smart city environments. *Int. J. Recent Technol. Eng.* **2019**, *8*, 7917–7921.
13. Sujata, J.; Abhijit, C.; Chanda, B. Impact of IOT technologies on smart city infrastructure: Current applications and future potential. *Int. J. Recent Technol. Eng.* **2019**, *8*, 2387–2391.
14. Sharma, S.; Nanda, M.; Goel, R.; Jain, A.; Bhushan, M.; Kumar, A. Smart cities using internet of things: Recent trends and techniques. *Int. J. Innov. Technol. Explor. Eng.* **2019**, *8*, 24–28.
15. Freudendal-Pedersen, M.; Kesselring, S.; Servou, E. What is smart for the future city? Mobilities and automation. *Sustainability* **2019**, *11*, 221. [CrossRef]
16. Sandoval, R.M.; Garcia-Sanchez, A.-J.; Garcia-Haro, J. Performance optimization of LoRa nodes for the future smart city/industry. *EURASIP J. Wirel. Commun. Netw.* **2019**, *2019*, 200. [CrossRef]
17. Andrade, R.O.; Yoo, S.G. A comprehensive study of the use of LoRa in the development of smart cities. *Appl. Sci.* **2019**, *22*, 4753. [CrossRef]

18. Jin, Y.; Yan, D.; Zhang, X.; Han, M.; Kang, X.; An, J.; Sun, H. District household electricity consumption pattern analysis based on auto-encoder algorithm. *IOP Conf. Ser. Mater. Sci. Eng.* **2019**, *609*, 072028. [CrossRef]
19. Zhou, S.; Zou, F.; Wu, Z.; Gu, W.; Hong, Q.; Booth, C. A smart community energy management scheme considering user dominated demand side response and P2P trading. *Int. J. Electr. Power Energy Syst.* **2020**, *114*, 105378. [CrossRef]
20. Ahmadian, E.; Byrd, H.; Sodagar, B.; Matthewman, S.; Kenney, C.; Mills, G. Energy and the form of cities: The counterintuitive impact of disruptive technologies. *Archit. Sci. Rev.* **2019**, *62*, 145–151. [CrossRef]
21. Wang, S.J.; Moriarty, P. Energy savings from Smart Cities: A critical analysis. *Energy Procedia* **2019**, *158*, 3271–3276. [CrossRef]
22. Costa, D.G.; Vasques, F.; Portugal, P.; Aguiar, A. A distributed multi-tier emergency alerting system exploiting sensors-based event detection to support smart city applications. *Sensors* **2020**, *20*, 170. [CrossRef]
23. Ilyas, N.; Shahzad, A.; Kim, K. Convolutional-neural network-based image crowd counting: Review, categorization, analysis, and performance evaluation. *Sensors* **2020**, *20*, 43. [CrossRef] [PubMed]
24. Hynes, M. Investigating Automobility and Its Implications for Local Urban Transport Policy Design. *Urban Sci.* **2017**, *1*, 14. [CrossRef]
25. Barthlemy, J.; Verstaevel, N.; Forehead, H.; Perez, P. Edge-computing video analytics for real-time traffic monitoring in a smart city. *Sensors* **2019**, *19*, 2048. [CrossRef]
26. Chong, X.J. Design and implementation of a self-powered smart water meter. *Sensors* **2019**, *19*, 4177. [CrossRef]
27. Mingaleeva, R.D.; Zajcev, V.S.; Bessel, V.V. Assessment of the technical potential of wind and solar energy in Russia. *Territ. Neftegaz* **2014**, *3*, 84–92.
28. Kuznetsov, P.N.; Lyamina, N.V.; Yuferev, L.Y. A device for remote monitoring of solar power plant parameters. *Appl. Sol. Energy* **2019**, *55*, 247–251. [CrossRef]
29. Cheboxarov, V.V.; Kuznetsov, P.N. Ways to solve the problem of instability in energy production of renewable energy plants. *Sci. Diary* **2019**, *4*, 60–66.
30. Kuznetsov, P.N.; Cheboxarov, V.V.; Yakimovich, B.A. Hybrid wind-solar power plants. *Bull. IzhSTU Named MT Kalashnikov* **2020**, *1*, 116–123. [CrossRef]
31. Kuznetsov, P.N.; Kuvshinov, V.V.; Issa, H.A.; Mohammed, H.J.; Al Barmani, A.G. Investigation of the losses of photovoltaic solar systems during operation under partial shading. *J. Appl. Eng. Sci.* **2020**, *18*, 313–320. [CrossRef]
32. Nazarov, B.I.; Saliev, M.A.; Mahmudov, A.N.; Abdullaev, S.F. The effect of aerosol pollution on the operation of solar receivers. *Rep. Acad. Sci. Repub. Tajikistan Dushanbe Presidium Acad. Sci. Repub.* **2016**, *49*, 206–213.
33. Savenko, A.E.; Osmanov, E.S. Calculation of realized capacity and power generation by a single-standing wind turbine. *Bull. Kazan State Energy Univ.* **2019**, *1*, 68–78.
34. Huld, T. PVMAPS: Software tools and data for the estimation of solar radiation and photovoltaic module performance over large geographical areas. *Sol. Energy* **2017**, *142*, 171–181. [CrossRef]
35. Tsai, H.L.; Tu, C.S.; Su, Y.J. Development of generalized photovoltaic model using MATLAB/SIMULINK. In Proceedings of the World Congress on Engineering and Computer Science, San Francisco, CA, USA, 22–24 October 2008; pp. 1–6.
36. Ebrahimpour, M.; Shafaghat, R.; Alamian, R.; Safdari Shadloo, M. Numerical investigation of the savonius vertical axis wind turbine and evaluation of the effect of the overlap parameter in both horizontal and vertical directions on its performance. *Symmetry* **2019**, *11*, 821. [CrossRef]
37. Kothe, L.B.; Möller, S.V.; Petry, A.P. Numerical and experimental study of a helical Savonius wind turbine and a comparison with a two-stage Savonius turbine. *Renew. Energy* **2020**, *148*, 627–638. [CrossRef]
38. Sahim, K.; Santoso, D.; Puspitasari, D. Investigations on the effect of radius rotor in combined Darrieus-Savonius wind turbine. *Int. J. Rotating Mach.* **2018**, *2018*, 3568542. [CrossRef]
39. Mandel, J. *The Statistical Analysis of Experimental Data*; Courier Corporation: New York, NY, USA, 2012.
40. Rabinovich, S.G. *Measurement Errors and Uncertainties: Theory and Practice*; Springer Science & Business Media: Berlin/Heidelberg, Germany, 2006.
41. Inoue, A.; Takahashi, R.; Murata, T.; Tamura, J.; Kimura, M.; Futami, M.O.; Ide, K. A calculation method of the total efficiency of wind generators. *Electr. Eng. Jpn.* **2006**, *157*, 52–62. [CrossRef]

applied sciences

MDPI

Article

Particle and Particle Agglomerate Size Monitoring by Scanning Probe Microscope

Pavel Gulyaev [1], Tibor Krenicky [2], Evgeny Shelkovnikov [1] and Aleksandr Korshunov [1,*]

[1] Udmurt Federal Research Centre of the Ural Branch of the Russian Academy of Sciences, Institute of Mechanics, 426067 Izhevsk, Russia; lucac@e-izhevsk.ru (P.G.); iit@udman.ru (E.S.)
[2] Department of Technical Systems Design and Monitoring, Faculty of Manufacturing Technologies with a Seat in Presov, Technical University of Kosice, Sturova 31, 080 01 Presov, Slovakia; tibor.krenicky@tuke.sk
* Correspondence: kai@udman.ru

Featured Application: The results presented in this paper can be used in microscopy and modern methods of particle inspection using images.

Abstract: In the present study, the use of a scanning probe microscope is described for monitoring the sizes of nanoparticles. Monitoring is the process of acquiring and analysing the set of overlapping images. The main analysis steps are image segmentation, determination of nanoparticles allocation and their sizes, determination of the overlap of images with one another, and exclusion of repeating measurements for the formation of the correct particle-size sampling. The thorough examination of commercial scanning probe microscopes, software, and image processing libraries showed that their capabilities are limited for image segmentation and determination of sizes in complex structured images. A method based on the surface curvature computation is proposed for the image segmentation (allocation of particles) and determination of particle sizes. The curvature is estimated using the surface area approximation with respect to the circumference. It is proposed to use sample displacement sensors as an aid for image stitching.

Keywords: scanning probe microscope; particle size monitoring; key points; descriptor; image stitching; segmentation; particle allocation

Citation: Gulyaev, P.; Krenicky, T.; Shelkovnikov, E.; Korshunov, A. Particle and Particle Agglomerate Size Monitoring by Scanning Probe Microscope. *Appl. Sci.* **2022**, *12*, 2183. https://doi.org/10.3390/app12042183

Academic Editor: Chih-Ching Huang

Received: 24 January 2022
Accepted: 16 February 2022
Published: 19 February 2022

1. Introduction

The physical and chemical properties of nanoparticles are mainly determined by their sizes. Size monitoring is an important component of the process of nanoparticles production [1,2]. The monitoring is the formation of the particle-size sampling followed by the analysis of the particle-size distribution. The analysis is of importance for the evaluation of the sizes of particles themselves and for the evaluation of particle-based products, for example, of the dispersity evaluation of powders containing particles under study. For the data acquisition and formation of the particle-size distribution such methods as laser diffraction [3,4], dynamic light scattering [5,6] and acoustic spectroscopy [7,8] are more widely used. The methods are mainly suitable for controlling particles in solutions. The disadvantage of these methods is measurement results distortion, which can occur with a significant difference in the particles size present in the solution.

In this study, we consider the case when nanoparticles are deposited on a substrate before research and size monitoring, carried out by images. An overview of deposition methods and a description of the simplest one can be found in [9]. Deposition is widely used for particles obtained in the liquid phase and in the form of powders. In the latter case, a suspension or colloidal solution is prepared from the particles. Traditional photography [10] or various microscopes and diffractometers [11–17] are widely used to obtain particles images on substrates. One of the most powerful tools among them is a scanning probe microscope (SPM). In comparison with other devices such as an electron microscope, for example, the SPM has the following advantages: variety of sample types, functionality in air and liquid, relative simplicity of sample preparation, direct topographic measurements and the presence of nanolithography functions permitting to mark studied regions during monitoring. Since the lifetime of particles on the substrate can be limited, and the execution time of one scan can reach 5–10 min, significant efforts of developers are aimed at time reduction and automating the monitoring of particle sizes. One of the ways to reduce monitoring time is to substitute sequential uniform scanning [18] with tracking scanning. Automation is possible due to additional image processing and the determination of tracking scan parameters based on the results of this processing.

Direct topographic measurements in SPM allow to use the conventional image processing methods. As a rule, SPMimages are processed with the use of conventional algorithms [19] presented in OpenCV and FLANN libraries [20–22]. Commercial universal SPM [23] software, as well as specialized software (ImageJ, MountainsSPIP) widely use the methods (binarization, segmentation, contour analysis) implemented in the mentioned libraries. However, it is further shown that in some cases it is reasonable to use specialized methods for image processing. There are various special-purpose SPMs that can be used for nanoparticle size monitoring such as a multi-probe scanning microscope (an AFM with parallel active cantilevers) and large-sample SPM. The multi-probe scanning microscope can be used for fast investigation (several minutes) of a large area (several mm^2). However, it has a low resolution of about 40–50 nm [24], which makes it impossible to monitor nanoparticles with a diameter smaller than 40–50 nm.

A large-sample SPM can also be used for nanoparticle size monitoring. Such an SPM is equipped with advanced sample-positioning systems, which leads to an increase in the effectiveness of size monitoring. The use of a large-sample SPM is economically sound for monitoring nano-objects in mass-production items such as silicon wafers, catalysts, membranes, or filters. However, when the parameters of an item are changed, for example, a decrease in the sizes of a sample and nanoparticles, the large-sample SPM capabilities are the same or even worse than those of a conventional laboratory SPM. Another type of SPM providing possibilities for monitoring nanoparticle sizes is a universal highly automated SPM, for example, an SPM Next II [23]. Such an SPM allows selecting an area for investigation, automatic positioning of a probe to a selected area, scanning several overlapping areas and forming a panoramic image. However, in the automatic mode of operation, the software of commercial microscopes is not always capable of distinguishing and correctly determining the sizes of polydisperse powders, particles with over-lapping contours or considerable oscillations of the brightness function.

Based on the above-mentioned, we can conclude that the search for means to modernize existing SPMs (for example, open-access platforms such as Bruker Dimension Icon) for the automation of the size monitoring is of current importance especially when these means facilitate the development of SPM in general. This is confirmed by examples of a meaningful (considering the features of the objects under study) implementation of the tracking scan concept, implemented in a homebuilt SPM [25].

In the present study, we consider the formation of sampling on the basis of the processing of the image set of nanoparticles and their agglomerates deposited on a substrate for monitoring. These particular cases are characterized by the presence of the following limitations:

- Impossibility of the formation of the monolayer of particles guaranteeing their absolute distinguishability;
- Difficulty in searching for particles on a substrate.

The objective of the present work is the development of methods for particle-size monitoring in images acceptable in the conditions of the afore-described limitations and suitable for individual particles and the particles in agglomerates and multilayer coatings. The SPM is well-suited for studying nanoparticles deposited on a substrate from a gaseous phase or a liquid. The most difficult case for monitoring is the deposition of nanoparticles from a liquid since it leads to the formation of multilayer, the relief irregularity of an obtained coating and the presence of agglomerates.

2. Materials and Methods

In the present study, there are examples of the SPM application for studying fullerene-containing powder (the particle sizes in the range of 0.7–100 nm) and nanoscale particles (the sizes in the range of 0.1–40 μm). The fullerene-containing powder was obtained in Udmurt Federal Research Centre (Izhevsk, Russia) by electrochemical deposition and dispersion at high pressure. The nanoscale particles were obtained in the multi-stage centrifugal mill designed and assembled in Udmurt Federal Research Centre. Highly oriented pyrolytic graphite (HOPG) and polished ceramic plates (1 × 1 cm) were used as substrates. The substrates were purchased from the NT-MDT company.

The electrochemical deposition of fullerene-containing powders on HOPG was performed from a fullerene solution (the concentration ~150 μg/g) in paraxylene with sodium metal at the voltage of 100–150 V and galvanic current of 10–30 μA. HOPG was preliminarily subjected to mechanical cleaning using scotch tape.The cooper particles dissolved in polystyrene were deposited on the plate surface by the spin-coating method. The plate was preliminarily cleaned with ethanol.

In order to automate the investigations, a microscope based on a commercial P47 SPM was used. The field of view of the microscope was 100 × 100 × 6 μm (±10%), and the minimum scanning step was 0.006 nm. It was upgraded as follows. Firstly, the manual positioning screw drivers were motorized. Secondly, the software designed for motion control and image processing was upgraded. To minimize any actions upon a surface the modes of tunnelling microscopy and semi-contact atomic force microscopy were used.

3. Size-Monitoring Method with the Use of the SPM

The nanoparticle-size monitoring with the use of the SPM can be described as follows. Firstly, a surface area is selected (for example, with the help of a video camera); after that, the selected area is scanned several times at high resolution. Let us consider the case when the size of the studied area exceeds the microscope field; it is rather common considering that the resolution of an optical microscope is 1 μm and the size of high-resolution scanners is 1–3 μm. Traditionally, the scanning of surface areas exceeding the field is performed using overlapped scans and a sample displacement. Before the displacement, the probe is taken away from the surface and brought back after the displacement. The scanning process is performed until there are a required number of particles on the acquired images, the sizes of which form the statistics for monitoring. The scans can be uniformly arranged over the specified area with a fixed overlap (uniform scanning, Figure 1a) or along the line with an increased concentration of controlled particles (tracking scan, Figure 1b). For example, when particles are deposited on the HOPG surface, the most probable site of their allocation is the boundaries of terraces. In this case, the tracking scan provides a considerable decrease in the monitoring time. At such scanning, the displacement of scans is performed by a scanner or motorized sample platform. Here it is necessary to consider that the error in the positioning by a motorized platform is significantly bigger than that in the case with a scanner. The acquired scans (images) are stitched into a single image before processing or are processed separately while the repeating results in the overlap regions are excluded.

The uniform scanning of a studied area is sufficiently elaborated by the microscope and software manufacturers. In the present study, we turn our attention to tracking scans. Tracking scanning allows to significantly save the time of scanning. The amount of savings provided in [25] amounted to up to 67% of the scanning time [25]. For example, in the case shown in Figure 1, the time savings is 5/9 or 55%. It should be noted that significant time savings are also achieved by automating the pre-processing of images and the selection of the trajectory of the displacement of the field of view, carried out by the experimenter manually.

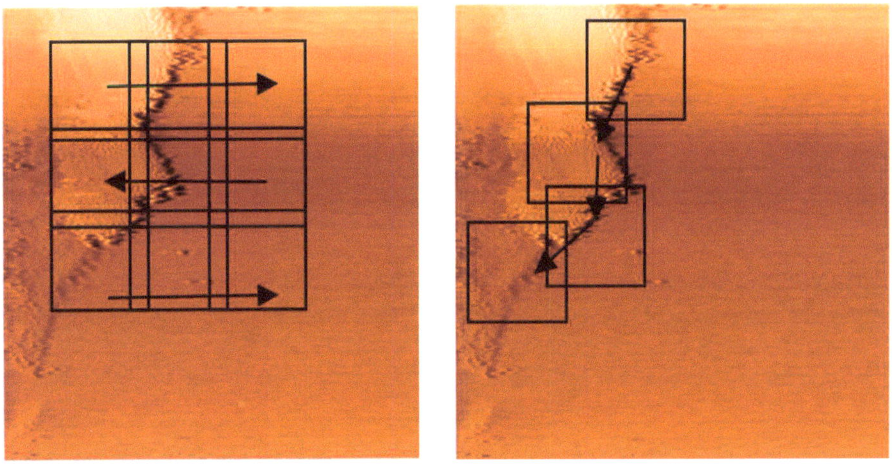

(a) (b)

Figure 1. Options of the SPM operation in the process of the size monitoring: (**a**) uniform scanning; (**b**) tracking scanning.

Our approach to particle size monitoring, founded on tracking scan, is based on solving the following tasks:

(1) Particle detection: image segmentation, particle allocation, the determination of the particle sizes and coordinates;
(2) Computation of the sample displacement value providing the coordinate coupling of two overlapping images and the continuation of the particle-size data acquisition;
(3) Computation of the "global" coordinates of the particle centres for excluding repeating measurements in the areas of the image overlapping.

To solve the first task, the modern SPM software contains the tools for particles detection and determination of their sizes, or such tools are in specialized software. For example, the NT-MDT microscopes (the software ImageAnalysis) contain a GrainAnalysis function, and in the specialized software MountainsSPIP there is a Grain/Pore Analysis function. The drawback of the above tools is the necessity of manual adjustment. In addition, these tools are based on the standard tools for image binarization and segmentation (Watershed [26], Otsu's method [27]), which decreases the accuracy of the size determination for superimposing particles and particles with different degrees of brightness. The difference in brightness is possible even when particles are deposited from a liquid solution on atomically smooth substrates. This is due to nano- and microparticles that tend to form polydisperse agglomerates and a surface with an irregular relief. There are more advanced methods for segmentation such as the parametric (with the estimation of the parameters of the grey-level distribution of each class of objects in an image [28,29]), variational [30], neuronet [31] and correlation methods [32]. All the above methods are limited in application and they are not always available in open-ended libraries such as OpenCV. In addition, the parametric and variational methods are distinguished by their considerable computational complexity and complexity of implementation; regarding the neuronet methods, sometimes it is difficult to provide a sufficient amount of the training material. The investigations of the segmentation methods with the use of open freeware (ImageAnalysis, ImageJ, WSxM) show the following: The images segmentation using conventional methods based on the intensity function is rather difficult for complex images with particle agglomerate. The difficulty of segmentation leads to the impossibility of accurately determining the size and coordinates of particles and the implementation of the proposed monitoring method. In the present study, we consider the approach of image segmentation based on the surface curvature function.

For solving the second task, motorized sample platforms are used; as a rule, microscopes for studying large-size samples are equipped with such platforms. In this case, the larger the platform travel range is, the larger the minimal step of positioning is (Table 1). Judging by the value of the minimal step, the sample-positioning platforms in a commercial SPM are designed for use with an optical microscope (resolution of about 1 μm). The use of such systems with a scanner having a travel range of less than 1 μm is difficult.

Table 1. Characteristics of probe microscopes.

SPM Model	Company	Range, mm	Minimal Step (Resolution), μm
C3M Next II [23]	NT MDT	5×5	0.3
AFM5500M [33]	Hitachi	100×100	2
Dimension Icon [34]	Brucker	180×150	2

The presence of an error in the positioning of the motorized platform causes an uncertainty in the value of the scan overlapping, which makes difficult the solution of the third task, the determination of the probe and particles coordinates in a single global coordinate system. To solve the task, in an SPM the sensor values or the tools for the overlapping-image processing can be used. When sensors are used, each image and each particle have their unique coordinates, which allows for excluding repeating results in the overlap region. In this case, the main problems are as follows:

- Sensors are not always present in the SPM positioning system, and sometimes their integration in this system is constructively impossible and impractical;
- The sensor accuracy may be insufficient for the unambiguous establishment of the correspondence of the centres of the particles.

Software (algorithm) for stitching or coordinate coupling of overlapping images can be an alternative to sensors. When algorithmic methods are used, the accuracy of the coordinate determination depends on the resolution of the scanner and image and not on the precision of the platform or sensor. When the images are stitched, a single image is formed, and each particle is unique within the image. At the coordinate coupling, the mutual shift (displacement) and rotation of the images are determined. Using these values, the local coordinates of particles in each image can be combined into a common coordinate system. The cases are known of the combined use of the sensor values and the images for high-precision probe positioning. For example, in the SPM NEXT II [23], the sample positioning by the motorized platform is additionally controlled by a high-resolution video camera.

Detectors and descriptors of key points are widely used for matching and stitching of images. Binary descriptors (BRIEF, ORB and BRISK) refer to low-level ones and they are sensitive to distortions associated with the change in scale, shape, and point of view [35] which are quite probable in an SPM. The SPMimages can contain distortions even after processing, which influence the result of the measurement of object sizes. For example, an image can contain lines significantly different from neighbouring lines due to the probe-surface interaction or 1/f noises [36]. There can also be brightness oscillations due to the final speed of the regulating system, surface contamination, vibroacoustic disturbances, and imperfect scanning system. Such disturbances influence the object shape and the position of key points. More complex, transformation invariant descriptors such as SIFT [37], SURF [38] can also be ineffective in the conditions of disturbances and distortions characteristic of the SPM.

The analysis shows that specialized methods should be used for image matching instead of conventional ones. For example, it is logical if the centres of key points coincide with the particle centres and the information on the particle size is included in the key point descriptor. It is also advisable to use the values of the platform sensors for determining the mutual orientation (displacement, angle of rotation) of the images (for example, in the topoStitch program [39], prior information on the image values can be used for image stitching).

Considering the difficulties in solving the problem of particle agglomerate size monitoring by conventional methods, we propose a new integrated approach based on the use of:

- The surface curvature detectors for correct image segmentation and particle detection;
- The detection results for overlapping images coordinate matching and sample displacement computation;
- The complexing of displacement sensors and the determined images overlap value for refinement of results.

4. Detection of Particles Using the Surface Curvature Function

For particle detection, the curvature function of surface or plane curves can be used. However, conventional methods for the curvature computation use derivatives; thus, the field of their application is limited to smooth surfaces [40]. Therefore, the development of new methods for the computation of the curvature of the SPMimages considering their peculiar features is of high practical value. Let us consider one of the possible realizations of the method for computing the surface curvature [41,42] and particles sizes in the SPMimages.

To determine the curvature the neighbourhood of each point in each line of the SPMimage Z(x) is approximated by the segment of the adjacent circle. The segment height $\varepsilon = AB$ is a constant value for the entire image, and the chord length $H = P_1P_2$ is a variable (Figure 2).

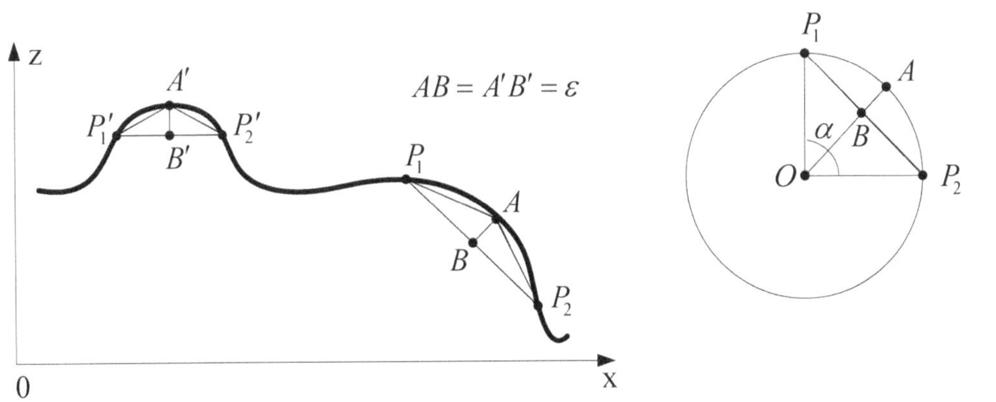

Figure 2. Schematic view of the curvature detector operation.

The radius R of the particle osculating circle is computed from the figure ΔP_1OB:

$$\left(R-\varepsilon^2\right)+\frac{H^2}{4}=R^2,\ R=\frac{H^2+4\varepsilon^2}{8\varepsilon} \quad (1)$$

The algorithm of the curvature computation is as follows. The points P_1 and P_2 are incrementally moved away from A with the step of 1 pixel along the x axis until $AB < \varepsilon$. After each step, the coordinates of points P_1 and P_2 are determined and the general equation of the line passing through these points is computed. The value AB is determined by the following formula:

$$AB = \left|\frac{ax + by + c}{\sqrt{a^2 + b^2}}\right|;\ AB = \left|\frac{c}{\sqrt{a^2 + b^2}}\right|_{(x=0,y=0)} \quad (2)$$

where a, b, and c are the coefficients of the general equation of the straight line drawn through P_1 and P_2. If the inequality $AB < \varepsilon$ ceases to hold, then the chord length is calculated. The H values for the image lines and columns form the functions $H(x)$ and $H(y)$.

For the images with a low level of distortions, the centre of a particle is the point of the coincidence of the local extrema $H(x)$ and $H(y)$. The particle radius is computed by Formula (1). In the presence of distortions, the particle centres can be determined by the local extrema of the function:

$$W^{+}(x, y) = \begin{cases} H_y(x) + H_x(y), & (H_y(x) \geq 0) \wedge (H_x(y) \geq 0), \\ C, & \text{otherwise}, \end{cases} \quad (3)$$

where $C \geq \max\ (H_y(x) + H_x(y))$is a constant. In the presence of several close-spaced extrema $W^{+}(x, y)$,the dominant extremum is chosen. The curvature detector is sensitive to image noise. Therefore, preliminary smoothing is required.

5. Determination of the Parameters of the Mutual Orientation of Overlapping Images

The existing methods of image stitching can require considerable (up to 50% [43]) overlapping of images which is necessary for the allocation of the sufficient number of common key points in the images Z_0 and Z. When particle sizes are determined with the use of several overlapping images, the result of the stitching process is not important. It is important that the parameters of the mutual orientation of overlapping images, displacement and rotation, are determined; this permits the determination of the overlap region and exclude the repeating results of the measurements relating to the region. The determination of the parameters is more exact when there is a sufficient number of common points. For the homography, the number of such points should not be less than 4 and at the use of the method of image matching by key points' pairs (IMBKPP) [44], it should be 3. Let us consider the realization of the IMBKPP method where the detected particle centres are taken as key points.

In order to compute the displacement and rotation of the images, we used the key points' pairs of the images, which are made up of the key points from the total set of key points. Each key points pair (line segment) is described by the coordinates of the ends (x_1, y_1) and (x_2, y_2) and the length l. For the computational shortcut, the displacement values Δ_x, Δ_y and the rotation angle φ are computed only for the pairs of points with the same length. The order of the computations is provided in [44]. The Δ_x and Δ_y values form a two-dimensional array $M(\Delta_x, \Delta_y)$ where the number of certain displacement values (Δ_x, Δ_y) is entered into the cell with the coordinate (Δ_x, Δ_y). The set of the values of the rotation angles forms a histogram of iteration $N(\varphi)$ for the range from $-180°$ to $180°$.The coordinates of the maxima in $M(\Delta_x, \Delta_y)$ and $N(\varphi)$ determine the displacement and rotation of the images.

During the size monitoring process, to minimize losses of time for scanning the same surface areas, it is practical to restrict the overlap region of the sequentially acquired images so that there would not be more than 3–5 particles. However, it can lead to a decrease in the overlapping coefficient by 10–15%. As the practice showed, the background elements (disturbances) cause the appearance of false key points and several local maxima in the array $M(\Delta_x, \Delta_y)$ which makes the computation of the mutual orientation parameters of two images difficult.

In the simplest case, a key points pair is described with the use of the descriptor D1 which represents the coordinates of the ends (x_1, y_1) and (x_2, y_2) and the length l of the line segment connecting them: $D1 = \{(x_1, y_1), (x_2, y_2), l\}$. To improve the accuracy of the pairs' correspondence, the more informative pair descriptor is recommended: D2 = {(x_1, y_1), (x_2, y_2), R_1, R_2, l}, where R_1 and R_2 are the radii of the particles, the centres of which coincide with key points. In this case, the correspondence search algorithm checks both the length equality of the pairs in the images and the validity of the condition:

$$\begin{gathered}(|R_1 - R_i + R_2 - R_j| < 2t)and(((|R_1 - R_i| < t)and \\ and(|R_2 - R_j| < t))or\ ((|R_1 - R_j| < t)and(|R_2 - R_i| < t)))\end{gathered} \tag{4}$$

where R_1 and R_2 are the radii of the curvature of the key points pair of the first image; R_i and R_j are the radii of the curvature of the key points pair of the second image; t is the tolerance value. It is also possible to exclude the key point pairs with a nonzero rotation angle from the correspondence analysis since the second image was acquired after the sample displacement in two orthogonal directions. The pair descriptor in this case has the form D3{(x_1, y_1), (x_2, y_2), R_1, R_2, l, φ}. Descriptors D1–D3 are sensitive to the scale of the image. Their use implies the maintenance of the field of view and the resolution of the microscope during the tracking scan.

6. Complexing of the Methods of Images Overlap Estimation

As was mentioned earlier [44], the IMBKPP method provides good results for high-quality images with a high overlapping coefficient and a considerable number of particles in the overlap region. When the overlapping coefficient is low, the number of common key points of the images decrease; the maximum value of the array M(Δ_x, Δ_y) may be small (only 1–3units), and the number of the equivalent maxima exceeds 1.

To avoid this, when choosing the offset of the subsequent scan, it is necessary to ensure that there are more than three particles in the overlap area. For example, we usually set the overlap region dimensions so that there were no fewer than 10 particles. The reason for this is that when the density of the similar-size particles is high, the probability of a false correspondence of key point pairs (especially short ones) increases. The technique for determining the overlap region is as follows: The overlap of 35% is set along X and Y axes (the area of the overlap region is 12.5%). If in the overlap region the number N_{OV} of particles, allocated in the overlap region, is smaller than required, the overlap region is gradually increased along one coordinate until the required value of N_{OV} is attained. When necessary, the overlap region can be increased along the other coordinate as well. Contaminations or image noise have an effect on the number of particles. If there is a chance of such contamination, then it is necessary to increase the overlap area.

Thus, to increase the reliability of the determination of the scan overlap parameters, several methods for estimating these parameters should be used [45]: first, algorithmic methods for the image processing; and second, the values of the sensors mounted on the sample-positioning platform or displacement values obtained on the basis of the signals controlling the sample-positioning platform drives. In our SPM we use the positioning platform where the leadscrews for manual control are equipped with inertia piezo-drives and sensors [45].

If there are several maxima in the array M(Δ_x, Δ_y), the sensor values limit the valid region of the M(Δ_x, Δ_y) and permit to select the truer extremum. In this case, the accuracy of the final result computed by the software is considerably higher than the sensor accuracy. The sensor resolution should correspond to the value of the linear scans overlapping along one of the axes and make up 10–15% of the scan size. The accuracy of the software method is 2–3 pixels, which is 0.8–1.1% for the scan of 256 × 256 in size.

7. Result and Discussion

We compare the proposed method of particle size monitoring with a GrainAnalysis function of a commercial P47 microscope. We use three pairs of agglomerate images: a test pair in Figure 3a,b; images of agglomerates on a flat surface in Figure 4a,b; images of agglomerates on a surface with a complex relief Figure 5a,b. Each test images contain 47 particles. The particles size on the images is 10 pixels. The images differ in the particle brightness. The test images were modelled, the images in Figures 4 and 5 were received in semi-contact mode (scanning time ~4 min).

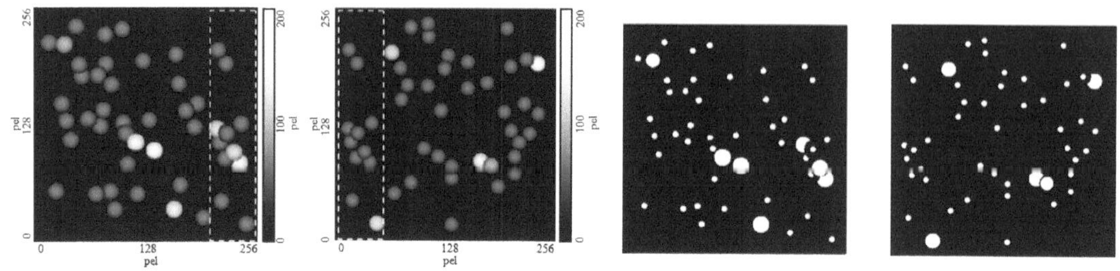

(a) (b) (c) (d)

Figure 3. Overlapping test images (256 × 256 pixels, Z range—200, x-axis offset 190 pixels): (**a**,**b**) source images with marked overlap area; (**c**,**d**) the results of segmentation by GrainAnalysis.

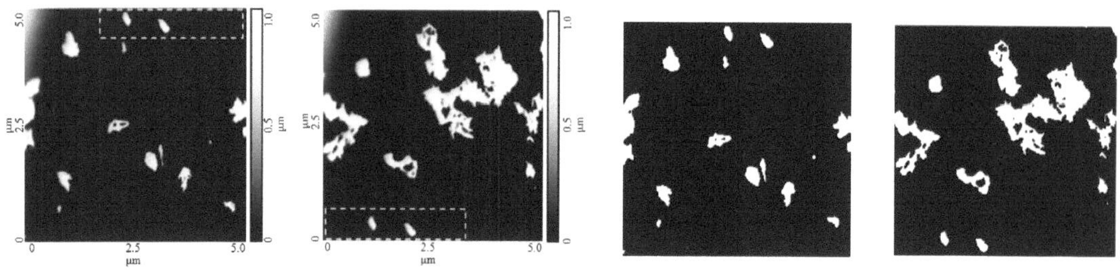

(a) (b) (c) (d)

Figure 4. Overlapping copper microparticles images (256 × 256 pixels): (**a**,**b**) source images with marked overlap area; (**c**,**d**) the results of segmentation by GrainAnalysis.

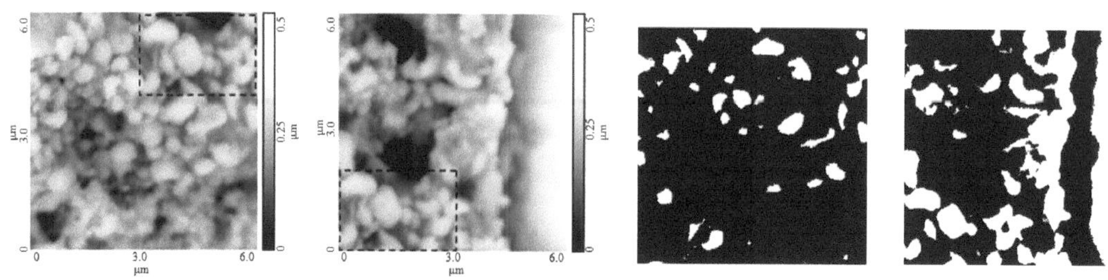

(a) (b) (c) (d)

Figure 5. Overlapping fullerene' images (256 × 256 pixels): (**a**,**b**) source images with marked overlap area; (**c**,**d**) the results of segmentation by GrainAnalysis.

The results of segmentation of test images obtained using the GrainAnalysis tool are shown in Figure 3c,d. For quantitative estimates, we used the number Nd of detected particles and their average size Rd, as well as the average size relative deviation ε. It follows from the comparison results (Table 2) that the conventional method did not allow the correct detection of agglomerate particles. At the same time, the separation of agglomerates into particles may be impossible or may be accompanied by a decrease in size. This is mainly due to the limited capabilities of conventional methods. The comparison of the segmentation results obtained by the curvature detector and conventional methods (Figure 6) shows the effectiveness of the curvature detector for the segmentation of agglomerate images. It detected all 47 particles in the test image, unlike the conventional methods. In addition, the average relative deviation of the particle sizes (in the test images) was no more than 3% at the same time. Unfortunately, it is difficult to assess the accuracy of the detector on more complex images (Figures 4 and 5a,b). The reason for this is the lack of prior reliable information about the sizes of particles and agglomerates.

Table 2. Comparison of segmentation results by the curvature Detector and the GrainAnalysis tool (the results for the Grain Analysis tool are given for Figure 3c,d in parentheses).

Parameter	Curvature Detector	Grain Analysis Tool
Nd	47	47(47)
Rd	9.84	8(9)
ε	0.016	0.2(0.1)

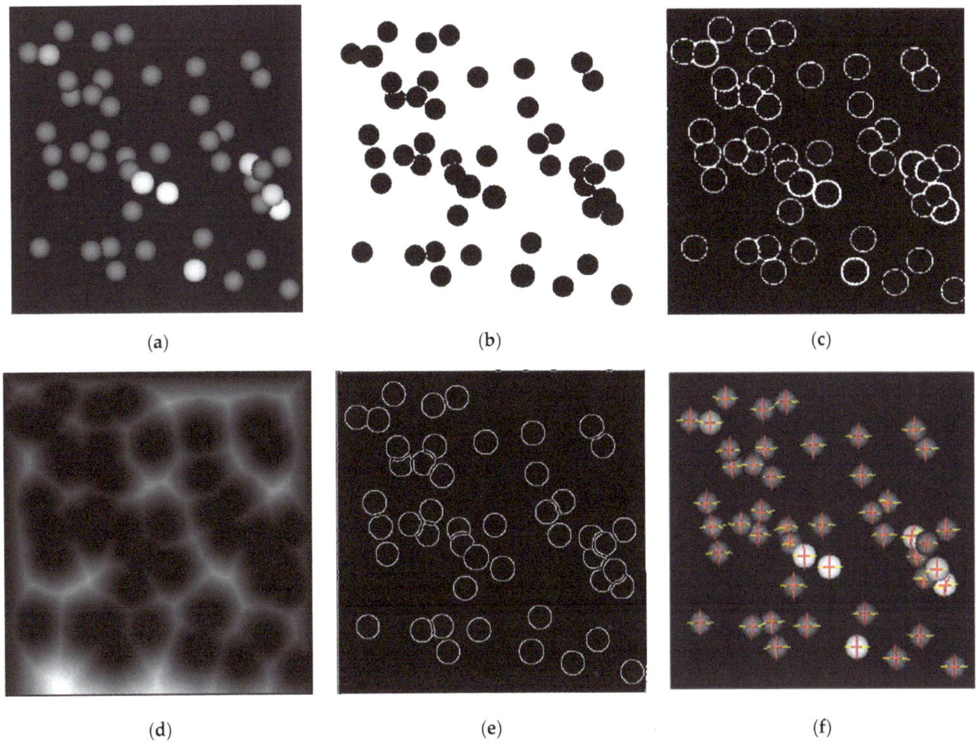

Figure 6. Comparative results of the segmentation on test image: (**a**) the initialimage (the scale according to Figure 3a), containing 47 particles with a radius of 10 pixels; (**b**) segmentation by Otsu's method; (**c**) segmentation with the use of the adaptive threshold technique; (**d**) segmentation by Watershed method; (**e**) segmentation by Canny method; (**f**) the centres of the particles, and the zones of maximum curvature, obtained using a curvature detector.

To compare the methods of particles monitoring on the complex images (Figures 4 and 5), we use threshold segmentation performed using the Grain Analysis tool. The threshold was adjusted manually to the value at which the agglomerate particles in the overlap area are separated from each other. A comparison of the segmentation results with the GrainAnalysis tool (Figures 4 and 5c,d) and the curvature detector (Figures 7–9) shows that the detector highlights more key points in the overlap zone, which makes it easier to overlap images in coordinate coupling. Figures 7, 8 and 9a,b shows two overlapping images and the computed results for the displacement and rotation. The scale of the images (Figures 7–9) corresponds to Figures 3–5. The array M is displayed in the form of the two-dimensional map of the displacement values (Figures 7, 8 and 9c: dot designates a non-zero element of the array M, and cross designates a maximum) and the array N(φ) is shown in the form of a diagram (Figures 7, 8 and 9d). The orientation parameters Δ_x, Δ_y and φ are determined from the positions of the maxima of the M and N arrays.

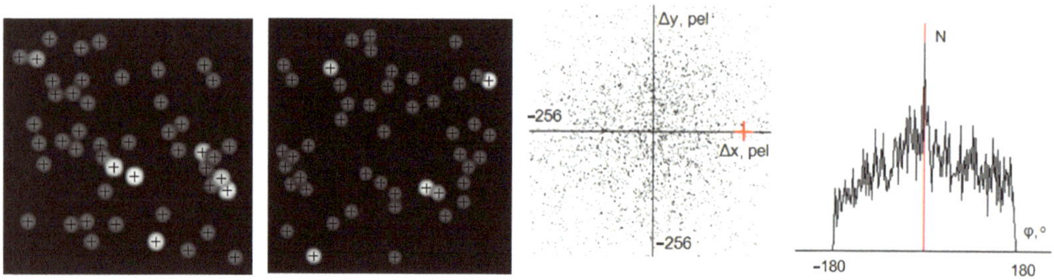

(**a**) (**b**) (**c**) (**d**)

Figure 7. Results obtained by the IMBKPP method for the test images: (**a**,**b**) source images with marked particle centres; (**c**) the map of displacement values; (**d**) the histogram of rotation angles.

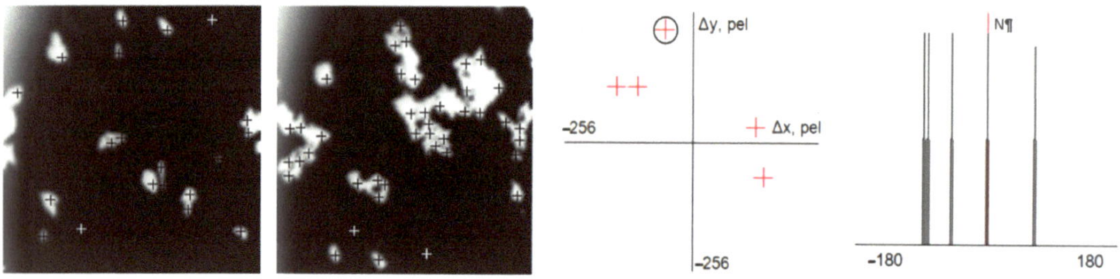

(**a**) (**b**) (**c**) (**d**)

Figure 8. Results obtained by the IMBKPP method for the copper microparticles' images: (**a**,**b**) source images with marked particle centres; (**c**) the map of displacement values; (**d**) the histogram of rotation angles.

Figures 7–9 present the curvature detector and IMBKPP which achieved satisfactory results (the specified *x*-axis offset is correctly determined) on images, on which the conventional methods (such as SURF descriptor and FLANN library) showed unsatisfactory results. The overlapping areas of the images (Figures 7–9) correspond to Figures 3–5. Figure 8 confirmed the operability of the proposed monitoring approach for images with low overlap, which required the use of a sensor to determine the true extreme of the displacement (Figure 8c, encircled). Figure 9 confirmed the operability of the proposed approach in images with complex relief. The detector's operability was determined by visual comparison of similar key points of overlapping images, as well as an assessment of their position on the image and histogram.

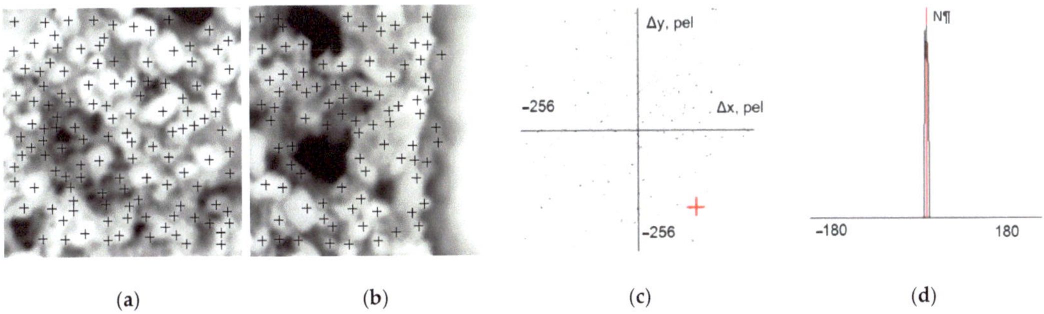

Figure 9. Results obtained by the IMBKPP method for fullerene-containing powder images: (**a**,**b**) source images with marked particle centres; (**c**) the map of displacement values; (**d**) the rotation angle histogram.

Figure 8 shows that the centres of several densely packed agglomerates, determined by the curvature detector, may not match the visually detected centres. This problem can be solved by increasing the image resolution or reducing the microscope field of view. The number of detected particles increases in proportion to their concentration if the image resolution is sufficient. Images of polystyrene spheres coated by gold (Figure 10), obtained from the scan gallery [46], and Table 3 confirm this fact. The size of the image elements that can be detected should not be less than 4–5 pixels. This is illustrated by Figure 11 [46], where some small elements, located at a short distance (4–5) pixels from other elements, were not detected.

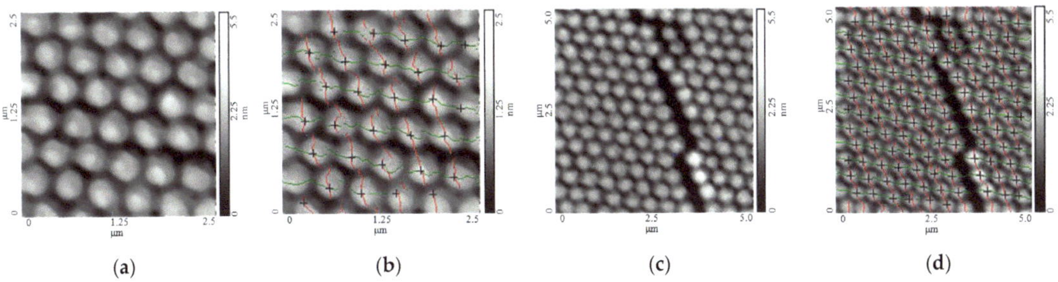

Figure 10. Images of polystyrene spheres coated by gold: (**a**) source images 2.5 × 2.5 µm; (**b**) images with marked centres; (**c**) source images 5 × 5 µm; (**d**) images with marked centres.

Table 3. Detection results on images with different particle concentrations.

Parameter	Figure 10a	Figure 10c
Nd	30	118
Rd	17.98	8.98

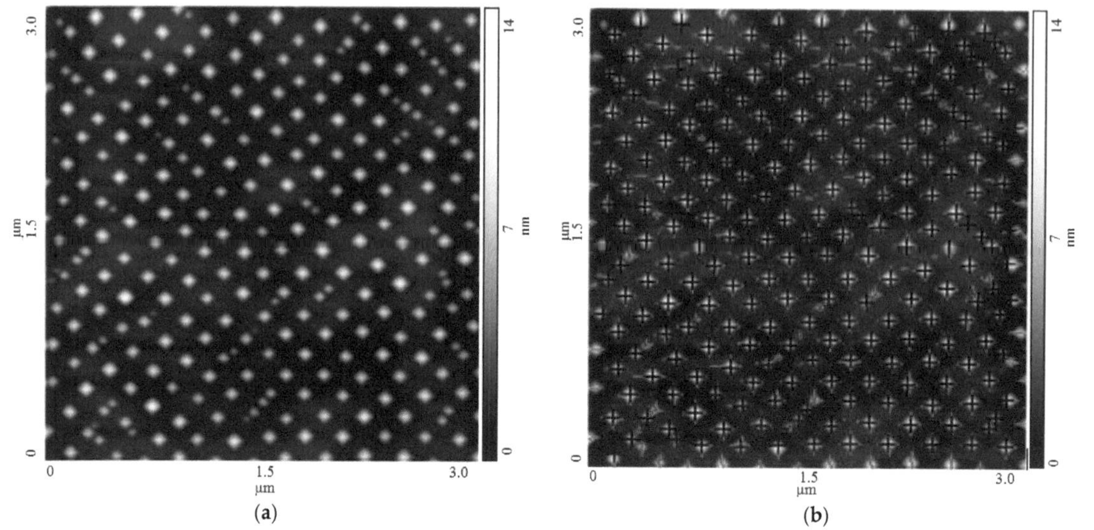

Figure 11. Array of ordered Ge(Si) pyramid islands, grown on relaxed SiGe buffer layer [46]: (**a**) source images 3 × 3 µm; (**b**) images with marked centres.

8. Conclusions

In order to automate the monitoring of agglomerate sizes, an approach to solving the task of surface tracking scan in probe microscopy is formulated. A feature of the proposed approach is the use of information about the particlesize and coordinates to determine the subsequentscan positions and generate statistics of the results.

The use of a surface curvature detector improved the accuracy of determining the coordinates and sizes of agglomerated particles. In combination with the use of motion sensors, this increases the reliability of stitching tracking images and allows to generate correct size statistics.

The proposed approach was tested on polystyrene particles, Ge(Si) islands, fullerene-containing powders, and agglomerates containing nanoscale diamonds and copper particles. The possibility of the separation of individual particles and the particles in the agglomerate composition was confirmed.

The results presented in the study can be used in other types of microscopy and inspection methods with the use of images.

An example of the implementation of algorithms in the Pascal (Delphi) is available at the link http://iam.udman.ru/ru/node/6653 (accessed 9 January 2022).

Author Contributions: Conceptualization, P.G. and A.K.; methodology, E.S. and T.K.; software, P.G. and E.S.; validation, P.G. and E.S.; formal analysis, P.G. and T.K.; resources, T.K.; data curation, P.G. and A.K.; writing—original draft preparation, P.G.; writing—review and editing, A.K. and T.K.; visualization, P.G. and E.S.; project administration, T.K. and A.K.; funding acquisition, T.K. All authors have read and agreed to the published version of the manuscript.

Funding: This work was supported by the Slovak Ministry of Education within project VEGA No. 1/0823/21 and by the Slovak Research and Development Agency under contract No. APVV-18-0316.

Institutional Review Board Statement: Not applicable.

Informed Consent Statement: Informed consent was obtained from all subjects involved in the study.

Data Availability Statement: In the paper the publicly archived gallery of scans [46] was used.

Conflicts of Interest: The authors declare no conflict of interest. The funders had no role in the design of the study; in the collection, analyses, or interpretation of data; in the writing of the manuscript, or in the decision to publish the results.

References

1. Merkus, H.G. *Particle Size Measurements*; Springer: Dordrecht, The Netherlands, 2009; 534p.
2. Raval, N.; Maheshwari, R.; Kalyane, D.; Youngren-Ortiz, S.R.; Chougule, M.B.; Tekade, R.K. Importance of physicochemical characterization of nanoparticles in pharmaceutical product development. In *Advances in Pharmaceutical Product Development and Research, Basic Fundamentals of Drug Delivery*; Tekade, R.K., Ed.; Academic Press: Cambridge, MA, USA, 2019; pp. 369–400. [CrossRef]
3. Jones, R.M. Particle size analysis by laser diffraction: ISO 13320, standard operating procedures, and Mie theory. *Am. Lab.* **2003**, *35*, 44–47.
4. Storti, F.; Balsamo, F. Particle size distributions by laser diffraction: Sensitivity of granular matter strength to analytical operating procedures. *Solid Earth* **2010**, *1*, 25–48. [CrossRef]
5. Frisken, B.J. Revisiting the method of cumulants for the analysis of dynamic Light-Scattering data. *Appl. Opt.* **2001**, *40*, 4087–4091. [CrossRef] [PubMed]
6. Kätzel, U.; Vorbau, M.; Stintz, M.; Gottschalk-Gaudig, T.; Barthel, H. Dynamic Light Scattering for the Characterization of Polydisperse Fractal Systems: II. Relation between Structure and DLS Results. *Part. Part. Syst. Charact.* **2008**, *25*, 19–30. [CrossRef]
7. Inam, M.; Frances, C. Effects of impurities on particle sizing by acoustic attenuation spectroscopy. *Powder Technol.* **2010**, *201*, 21–26. [CrossRef]
8. Dukhin, A.S.; Goetz, P.J. Applications for emulsions and other soft particles. In *Studies in Interface Science*; Elsevier: Amsterdam, The Netherlands, 2010; Volume 24, pp. 369–400. [CrossRef]
9. Markelonis, A.R.; Wang, J.S.; Ullrich, B.; Wai, C.M.; Brown, G.J. Nanoparticle film deposition using a simple and fast centrifuge sedimentation method. *Appl. Nanosci.* **2015**, *5*, 457–468. [CrossRef]
10. Shanthi, C.; Porpatham, R.; Pappa, N. Image Analysis for Particle Size Distribution. *Int. J. Eng. Technol.* **2014**, *6*, 1340–1345.
11. Mulligan, S.; Ojeda, J.; Kakonyi, G.; Thornton, S.; Moharamzadeh, K.; Martin, N. Characterisation of Microparticle Waste from Dental Resin-Based Composites. *Materials* **2021**, *14*, 4440. [CrossRef]
12. Gazi, M.; Apu, S.; Sharmili, N.; Rahman, M. Origin and characterization of clay deposits in the DupiTila Formation of the Bengal Basin, Bangladesh. *Solid Earth Sci.* **2021**, *6*, 317–327. [CrossRef]
13. Couteau, O.; Roebben, G. Measurement of the size of spherical nanoparticles by means of atomic force microscopy. *Meas. Sci. Technol.* **2011**, *22*, 065101. [CrossRef]
14. Grobelny, J.; DelRio, F.W.; Pradeep, N.; Kim, D.I.; Hackley, V.A.; Cook, R.F. Size Measurement of Nanoparticles Using Atomic Force Microscopy. *Methods Mol. Biol.* **2011**, *697*, 71–82. [CrossRef] [PubMed]
15. Misumi, I.; Sugawara, K.; Takahata, K.; Takahashi, K.; Ehara, K. Size measurements of standard nanoparticles using metrological atomic force microscope and evaluation of their uncertainties. *Precis. Eng.* **2017**, *51*, 691–701. [CrossRef]
16. Dorofeev, G.; Streletskii, A.; Povstugar, I.; Protasov, A.; Elsukov, E. Determination of nanoparticle sizes by X-ray diffraction. *Colloid J.* **2012**, *74*, 678–688. [CrossRef]
17. Linsinger, T.; Roebben, G.; Gilliland, D.; Calzolai, L.; Rossi, F.; Gibson, N.; Klein, C. *Requirements on Measurements for the Implementation of the European Commission Definition of the Term 'Nanomaterial'*; Publications Office of the European Union: Luxembourg, 2012. [CrossRef]
18. Sun, C.; Beare, R.; Hilsenstein, V.; Jackway, P. Mosaicing of microscope images. In Proceedings of the Digital Imaging Computing: Techniques and Applications, Cairns, Australia, 6 December 2005; p. 50. [CrossRef]
19. ISO 13322-1:2014. *Particle Size Analysis—Image Analysis Methods—Part 1: Static Image Analysis Methods*, 2nd ed.; International Organization for Standardization: Geneva, Switzerland, 2014.
20. ISO 13322-1:2016. *Particle Size Analysis—Image Analysis Methods—Part 2: Dynamic Image Analysis Methods*, 2nd ed.; International Organization for Standardization: Geneva, Switzerland, 2016.
21. OpenCV. Available online: https://opencv.org/ (accessed on 7 September 2021).
22. Feature Matching with FLANN. Available online: www.docs.opencv.org/3.4/d5/d6f/tutorial_feature_flann_matcher.html (accessed on 7 September 2021).
23. SPM NEXT II. Available online: https://www.ntmdt-si.com/products/automated-afm/next-ii (accessed on 11 February 2021).
24. Somnath, S.; Kim, H.J.; Hu, H.; William, W.P. Parallel nanoimaging and nanolithography using a heated microcantilever array. *Nanotechnology* **2013**, *25*, 014001. [CrossRef]
25. Yongbing, W.; Jianmin, S.; Xinjian, F.; Danish, H.; Hao, Z.; Hui, X. Fast Specimen Boundary Tracking and Local Imaging with Scanning Probe Microscopy. *Scanning* **2018**, *2018*, 3979576. [CrossRef]
26. Beucher, S.; Lantuéj, C. Use of whatersheds in contuor detection. In Proceedings of the Workshop on Image Processing, Real-Time Edge and Motion Detection, Rennes, France, 17–21 September 1979; Available online: http://cmm.ensmp.fr/~{}beucher/publi/watershed.pdf (accessed on 11 February 2021).
27. Otsu, N. A Threshold Selection Method from Gray-Level Histograms. *IEEE Trans. Syst. Man Cybern.* **1979**, *9*, 62–66. [CrossRef]

28. Navon, E.; Miller, O.; Averbuch, A. Color image segmentation based on adaptive local thresholds. *Image Vis. Comput.* **2005**, *23*, 69–85. [CrossRef]
29. Abdullah, L.S.; Hambali, H.A.; Jamil, N. Segmentation of Natural Images Using an Improved Thresholding-Based Technique. *Procedia Eng.* **2012**, *41*, 938–944. [CrossRef]
30. Bui, K.; Fauman, J.; Kes, D.; Mandiola, L.T.; Ciomaga, A.; Salazar, R.; Bertozzi, A.L.; Gilles, J.; Goronzy, D.P.; Guttentag, A.I.; et al. Segmentation of scanning tunneling microscopy images using variational methods and empirical wavelets. *Pattern Anal. Appl.* **2020**, *23*, 625–651. [CrossRef]
31. Ronneberger, O.; Fischer, P.; Brox, T. U-Net: Convolutional networks for biomedical image segmentation. In Proceedings of the International Conference on Medical Image Computing and Computer-Assisted Intervention, Munich, Germany, 5–9 October 2015; pp. 234–241. [CrossRef]
32. Gulyaev, P.V. Reference marks application for surface coordinate binding in scanning probe microscopy. *Comput. Opt.* **2020**, *44*, 420–426. [CrossRef]
33. Scanning Probe Microscope AFM5500M. Available online: https://www.hitachi-hightech.com/global/product_detail/?pn=em-afm5500m) (accessed on 11 February 2021).
34. Atomic Force Microscope Dimension Icon. Available online: https://www.bruker.com/en/products-and-solutions/microscopes/materials-afm/dimension-icon-afm.html (accessed on 11 February 2021).
35. Krasnabayeu, Y.A.; Chistabayeu, D.V.; Malyshev, A.L. Comparison of binary feature points descriptors of images under distortion conditions. *Comput. Opt.* **2019**, *43*, 434–445. [CrossRef]
36. Stoll, E.; Marti, O. Restoration of Scanning Tunnelling Microscope data, blurred by limited resolution, and hampered by 1/f-like noise. *Surf. Sci.* **1987**, *181*, 222–229. [CrossRef]
37. Rublee, E.; Rabaud, V.; Konolige, K.; Bradski, G. ORB: An efficient alternative to SIFT or SURF. *IEEE Int. Conf. Comput. Vis.* **2011**, *58*, 2564–2571. [CrossRef]
38. Herbert, B.; Ess, A.; Tuytelaars, T.; Van Gool, L. SURF: Speeded up robust features. Computer vision and image understanding. In *Lecture Notes in Computer Science*; Springer: Berlin/Heidelberg, Germany, 2006; Volume 3951. [CrossRef]
39. Topostitch. Available online: https://www.topostitch.com/ (accessed on 11 February 2021).
40. Golkar, E.; Prabuwono, A.S.; Patel, A. Real-time curvature defect detection on outer surfaces using best-fit polynomial interpolation. *Sensors* **2012**, *12*, 14774–14791. [CrossRef] [PubMed]
41. Han, J.H.; Poston, T. Chord-to-point distance accumulation and planar curvature: A new approach to discrete curvature. *Pattern Recognit. Lett.* **2001**, *22*, 1133–1144. [CrossRef]
42. Gulyaev, P.V. Measurement of the length of objects on scanning probe microscope images using curvature detectors. *Meas. Tech.* **2021**, *64*, 21–27. [CrossRef]
43. Chen, C.; Klette, R. An image stitcher and its application in panoramic movie making. In Proceedings of the DICTA'97, Auckland, New Zealand, 22 April 1997; pp. 101–106.
44. Shelkovnikov, Y.K.; Gafarov, R.M.; Gulyaev, P.V. Surface images creation at the multiscan mode of the scanning tunnel microscope. *Chem. Phys. Mesoscopy* **2008**, *10*, 514–520.
45. Gulyaev, P.V.; Shelkovnikov, E.Y.; Tyurikov, A.V. Software-Hardware Systems for Measurement of Sample Displacements in Probe Microscopes. *Meas. Tech.* **2019**, *61*, 987–991. [CrossRef]
46. NT-MDT. Spectrum Instruments. Scan Gallery. Available online: https://www.ntmdt-si.ru/resources/scan-gallery (accessed on 17 February 2022).

applied sciences

MDPI

Article

Center of Gravity Coordinates Estimation Based on an Overall Brightness Average Determined from the 3D Vision System

Radovan Holubek [1,*] and Marek Vagaš [2]

1 Department of Production Devices and Systems, Faculty of Materials Science and Technology, Slovak University of Technology, Bottova 25, 91724 Trnava, Slovakia
2 Department of Industrial Automation and Mechatronics, Faculty of Mechanical Engineering, Technical University of Kosice, Letná 9, 04200 Kosice, Slovakia; marek.vagas@tuke.sk
* Correspondence: radovan.holubek@stuba.sk; Tel.: +421-908-647-132

Abstract: In advanced manufacturing technologies (including complex automated processes) and their branches of industry, perception and evaluation of the object parameters are the most critical factors. Many production machines and workplaces are currently equipped as standard with high-quality special sensing devices based on vision systems to detect these parameters. This article focuses on designing a reachable and fully functional vision system based on two standard CCD cameras usage, while the emphasis is on the RS 232C communication interface between two sites (vision and robotic systems). To this, we combine principles of the 1D photogrammetric calibration method from two known points at a stable point field and the available packages inside the processing unit of the vision system (as filtering, enhancing and extracting edges, weak and robust smoothing, etc.). A correlation factor at camera system (for reliable recognition of the sensed object) was set from 84 to 100%. Then, the pilot communication between both systems was proposed and then tested through CREAD/CWRITE commands according to protocol 3964R (used for the data transfer). Moreover, the system was proven by successful transition of the data into the robotic system. Since research gaps in this field still exist and many vision systems are based on PC processing or intelligent cameras, our potential research topic tries to provide the price–performance ratio solution for those who cannot regularly invest in the newest vision technology; however, they could still do so to stay competitive.

Keywords: processing in automotive industries; implementation of non-conventional technology; advanced vision systems for sensing; determination of chosen object characteristic; experimental analysis of manufacturing processes

Citation: Holubek, R.; Vagaš, M. Center of Gravity Coordinates Estimation Based on an Overall Brightness Average Determined from the 3D Vision System. *Appl. Sci.* **2022**, *12*, 286. https://doi.org/10.3390/app12010286

Academic Editor: Andrea Prati

Received: 29 October 2021
Accepted: 24 December 2021
Published: 28 December 2021

1. Introduction

The vision system is a crucial sensor implemented in manufacturing processes to enlarge its peripheral abilities and possibilities to the Industry 4.0 concept inclusion [1]. Presently, current research in this field shows that the determination of the object's parameters (with 3D view orientation) is possible with the help of at least two single-camera sensors. The advantage of this principle is the easy implementation to industrial purposes in order to sense, record, and transmit the obtained picture by an interface such as serial interface with the connection of a PC, or an additional processing system. Images that we receive in such a way are subsequently evaluated concerning the necessary information extraction of the relevant sensed product [2]. Usually, after this process, we can subsequently determine and compile the necessary process data of the sensed object.

This study investigates the classical approaches and techniques for determining necessary sensing object characteristics when randomly oriented products are consistently moving, for example, by a conveyor at an automated workplace. The principles of calculating the displacements of two captured pictures using the DIC technique is also known. It is essential to note that this method is sensitive to the average gray intensity of the images [3].

However, the intelligent cameras still further support the manufacturing engineering and simplify it because the image analysis is run directly inside the CCD system [4]. Their central part usually consists of the processor and the algorithm setting necessary for an object's properties determination. As part of the CCD, the system also consists of the image sensor that can be described as: "the high-quality device that can recognize monochromatic pictures in resolution, what we need, and with speed what is necessary" (60 per sec, for example) [5].

Meanwhile, we can produce sharp pictures that increase algorithms' precision, such as "edge detection" and "pattern recognition". The next step consists of reliable information providing (target coordinates for robotic arm) that are sent to the advanced control system of the workplace. The final goal is to monitor and ensure the product's correct grasp. Moreover, the vision systems demonstrate their irreplaceability when it is necessary to implement overall correctness for each sensed product (TQM). A good example is implementing an infrared camera system for temperature distribution of the composites [6]. Inspection of the objects, finding the parameters such as color, shape, and so on is their primary purpose, see Figure 1.

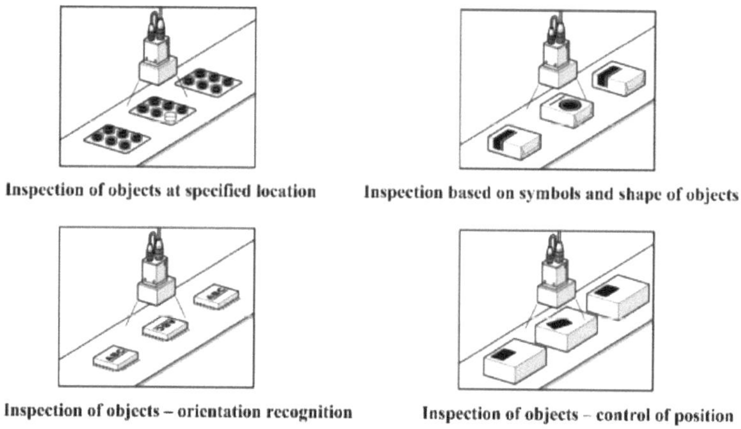

Figure 1. Purpose of cameras.

Due to the application attractiveness, many intelligent camera manufacturers have broadly comparable parameters [7]. The gradually increasing deployment of vision systems under the pressure of the emerging market has caused this issue to have been studied widely [8]. Yet, over the past few decades, their complexity and ability to control processes and monitor products in an automated production workplace have become attractive to many global manufacturers [9]. In recent years, the vision systems market has focused mainly on intelligent cameras with a high frame rate, excellent electrical coverage, and a powerful digital signal processor in one compact unit. Their advantage is, in particular, a more straightforward implementation of the application and their increased robustness. Parameters such as flux viscosity, dipping acceleration, dipping time, and flux stable time can be investigated using a high-speed camera system and MATLAB data processing [10].

Unfortunately, there are still few solutions focused on affordability and applicability to the situations where the pallet deployment cannot be used in industry [11]. Our current research is based in this area, where we try to process the acquired image using a comprehensive vision system OMRON F150-3. This system consists of two CCD sensors, which are placed next to each other [12]. The aim is to improve the existing classical approaches and techniques by determining some necessary scanning object characteristics. Such a solution could be developed by us and considered as a price-affordable complex vision system (compared with company solution) [13].

This paper intends to show an approach for a simple vision system solution in terms of price and applicability to several cross-sectoral areas (manufacturing, assembly, and automotive). Contrary to passive and closed intelligent camera systems, embedded complex vision systems are significant for most machine vision applications in a small footprint. They are usually provided as units that deliver extensive processing power suited for industrial applications [14]. As a result, identification and subsequent navigation of the robotic arm become optimal and are maintained throughout the entire smooth automated operation. Nevertheless, some research and primary praxis demonstrated that the CCD sensor itself usually has, around the lens, a ring of red LEDs that serve as a passive light source [15]. Ho6ever, additional lighting makes the solution more expensive on the one hand but increases the versatility and performance of work in the workplace. Despite this factor, we decided not to use the additional lightning and realize a solution based on the serial communication interface connected to the robotic arm control system. Different lighting was not considered, and the system compares the current and saved scene in real time according to the relevant lighting conditions. One crucial issue in the research of vision system usage is applied to the smartness, and the fast calibration is mainly followed for 3D measurement [16]. Therefore, this paper proposes a reachable vision system that will determine sensed object properties via communication between two CCD sensors, a processing unit, and a robotic design.

This paper is divided into seven sections: Section 1 offers a review of the current deployment, research, and customer requirements for the vision system in general, as well as provides examples for their theoretic implementation. Existing technical solutions in this area focus on the progressive trends oriented on the various scanned parts to the sensing, following, and control as a whole. Section 2 deals with the available methods and procedures responsible for image processing followed by mathematical determination (epipolar geometry) regarding using 2D or 3D vision systems. Equally, in Section 3, there is an explanation of the proposed vision system that includes available communication interfaces, data transfer, and setup of the vision system itself. Finally, Section 4 consists of the framework for communication between the vision and control systems of the robotic arm, based on the communication RS232C channel. Part of this section is also the configuration and setting for the communication protocols, its composition, and an example for mutual communication between both sides (vision system and control system of robotic arm).

Section 5 deals with the experimental setup of a technical solution based on two CCD systems, determinates its variables, and provides us with the experimental measurement for image processing. This section also includes the necessary steps for obtaining the object coordinates and the center of gravity from a sensed object that is also one of the leading research goals. The verification of the proposed solution is proven by the graphical and theoretical test of normality for each object coordinates. Section 6 focuses on summarizing and comparing our achieved results with actual relevant research in vision systems. This part of the article is focused on explaining the advantages and disadvantages of this presented solution with emphasis on its applicability and further development. At the same time, we present some professional solutions compared with our research in terms of usability and deployment in actual conditions. Finally, the last section summarizes realized development and research in this field and the recommendations for future work and improvements.

2. Materials and Methods

The methodology for processing the image information obtained from the image is currently focused on determining the position of individual parameters of the sensed product and its comparison with the reference values of the correct product [17]. After that, we can evaluate a geometric imagination of the object being sensed and its parameters [18]. The next step in the process supposes the movement of the reference plane to various spacing horizontally in "x" and simultaneous measurement at the equal point with the assignment of correct values for the translation (in pixels). In additional calculations, we can also state the precise value between the sensing device (camera) and our sensed product. The correctness of this methodology is contingent on the fixation of the coordinate "x" and "z" with the usage of (at least) two sensors that are fixed alongside. Finally, the targeted product will be sensed, captured, and evaluated with the condition to the intersection of sensed directions while its axes are parallel [19]. The sensor spacing between the sensed product can be thus determined as a ratio between the spacing of the cameras "b" and spacing from its reference plane "r" to the absolute number between the measured translation "a" and "c" as it is given by the Equation (1):

$$Z = \frac{(b \times r)}{|a - c|}, \tag{1}$$

were,

b—Spacing between the sensors,
r—Spacing of the reference and ocular plane,
a—Value measured for sensor 1,
c—Value measured for sensor 2.

We must also consider the facts during the image processing:

Resolution of the sensed product,
Value of the images per second,
Picture intensity, color, and persistence.

Based on this, we can identify two basic principles for *3D* visioning.

2D sensors,
3D sensors.

2.1. 2D Sensors

Initial calibration of both 2D sensors is needed because it is only possible to state the relevant values for the sensed products successfully. We can achieve this by setting the 2D sensors values (calibration matrix, distortion factor) and "mutual spacing and 2D sensors orientation". Furthermore, epipolar geometry (which addresses two sensor issues) is needed for the statement of the number of 3D points that we can estimate based on the positions at 2D sensors [20]. During the determination of the search points $m = [x', y', 1]$ using the obtained picture, the following equation applies:

$$m = P \times M \tag{2}$$

where, the dot "m" is the picture of point "M"in the surface for screening (therefore, in the surface that is sensed by the camera system for capture). Value "P" for the projection matrix dimension has a state 3×4 and shows the dependence of the actual dot with the sensed dot as:

$$P = K \times [I, 0] \tag{3}$$

where, dimension calibration matrix "K" has a state of 3×3 and describes the values of the environment and values of the 2d sensors. The followed matrix is "I", 3×3 is the dimension, and "0" is zero columns.

$$K = \begin{bmatrix} a_x & c & t_x \\ 0 & a_y & t_y \\ 0 & 0 & 1 \end{bmatrix} \tag{4}$$

a—trend is scaling in axes "x" and "y" related to the 2d sensors.
t—intersection of sensed surface and visual surface.
c—Distortion parameter (*usually* = 0).

Matrix "P" can be other than the higher mentioned and written representation for the value of translation and orientation to the coordinate system. Regarding this assumption, we can add the transformation matrix for the 2D sensor coordinate and environment surface:

$$P = K \times [I, 0] \times \begin{bmatrix} R & RC' \\ 0 & 1 \end{bmatrix} \tag{5}$$

were,

R—this is the matrix 3×3 and explains the rotation.
C'—translation vector to the environment surface.

In such a way, we can state the value between dot "m" from the first sensed product, allocating dot "m'" from the second sensed product. In the case of these two dots, we propose the matrix "F" with 3×3:

$$m'^T \times F \times m = 0 \tag{6}$$

In the next step, we suggest that one of the projection matrices is allocated at the start of the coordinate system, so that no additional activities are needed. The satisfactory result lies in the determination of the second matrix using the dual sensors:

$$W = \begin{bmatrix} 0 & -1 & 0 \\ 1 & 0 & 0 \\ 0 & 0 & 1 \end{bmatrix} \tag{7}$$

To achieve the success of this methodology, we need to test projection matrices at any dot and choose the correct one. This means our selected dot must be positioned in front of the 2D sensors. After this, we can state the coordinates of the dots (with help from the linear method for the triangulation) and the relevant equations:

$$\begin{bmatrix} xp^{3T} & -p^{1T} \\ yp^{3T} & -p^{2T} \\ x'p'^{3T} & -p'^{1T} \\ y'p'^{3T} & -p'^{2T} \end{bmatrix} \tag{8}$$

By using the least square methods, we can state a resultant *3D* coordinate of point "X". Variable's "x" and "y" are a coordinate "p^i", which is an i-th row of matrix "P".

A sensed product and its 3D position are crucial for picture transformation, and every individual dot realizes it compared with the same "y" coordinate. Therefore, after we found the corresponding dot and its row (for the second sensor), we were advised to firstly realize the preprocessing, and only after discovering the processing process of an obtained image [21].

2.2. 3D Sensors

A sensing system for capturing images with the help of one 3D sensor as advanced and complex as a 3D camera system usually consists of two side-by-side cameras in a single housing (3D-A5000 Cognex 3D camera, SICK IVC-3D, OMRON Xpectia FZD, etc.) [22]. The aim of such devices is image obtaining and its transformation into separate parameters (with the help of the decomposition process) to obtain the relevant information about a product [23]. The data that we capture in such a way allows us to achieve the advanced calculation of 3D data. We can evaluate the sensing object's dimensions, thickness, position, and orientation based on this data. However, this method has disadvantages. For example, sometimes we obtain the deformation image from the sensors, but we can measure the distorted images and repair it. Then, we evaluate the modification with the help of calibration. Usually, these types of 3D camera systems utilize algorithms for autocalibration or particular types of calibration, depending on the camera model, which includes the influence of its actual lenses [24]. So, as we know, the dependence between geometric and the sensed product exists here; by comparing with the 3D CAD model, we can state the correct positioning and orientation [25].

3. Setup of the Vision System

To ensure the communication quality of both the robotic system and sensing device, they need to be proven in conjunction with standard known protocols. Currently, there are several interfaces with principle and topology on the Ethernet, RS232, and PROFIBUS systems, and so on. In addition, the control system of the robotic arm needs the relevant libraries and functions to understand the commands.

The composition of the sensing principle is shown in Figure 2. The sensing process continues with calibration. Its purpose is to determine the mathematical values inside and outside for the lens distortion. This principle helps us state the correct dots, their selection, and extraction from the obtained images. Transformation of the coordinate systems precedes this step (robotic arm and sensing device) [26].

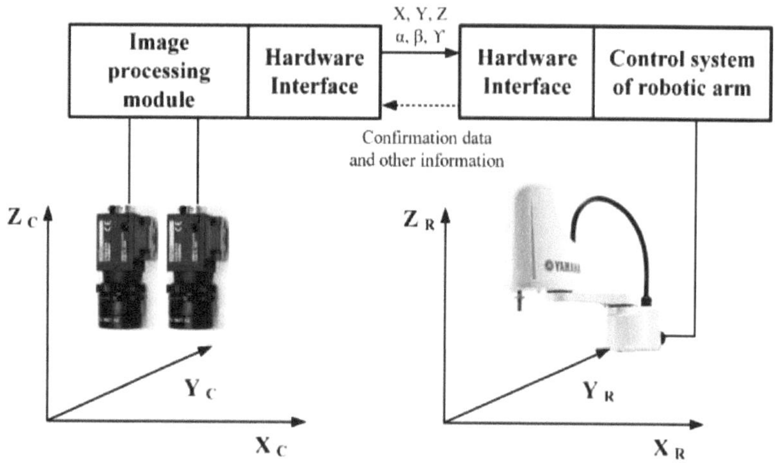

Figure 2. Composition of the sensing principle.

4. Verification of Data Transfer (Communication)

Data transfer between the robotic arm and the vision system is via the RS232C communication channel. Communication at both sides is supposed (to the serial communication channel) followed by configuration and specification of transmitting processes [27]. In the next step, this configuration process is followed by the cold startup of the control system of the robotic arm. Finally, the serial interface must be assigned to the operating system for the transmission with *CREAD/CWRITE* usage, see Tables 1–3. Usually, this can be configured for communication through *CREAD/CWRITE* according to the following scheme shown in Figure 3.

Table 1. Configuration of the serial interface.

1.	(COM3)	
2.	BAUD = 9600	110, 150, 300, 600, 1200, 2400, 4800, 9600, 19200, 38,400, 57,600
3.	CHAR_LEN = 8	7, 8
4.	STOP_BIT = 1	1, 2
5.	PARITY = 2	EVEN = 2, ODD = 1, NONE = 0
6.	PROC = 1	3964R = 1, SRVT = 2, WTC = 3, XON/XOFF = 4

Table 2. Configuration of transmission protocol 3964R for the data transfer.

1.	(3964R)	
2.	CHAR_TIMEOUT = 500	msec, max. interval between two symbols
3.	QUITT_TIMEOUT = 500	msec, max. waiting time at control system of a robotic arm to symbol DLE
4.	TRANS_TIMEOUT = 500	; msec,
5.	MAX_TX_BUFFER =2	1.5, max. value of cache output
6.	MAX_RX_BUFFER = 10	1.20, max. value of cache input
7.	SIZE_RX_BUFFER = 1.2048	dimension of receiving memory input (in bytes)
8.	PROTOCOL_PRIOR = 1	HIGH = 1, LOW = 0, priority

Table 3. An example of successful communication between both systems (vision system and control system of robotic arm).

1.	DEFDAT SEND
2.	DECLARATION
3.	INT HANDLE
4.	DECL STATE_T SW_T, SC_T
5.	DECL MODUS_T, MW_T
6.	ENDDAT
7.	DEF SEND ()
8.	INITIALIZATION
9.	MW_T = #SYNC
10.	INSTRUCTION
11.	OPEN_P ()
12.	WRITE ()
13.	CLOSE_P ()
14.	END

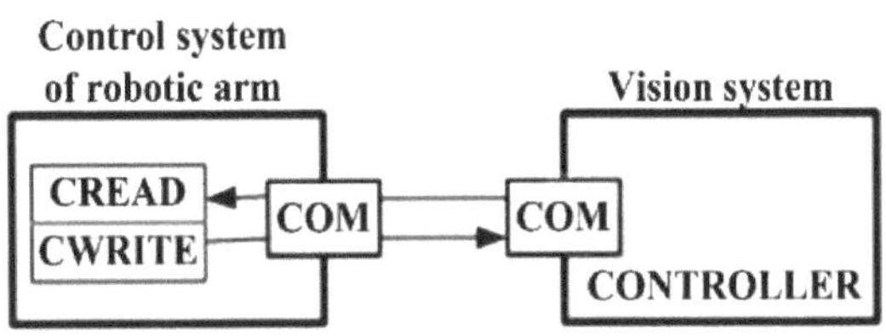

Figure 3. The setting of serial interface communication.

5. Experimental Testing (Setup)

The main aim of the vision system OMRON F150-3 is its experimental testing to verify and prove the possibilities and abilities of such a device, primarily to determine the chosen object characteristics. This system belongs to the so-called compact vision systems, consisting of an image information processing unit (independent of the PC), an externally connected CCD camera, or several CCD cameras. Image information is processed and evaluated in an external unit outside the cameras. The unit's architecture is usually similar to an industrial or embedded PC built on a powerful processor. This system makes it possible to decide on a sensing object position in a working envelope followed by the parameters of the center of gravity (x, y, and z), as one of the goals in this paper. Such obtained information follows the control system of the robotic arm for their successive transposition [28]. The source point for the introduced solution consists of the realized technical solution with the object coordinates measuring to the object to determine its center of gravity in 3D space. By step sequencing together with the help of two CCD camera systems, precise methodological verification of measurement was reached. The vision systems' key parameters (OMRON F150-3) taken into account for the present solution are the resolution (512 × 484), the field of view (in our case = 0 because we did not use an additional source of light), and focus (35 mm). The auxiliary light source usage makes it possible to determine sensing objects up to 50 × 50 mm and distance up to 76 mm. Experimental measurement requires implementing the complex control, input–output, and communication peripherals, mainly: control system for post-processing, CCD camera systems with 35 mmm lens, console, monitor, and corresponding cables. In addition, CCD cameras fixation directly to the aluminum profile with modular construction is required (see Figure 4).

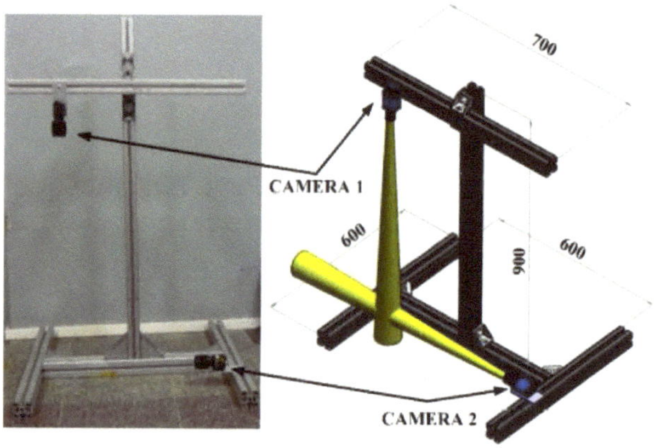

Figure 4. Realized technical solution experimental measurement stand with CCD cameras.

The presented and realized technical solution with the two CCD cameras assumes such disposition to capture all object data coordinates. Thus, we try to avoid a technical solution and such disposition where the "z" coordinate axis will not be sufficiently able to determine shape, depth, and height of the sensing object [29]. Concerning this requirement, the CCD cameras' fixation at a right angle (in the mutual one) was also considered. From the vertical CCD camera system regarding the proposed realized technical solution, it is possible to determine the object coordinates in "x" and "y". We evaluate the coordinate "z" as the third supplementary axis with the help of a horizontal camera system. Table 4 shows an example of a symmetric object tested to obtain its coordinates (and subsequently the center of gravity). Correct and incorrect sensed objects are evaluated by the correlation factor of the camera system (inside the processing unit with a value from 84 to 100). When results under this value were obtained, the measurement process of the camera system was considered to be incorrect (faulty) and repeating the test was required

Table 4. Specification of the sensed object.

3D Object	Ground Projection	Correlation Factor
	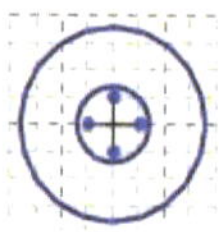	84–100

Suppose we need to obtain relatively high accuracy of the scanned object position. In that case, we need to eliminate the contrast factor of the object within the measured scene as much as possible. Thus, poorly sensed objects would hardly be identifiable, and the system would be unstable to accurately determine their shape, as shown in the left part of Figure 5.

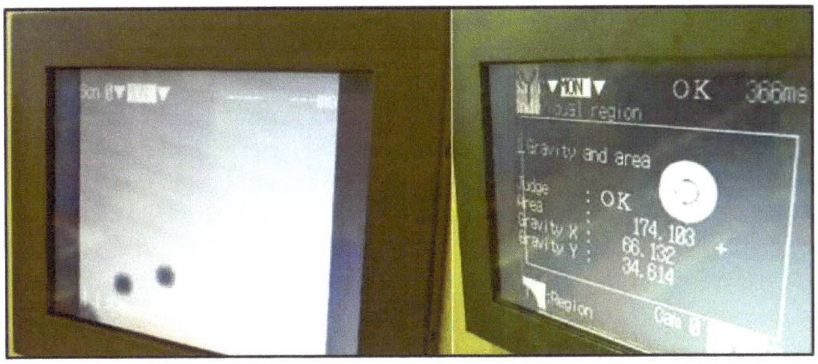

Figure 5. Incorrect procedures for determination of the sensed objects at the measured scene.

Obtaining the "x" and "y" coordinates is dependent on the calibration of the CCD cameras. Their calibration is necessary and required, especially in the usage of the triangular method. Their aim covers the reconstruction of sensed objects in order to obtain their shape properties. The example of non-calibrated CCD cameras can be seen in the right part of Figure 5. In the case of a sensed object moving, CCD cameras cannot determine the object's correct position. Therefore, it is appropriate to use the information in the form of optical calibration tag markers that identify the sensed scene. Markers should be identifiable in the measurement scene by their brightness profile related to the correlation factor. Using the simple thresholding method, we assume the intensity of the marker contrast will be higher than the contrast of all other points on the background of the measured scene [30]. A certain degree of change in the area of the marker depending on its position is natural and results from the physical nature of the method used. Our presented solution uses a 1D photogrammetric calibration method from two known points at a stable point field, as shown in Figure 6 (left part of the picture with two white pins).

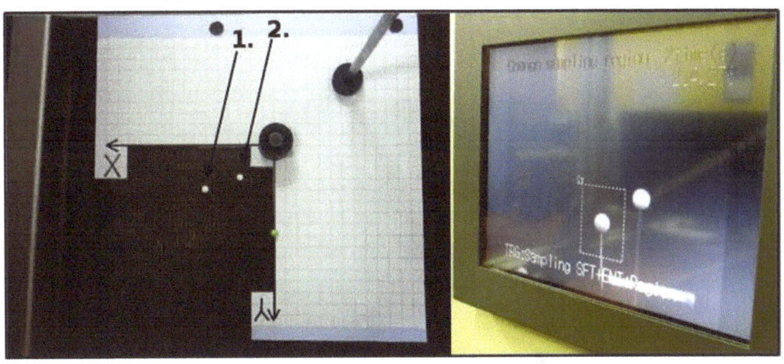

Figure 6. Initial calibration of CCD cameras.

Necessary Steps for Coordinates of the Sensed Object

The following steps aim to describe and present the necessary activities and functions needed for the measurement realization. Firstly, both CCD camera system (initial) calibrations are essential, as illustrated in Figure 6. This step contains several subactivities such as the vision system startup followed by CCD camera system registration and the additional light source specification. After these subactivities, it is also necessary to realize and specify the processing mode type required for the following processing. Finally, by setting the CCD camera system exposure time (in the case of more extended exposure time, the setting will be monitor blank), we achieve picture focusing using the actual light source conditions available in the working environment (together with the sensing object).

With the help of an accurate digital measurement instrument (in the form of the digital caliper), we can determine the coordinates for the first calibration point, i.e.,: (x = 25 mm, y = 10 mm, z = 15 mm). Similarly, for the second calibration point the coordinates were: (x = 30 mm, y = 15 mm, z = 15 mm). In terms of calibration, two calibration points are sufficient because the magnification scale in axes "x" and "y" is the same. The additional step of the CCD camera system setting is object presence detection (if it is in the working envelope for both CCD cameras); see the left part of Figure 7. We use the density averaging function (that is inside the vision system) to determine the measurement area [31]. This function compares the two pictures' brightness and evaluates them based on the greyscale. Following this, we determine an overall brightness average with the measurement (firstly, for the empty area, and second for the area with the sensing object).

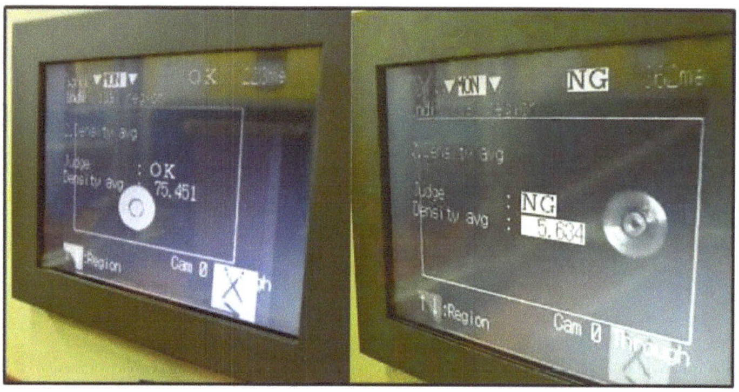

Figure 7. Detection of sensed object presence.

The measurement realization verifies the sensing object detection (concerning the actual source light conditions). First, it evaluates an overall density average at the selected area (rectangle shape) that specifies a value of 74.451. The setting of the limit values for the density averaging function is carried out after this process is realized (based on the reliability of the sensing object presence or absence evaluation) [32]. Second, detection of the sensed object gives us qualitative information about errors from the measurement by the vision system. A failed measurement leads us to return to the initial calibration to perform recalibration; see failed detection example of a sensed object shown in the right part of Figure 7. An identification of coordinate axes "x" and "y" for the sensing object is demonstrated in the next step (by using the gravity and area function). The following fundamental step consists of calculation of the area of the sensing object that we can view from the top. The sensing object is evaluated by white pixels (mainly, according to the light source), and the other area is black (see left part of Figure 8).

The next step consists of the "z" axis determination of the sensing object's shape, depth, and height (see right part of Figure 8). Subactivities of this step also consist of a post-processing process applied to the captured picture from the vision system in a way in which the edge position function successfully realizes sensing the object [33]. This function detects the sensing object's edges, allowing sufficient contrast between the object and the environment. By applying this process, we can evaluate (in the appropriate direction) the edges of a sensing object in relation to the simultaneous usage of the functions "light to dark" (the sensing object is dark, and the background is bright) or "dark to light" (the sensing object is bright, and the background is dark).

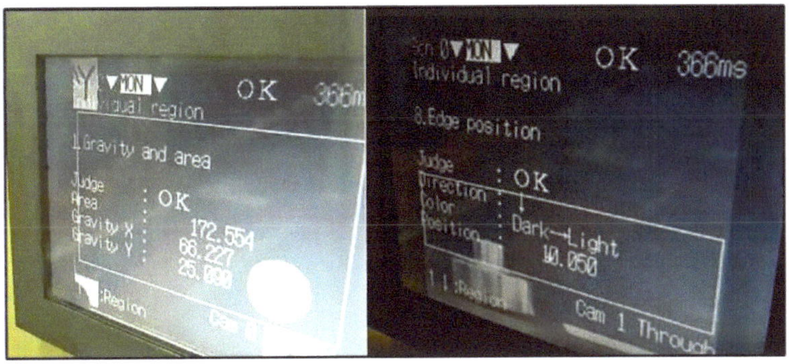

Figure 8. Determination of the object's coordinates.

Finally, the last step consists of determining the center of gravity parameters for the sensing object, which we choose by a simple mathematical operation (x, y, $z/2$) because we consider the center of gravity to be at the middle of the sensed object. Information obtained by this principle is then transferred into the computer via the RS232C serial interface for advanced processing (transformation of the coordinates from the external coordinate system to the coordinates of the robotic arm control system) [34]. Subactivities of this step include transmitting coordinate information directly to the control system of the robotic arm. We can state that a major influence on the captured results is the lighting conditions (light uniformity) or vertical camera system distance from the sensing object. Consequently, to verify obtained results, we carried out a test of normality [35]. We used a graphical test of normality to verify the normality, so-called quantile–quantile plot (*Q-Q plot*). If the obtained data are from the normal distribution, the points lie on a straight line for the coordinate axis "x" to be fulfilled. The graph is shown in Figure 9.

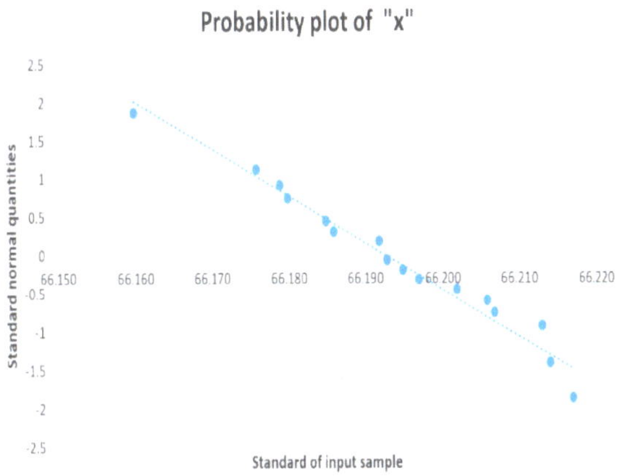

Figure 9. Normality test for "x" axis of sensing object.

Experimental measurements were performed for each axis of the sensing object 20 times. The measured values were statistically processed, and the normality of the data of individual coordinates of the sensing object was verified first. Results are shown in the graphs in Figures 9–11, where we can see that the unique points are, for the most part, at a sufficiently small distance from the line. It, therefore, implies the normality of the captured data.

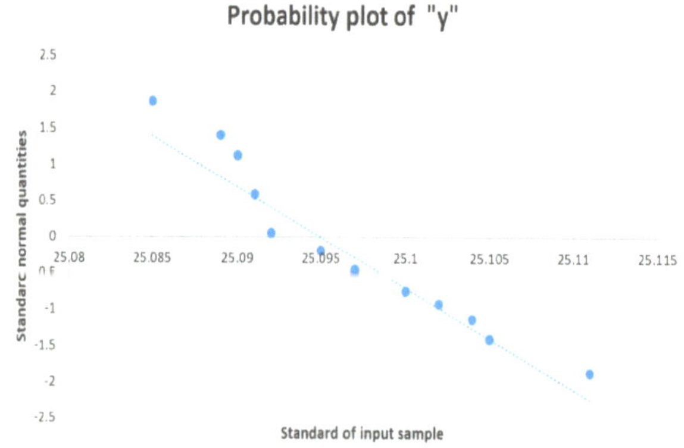

Figure 10. Normality test for "*y*" axis of sensing object.

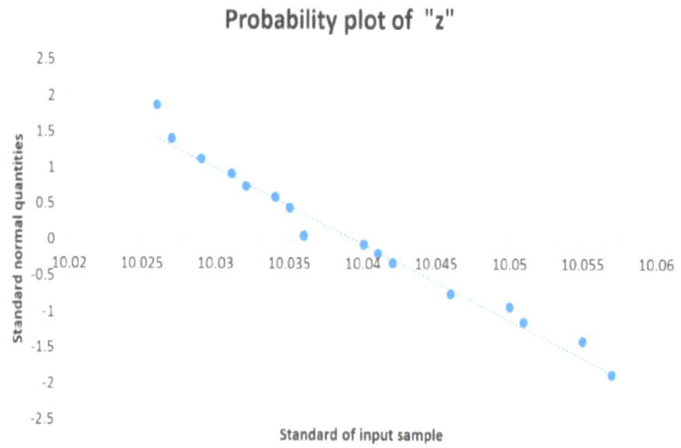

Figure 11. Normality test for "z" axis of sensing object.

We perform the test of normality based on the hypothesis that the sample file with the range $n = 20$ comes from the normal distribution. After the graphical test of normality verification, we chose the level of significance $\alpha = 0.05$. Then, we tested it with the zero hypothesis H0, which means that the sample file is from the normal distribution opposite to the alternative hypothesis H_1, which comes from another distribution. Finally, the results based on the S.W. test confirms goodness, so the selected sample file comes from the normal distribution probability (value of tested statistics = 0.9726, p-value = 0.4335).

6. Discussion

The novelty of our proposed visioning system (compared with other available solutions) combines principles of 1D photogrammetric calibration methods from two known points at a stable point field and the use of the available packages inside the processing unit of the vision system (such as filtering, enhance and extract edges, weak and robust smoothing, etc.). Combining these two approaches makes it possible to create modularly oriented recognition. We want to mention that the parameters we know are entirely sufficient for the 1D photogrammetric calibration method. However, this method allows more than two points (e.g., three, four, or more). Traditionally, 2D or 3D calibration can be used depending on the object's sensed size. These methodologies usually use a specific type of calibration field (e.g., checkerboard for 2D) or 3D calibration objects (including a spatial object used to calibrate the camera system). The 1D photogrammetric method was chosen because of the efficiency in calibrating our type of camera system, the known reference coordinates of two points (point field), and the possibility of calibrating objects of our shape and size.

The ultimate goal is to determine the necessary properties of the sensed object. The experiments found that a sufficient correlation factor that guarantees reliable recognition of the sensed object should be a value between 84% and 100%. Pilot communication between both systems (vision and robotic systems) was proposed and then tested through CREAD/CWRITE commands according to the transmit protocol 3964R used for the data transfer at our solution. It was verified and approved the successful transmission of data after the object sensing process was submitted into the robotic system (see Table 3 with successful and verified communication commands). The reliable usage of the assistive commands inside the processing unit was most clearly illustrated by the clear and understandable image of the obtained sensed object. Furthermore, such obtained results at the moment are well prepared for their further and deeper processing during the next phase of development.

Currently, many vision systems are based on PC processing, complex vision systems, or intelligent cameras. Due to the growth in popularity of advanced vision systems usage together with their rapid advance, we attempted to develop an available solution to meet the requirement of (trouble-free) implementation into the (mainly, but not only) automated manufacturing workplaces and automotive industry [36]. The majority of current vision systems consist of closed professional processing algorithms, and as such they cannot be customized to specific situations or specific working conditions. Consequently, there is a growing demand for price-affordable vision systems, such as the one proposed in this paper. On the other hand, weaknesses should also be noted and mentioned in this project. Although encouraging results are obtained from this research, there are still some significant limitations to consider. Hopefully, the potential of new modularities for our application will not be limited by robotic system selection. Firstly, this project will be tested at the KUKA robotic system available at our laboratory.

To reduce single-purpose usage, it is necessary to also verify our proposed vision system at another robotic system (to further determine reliable communication and recognition of sensed objects). Secondly, the proposed vision system is significantly dependent on the existing light conditions of the location where it will be implemented because we do not use additional lightning in our proposal, therefore, it is also worth considering this aspect, as lighting is an essential factor. Thirdly, the serial communication interface should be replaced in the future with a more advanced system, such as ETHERNET, OPC Communication, FIREWARE, USB, etc. The configuration process for these newer communication protocols is much simpler because they are based on open communication standards. Usually, the user is required to enter the IP address of the device and the appropriate port with which they want to communicate. In addition, serial communication has a low baud rate and is used over shorter distances.

Last but not least, we only performed experiments inside one of our laboratories. More experiments have to be realized to compare use, further demonstration, and the functionality of the proposed project in a diversity of conditions.

7. Conclusions

This paper proposes a reachable and fully functional vision system based on two standard CCD cameras to provide a price–performance ratio solution for those who cannot often invest in the newest vision technology at their workplaces. The innovation of our proposed visioning system (compared with other available solutions) combines the principles of the 1D photogrammetric calibration method, vision system, and standard RS 232C communication to determine the necessary properties of the sensed object. In addition, this approach makes it possible to create modularly oriented recognition (depending on the built-in functions of the camera system's processor unit, of course). Furthermore, with the continued development of vision systems in advanced manufacturing technologies on the one hand, and to stay competitive on the other hand, we prove the elimination of the weakness of an object´s recognition by using the available packages allocated inside the processing unit of the vision system (such as filtering, enhancing and extracting edges, or weak and robust smoothing) [37]. Moreover, our new findings were validated by reliable testing of pilot communication through CREAD/CWRITE commands (according to the transmit protocol 3964R), determining the sufficient correlation factor that guarantees reliable recognition of the sensed object (84% to 100%) and by probability plots. Furthermore, the results of our work are illustrated by experimental tests, where the feasibility of the design is proved by obtainment of clear and stable images of the sensed object [38].

Future work consists of further and deeper image processing to determine advanced properties (information) about sensed objects, which will probably greatly influence the vision system performance. Consequently, exciting areas of additional research include reducing usage of the single-purpose serial communication system and testing other possible communication protocols (ETHERNET IP and OPC communication) with the KUKA robotic system to verify modularity and identify additional limitations of our proposed design.

Author Contributions: All authors discussed and agreed upon the idea and made scientific contributions: writing—original draft preparation, R.H. and M.V.; experiment designing, M.V. and R.H.; experiment performing, M.V.; data analysis and writing—review and editing, M.V. and R.H. All authors have read and agreed to the published version of the manuscript.

Funding: This research was supported by research grant VEGA 1/0330/19; research and design of algorithms and systems for the fusion of heterogeneous data in multisensor architectures, KEGA 044TUKE-4/2021; remote access to laboratory exercises for industrial automation, erasmus+ 2019-1-RO01-KA203-063153; and development of mechatronics skills and innovative learning methods for Industry 4.0.

Institutional Review Board Statement: Not applicable.

Informed Consent Statement: Not applicable.

Data Availability Statement: Not applicable.

Conflicts of Interest: The authors declare no conflict of interest.

References

1. Distante, A.; Distante, C. *Handbook of Image Processing and Computer Vision*; Springer Nature: Berlin, Germany, 2020; 448p, ISBN 978-3-030-42373-0.
2. Hasegawa, Y.; Shimon, Y.N. *Springer Handbook of Automation*; Springer: Berlin, Germany, 2020; 1812p, ISBN 978-3-540-78830-0.
3. Zhang, W.; Gu, X.; Zhong, W.; Ma, Z.; Dong, X. A review of transparent soil model testing technique in underground construction: Ground visualization and result digitalization. *Undergr. Space* **2020**. [CrossRef]
4. Xiang, Y.; Liu, H.; Zhang, W.; Chu, J.; Zhou, D.; Xiao, Y. Application of transparent soil model test and DEM simulation in study of tunnel failure mechanism. *Tunn. Undergr. Space Technol.* **2018**, *74*, 178–184. [CrossRef]
5. Kahily, H.M.; Sudheer, A.P.; Narayanan, M.D. RGB-D sensor-based human detection and tracking using an armed robotic system. In Proceedings of the 2014 International Conference on Advances in Electronics Computers and Communications, Bangalore, India, 10–11 October 2014; pp. 1–4.
6. Zhan, L.; Junhui, L.; Xiaohe, L. Novel Functionalized BN Nanosheets/Epoxy Composites with Advanced Thermal Conductivity and Mechanical Properties. *ACS Appl. Mater. Interfaces* **2020**, *12*, 6503–6515. [CrossRef]
7. Saukkoriipi, J.; Heikkilä, T.; Ahola, J.M.; Seppälä, T.; Isto, P. Programming and control for skill-based robots. *Open Eng.* **2020**, *10*, 368–376, ISSN 2391-5439. [CrossRef]
8. Favorskaya, M.N.; Jain, L.C. *Computer Vision in Control Systems-4: Real-Life Applications*; Springer International Publishing AG: Berlin/Heidelberg, Germany, 2018; p. 317; ISBN 978-3-319-67993-8.
9. Shallari, I.; O'Nils, M. From the Sensor to the Cloud: Intelligence Partitioning for Smart Camera Applications. *Sensors* **2019**, *19*, 5162. [CrossRef] [PubMed]
10. Li, J.; Xia, Y.; Wang, W.; Zhang, W.; Zhu, W. Dipping Process Characteristics Based on Image Processing of Pictures Captured by High-speed Cameras. *Nano-Micro Lett.* **2015**, *7*, 11. [CrossRef] [PubMed]

11. Tong, M.; Fan, L.; Nan, H.; Zhao, Y. Smart Camera Aware Crowd Counting via Multiple Task Fractional Stride Deep Learning. *Sensors* **2019**, *19*, 1346. [CrossRef] [PubMed]
12. Frontoni, E.; Mancini, A.; Zingaretti, P. Embedded Vision Sensor Network for Planogram Maintenance in Retail Environments. *Sensors* **2015**, *15*, 21114–21133.
13. Popovic, V.; Seyid, K.; Cogal, Ö.; Akin, A.; Leblebici, Y. *Design, and Implementation of Real-Time Multi-Sensor Vision Systems*; Springer International Publishing AG: Berlin/Heidelberg, Germany, 2017; p. 257; ISBN 978-3-319-59056-1.
14. Campbell, F.C. *Inspection of Metals: Understanding the Basics*; ASM International: Almere, The Netherlands, 2013; p. 487; ISBN 978-1-62708-000-2.
15. Garbacz, P.; Burski, B.; Czajka, P.; Mężyk, J.; Mizak, W. The Use of 2D/3D Sensors for Robotic Manipulation for Quality Inspection Tasks. *Solid State Phenom. Adv. Manuf. Eng. II* **2015**, *237*, 77–82, ISSN 1662-9779.
16. Sinha, P.K. *Image Acquisition and Preprocessing for Machine Vision Systems*; SPIE Press: Bellingham, WA, USA, 2012; p. 750; ISBN 9780819482037.
17. Chen, J.; Jing, L.; Hong, T.; Liu, H.; Glowacz, A. Research on a Sliding Detection Method for an Elevator Traction Wheel Based on Machine Vision. *Symmetry* **2020**, *12*, 1158, ISSN 1424-8220. [CrossRef]
18. Sukop, M.; Hajduk, M.; Baláž, V.; Semjon, J.; Vagaš, M. Increasing degree of automation of production systems based on intelligent manipulation. *Acta Mech. Slov.* **2011**, *15*, 58–63, ISSN 1335-2393. [CrossRef]
19. Tannoury, A.; Darazi, R.; Makhoul, A.; Guyeux, C. Wireless multimedia sensor network deployment for disparity map calculation. In Proceedings of the IEEE Middle East and North Africa Communications Conference (MENACOMM), Jounieh, Lebanon, 18–20 April 2018; pp. 1–6.
20. Mohamed, A.; Culverhouse, P.; Cangelosi, A.; Yang, C. Active stereo platform: Online epipolar geometry update. *EURASIP J. Image Video Processing* **2018**, *2018*, 54. [CrossRef]
21. Deepa; Jyothi, K. A robust and efficient preprocessing technique for stereo images. In Proceedings of the International Conference on Electrical, Electronics, Communication, Computer, and Optimization Techniques (ICEECCOT), Mysuru, India, 15–16 December 2017; pp. 89–92.
22. Van Baer, T.; Introduction to Machine Vision. Cognex Confidential Webinar. 2018, p. 49. Available online: https://www.cognex.com/library/media/ondemandwebinars/slides/2018_intro_machine_vision-webinar.pdf (accessed on 26 August 2021).
23. Collado, J.F. New Methods for Triangulation-based Shape Acquisition Using Laser Scanners. Ph.D. Thesis, Department d'Electronica, Informatica i Automatica, Universitat de Girona, Girona, Italy, 2004; ISBN 84-689-3091-1.
24. Gil Ruiz, A.; Victores, J.G.; Łukawski, B.; Balaguer, C. Design of an Active Vision System for High-Level Isolation Units through Q-Learning. *Appl. Sci.* **2020**, *10*, 5927. [CrossRef]
25. Lv, Z.; Zhang, Z. Build 3D Scanner System based on Binocular Stereo Vision. *J. Comput.* **2012**, *7*, 399–404. [CrossRef]
26. Novák, P.; Špaček, P.; Mostýn, V. Stereovision system—Detection of the correspondings points. In Proceedings of the ICMT '11—International Conference on Military Technologies; University of Defence: Brno, Czech Republic, 2011; pp. 961–968, ISBN 978-80-7231-787-5.
27. Vagaš, M.; Sukop, M.; Baláž, V.; Semjon, J. The calibration issues of the 3D vision system by using two 2D camera sensors. *Int. Sci. Her.* **2012**, *3*, 234–237, ISSN 2218-5348.
28. Schneider, M.; Machacek, Z.; Martinek, R.; Koziorek, J.; Jaros, R. A System for the Detection of Persons in Intelligent Buildings Using Camera Systems—A Comparative Study. *Sensors* **2020**, *20*, 3558, ISSN 1424-8220. [CrossRef] [PubMed]
29. Olivka, P.; Mihola, M.; Novák, P.; Kot, T.; Babjak, J. The 3D laser range finder design for the navigation and mapping for the coal mine robot. In Proceedings of the 17th International Carpathian Control Conference (ICCC), High Tatras, Slovakia, 29 May–1 June 2016; pp. 533–538.
30. Davies, R.E. *Computer Vision: Principles, Algorithms, Applications, Learning*; Academic Press: Cambridge, MA, USA, 2018; p. 900; ISBN 978-0-12-809284-2.
31. Du, Y.C.; Muslikhin, M.; Hsieh, T.H.; Wang, M.S. Stereo Vision-Based Object Recognition and Manipulation by Regions with Convolutional Neural Network. *Electronics* **2020**, *9*, 210, ISSN 2079-9292. [CrossRef]
32. Matusek, O.; Zdenek, V.; Hotar, V. Detection of glass edge corrugation for cutting distance optimization. *MM Sci. J.* **2017**, 1734–1737. [CrossRef]
33. Vachálek, J.; Čapucha, L.; Krasňanský, P.; Tóth, F. Collision-free manipulation of a robotic arm using the MS Windows Kinect 3D optical system. In Proceedings of the 20th International Conference on Process Control, Strbske Pleso, Slovakia, 9–12 June 2015; pp. 96–106.
34. Holubek, R.; Ružarovský, R.; Delgado Sobrino, D.R. An innovative approach of industrial robot programming using virtual reality for the design of production systems layout. In *Advances in Manufacturing*; Springer Nature: Cham, Switzerland, 2019; pp. 223–235; ISBN 978-3-030-18714-9.
35. Xu, X.; Yang, H. Vision Measurement of Tunnel Structures with Robust Modelling and Deep Learning Algorithms. *Sensors* **2020**, *20*, 4945. [CrossRef] [PubMed]
36. Hain, J. Comparison of Common Tests for Normality. Ph.D. Thesis, Julius-Maximilians-Universitat Wurzburg Institut fur Mathematik und Informatik, Wurzburg, Germany, 2010; 235p.

37. Pérez, L.; Rodríguez, Í.; Rodríguez, N.; Usamentiaga, R.; García, D.F. Robot Guidance Using Machine Vision Techniques in Industrial Environments: A Comparative Review. *Sensors* **2016**, *16*, 335, ISSN 1424-8220. [CrossRef] [PubMed]
38. Halenár, I.; Juhás, M.; Juhásová, B.; Vladimirovič, D.B. Virtualization of production using digital twin technology. In Proceedings of the 17th International Carpathian Control Conference (ICCC), Krakow-Wieliczka, Poland, 26–29 May 2019; pp. 1–5, ISBN 978-1-7281-0702-8.

applied sciences

MDPI

Article

Experimental Determination of the Coefficient of Friction on a Screw Joint

Jozef Mascenik * and Tomas Coranic

Department of Design and Monitoring of Technical Systems, Faculty of Manufacturing Technologies with a Seat in Presov, Technical University of Kosice, Bayerova 1, 080 01 Presov, Slovakia
* Correspondence: jozef.mascenik@tuke.sk; Tel.: +421-55-602-6337

Featured Application: The results of this paper can help in lubricant selection when frequently disassembling and assembling a screw joint or when using a screw joint with fine mechanics. By applying the results in practice, it is possible, in certain cases, to minimize the tightening torque as the main parameter for screw joints.

Abstract: This paper deals with the coefficient of determination of screw connection friction between the thread and the matrix. The coefficient of friction was measured using a laboratory device with an M20 screw connection without any grease and, subsequently, plastic grease was added (CX80 silicone, lithium, and copper grease). When grease is added, the friction in the threads and screw heads is limited and consistently retained. When tightening by torque, which represents the prevailing assembly method in standard screwing practice, only part of the torque is effectively used to create axial force and pre-stress. The rest of the torque is employed in friction suppression between threads and converted into heat. In general, the coefficient of friction depends on diverse factors such as the roughness of the thread surface, the gradient angle of the helix, and the grease properties. The tightening torque represents a primary parameter in the experimental measurements, monitored using a digital torque spanner, and generates an axial force in the screw. Based on the aforementioned parameters, the objective of this paper was to monitor changes in the coefficient of friction between the thread of the screw and the matrix in the case of different grease types. The actual coefficient of friction was calculated through the exponential equation of the torque balance. First measured was the load of the bolted joint without the use of grease, where the average value of the coefficient of friction was 0.44732; this value served as a benchmark for comparison to the measurements with the use of grease. The measurements showed that the value of the friction coefficient was reduced by 30.57% when using lithium grease, by 40.56% when using silicone grease, and by 47.64% when using copper-based grease, making the latter the most suitable for the application. Without appropriate greasing, friction suppression was accompanied by extremely high torques, which resulted in insufficient screw prolongation.

Keywords: bolted joint; load; axial force; coefficient of friction; grease

Citation: Mascenik, J.; Coranic, T. Experimental Determination of the Coefficient of Friction on a Screw Joint. *Appl. Sci.* **2022**, *12*, 11987. https://doi.org/10.3390/app122311987

Academic Editor: Jacek Tomków

Received: 21 October 2022
Accepted: 14 November 2022
Published: 23 November 2022

1. Introduction

Currently, only minor attention is paid to the given issue; thus, few papers have been published in this field. Most of the friction coefficient research focuses on ball and trapezoidal screws, which are used in machine tool drives. The findings have led authors to design a measuring stand for testing bolted joints, where the primary monitored parameter is the coefficient of friction between the threads.

Bolted joints are a type of dismountable joint, designed to connect diverse components. From a technical point of view, their function is based on threaded coupling. Threads are referred to as functional elements, and they are standardized to a high degree.

The current situation and level of technology in the mechanical coupling of components with thread fasteners are characterized as follows:

- The prevailing results are generally underestimated, based on the fact that the screws and matrices are widely used and, compared to other structural elements, they are the cheapest available.
- Double, even triple, dimensioning, which represents the weakest point of current screwing technology. On the one hand, the structural designer is frightened of bearing potential responsibility for a possible accident, and on the other hand, there is the issue of mastering the theory and practice of screwing.
- Economic pressure to build light structures to achieve material savings [1].

The causes of failures in screw connections have become a subject of interest in many companies and relevant workplaces. The well-known SFK company created a hierarchy of failure causes, showing that the highest number of failures occur in assembly. This is a logical outcome of imprecise tightening.

Bolted joint design includes designing materials and strength dimensioning. At the same time, requirements related to operational load must be taken into consideration. A bolted joint functions as required only if all of the mechanical properties of the individual elements are preserved, along with the degree of precision of the defined tolerated dimensions. In practice, bolted joints are subjected to load tests according to specified calculations. These tests verify the efficiency, different deformations, thread strength, etc., of bolted joints.

All inevitable forces are generated when the respective screw or bolt nut is tightened with the predefined torque. Apart from the force acting in the direction of the screw axis, bolted joints induce other forces perpendicular to their axis, which are caused by friction forces. These forces especially occur in abutting areas of the connected parts. Friction force must act adequately to prevent a shift in the connected parts and consequent shear stress on the screw [2].

From the point of view of modern design engineering, assessment of the carrying capacity, the service life, and the fulfillment of the functional requirements of bolted joints is significantly influenced by the correct definition of axial force F_0 when a screw or a bolt nut is tightened, as well as by practical procedures necessary for their achievement.

If the force is extremely low in comparison to the tension tolerated in the screw core, the maximum technical parameters of the structure cannot be achieved. In case of high values, when the superposition of the force, along with the effects of other forces, reach characteristic stress in the thread core due to operational stress, the bolted joint is released or the threads are cut, which represents an unacceptable condition (tightness is insufficient, abutment joints are released, etc.).

Therefore, close attention has been paid to the experimental processes, assuring the pre-stress of screws as required. These processes include the following:

- The use of negative dilatation of a loosely tightened pre-heated screw, which assures pre-stress and the required properties when it is cooled down;
- The application of ultrasound (change in the wave speed along the screw);
- The application of the screw with a dynamometer element (e.g., a resistance strain gauge);
- The application of dynamometer power boards;
- Tightening of the regulated torque, etc.

2. Bolted Joint Design and Testing

Nowadays, a structural designer disposes of a broad assortment of decision-making tools, and just their knowledge and experience are relevant in decision making regarding which variant they opt for. Certainly, a screw connection can serve its function only when correctly installed, i.e., only when tightened to a value prescribed for the tightening torque. As soon as the tightening starts, it changes into a tribological node, in which case, in addition to the mechanical properties, friction plays an important role. Particular attention must be paid to the screw connections for a low coefficient of friction in the course of tightening and for high friction when assembly is completed in order to prevent disengagement of the connection.

Despite the fact that there exist various tables containing recommended torques determined through calculations of material hardness, the final decision is made by a structural designer. The reason is simple: no theoretical calculation can take into consideration the influence of the operating conditions, i.e., the type, magnitude, sense, and direction of stress of the respective structural node during operation. Based on these theoretical facts, experiments were conducted to contribute to the knowledge related to screw connections in practice.

2.1. *Ratio of the Forces in Bolted Joints*

The ratio of the forces in a screw is the ratio of the forces on the inclined plane. It is the relationship between the axial force F_Q, driving force F_1, and other thread parameters (Figure 1).

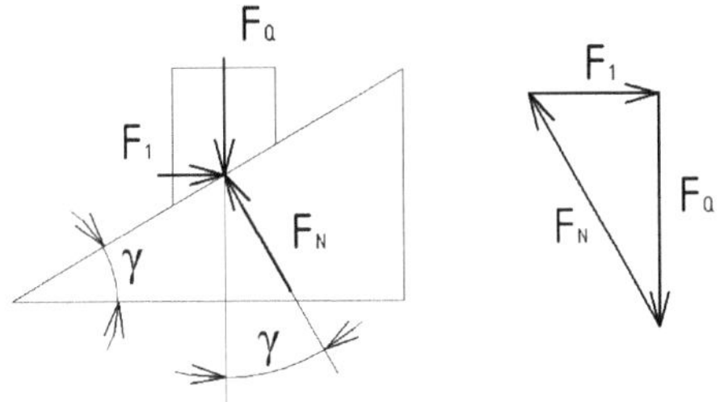

Figure 1. Ratio of the forces for frictionless motion.

The flat thread premise is accepted when the following conditions are met:

- A thread is perpendicular to an axis, causing forces to act tangentially;
- An axis load is evenly distributed along all threads;
- During stroking, the screw rotates and the bolt nut is in the fixed position;
- A calculation is performed with the mean diameter d_2 or d_S.

2.1.1. Frictionless Motion

$$F_1 = F_Q \cdot tg\gamma \tag{1}$$

Here, F_1 is the tangential force acting on the mean diameter of the thread; F_Q is the axial force; and F_N is the normal force.

2.1.2. Frictionless Motion—Tightening

$$F_T = F_N \cdot f,\ f_{\text{–coefficient of friction}} \tag{2}$$

$$\begin{gathered}\sum F_{ix} = 0 \Rightarrow F_1 = F_N \cdot \sin\gamma + f \cdot F_N \cdot \cos\gamma \\ \sum F_{iy} = 0 \Rightarrow F_Q = F_N \cdot \cos\gamma - f \cdot F_N \cdot \sin\gamma \\ \frac{F_1}{F_Q} = \frac{F_N \cdot \sin\gamma + f \cdot F_N \cdot \cos\gamma}{F_N \cdot \cos\gamma - f \cdot F_N \cdot \sin\gamma} \Rightarrow \\ F_1 = F_Q \cdot \frac{\sin\gamma + f \cdot \cos\gamma}{\cos\gamma - f \cdot \sin\gamma}\end{gathered} \tag{3}$$

When modified (Figure 2),

$$F_1 = F_Q \cdot \frac{tg\gamma + f}{1 - f \cdot tg\gamma}\ resp.\ F_1 = F_Q \cdot \frac{tg\gamma + tg\phi}{1 - tg\gamma \cdot tg\phi} = F_Q \cdot tg(\gamma + \phi) \tag{4}$$

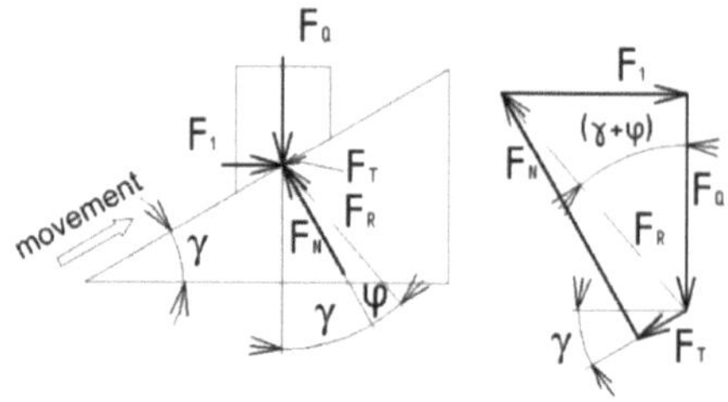

Figure 2. Ratio of the forces for motion with friction (tightening).

2.2. Influencing Factors

The factors influencing the relationship between tightening torque and axial force in a screw can be classified as follows:

- Geometrical—thread shape and contact area between reciprocally rotating parts;
- Tribological—contact area friction [3–5].

The value of the coefficient of friction between the area of the matrix thread and the area of the screw thread is influenced by many factors, such as the characteristics of the surface and of the thread grease, as well as the magnitude of the angle of the teeth's sides. At the same time, the coefficient of friction is a function of the heat treatment of screws and it affects the axial force value. Apart from the aforementioned heat, the friction that occurs in threads causes stress on the screw connection upon torsion, especially in the course of assembling and disassembling, in which case, they become corroded under the long-term influence of a harsh environment [6,7].

2.3. Coefficient of Friction of the V-Thread

As with friction, the flat thread premise is accepted; from it, the moment equation stems, in which φ or f is substituted by φ' or f'.

$$\begin{aligned} T_A - F_O \cdot tg(\gamma + \phi\prime) \cdot \tfrac{d_s}{2} = 0 \\ T_A - F_O \cdot (tg\gamma + f\prime) \cdot \tfrac{d_s}{2} = 0 \end{aligned} \tag{5}$$

The axial force in the screw is as follows:

$$F_O = \frac{2 \cdot \pi \cdot F_1 \cdot l}{\pi \cdot f\prime \cdot d_s + s} \tag{6}$$

The coefficient of friction in the threads regardless of the inclination angle of the thread is:

$$f\prime = \frac{2\pi \cdot T_A - F_O \cdot s}{\pi \cdot F_O \cdot d_s} \tag{7}$$

The coefficient of friction of the V-thread with α standing for the inclination angle of the thread is:

$$f\prime = \frac{f}{\cos\frac{\alpha}{2}} \Rightarrow f = f\prime \cdot \cos\frac{\alpha}{2} \tag{8}$$

The influence of the α angle results in an increase in the coefficient of friction f' in the case of the V-thread with an increasing α value. In this paper, we used the magnitude of the coefficient of friction $f\prime$.

3. Stand and Elements Essential for the Experiment

The experimental monitoring and the measurement of the friction coefficient in the case of the threads of the screw connection were monitored loading (tightening torque) and emerging axial force. In the case of the laboratory measurement, an M20 × 100 STN EN 24,017 bolt was used, due to the current design and capabilities of the measuring equipment.

Measurement was preceded by degreasing the screw to prevent errors when measuring the coefficient of friction.

The emerging axial force *Fo* was monitored in the experimental measurement through controlled tightening (axial) torque. The bolted joint load was monitored using degreased meshing threads and threads greased with MOL Alubia AK2.

A stand (Figure 3) was used for the practical part, designed for monitoring the axial force in the bolted joint. Measurement was performed in the Testing and Monitoring Centre of Technical Systems in the Department of Technical Systems Design and Monitoring in the Faculty of Manufacturing Technologies, Technical University of Kosice. Measurements were performed with M20 × 1.5 threaded bolts; a total of 10 bolts were used and five measurements were carried out for each alternative of the experiment. The measurement required preparation of a measuring stand, including pre-specified parts, the absence of which would make the measurement impossible [8,9]. Measurement was performed with pre-defined values of axial torque. Therefore, the computer software displayed a graph that allowed reading of the axial force with individual axial torque values.

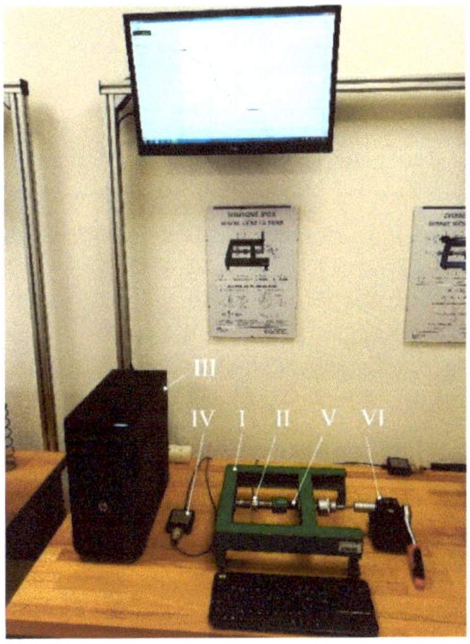

Figure 3. The stand designed for monitoring bolted joints during controlled loading: I—Basic structure; II—membrane force sensor; III—computer; IV—A/D converter; V—tested screw; VI—torque measuring instrument.

The system designed for monitoring and determining the coefficient of friction in the case of screw connections consisted of a basic structure (I), a membrane force sensor (II), and connection to a computer (III) by means of a converter (IV). The equipment included a replaceable tested screw (V), loaded via tightening torque and monitored by the measuring instrument (VI). Loading using the required torque was a manual process, which consequently generated loading of the screw connection by axial force. Monitoring of the axial force was conducted by means of EMS Center software from Emsyst.

3.1. Digital Electronic Converter EMS650

Laboratory measurement was performed using a digital electronic converter, Emsyst EMS650 (Figure 4), with a USB 2.0 interface. EMS650 is a digital converter designed to process the electrical signals produced by a tensometric sensor of force (e.g., tensile and compression force and torque). The signal produced by the sensor is intensified, adequately modified and driven to the analogue input of the microprocessor. Therefore, it is converted to a digital number and sent to the computer [10] through the USB interface. The converter does not dispose of any control elements, as it has two connectors only. The sensor is connected to a Cannon type connector, and the other connector (micro-USB) is used to connect the converter to the computer. All converter functions are controlled solely through the USB interface [11].

Figure 4. Emsyst EMS650 converter.

3.2. *Tensometric Force Sensor*

The experimental measurements were performed by means of a universal membrane sensor EMS20 (Figure 5), which was used for sensing the axial force within a range from 0 to 5 kN. This sensor can be used to measure both the tensile and the compression force acting in the direction of the sensor axis, and external threads are employed for attaching [12]. The sphere of application consisted of Testing, the measurement of forces on the machines, dynamometers and electronic scales. The recommended loading mode is tensile or compressive loading with a minimum distance of 2 mm between the tensometric sensor and an articulated head or a silent block [13]. The sensitivity of the sensor was $1.0 \pm 2\%$ mV/V, and the temperature coefficient was 0.1% Fn/10 °C. The sensor could be used within a temperature range from −10 °C to 70 °C. The insulation resistance exceeded 5000 Ω, and the bridge resistance was $395 \pm 10\%$ Ω at the input and $350 \pm 5\%$ Ω at the output.

Figure 5. Tensometric force sensor EMS20.

3.3. *Digital Torquemeter WRG3-135*

A small digital torquemeter was designed for fast measurement and force control during screw tightening. The torquemeter was installed directly onto a tightening key. The torquemeter was equipped with a digital display, which shows the actual measured values of the axial torque or the maximal value [14] in the case of maximum loading. The digital torquemeter WRG3-135, with a measurement range from 6.8 to 135 Nm (Figure 6), had an accuracy of ±2%. The meter was equipped with a sound signalization unit that was activated upon reaching a pre-specified axial torque value. The meter enabled measuring of the screws with either the right or left thread.

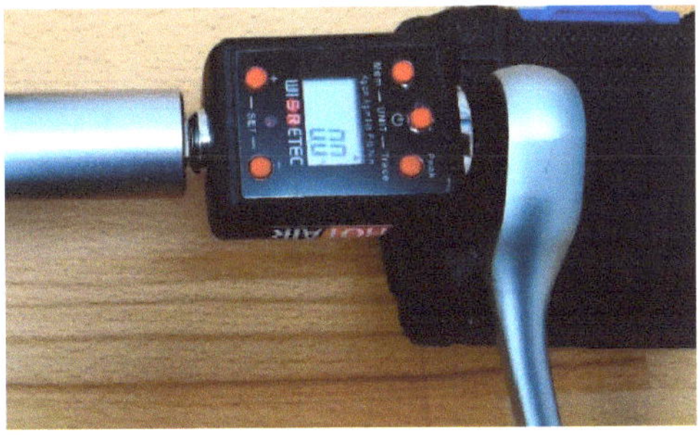

Figure 6. Torquemeter WRG3-135 (from 6.8 to 135 Nm).

3.4. Software EMS Center

All laboratory measurement results of the controlled axial torque loading and of the monitoring of the emerging axial force in the case of particular bolted joints were evaluated via the EMS Center software. The software version was V1.02, which supports the Windows 7, 8, and 10 operating systems. The software was primarily designed for data collection by means of digital electric units using EMSYST. The software offers graphical plotting of the selected developments and their subsequent storage in files of desired formats [15].

The software includes a record data function, which is activated when the record data button is pressed. The length of individual records is selected according to the number of samples or record time in seconds. All recorded parameters and data can be processed further in Excel, for instance. Therefore, it is advisable to record data in the txt.format. The software includes basic buttons—Start, Stop, and Zero. When the Zero button is pressed, the display is reverted to a zeroized state [16]. In order to create a new profile in the given program, the display must be put in a zeroized state. The Start button activates the measurement, while the Stop button discontinues the measurement.

3.5. Lubricants Used in the Experiment

The bolted joint load was monitored using the degreased meshing threads and the threads greased with MOL Alubia AK2. Samples of the lubricants are shown in Figure 7.

Figure 7. Visual comparison of the used lubricants.

4. Experimental Measurements

4.1. Center Dry Abrasion

The first experimental measurements and the subsequent friction coefficient determination in the case of screw connections were realized without any grease or oils. In the event of such direct contact between the screw threads and the matrix, the coefficient of friction should achieve the highest value, contrary to values from the following measurements in which plastic grease was used. Table 1 displays the measured values, which were processed and are shown in Figure 8.

Table 1. Measured values in dry abrasion.

No. of Measurements	1	2	3	4	5
Loading—tightening torque M_T (Nm)	4.0	8.0	12.0	16.0	20.0
Emerging axial force F_O (N)	624.8	1218.2	1789.4	2326.7	2792.8

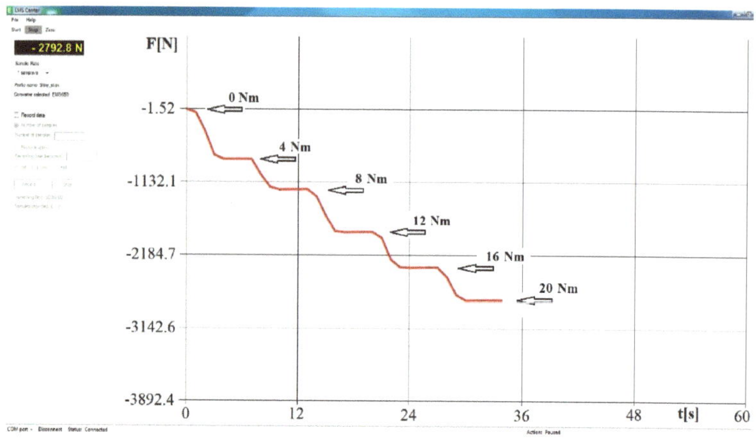

Figure 8. Dependence of the emerging axial force F_O on time from the EMS Center software.

The calculated coefficients of friction f' for the individual tightening torques in dry abrasion are as follows:

$$
\begin{aligned}
f'_1 &= \frac{2\pi \cdot M_{T1} - F_{O1} \cdot s}{\pi \cdot F_{O1} \cdot d_s} = \frac{2\pi \cdot 4\ Nm - 624.8\ N \cdot 0.0025\ m}{\pi \cdot 624.8\ N \cdot 0.018376\ m} = 0.6534 \\
f'_2 &= \frac{2\pi \cdot M_{T2} - F_{O2} \cdot s}{\pi \cdot F_{O2} \cdot d_s} = \frac{2\pi \cdot 8\ Nm - 1218.2\ N \cdot 0.0025\ m}{\pi \cdot 1218.2\ N \cdot 0.018376\ m} = 0.6714 \\
f'_3 &= \frac{2\pi \cdot M_{T3} - F_{O3} \cdot s}{\pi \cdot F_{O3} \cdot d_s} = \frac{2\pi \cdot 12\ Nm - 1789.4\ N \cdot 0.0025\ m}{\pi \cdot 1789.4\ N \cdot 0.018376\ m} = 0.6865 \\
f'_4 &= \frac{2\pi \cdot M_{T4} - F_{O4} \cdot s}{\pi \cdot F_{O4} \cdot d_s} = \frac{2\pi \cdot 16\ Nm - 2326.7\ N \cdot 0.0025\ m}{\pi \cdot 2326.7\ N \cdot 0.018376\ m} = 0.7051 \\
f'_5 &= \frac{2\pi \cdot M_{T5} - F_{O5} \cdot s}{\pi \cdot F_{O5} \cdot d_s} = \frac{2\pi \cdot 20\ Nm - 2792.8\ N \cdot 0.0025\ m}{\pi \cdot 2792.8\ N \cdot 0.018376\ m} = 0.7361
\end{aligned} \tag{9}
$$

The actual coefficients of friction f for the individual tightening torques in dry abrasion are as follows:

$$
\begin{aligned}
f_1 &= f'_1 \cdot \cos\frac{\alpha}{2} = 0.6534 \cdot \cos\frac{60^\circ}{2} = 0.4233 \\
f_2 &= f'_2 \cdot \cos\frac{\alpha}{2} = 0.6714 \cdot \cos\frac{60^\circ}{2} = 0.4349 \\
f_3 &= f'_3 \cdot \cos\frac{\alpha}{2} = 0.6865 \cdot \cos\frac{60^\circ}{2} = 0.4448 \\
f_4 &= f'_4 \cdot \cos\frac{\alpha}{2} = 0.7051 \cdot \cos\frac{60^\circ}{2} = 0.4568 \\
f_5 &= f'_5 \cdot \cos\frac{\alpha}{2} = 0.7361 \cdot \cos\frac{60^\circ}{2} = 0.4768
\end{aligned} \tag{10}
$$

4.2. Abrasion with the Use of Lubricating CX80 Silicone Grease

Colorless lubricant with high viscosity creates a protective film with excellent resistance; hence, coverage of the friction surfaces with a thin layer of silicone grease leads to an effective reduction in friction and provides long-lasting protection of the equipment, suitable for use on contact surfaces of plastic, metal, ceramics, rubber, etc. It does not stain, does not adhere to dust, and is non-adhesive, facilitating the removal of forms, with an operating temperature of −60 to +230 °C.

It is compatible with all kinds of sanitary equipment, including conventional and lever (gas) valves, seals, connectors, fittings, valves, and joints. It can also be used in the clean lubrication of equipment in the textile, paper, and food industries; electric and power distribution equipment (safety protection of electric circuits); power cases that work under extreme temperatures; and the lubrication of links, locks, hinge screws, and sleeves in the automotive industry. Moreover, it protects against corrosion in printing machines where the lubricant comes into contact with paper [17]. It has also been approved by NSF for use in the food industry. The measured values with silicone grease are presented in Table 2 and Figure 9.

Table 2. Values determined via measurement with lubricant CX80—silicone grease.

No. of Measurements	1	2	3	4	5
Loading—tightening torque M_{TS} *(Nm)*	4.0	8.0	12.0	16.0	20.0
Emerging axial force F_{OS} *(N)*	974.2	1945.6	2874.6	3823.2	4681.6

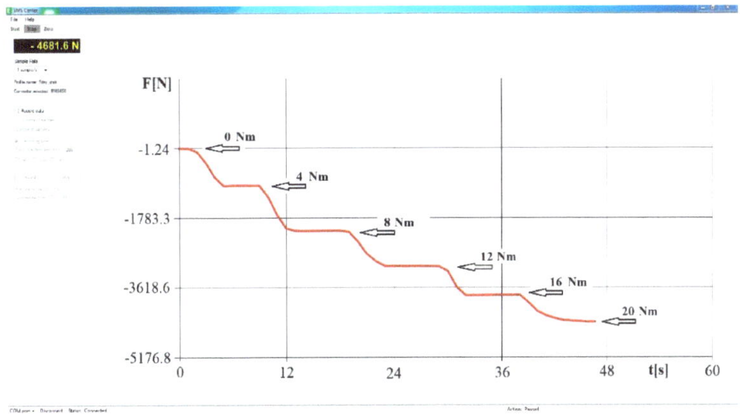

Figure 9. Dependence of the emerging force F_{OS} on time.

The calculated coefficients of friction f' for the individual tightening torques in abrasion with lubricating silicone grease CX80 are as follows:

$$
\begin{aligned}
f'_{S1} &= \frac{2\pi \cdot M_{TS1} - F_{OS1} \cdot s}{\pi \cdot F_{OS1} \cdot d_s} = \frac{2\pi \cdot 4\ Nm - 974.2\ N \cdot 0.0025\ m}{\pi \cdot 974.2\ N \cdot 0.018376\ m} = 0.4035 \\
f'_{S2} &= \frac{2\pi \cdot M_{TS2} - F_{OS2} \cdot s}{\pi \cdot F_{OS2} \cdot d_s} = \frac{2\pi \cdot 8\ Nm - 1945.6\ N \cdot 0.0025\ m}{\pi \cdot 1945.6\ N \cdot 0.018376\ m} = 0.4042 \\
f'_{S3} &= \frac{2\pi \cdot M_{TS3} - F_{OS3} \cdot s}{\pi \cdot F_{OS3} \cdot d_s} = \frac{2\pi \cdot 12\ Nm - 2874.6\ N \cdot 0.0025\ m}{\pi \cdot 2874.6\ N \cdot 0.018376\ m} = 0.4110 \\
f'_{S4} &= \frac{2\pi \cdot M_{TS4} - F_{OS4} \cdot s}{\pi \cdot F_{OS4} \cdot d_s} = \frac{2\pi \cdot 16\ Nm - 3823.2\ N \cdot 0.0025\ m}{\pi \cdot 3823.2\ N \cdot 0.018376\ m} = 0.4121 \\
f'_{S5} &= \frac{2\pi \cdot M_{TS5} - F_{OS5} \cdot s}{\pi \cdot F_{OS5} \cdot d_s} = \frac{2\pi \cdot 20\ Nm - 4681.6\ N \cdot 0.0025\ m}{\pi \cdot 4681.6\ N \cdot 0.018376\ m} = 0.4216
\end{aligned}
\tag{11}
$$

The actual coefficients of friction f for the individual tightening torques in abrasion with lubricating silicone grease CX80 are as follows:

$$\begin{aligned} f_{S1} &= f\prime_{S1}\cdot \cos\tfrac{\alpha}{2} = 0.4035\cdot \cos\tfrac{60^\circ}{2} = 0.2614 \\ f_{S2} &= f\prime_{S2}\cdot \cos\tfrac{\alpha}{2} = 0.4042\cdot \cos\tfrac{60^\circ}{2} = 0.2618 \\ f_{S3} &= f\prime_{S3}\cdot \cos\tfrac{\alpha}{2} = 0.4110\cdot \cos\tfrac{60^\circ}{2} = 0.2662 \\ f_{S4} &= f\prime_{S4}\cdot \cos\tfrac{\alpha}{2} = 0.4121\cdot \cos\tfrac{60^\circ}{2} = 0.2670 \\ f_{S5} &= f\prime_{S5}\cdot \cos\tfrac{\alpha}{2} = 0.4216\cdot \cos\tfrac{60^\circ}{2} = 0.2731 \end{aligned} \quad (12)$$

Based on the given calculations, the coefficients of the friction values were determined. Hence, based on the acquired data, the actual coefficients of friction for the individual tightening torques were calculated.

4.3. Abrasion with the Use of Lubricating Lithium Grease CX80

This is a multi-purpose product; thus, it is used for the lubrication of industrial and agricultural systems and for the systems of motor vehicles. It ensures excellent lubrication, even under the toughest conditions.

It is resistant to water and oxidation, protects against attrition and corrosion, extends the life of bearings [4,5], contains EP additives, is mechanically stable, and, due to its soft consistency, is easy to apply. It passes the SKF EMCOR test, with operating temperatures ranging from −20 to +130 °C. Its density is 0.807 g/cm at 20 °C. The measured values with lithium grease are presented in Table 3 and in Figure 10.

Table 3. Values determined via measurement with lubricant CX80—lithium grease.

No. of Measurements	1	2	3	4	5
Loading—tightening torque M_{TL} (Nm)	4.0	8.0	12.0	16.0	20.0
Emerging axial force F_{OL} (N)	836.2	1678.5	2518.4	3322.8	4092.3

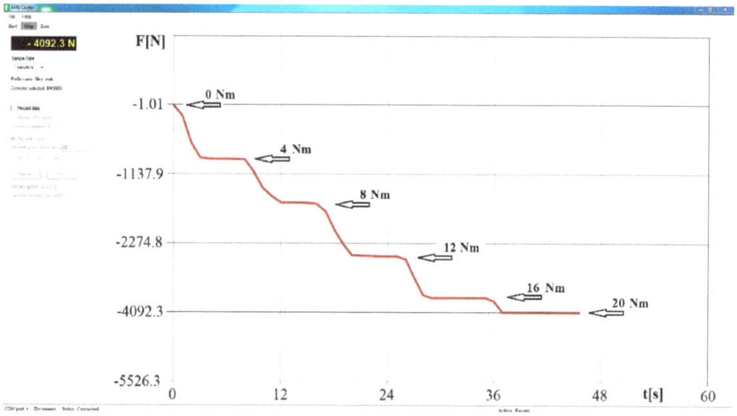

Figure 10. Dependence of the emerging force F_{OL} on time.

The calculated coefficients of friction f' for the individual tightening torques in abrasion with lubricating lithium grease CX80 are as follows:

$$\begin{aligned}
f'_{L1} &= \frac{2\pi \cdot M_{TL1} - F_{OL1} \cdot s}{\pi \cdot F_{OL1} \cdot d_s} = \frac{2\pi \cdot 4\ Nm - 836.2\ N \cdot 0.0025\ m}{\pi \cdot 836.2\ N \cdot 0.018376\ m} = 0.4773 \\
f'_{L2} &= \frac{2\pi \cdot M_{TL2} - F_{OL2} \cdot s}{\pi \cdot F_{OL2} \cdot d_s} = \frac{2\pi \cdot 8\ Nm - 1678.5\ N \cdot 0.0025\ m}{\pi \cdot 1678.5\ N \cdot 0.018376\ m} = 0.4754 \\
f'_{L3} &= \frac{2\pi \cdot M_{TL3} - F_{OL3} \cdot s}{\pi \cdot F_{OL3} \cdot d_s} = \frac{2\pi \cdot 12\ Nm - 2518.4\ N \cdot 0.0025\ m}{\pi \cdot 2518.4\ N \cdot 0.018376\ m} = 0.4752 \\
f'_{L4} &= \frac{2\pi \cdot M_{TL4} - F_{OL4} \cdot s}{\pi \cdot F_{OL4} \cdot d_s} = \frac{2\pi \cdot 16\ Nm - 3322.8\ N \cdot 0.0025\ m}{\pi \cdot 3322.8\ N \cdot 0.018376\ m} = 0.4807 \\
f'_{L5} &= \frac{2\pi \cdot M_{TL5} - F_{OL5} \cdot s}{\pi \cdot F_{OL5} \cdot d_s} = \frac{2\pi \cdot 20\ Nm - 4092.3\ N \cdot 0.0025\ m}{\pi \cdot 4092.3\ N \cdot 0.018376\ m} = 0.4886
\end{aligned} \tag{13}$$

The actual coefficients of friction f for the individual tightening torques in abrasion with lubricating lithium grease CX80 are as follows:

$$\begin{aligned}
f_{L1} &= f'_{L1} \cdot \cos\tfrac{\alpha}{2} = 0.4773 \cdot \cos\tfrac{60^\circ}{2} = 0.3092 \\
f_{L2} &= f'_{L2} \cdot \cos\tfrac{\alpha}{2} = 0.4754 \cdot \cos\tfrac{60^\circ}{2} = 0.3080 \\
f_{L3} &= f'_{L3} \cdot \cos\tfrac{\alpha}{2} = 0.4752 \cdot \cos\tfrac{60^\circ}{2} = 0.3079 \\
f_{L4} &= f'_{L4} \cdot \cos\tfrac{\alpha}{2} = 0.4807 \cdot \cos\tfrac{60^\circ}{2} = 0.3114 \\
f_{L5} &= f'_{L5} \cdot \cos\tfrac{\alpha}{2} = 0.4886 \cdot \cos\tfrac{60^\circ}{2} = 0.3165
\end{aligned} \tag{14}$$

4.4. Abrasion with the Use of Lubricating CX80 Copper Grease

This is an anti-adhesive, multi-purpose product for use under the most difficult conditions and within a high temperature range from −30 to 1200 °C. It ensures excellent lubrication, safeguards against attrition, corrosion, and high temperatures, and has a wide temperature range (−30 to +1200 °C). Meanwhile, its density is 0.9 g/cm^3 [18].

It is compatible with screws and nuts, threads and all connected surfaces, fittings, spark plugs, wheel and brake discs, battery joints, port installations, revolving discs, highly stressed bearings of all kinds, elements in the industrial and construction fields, other elements of industrial systems, construction, and agricultural and motor vehicles [19,20]. The measured values with copper grease can be seen in Table 4 and Figure 11.

The calculated coefficients of friction f' for the individual tightening torques in abrasion with lubricating copper grease CX80 are as follows:

$$\begin{aligned}
f'_{C1} &= \frac{2\pi \cdot M_{TC1} - F_{OC1} \cdot s}{\pi \cdot F_{OC1} \cdot d_s} = \frac{2\pi \cdot 4\ Nm - 1076.4\ N \cdot 0.0025\ m}{\pi \cdot 1076.4\ N \cdot 0.018376\ m} = 0.3611 \\
f'_{C2} &= \frac{2\pi \cdot M_{TC2} - F_{OC2} \cdot s}{\pi \cdot F_{OC2} \cdot d_s} = \frac{2\pi \cdot 8\ Nm - 2154.3\ N \cdot 0.0025\ m}{\pi \cdot 2154.3\ N \cdot 0.018376\ m} = 0.3608 \\
f'_{C3} &= \frac{2\pi \cdot M_{TC3} - F_{OC3} \cdot s}{\pi \cdot F_{OC3} \cdot d_s} = \frac{2\pi \cdot 12\ Nm - 3236.9\ N \cdot 0.0025\ m}{\pi \cdot 3236.9\ N \cdot 0.018376\ m} = 0.3601 \\
f'_{C4} &= \frac{2\pi \cdot M_{TC4} - F_{OC4} \cdot s}{\pi \cdot F_{OC4} \cdot d_s} = \frac{2\pi \cdot 16\ Nm - 4315.7\ N \cdot 0.0025\ m}{\pi \cdot 4315.7\ N \cdot 0.018376\ m} = 0.3601 \\
f'_{C5} &= \frac{2\pi \cdot M_{TC5} - F_{OC5} \cdot s}{\pi \cdot F_{OC5} \cdot d_s} = \frac{2\pi \cdot 20\ Nm - 5318.6\ N \cdot 0.0025\ m}{\pi \cdot 538.6\ N \cdot 0.018376\ m} = 0.3659
\end{aligned} \tag{15}$$

Table 4. Values determined via measurement with lubricant CX80—copper grease.

No. of Measurements	1	2	3	4	5
Loading—tightening torque M_{TC} *(Nm)*	4.0	8.0	12.0	16.0	20.0
Emerging axial force F_{OC} *(N)*	1076.4	2154.3	3236.9	4315.7	5318.6

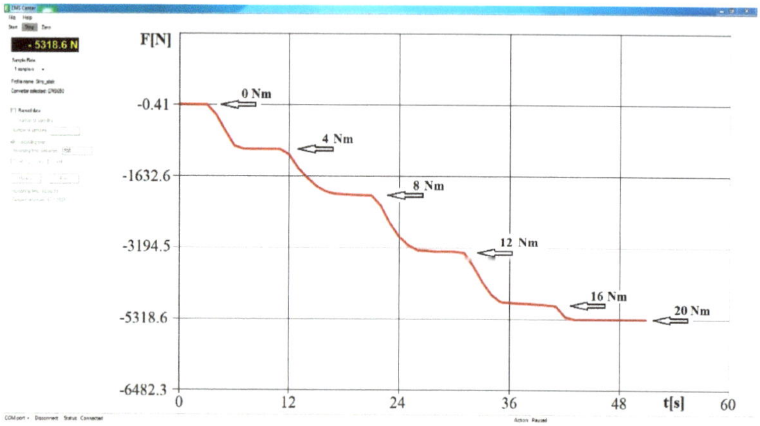

Figure 11. Dependence of emerging force F_{OC} on time.

The actual coefficients of friction f for the individual tightening torques in abrasion with lubricating copper grease CX80 are as follows:

$$\begin{aligned}
f_{C1} &= f\prime_{C1}\cdot \cos\tfrac{\alpha}{2} = 0.3611\cdot \cos\tfrac{60^\circ}{2} = 0.2339\\
f_{C2} &= f\prime_{C2}\cdot \cos\tfrac{\alpha}{2} = 0.3608\cdot \cos\tfrac{60^\circ}{2} = 0.2337\\
f_{C3} &= f\prime_{C3}\cdot \cos\tfrac{\alpha}{2} = 0.3601\cdot \cos\tfrac{60^\circ}{2} = 0.2333\\
f_{C4} &= f\prime_{C4}\cdot \cos\tfrac{\alpha}{2} = 0.3601\cdot \cos\tfrac{60^\circ}{2} = 0.2333\\
f_{C5} &= f\prime_{C5}\cdot \cos\tfrac{\alpha}{2} = 0.3659\cdot \cos\tfrac{60^\circ}{2} = 0.2370
\end{aligned} \quad (16)$$

5. Results and Discussion

Figure 12 is a graphical representation of the dependence of the emerging axial force in the bolt on the tightening torque for tests without lubricating grease and with different alternatives of the tested grease.

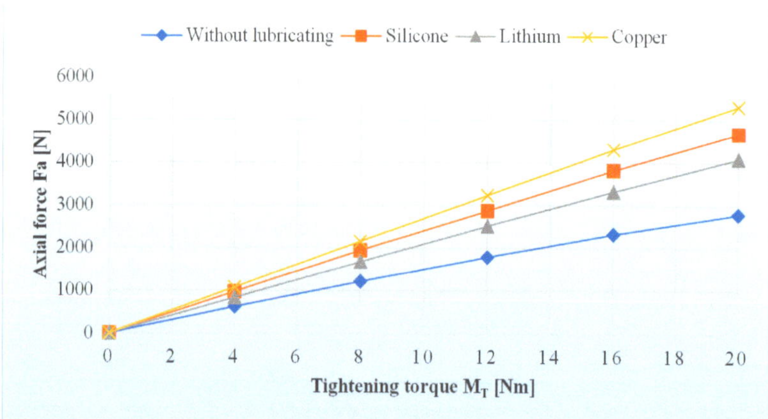

Figure 12. Dependence of the loading tightening torque on the emerging axial force with and without lubricating grease.

The dependence of the calculated actual friction coefficients on the tightening torque for the different test variants with or without lubricating grease is shown in Figure 13.

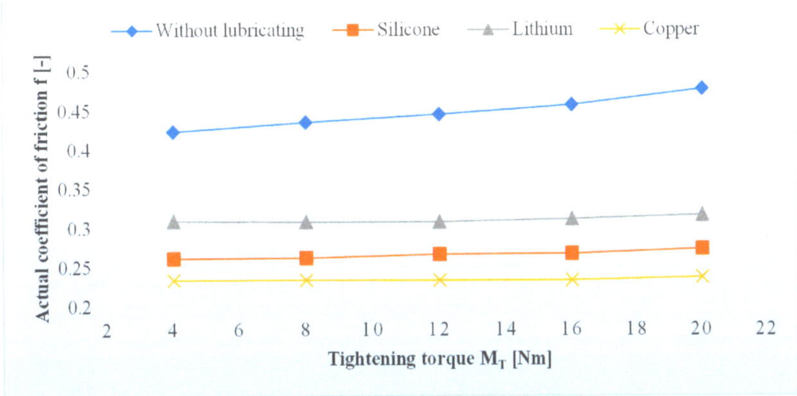

Figure 13. Dependence of the tightening torques on the actual calculated coefficients of friction with and without lubricating grease.

The calculated coefficients of friction in the thread were slightly distorted. This could be due to many factors, including the running-in of the screw connection. This caused an increase in the axial force in the screw threads, given by law, which was the main cause of the scatter of the measured values. Therefore, after each measurement, we completely unscrewed the screw and screwed it in again so that the same threads did not meet again.

Unlike linear theory, the actual dependence of the axial force on the tightening torque is progressive.

Based on the results of the verification experiments, we recommend a linear formula for estimating the lower limit of the axial force F_O when applying the tightening torque M_T, considering the following:

- The actual F_O value can be up to 50% higher, especially at large M_T values, which are suitable for screws with a high yield strength, because the thread surfaces grind against one another. The actual course of F_O is progressive.
- In order to achieve reproducible values on a set of the same screws, attention must be paid to the same method of lubrication [21].

We determined the coefficients of friction in the thread during dry and lubricated friction. By using better lubricants, the value of the coefficient can be minimized to values close to 0.2, which is the standard value of the steel-steel shear coefficient.

In conclusion, it is advisable to use modern hydraulic tightening machines for demanding applications due to their better accuracy and capability for operation without harmful effects on the operator.

For better-quality measurement, it would be competent to use modern hydraulic tightening machines to derive a more accurate value of the tightening torque. These machines have a small deviation value [22].

Due to the dimensional diversity of the screw connection, it would also be appropriate to adapt the measuring device to a more universal format:

- By making inserts for fastening screws of smaller dimensions. The design of the inserts is of great importance. By making them, the coefficient of friction could also be measured in other types of threads.
- By modifying the device so that it is possible to carry out a second method of measurement, in which the nut under the washer is tightened. With this measurement, another quantity would enter our calculations—the friction coefficient between the

nut and the washer. In this method, it is necessary to ensure that the screw head is guided to ensure that the screw does not rotate but only moves in the axial direction.
- By adjusting the device frame or its parts so that screws with smaller and larger lengths can be used.

6. Conclusions

The characteristic of the influence of geometrical factors such as thread shape is predominantly deterministic; however, tribological factors, which include friction occurring in the areas of contact, cause particular variance in the result. Therefore, it is necessary to pay attention to its global effects on screw connections from an experimental point of view. A derived semi-empirical formula appeared to be appropriate for those screws with a metric thread.

The laboratory measurements conducted using a testing stand proved that the coefficient of friction is considerably influenced by the loads of the factors, along with well-run screw connections. This causes axial force enlargement after tightening several times and represents the main reason for the variance in the measured values. Therefore, each measurement was followed by complete unscrewing and re-screwing of the screw to avoid encountering identical threads. Contrary to the linear theory, the actual dependence of the axial force on the tightening torque has a progressive characteristic. At the same time, to achieve reproducible values in a set of uniform screws, it is necessary to pay particular attention to ensuring their identical greasing. The results of the experimental measurements and necessary calculations were presented and plotted in individual figures and graphs.

Under laboratory conditions, the copper-based lubricant had the greatest effect on the coefficient of friction between the threads, where the decrease compared to the measurements without lubricating grease was 47.64%. An important factor in the lubrication of bolted joints is the service life or degradation of the lubricating grease. Lubricant degradation is mainly influenced by ambient conditions such as weathering, temperature gradients, and environmental cleanliness. Therefore, further experiments in this field are needed to verify the achieved laboratory results in technical practice.

Author Contributions: Conceptualization, J.M. and T.C.; methodology, J.M. and T.C.; software, T.C.; formal analysis, J.M. and T.C.; resources, T.C.; data curation, J.M. and T.C.; writing—original draft preparation, T.C.; writing—review and editing, J.M.; visualization, T.C.; project administration and funding acquisition, J.M. All authors have read and agreed to the published version of the manuscript.

Funding: This work was supported by the Slovak Ministry of Education within project KEGA 017TUKE-4/2021 and by the Slovak Research and Development Agency under contract No. APVV-18-0316.

Institutional Review Board Statement: Not applicable.

Informed Consent Statement: Informed consent was obtained from all subjects involved in this study.

Conflicts of Interest: The authors declare no conflict of interest.

References

1. Gaspar, S.; Coranic, T.; Pasko, J. Influence of die casting factors and of hardening temperature upon mechanical properties of a die cast. *MM Sci. J.* **2018**, *12*, 2639–2642. [CrossRef]
2. Coranic, T.; Gaspar, S.; Pasko, J. Utilization of Optimization of Internal Topology in Manufacturing of Injection Moulds by the DMLS Technology. *Appl. Sci.* **2021**, *11*, 262. [CrossRef]
3. Cacko, P.; Krenicky, T. Impact of lubrication interval to operating status of bearing. *Appl. Mech. Mater.* **2014**, *616*, 151–158. [CrossRef]
4. Cacko, P.; Krenicky, T.; Dobransky, J. Impact of an Excessive Wear of Bearing on the Mechatronic Devices. *Appl. Mech. Mater.* **2014**, *460*, 99–106. [CrossRef]
5. Majernik, J.; Gaspar, S.; Gryc, K.; Socha, L. Changes in Eutectic Silumin Structure Depending on Gate Geometry and Its Effect on Mechanical Properties of Casting. *Manuf. Technol.* **2018**, *18*, 439–443. [CrossRef]
6. Mascenik, J. Experimental determination of cutting speed influence on cutting surface character in material laser cutting. *MM Sci. J.* **2016**, *3*, 960–963. [CrossRef]

7. Trojanova, M.; Cakurda, T.; Hosovsky, A.; Krenicky, T. Estimation of Grey-Box Dynamic Model of 2-DOF Pneumatic Actuator Robotic Arm Using Gravity Tests. *Appl. Sci.* **2021**, *11*, 4490. [CrossRef]
8. Mascenik, J.; Pavlenko, S. Gear production using nonconventional technologies. *Tech. I Technol. Innovacii I Kačestvo Baranoviči* **2007**, 55–57.
9. Straka, L.; Hasova, S. Prediction of the heat-affected zone of tool steel EN X37CrMoV5-1 after die-sinking electrical discharge machining. *Proc. Inst. Mech. Eng. Part B-J. Eng. Manuf.* **2018**, *232*, 1395–1406. [CrossRef]
10. Trojanowska, J.; Pająk, E. Using the Theory of Constraints to Production Processes Improvement. In Proceedings of the 7th International Conference of DAAAM Baltic Industrial Engineering, Tallinn, Estonia, 22–24 April 2010; Volume 1, pp. 322–327.
11. BUŃ, P.; Trojanowska, J.; Ivanov, V.; Pavlenko, I. The Use of Virtual Reality Training Application to Increase the Effectiveness of Workshops in the Field of Lean Manufacturing. In Proceedings of the 4th International Conference of the Virtual and Augmented Reality in Education, Budapest, Hungary, 17–19 September 2018; pp. 65–71.
12. Husar, J.; Knapcikova, L.; Balog, M. Implementation of Material Flow Simulation as a Learning Tool. In *Design, Simulation, Manufacturing: The Innovation Exchange*; Springer: Cham, Switzerland, 2019; pp. 33–41.
13. Bun, P.; Gapsa, J.; Husar, J.; Kascak, J. Mixed Reality Training in Electrical Equipment Operating Procedures. In *International Scientific-Technical Conference MANUFACTURING*; Springer: Cham, Switzerland, 2022; pp. 306–316.
14. Kascak, J.; Husar, J.; Knapcikova, L.; Trojanowska, J.; Ivanov, V. Conceptual Use of Augmented Reality in the Maintenance of Manufacturing Facilities. In *International Scientific-Technical Conference MANUFACTURING*; Springer: Cham, Switzerland, 2022; pp. 241–252.
15. Majernik, J.; Gaspar, S.; Pasko, J.; Coranic, T. Design of Casting and Gating Systems. *Lüdenscheid: RAM-Verl.* **2021**, 108.
16. Straka, L.; Pitel, J.; Corny, I. Influence of the main technological parameters and material properties of the workpiece on the geometrical accuracy of the machined surface at wedm. *Int. J. Adv. Manuf. Technol. Springer Int. Publ.* **2021**, *115*, 3065–3087. [CrossRef]
17. Straka, L.; Panda, A. Verification of the non-contact method of measuring roughness parameters on surface machined by WEDM technology. *Stud. I Mater.* **2021**, *41*, 32–37.
18. Majernik, J.; Podaril, M.; Gojdan, D. Influence of overflow connecting channel cross-section design on selected parameters of high pressure die casting. *Arch. Foundry Eng.* **2021**, *21*, 75–80.
19. Anisimov, V.M.; Anisimov, V.V.; Krenicky, T. Properties Prediction of Linear Block-polyurethanes Based on the Mixtures of Simple Oligoethers. *Manag. Syst. Prod. Eng.* **2019**, *27*, 217–220. [CrossRef]
20. Szczepaniak, M.; Trojanowska, J. Methodology of Manufacturing Process Analysis. *Adv. Manuf. II Prod. Eng. Manag.* **2019**, *2*, 281–294.
21. Ivanov, V.; Pavlenko, I.; Liaposhchenko, O.; Gusak, O.; Pavlenko, V. Determination of contact points between workpiece and fixture elements as a tool for augmented reality in fixture design. *Wirel. Netw.* **2021**, *27*, 1657–1664.
22. Kascak, J.; Gaspar, S.; Pasko, J.; Knapcikova, L.; Husar, J.; Baron, P.; Torok, J. Design of an Atypical Construction of Equipment for Additive Manufacturing with a Conceptual Solution of a Printhead Intended for the Use of Recycled Plastic Materials. *Appl. Sci.* **2021**, *11*, 2928. [CrossRef]

MDPI AG
Grosspeteranlage 5
4052 Basel
Switzerland
Tel.: +41 61 683 77 34

Applied Sciences Editorial Office
E-mail: applsci@mdpi.com
www.mdpi.com/journal/applsci

www.ingramcontent.com/pod-product-compliance
Lightning Source LLC
Chambersburg PA
CBHW041428210926
51850CB00013B/3381
* 9 7 8 3 7 2 5 8 4 0 4 4 1 *